2011 22nd Annual IEEE/SEMI Advanced Semiconductor Manufacturing Conference

(ASMC 2011)

Saratoga Springs, New York, USA
16-18 May 2011

IEEE Catalog Number: CFP11ASC-PRT
ISBN: 978-1-61284-408-4

Copyright © 2011 by the Institute of Electrical and Electronic Engineers, Inc
All Rights Reserved

Copyright and Reprint Permissions: Abstracting is permitted with credit to the source. Libraries are permitted to photocopy beyond the limit of U.S. copyright law for private use of patrons those articles in this volume that carry a code at the bottom of the first page, provided the per-copy fee indicated in the code is paid through Copyright Clearance Center, 222 Rosewood Drive, Danvers, MA 01923.

For other copying, reprint or republication permission, write to IEEE Copyrights Manager, IEEE Service Center, 445 Hoes Lane, Piscataway, NJ 08854. All rights reserved.

***This publication is a representation of what appears in the IEEE Digital Libraries. Some format issues inherent in the e-media version may also appear in this print version.

IEEE Catalog Number:	CFP11ASC-PRT
ISBN 13:	978-1-61284-408-4
ISSN:	1078-8743

Additional Copies of This Publication Are Available From:

Curran Associates, Inc
57 Morehouse Lane
Red Hook, NY 12571 USA
Phone: (845) 758-0400
Fax: (845) 758-2633
E-mail: curran@proceedings.com
Web: www.proceedings.com

ASMC 2011 - Table of Contents

These technical proceedings contain copyrighted manuscripts from the 2011 22nd Annual IEEE/SEMI Advanced Semiconductor Manufacturing Conference (ASMC). Every effort has been made to ensure accuracy. However, IEEE and SEMI cannot be held responsible for errors or omissions. The listings and contents of these proceedings are proprietary and cannot be reproduced in part or in whole without the express written consent of IEEE, SEMI and the author(s).

Session 1: Factory Optimization I

1.1 **The Global Supply Chain is Our New Fab: Integration and Automation Challenges 001**
Hans Ehm, Thomas Kaufmann, Thomas Ponsignon, Infineon Technologies AG

1.2 **Ambient Persuasion in the Factory: The Case of the Operator Guide 007**
Alexander Meschtscherjakov, Patricia Kluckner, Florian Pöhr, Wolfgang Reitberger, Astrid Weiss, Manfred Tscheligi, University of Salzburg; Karl Horst Hohenwarter, Peter Oswald, Infineon Technologies AG

1.3 **Looking Towards a Sustainable and Green Future 013**
Sumita Basu, Sandra Viarengo, Intel Corporation

1.4 **Dynamic Management of Controls in Semiconductor Manufacturing 018**
Justin Nduhura Munga, Stéphane Dauzère-Pérès, Claude Yugma, Ecole des Mines de Saint-Etienne; Philippe Vialletelle, STMicroelectronics;

1.5 **Characterizing the Value of Technological Knowledge for Lean Manufacturing 024**
Charles M. Weber, Portland State University

Session 2: Advanced Metrology

2.1 **In-Line Metrology of High Aspect Ratio Structures with MBIR Technique 032**
Delphine Le Cunff, STMicroelectronics; Leif Jonny Höglund, Nicolas Laurent, SEMILAB AMS

2.2 **Full Automatic on the Fly Optical Macro Wafer Edge Inspection System 037**
Heiko Fröhlich, Infineon Technologies; Denis Kirmizigül, Dresden University of Technology

2.3 **Application of CGS Stress Metrology to Advanced Process Control & Monitoring 040**
David Owen, Jeff Hebb, Ultratech, Inc.; Christian Otten, Texas Instruments Inc

2.4 **Inline Control of an Ultra Low-k ILD Layer using Broadband Spectroscopic Ellipsometry 046**

Ronny Haupt, Jiang Zhiming, Leander Haensel, KLA-Tencor; Ulf Peter Mueller, Ulrich Mayer, GLOBALFOUNDRIES

2.5 **Advanced Overlay Control in Volume Manufacturing 052**

Timothy Wiltshire, Christopher Ausschnitt, Nelson Felix, Emily Huwang, Michael Pike, Allen Gabor, IBM Corporation; Moshe Preil, Vincent Couraudon, James Schreiber, Timon Fliervoet, Geert Simons, Erica Rottenkolber, ASML

Session 3: Advanced Equipment, Materials & Processes

3.1 **On the Technology and Ecosystem of 3D / TSV Manufacturing 058**

Klaus Hummler, Larry Smith, Raymond Caramto, Robert Edgeworth, Stephen Olson, Daniel Pascual, Jamal Qureshi, Andy Rudack, Roger Quon, Sitaram Arkalgud, SEMATECH

3.2 **Parasitics Extraction, Wideband Modeling and Sensitivity Analysis of Through-Strata-Via (TSV) in 3D Integration/Packaging 064**

Zheng Xu, James Jian-Qiang Lu, Rensselaer Polytechnic Institute; Xiaoxiong Gu, IBM T.J. Watson Research Center

3.3 **Scaling Of Copper Seed Layer Thickness Using Plasma-Enhanced ALD and an Optimized Precursor 070**

Jiajun Mao, Eric Eisenbraun, College of Nanoscale Science and Engineering, The University at Albany; Vincent Omarjee, Andrey Korolev, Christian Dussarrat, American Air Liquide

3.4 **Investigation of the Structural and Electrical Characterization on ZrO_2 Addition for ALD HfO_2 with La_2O_3 Capping Layer Integrated Metal-oxide Semiconductor Capacitors 074**

C.K. Chiang, J.C. Chang, W.H. Liu, C.C. Liu, J.F. Lin, C.L. Yang, J.Y. Wu, United Microelectronics Corporation; S.J. Wang, National Cheng Kung University

3.5 **Low-k Etching Using CF_3I, A Path To Overcome Current BEOL Integration Issues 078**

Adam G. Gildea, Justin C. Long, Eric Eisenbraun, College of Nanoscale Science and Engineering, The University at Albany; Vincent Omarjee, Nathan Stafford, François Doniat, Christian Dussarrat, American Air Liquide

Session 4: Interactive Session

4.1 **Advanced Elemental Analysis Methods for Sub 30nm Defects in Defect Review SEM 083**

Dror Shemesh, Adi Boehm, Ofir Greenberg, Kfir Dotan, Applied Materials

4.2 **Advanced Excursion Control and Diagnostics for CMP Process Monitoring 087**

Andrew Stamper, IBM Corporation; Gangadharan Sivaraman, Ravi Sankar, KLA-Tencor

4.3 **Advanced Floating Gate CD Uniformity Control in the 75nm Node NOR Flash Memory 090**

Sheng-Yuan Chang, Yu-Chung Chen, An Chyi Wei, Hong-Ji Lee, Nan-Tzu Lian, Tahone Yang, Kuang-Chao Chen, Chih-Yuan Lu, Macronix International Co., Ltd.

4.4 Automated Systematic Discovery for Development and Production 094

Brad W. Austin, IBM Corporation; Andrew Cross, Marcus Liesching, KLA-Tencor

4.5 Data Mining Using PLS-Trees and other Projection Methods 099

Tamara Byrne, Umetrics / MKS Instruments; Svante Wold, Umea University

4.6 The Effect of Bevel Film Removal on Wafer Warpage and Film Stress 104

Keechan Kim, Kwanwook Kwon, YS Kim, Lam Research; Russ Dudley, David Marx, Tamar Technology

4.7 Eliminating a Polysilicon Hole Defect Created during oxide removal 108

Ikhoon Shin, Jason Doub, Keith Mortesen, Raymond Lappan, ON Semiconductor

4.8 Establishing Continuous Flow Manufacturing in a Wafertest-Environment Via Value Stream Design 112

Sophia Keil, Germar Schneider, Dietrich Eberts, Infineon Technologies Dresden GmbH; Kristina Wilhelm, Ingo Gestring, University of Applied Sciences Dresden; Rainer Lasch, Technical University Dresden; Arthur Deutschländer, University of Applied Sciences Stralsund

4.9 FMEA for Lean Manufacturing 119

Michael E. Lombardi, Intel Corp

4.10 Laser Spike Annealing For Nickel Silicide Formation 121

Jeffrey Hebb, Yun Wang, Shrinivas Shetty, Jim McWhirter, David Owen, Michael Shen, Van Le, Jeffrey Mileham, David Gaines, Serguei Anikitchev, Shaoyin Chen, Paul Bischoff, Ultratech, Inc.; Joe Lee, LeeDAC Consulting

4.11 Mechanical Properties of Si-C-O-H Low-k Dielectrics Prepared by Plasma Enhanced Chemical Vapor Deposition 127

Peter Woytowitz, Sassan Roham, Dong Niu, Haiying Fu, Novellus Systems, Inc.

4.12 New Methods for Improved SRAM Detection through Scattered Light Collection 133

Reuven Barel, Keren Shachar, Yakir Bechler, Nir Horesh, Applied Materials; Hsien-Tsung Chiang, To-Yu Chen, Taiwan Semiconductors Manufacturing Company (TSMC)

4.13 Reduction of CMP-Induced Wafer Defects through *In-Situ* Removal of Process Debris 136

S.J. Benner, G. Perez, D. W. Peters, Confluense ; K. Hue, P.O'Hagan, Particle Sizing Systems

4.14 Sampling Process Information from Unstructured Data 140

Jens Popp, Dirk Ortloff, Process Relations GmbH; Thilo Schmidt, Kai Hahn, Matthias Mielke, Rainer Brück, University of Siegen

4.15 Substrate Cleaning using Ultrasonics/Megasonics 146

Mohammad Kazemi, Helmuth Treichel, Rito Ligutom, Xyratex International

4.16 Use of Neural Network to Model the FTIR Spectra of PECVD Silicon Nitride Films for Cadiovascular Pressure Sensor Applications 152

Thongchai Thongvigitmanee, Wisut Titiroongruang, King Mongkut's Institute of Technology Ladkrabang; Archkom Srihapat, Amporn Poyai, Thai Microelectronics Center

4.17 Virtual Metrology for Predicting Average PECVD Oxide Film Thickness 157

Ariane Ferreira, Agnés Roussy, Christelle Kernaflen; Ecole Nationale Supérieure des Mines de Saint-Etienne; Dietmar Gleispach, Günter Hayderer, austriamicrosystems AG; Hervé Gris, Jérôme Besnard, PDF Solutions

4.18 Wet Etch Step Modeling for Shallow Trench Isolation Module Control 163

Agnés Roussy, M. Gedion,, EMSE-CMP; N. Crousier, J. Pinaton K. Labory, STMicroelectronics

4.19 Yp - Ypk: Product Test Yield and Yield Dispersion Indicators 169

Matthias Thomas Bostelmann, ALTIS Semiconductor

Session 5: Advanced Process Development and Control

5.1 Neural Network Modeling of Fabrication Yield Using Manufacturing Data 173

Zubin Mevawalla, Gary S. May, Georgia Institute of Technology; M. Honjo, M.W. Kiehlbauch, Micron Technology

5.2 Reducing Environmentally Induced Defects While Maintaining Productivity 179

R. van Roijen, S. Conti, R. Keyser, R. Burda, J. Ayala, J. Maxson, R. Henry, E. Meyette, W. Steer, K. Tabakman, C. Yu, IBM Systems and Technology

5.3 Cost Effective and Robust Nickel Silicidation Process Qualification and Chamber Matching in Rapid Thermal Processing Tools 184

Weihua Tong, K. Suresh, Miowchin Tan, Peter Benyon, Vish Srinivasan, Jinping Liu, GLOBALFOUNDRIES

5.4 A New Device for Highly Accurate Gas Flow Control With Extremely Fast Response Times 188

Kevin Boyd, IBM Semiconductor Research and Development Center; Adam Monkowski, Jialing Chen, Tao Ding, Joseph Monkowski, Pivotal Systems

5.5 Thermal Budget Reduction and Throughput Enhancement for CMOS Epi Stressors via Wet Clean Interface Contamination Evaluation and Control 193

Paul Brabant, Keith Chung, Manabu Shiniriki, Scott Hasaka, Dane Scott, Mark Wirzbibki, Terry Francis, MathesonGas; Hong He, Devendra K. Sadana, IBM Research at Albany NanoTechnology Center and Technology Group

Session 6: Advanced Lithography

6.1 Optimization of Pitch-Split Double-Patterning Photoresist for Applications at 16 nm Node 196

Steven J. Holmes, Yunpeng Yin, Chiew-seng Koay, Shyng-Tsong Chen, Karen Petrillo, Guillaume Landie, Scott Halle; Sean Burns, John C. Arnold, Terry Spooner, Matthew Colburn, Rex Chen, S. Liu, R. Varanasi, IBM; Cherry Tang, Nicolette Fender, Brian Osborn, Lovejeet Singh, Mark Slezak, JSR Micro; Sumanth Kini, KLA-Tencor; Hideyuki, Tomizawa, Toshiba; Shannon Dunn, Jason Cantone, David Hetzer, Shinichiro Kawakami, Tokyo Electron Technology

6.2 Investigation of Noise Sources in the Focus Control Process for Immersion Lithography 204

Jasper Paul Munson, Jay Brown, Nikon Precision

6.3 Increase Fab Capacity with Predictive Short-Interval Scheduling 211

David Hanny, Applied Materials

6.4 Extendible Scanner Platforms for Mass Production, Now and In the Future 215

Hamid Khorram, Nikon Precision

6.5 Strategies for Single Patterning of Contacts for 32 and 28 nm Technology 222

Bradley Morgenfeld, Ian Stobert, Ju J An, Massud Aminpur, Colin Brodsky, Alan Thomas, IBM Corporation; Hideki Kanai, Toshiba America; Norman Chen, GLOBALFOUNDRIES; Henning Haffert, Martin Ostermayr, Infineon Technologies NA

Session 7 Factory Optimization II:

7.1 Non-Contact-Handling and Transportation for Substrates and Microassembly using Ultrasound-Air-Film-Technology 230

Gunther Reinhart, Michael Heinz, Johannes Stock, Technische Universität München; Josef Zimmerman, Michael Schilp, Adolf Zitzmann, Jens Hellwig, Zimmermann & Schilp Handhabungstechnik GmbH

7.2 Lithography Cost Savings Through Resist Reduction and Monitoring Program 236

Terri Couteau, Scott Lindauer, Chris Stewart, Spansion; Jennifer Braggin, Brent Bjornberg, Entegris

7.3 Improvements Wafer Placement Repeatibility and Robot Speed Improvements for Bonded Wafer Pairs Used in 3D 240

Andrew Rudack, SEMATECH; Michael Dailey, Fabworx Solutions, Inc..

7.4 200mm Fab AMHS Improvement During Aggressive Ramp 245

Sylvain Bouhnik, Micron Semiconductor

Session 8: Defect Inspection and Yield Optimization

8.1 Automated SEM Offset Using Programmed Defects 249

Oliver Patterson, Andrew Stamper, IBM Semiconductor Research and Development Center; Roland Hahn, KLA-Tencor

8.2 Post Etch Killer Defect Characterization and Reduction in a Self-aligned Double Patterning Technology 254

Hong-Ji Lee, Sun-Yi Lin, I-Ting Lin, Kuo-Liang Wei, Shen-Yuan Chang, Nan-Tzu Lian, Tahone Yang, Kuang-Chao Chen, Chih-Yuan Lu, Macronix International

8.3 A System to Optimize Inline Defect Detection Using Short Loop Testchips Leading to Faster Yield Learning 258

Tanya Yang, Hun Chow Lee, Victor Lim, Fang Hong Gn, Tri Mardiyono, Qionghan Wang, Long Phan Nguyen, GLOBALFOUNDRIES; Fei Li, Sa Zhao, Anand Inani, PDF Solutions

8.4 A Quality Metric for Defect Inspection Recipes 262

Ralf Buengener, GLOBALFOUNDRIES; Julie L. Lee, Brian M. Trapp, John A. Rudy, IBM Microelectronics

Session 9: Data Management

9.1 Managing Data for a Zero Defect Production 269

Gottfried Schmid, Infineon Technologies AG; Tilmann Hanitzsch, Senior Management Consultant

9.2 The Deployment Page: Integrated Real Time Views of Tools, Operations, and Lots 274

Henry Antonovich, IBM Microelectronics

9.3 Cycle Time Prediction in Wafer Fabrication Line by applying Data Mining Methods 280

Israel Turkel, Ben-Gurion University of the Negev

9.4 Recent Innovations in CMOS Image Sensors 285

Ray Fontaine, Chipworks, Inc.

Session 10: Defect Inspection and Yield Optimization

10.1 Parametric Composite Limited Yield Index for Functional Circuits Yield Prediction 290

Jiun-Hsin Liao, Ishtiaq Ahsan, Ronald Logan, George Rudgers, Fred Towler, IBM Microelectronics

10.2 Embedded Memory Fail Analysis for Production Yield Enhancement 294

Youssef Baltagi, Daniele Li Rosi, Vincenzo Tancorre, Christophe Garagnon, Eric Faehn, Mario Barone, Davide Appello, STMicroelectronics; Christophe Suzor, Synopsys

10.3 45nm Yield Model Optimization 299

Brian L. Walsh, John Colt, Jr., Daniel Poindexter, Thomas Joseph, IBM Systems and

Technology Group

10.4 Yield Optimization for Third Party Libraries Elements 303

Jeanne Paulette Bickford, Francis Chan, Mark Styduhar, Lee Wang, Robert Arelt, Ioanoa Graur, Steven Parker, Deborah Ryan, Tina Wagner, IBM Systems and Technology Group; Anand Kumaraswamy, IBM Corporation

2011 ASMC ORGANIZING COMMITTEE

Our special appreciation to ASMC committee members, who together volunteered countless hours to the organization of the 22nd Annual IEEE/SEMI Advanced Semiconductor Manufacturing Conference.

CONFERENCE Co-CHAIRS: Scott Lantz, Intel Corporation
Holly Magoon, Nikon Precision

GENERAL Co-CHAIRS: Jeff Barnum, KLA-Tencor
Daniel Maynard, IBM Microelectronics

ASMC 2011 COMMITTEE:

Sumita Basu, Intel Corporation
Duane Boning, PhD., Massachusetts Institute of Technology
Jennifer Braggin, Entegris, Inc.
Thanas Budri, National Semiconductor Corporation
Thomas Carbone, Fairchild Semiconductor
Christian Carlier, STMicroelectronics
Shi-Chung Chang, National Taiwan University
John Conway, Intel Corporation
Alain Diebold, PhD., University at Albany
Russell Dover, Brion Technologies (an ASML company)
Eric T. Eisenbraun, PhD., University at Albany
Eric Englhardt, PhD., Applied Materials
Patrice T. Flack, PhD.
Rainer Gehres, IBM Systems & Technology Group
Otto Graf, Infineon Technologies
Gary Green, Green Consulting
Dave Gross, GLOBALFOUNDRIES
Christopher Hess, PDF Solutions, Inc.
Grant Imper, JDS Uniphase
Dick James, Chipworks Inc.
Franz-Josef Kahlen, University of Cape Town
Fourmun Lee, ASM America, Inc.
John Lin, Taiwan Semiconductor Manufacturing Company, Ltd.
Christopher Long, IBM Research

Scott McClure, IBM Systems & Technology Group
Mike McIntyre, GLOBALFOUNDRIES
Winfried Meier, Nikon Precision Europe GmbH
Hanno Melzner, Infineon Technologies AG
William Miller, IBM Microelectronics
Kevin Nason, Fairchild Semiconductor
Kazunori Nemoto, Ph.D., Hitachi High Technologies
Viraj Pandit, Novellus Systems
Oliver D. Patterson, PhD., IBM Microelectronics
Larry Pulvirent, GLOBALFOUNDRIES
Huatan Qiu, Novellus Systems
Stefan Radloff, Intel Corporation
Dieter Rathei, D R YIELD Software & Solutions
Ron Remke, International SEMATECH
Theresa Roeder, PhD., San Francisco State University
Leonard Rubin, Ph.D., Axcelis Technologies
Arthur Tay, PhD, National University of Singapore
Thuy Tran-Quinn, IBM Corporation
Helmuth Treichel
Jacek Tyminski, PhD., Nikon Precision Inc.
Alok Vaid, GLOBALFOUNDRIES
Aarthi Venkateshan, Ph.D., Canon Anelva Corporation

James Lu, PhD., Rensselaer Polytechnic
Institute
Johann Massoner, Infineon Villach

Paul Werbaneth, MEMS Investor Journal, Inc
Brett Williams, ON Semiconductor
Naomi Yoshida, Applied Materials

The Global Supply Chain Is Our New Fab: Integration and Automation Challenges

Hans Ehm, Thomas Ponsignon, Thomas Kaufmann

Corporate Supply Chain
Infineon Technologies AG
Neubiberg, Germany
hans.ehm@infineon.com, thomas.ponsignon@infineon.com, thomas.kaufmann1@infineon.com

Abstract—**The globalization of the world economy as well as progresses in information technology made global supply chains a new paradigm for high-tech and semiconductor manufacturers like Infineon Technologies. Consequently supply chain operational excellence has become a key competitive advantage. Along with it comes the need for an agile, adaptable, and aligned global manufacturing network for mastering the volatile market demand – known as the triple-A challenge. A high degree of integration and automation is required across all stages of the value chain: equipment, factory, and supply chain levels. In this paper we describe several factors from shop-floor to corporate level addressing this challenge, and we outline a successful example of frontend-backend integration.**

Global Supply Chain; Manufacturing Network; Triple-A Challenge; embedded Wafer Level Ball;

I. SUPPLY CHAIN OPERATIONAL EXCELLENCE: A KEY COMPETITIVE ADVANTAGE IN A GLOBALIZED WORLD

In the last decades global supply chains have become a standard due to diminished trade barriers that made resource sharing attractive. Also tremendous progresses in information technology now allow managing and controlling complex systems (Stadler and Kilger, 2007). As a result the scope of operational excellence is no longer confined to single fabs, but it encompasses nowadays the concept of manufacturing network: the global supply chain is our new fab (Chien, 2007; Chien *et al.*, 2008).

Achieving an efficient utilization of the resources of the global supply chain is a must for those surviving in the semiconductor business but it remains quite challenging due to several factors. First, it usually involves dozens of in-house frontend and backend fabrication sites plus silicon foundries and subcontractors organized as a dynamic network that is spread all over the world. Secondly, it has to be dealt with long lead times while products tend towards shorter life cycles and steeper production ramp ups. In addition, semiconductors are certainly in one of the most volatile markets with cyclic up- and downturn phases, which make the demand very difficult to predict. Finally, capacity expansions typically take long times and are expensive, which comes in contradiction with the need for flexibility (Uzsoy *et al.*, 1992; Gupta *et al.*, 2006). Consequently, supply chain operational excellence along the Plan, Source, Make, Deliver, and Return processes (Supply Chain Council, 2011) has become a key competitive advantage

in the globalized world of semiconductors. To overcome and also to take advantage of the characteristics of the semiconductor industry an Agile, Adaptable, and Aligned supply chain is required. It has been made famous as the Triple-A Challenge (Lee, 2004). Agility describes the ability to deal with demand and supply uncertainties, Adaptability depicts a dynamic supply chain that copes with shortening product and technology cycles, and Alignment balances the interests of multiple players in the supply chain.

Looking at the flows of materials that occur between the workstations of a wafer fabrication site (Hopp and Spearman, 1997), similar complexity can be observed between the locations of a semiconductor network (Hopp, 2008). In the following we show an example of a successful Agile, Adaptable, and Aligned global supply chain at Infineon Technologies of a highly integrated chip for a platform chipset. The product is initially manufactured – not prototyped – at the only technically qualified factories, i.e. wafer fabrication in Germany, bumping in Taiwan, testing back in Germany, assembly in Korea, final test back in Germany. Whilst the product matures and penetrates the market, it requires more capacity and new routing opportunities – including production partners – since the manufacturing cost begins to dominate further growth. Thus, after a year of booming demand the chip has successively used more than fifteen different supply chains (Fig. 1).

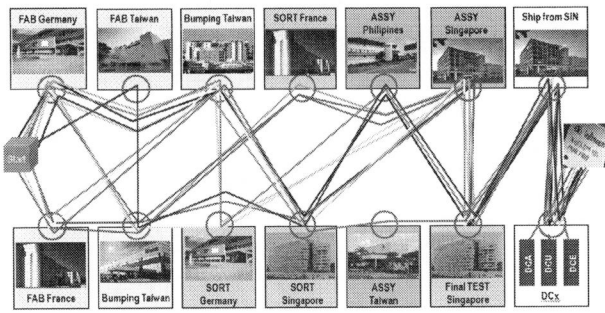

Figure 1. Example of a Triple-A supply chain, i.e. Agile, Adaptable and Aligned, of a highly integrated chip for a platform chipset at Infineon Technologies: the chip went through more than fifteen successive supply chains in a one year time period to be able to satisfy booming market demand.

978-1-61284-408-4/11 $26.00 © 2011 IEEE

Each new route allowed either increasing throughput or decreasing cost (Fig. 2). Obviously, the cost would have been even lower using one single low cost production site for the entire value chain, but it is technically hardly achievable within the short product life cycle, also the risk would be too high since the manufacturing flexibility is lost, and the learning from best practices of neighbor routes would be lost. Hence, by harvesting the opportunities of a global network, cost reduction of mid double digit percent ranges within one year can be reached.

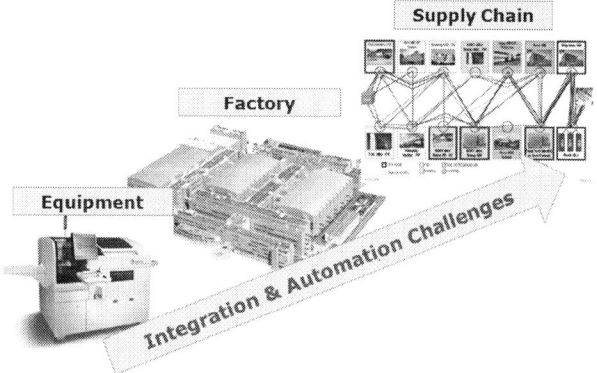

Figure 2. On the example of the chip from Fig. 1, the Triple-A supply chain allowed a much steeper production ramp up than initally forecasted, i.e. +100% throughput after four quarters, while achieving drastic cost reductions.

II. CHALLENGES FROM EQUIPMENT TO SUPPLY CHAIN: INTEGRATION AND AUTOMATION

Mastering the triple-A challenge as described above requires a certain degree of automation for dealing with flows of goods and information at each level of the global manufacturing network as well as an enhanced integration across all stages of the value chain. In the following we outline several key factors from equipment, factory, and supply chain levels, which contribute to the operational excellence of Infineon Technologies (Fig. 3).

Figure 3. Integration and automation challenges need to be tackled across all levels, i.e. from equipment, to factory, to supply chain levels, in order to achieve operational excellence.

The best example of operational excellence in the recent past is probably the management of the threefold increase of the factory loading in certain technology nodes in only 12 month time after the "cold steel phase" in 2008. The steep production ramp up was of major importance for a timely increase of the market share. The practical consequences imply for example integrating new equipment on the shop-floor in short delays, dealing with a tremendous amount of data at each fabrication site, and keeping track of lots during intra- and inter-site movements.

A. Equipment and Single Process Level

To guarantee the quality of the chips under the constraint of dynamic, complex, and globally distributed means of production, advanced information flow strategies are required. It is achieved by implementing Statistical Process Control (SPC) as well as Advanced Process Control (APC) techniques directly at the equipment level (Montgomery, 2001; Moyne *et al.*, 2001). A fully automated data acquisition is a pre-requisite to be able to handle the important amount of data that is needed. SPC and APC methods are part of an approach that enables to specify and improve quality metrics of general application (Ryan, 2000), to detect violations and process trends in time, to react quickly, and to prevent future abnormal occurrences. Among others Western Electric rules allow reacting before an impending quality metric violation could stop the material flow (Western Electric Co., 1985). By means of these techniques a consistent monitoring report of the Overall Equipment Efficiency (OEE) across the shop-floor is provided to planning organizations. Consequently, production recipes and test programs are adjusted accordingly.

Furthermore, the mentioned well-known best practices are extendable to other levels of the manufacturing network for documenting the quality of all processes, not only for the fabrication but also supply chain planning processes. (Russland *et al.*, 2011) shows an example on how to monitor demand forecasts with the help of SPC methods.

Finally, the reliable control of processes on the equipment level and single process level with the help of SPC and APC methods is the foundation for flexibility options at higher levels.

B. Factory Level

Paperless manufacturing is becoming standard in semiconductor manufacturing. The main challenge not only consists in having it done at the start of a manufacturing line; more often the challenge comes when flexible processing is required as IT systems have to prove that they are able to organize routine processes as well as to handle agile production. Besides others, this is made possible at Infineon Technologies via an indoor GPS system for easy and reliable material identification and also with an access control system for operators, i.e. only those who are qualified to run a certain type of equipment or process are allowed to operate it.

The factory challenge in semiconductor manufacturing has always been to achieve both high utilization rates and fast processing times (Leachman *et al.*, 2007). The operating curve management is the toolset that is used at Infineon Technologies

978-1-61284-408-4/11 $26.00 © 2011 IEEE

to reduce the alpha value (Fig. 4). Indeed, alpha is the measurement of the variability, which means, a low alpha value enables higher capacity utilization and higher throughput for a given X-factor value, i.e. the ratio between cycle time and raw process time. State-of-the-art semiconductor manufacturing fabs typically run their production with an X-factor value between two and three (Fowler and Robinson, 1995; Robinson *et al.*, 2003).

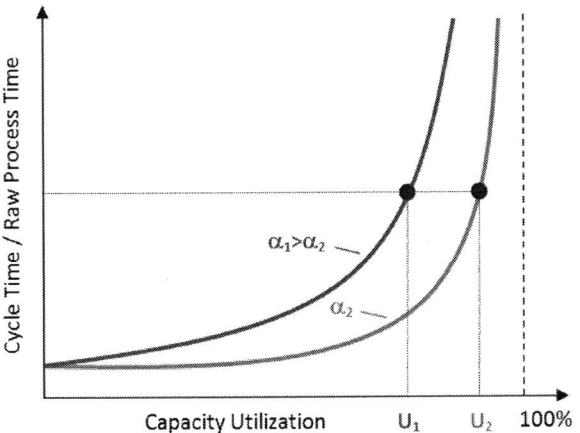

Figure 4. For a given ratio between cycle time and raw process time, known as the X-factor, a higher capacity utilization $U_2 > U_1$ is reached via lower variability $\alpha_2 < \alpha_1$.

In order to decrease the overall alpha value of the manufacturing network, the variability within each fab needs to be diminished. Variability may be due to equipment downtimes (scheduled or unscheduled), operator availability (physical presence, trainings...), process issues, cleanroom disturbances, raw material availability. All these disruption parameters need to be as low as possible, and those which are unavoidable should be synchronized. Thus, the synchronization can be done using the 4- (or X-) partner method (Fig. 5). It is shown that the variability of the entire system goes down when synchronization is achieved, e.g. the break of an operator occurs at the same time as a scheduled maintenance of the equipment and the scheduling system takes into account that the work-in-progress in front of the machine has to be processed before the maintenance starts.

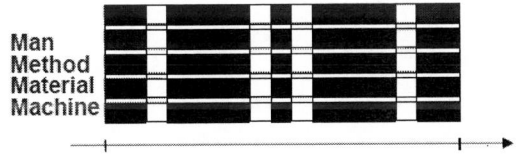

Figure 5. The 4-partner method is used to synchronize disruptions that occur during the processing; it allows decreasing the variability (i.e. alpha).

In addition, Automated Material Handling Systems (AMHS) as well as automated cleanroom monitoring are further enablers for stabilized processes in manufacturing areas.

C. Global Supply Chain Level

Having statistical control for equipment and process as one pillar, and operating curve management, X-factor and X-partner method as a second pillar on the factory level, the top area is now the supply chain level. Hence, the challenges faced in each factory are now transposed on a global scale. In other words, equipment on factory level becomes an entire fabrication site among the whole production network, and a manufacturing route linking equipment on shop-floor with respect to the fabrication recipe becomes a global supply chain that describes the sequence of fabrication sites being used.

The challenge on global supply chain level becomes integrating all elements of a supply chain. The material flow, the information flow, and the value flow need to be synchronized. For a semiconductor company like Infineon Technologies, where almost every produced chip is traveling once around the world from raw wafer until entering the production site of the customer, the value flow managed in all aspects is one of the key enablers in order to always fulfill custom and financing requirements and for taking full benefit of the globalization.

Global process change management (including qualifications), transit time measurements for regular flows, as well as standardized emergency shipment procedures for latest demand changes are other aspects of managing global supply chains. In addition, a master data management that is globally harmonized and interlinked (Fig. 6) is a mandatory enabler to support blended best-of-breed IT Tools (Fig. 7).

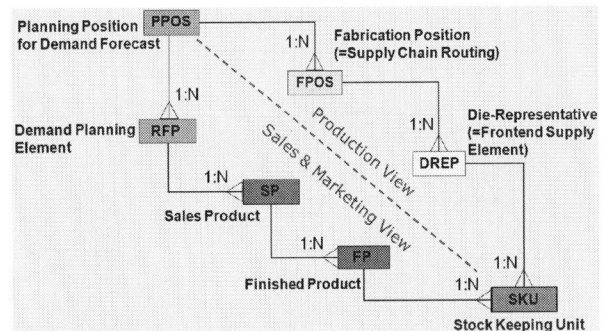

Figure 6. An effective master data management based on a consistent structure is necessary to support suply chain processes and IT tools.

Enablers of the enabler are the skills needed for the human interactions. Indeed one common "language" based on the Supply Chain Operations Reference (SCOR) model (Supply Chain Council, 2011) is required to allow having an aligned supply chain. Furthermore, a global online training system – called the Supply Chain Academy – is also very beneficial to keep the same level of knowledge across all worldwide sites.

978-1-61284-408-4/11 $26.00 © 2011 IEEE

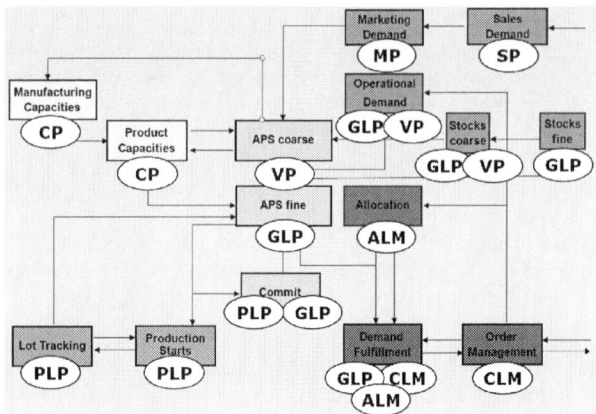

Figure 7. Various planner communities are involved to perform the different planning processes as shown on this simplified representation of Infineon's planning landscape (SP: Sales Planner; MP: Marketing Planner; VP: Volume Planer; CP: Capacity Planner; GLP: Global Logistics Planner; PLP: Production Logistics Planner; CLM: Customer Logistics Manager; ALM: Allocation Manager). A common understanding and level of knowledge is required company-wide to achieve an aligned supply chain.

III. A SUCCESSFUL EXAMPLE OF INTEGRATION AND AUTOMATION: BACKEND MEETS FRONTEND

One of Infineon's successful examples of integration and automation among all three levels is the introduction of the embedded Wafer Level Ball (eWLB) technology. This wafer level packaging solution uses a combination of traditional frontend and backend manufacturing techniques (Fig. 8), which has the potential to reduce the package size, to diminish manufacturing costs, and to speed up production since the backend process is literally integrated into frontend manufacturing steps (STATSChipPAC, 2007).

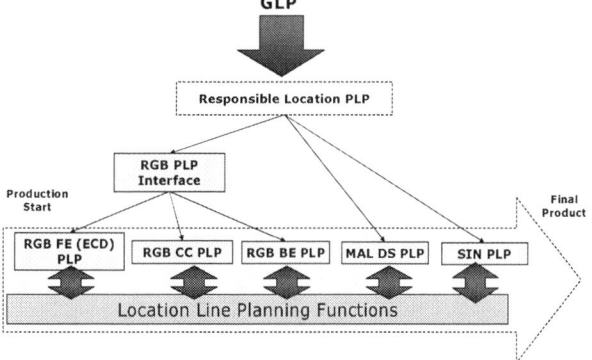

1) **Reconstituted wafer**
   - Wafer saw and pick-and-place from incoming wafer
   - Probed good die
   - Molded reconstituted wafer using proven materials
   - Molded artificial wafer starting point for thin film technology
2) **Redistribution**
   - Thin film technology with advanced design rules
   - Standard thin film equipment
   - Proven and reliable material set
3) **Ball Mount and Singulation**
   - Standard back-end assembly flow (and equipment)
4) **Test, Mark, Scan, Pack**
   - Standard or wafer level-based test flow
   - Standard assembly

Figure 8. The introduction of the embedded Wafer Level Ball (eWLB) technology was and is still challenging from equipment, to factory, to supply chain level due to its modified process flow.

The implementation of the new production process was quite challenging across all levels due to its modified process flow. In the following, we describe the difficulties encountered during the eWLB Supply Chain Task Force focusing mainly on

chipsets for the wireless communication market segment. Note that Infineon Technologies sold its Wireless business division to Intel Corporation in the first quarter of 2011 (Intel and Infineon Technologies, 2010).

A. Equipment Level

The biggest challenge we faced at equipment level while introducing eWLB was to enable the traceability of the chips up to the silicon wafers. Traditionally, the necessary data are lot and wafer identification numbers. However, during the eWLB reconstitution process each silicon wafer is cut into dices that are picked and placed onto the mold wafers. The final test maps for silicon wafers can only be reconstructed from information available from mold wafer test maps. Therefore, a dedicated database has been created with data from die placement equipment – i.e. identification numbers of silicon and mold wafers and x,y-coordinates of each die on these wafers – and from test equipment – i.e. identification numbers and x,y-coordinates for mold wafers. This single die tracking concept guarantees the chip traceability.

B. Factory Level

The manufacturing sequence of eWLB gets beyond the traditional distinction between frontend and backend since frontend equipment is being used for the redistribution process. The resource sharing brought challenges on factory level.

First, in terms of production planning and scheduling a logistics interface organization dedicated to eWLB production control has been introduced along with the creation of a responsible location Production Logistics Planner role that coordinates the utilization of frontend and backend shared capacities (Fig. 9).

Figure 9. The responsible location Production Logisitcs Planner (PLP) coordinates the utilization of frontend and backend shared resources for eWLB manufacturing (GLP: Global Logistics Planner; RGB: Regensburg prod. site; SIN: Singapore prod. site; MAL: Malacca prod. site; CC: Chip Card product segment; DS: discrete product segment).

Secondly, for lot tracking purpose we needed visibility of the frontend redistribution manufacturing step in the backend tracking. This was only possible by re-programming the lot tracking tool, the yield tracking tool, and the traceability tool in order to work in a "pool concept", which means, to give

978-1-61284-408-4/11 $26.00 © 2011 IEEE

visibility simultaneously to frontend and backend tools (Fig 10).

Thirdly, in terms of ownership of yield failures, a deficiency happening during the redistribution process is due to a frontend technology but is detected during the backend manufacturing stage. Hence, for the very first time the collaboration between both frontend and backend quality department was needed to identify and assign frontend failures to the backend production. This is the only way to guarantee the outmost quality of the products.

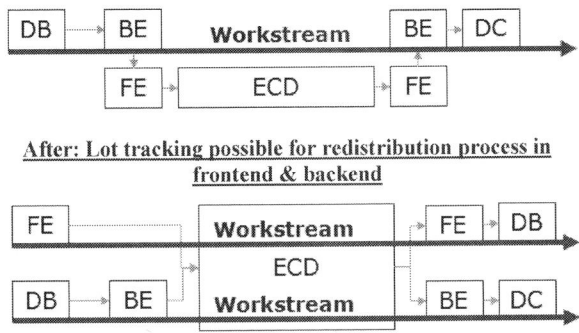

Figure 10. With the tools available the lot tracking during the redistribution process was not possible in backend. Thanks to the re-programmation of the tools to a pool concept approach, visibility was enabled in both frontend and backend (ECD: Lithography and Electrical & Chemical Deposition equipment; Workstream is the tool being used for WIP monitoring)

C. Global Suply Chain Level

On supply chain level, the first challenge encountered was the definition of a new product structure and the corresponding data handing process when creating a new product. Traditional assembly specifications (e.g. wafer data and technology, packing sequence, package type, ball apply material, back side protection material...) were extended to include both reconstitution (e.g. molding material) and redistribution process data (e.g. process line, pad metallization material...). Also, the release of this new assembly specification implied the creation of a new assembly data creation workflow involving nine departments in three locations.

Another challenge was to agree on the splitting and merging rules for the reconstitution process since it impacts the extent of a quality issue, i.e. the number of lots that need to be blocked or the number of wafers that need to be scrapped.

We also needed to agree on the Country Of Assembly (COA) since the assembly process steps (i.e. dicing of the silicon wafer, reconstitution, rounding, redistribution, marking, ball apply, and dicing of mold wafer) are physically done in multiple locations, but for legal purpose the main assembly location has to be defined. Different approaches have been considered (e.g. looking at cycle time, cost...). The final decision has been taken to align with other products and

to choose the ball apply production site as the main assembly location.

Another hurdle was to adapt Infineon's key performance indicators (i.e. confirmed line item performance, yield control, WIP and cycle time monitoring...) in order to include both reconstitution and redistribution process; e.g. changes to the yield reporting structure had an impact on all Infineon planning processes from the calculation of wafer starts for a single process step up to the monthly volume rolling forecast process on corporate level.

CONCLUSION

In this paper we have shown that today's progress in semiconductor manufacturing goes beyond the strict control of physical and chemical processes in equipment and single process steps controlled via Statistical Process Control. It even goes beyond the particle-free manufacturing in fabs that are continuously optimized via operating curve management. Semiconductor manufacturing benefits from the flexibility options of global supply chains more than other industries since transportation costs are more negligible than for most of other high-tech products. Agile and adaptable supply chains emerge when the complexity in material, information, and value flows is understood, reduced to a minimum, and is perfectly managed. If those that work in semiconductor supply chains can be used in aligned end-to-end supply chains from suppliers' suppliers to customers' customers, then we are very close to the saying of Dr. Hau Lee: "instead of company to company competition, we are now in an era of supply chain to supply chain competition" (Lee, 2010).

ACKNOWLEDGMENTS

The authors would like to thank Geraldine Yachi for her contribution to the third part of this paper.

REFERENCES

[1] C.-F. Chien, "Made in Taiwan: shifting paradigms in high-tech industries," Industrial Engineer, vol. 39, no. 2, pp. 47–49, February 2007.

[2] C.-F. Chien, S. Dauzère-Pérès, H. Ehm, J. W. Fowler, Z. Jiang, S. Krishnaswamy, L. Mönch, and R. Uzsoy, "Modeling and analysis of semiconductor manufacturing in a shrinking world: challenges and successes," Proceedings of the Winter Simulation Conference, pp. 2093–2099, December 2008.

[3] J. W. Fowler and J. K. Robinson, "Measurement and Improvement of Manufacturing Capacity (MIMAC) – Designed experiment report," SEMATECH Technology Transfer #95062861A-TR, July 1995.

[4] J. N. D. Gupta, R. Ruiz, J. W. Fowler, and S. J. Mason, "Operational planning and control of semiconductor wafer fabrication," Production Planning and Control, vol. 17, no. 7, pp. 639–647, October 2006.

[5] W. J. Hopp, Supply Chain Science, New York: McGraw-Hill, 2008.

[6] W. J. Hopp and M. L. Spearman, Factory Physics, Yearbook of Science and Technology, New York: McGraw-Hill, 1997.

[7] Intel and Infineon Technologies, "Intel to acquire Infineon's wireless solutions business," Joint news release, August 2010.

[8] R. C. Leachman, S. Ding, and C.-F. Chien. "Economic efficiency analysis of wafer fabrication," IEEE Transactions on Automation Science and Engineering, vol. 4, no. 4, pp. 501–512, October 2007.

[9] H. L. Lee, "The triple-A supply chain," Harvard Business Review, vol. 82, no. 10, pp. 102–112, October 2004.

978-1-61284-408-4/11 $26.00 © 2011 IEEE

[10] H. L. Lee, "Matching supply and demand with 'sensible sense' and 'responsive response'," SCM World Live: RaptureWorld webinar, February, 2010.

[11] D. C. Montgomery, Introduction to Statistical Quality Control, 4th ed. New York: Wiley, 2001.

[12] J. Moyne, E. Del Castillo, and A. M. Hurwitz, Run-To-Run Control in Semiconductor Manufacturing, Boca Raton: CRC Press, 2001.

[13] J. K. Robinson, J. W. Fowler, and E. Neacy, "Capacity loss factors in semiconductor manufacturing," working paper, March 2003.

[14] T. P. Ryan, Statistical Methods for Quality Improvement, Wiley Series in Probability and Statistics, 2nd ed. New York: Wiley, 2000.

[15] T. Russland, H. Ehm, and T. Ponsignon, "A worflow management system for cross-system processes on the example of Infineon Technologies AG," unpublished, Feburary 2011.

[16] H. Stadler and C. Kilger, Supply Chain Management and Advanced Planning: Concepts, Models, Software, and Case Studies, 4th ed. Berlin: Springer, 2007.

[17] STATSChipPAC, "Embedded wafer level ball grid array," http://www.statschippac.com/services/packagingservices/waferlevelprod ucts/ewlb.aspx, February 2011.

[18] Supply Chain Council, "Supply Chain Operations Reference (SCOR) model," http://supply-chain.org/scor, February 2011.

[19] R. Uzsoy, C.-Y. Lee, and L. A. Martin-Vega, "A review of production planning and scheduling models in the semiconductor industry – part I: system characteristics, performance evaluation and production planning," IIE Transactions, vol. 24, no. 4, pp. 47–60, September 1992.

[20] Western Electric Co., Statistical Quality Control Handbook, 11th ed. DC: Delmar Printing Company, 1985.

Ambient Persuasion in the Factory:
The Case of the Operator Guide

Alexander Mechtscherjakov, Patricia Kluckner,
Florian Pöhr, Wolfgang Reitberger, Astrid Weiss,
Manfred Tscheligi
CDL on Contextual Interfaces
ICT&S Center, University of Salzburg
Salzburg, Austria
firstname.lastname@sbg.ac.at

Karl Horst Hohenwarter,
Peter Oswald

CDL on Contextual Interfaces
Infineon Technologies Austria AG
Villach, Austria
firstname.lastname@infineon.com

Abstract—Increasing the efficiency in production while maintaining a low error rate is one of the key goals of semiconductor factories. Often this is achieved on a technical level but there is significant potential on the human side as well. By reducing the complexity of the work environment and presenting the right information at the right time and place operators can be supported in their decision making process and therefore the productivity can be increased. Ambient Persuasion, which combines ambient intelligence and persuasive technology, is a promising approach in this direction. In this paper we present the "Operator Guide", an ambient persuasive system that aims to improve work efficiency and to reduce error response times of operators by showing the next best working steps. We provide a detailed description of the Operator Guide along with an outlook of the next steps within a user-centered design approach.

Keywords-ambient persuasion; ambient intelligence; persuasive technology; operator guide

I. INTRODUCTION

From a semiconductor company's perspective the most important goals within the production cycle are twofold. On the one hand the goal is—as in most commercial areas—the improvement of overall productivity with as little resources as possible leading to a possible optimal efficiency. On the other hand a low error rate is indispensable to provide perpetually high product quality and reduce the loss of defective products. The supreme goal of all is a zero-defect production, since in semiconductor factories produced goods are very expensive. In order to reach these goals one strategy could be a full automation of the manufacturing process to eliminate human errors. Since short-term changes in an automated manufacturing process are expensive or even impossible full-automation is not always desirable. More flexibility can be provided by the employment of human operators. They are able to react to problems immediately. A possible solution to reduce errors and simultaneously keep flexibility at a high level is for instance the "smart factory" strategy. This approach consists of a prudent interplay between human operators and manufacturing equipments [1].

We propose that "ambient persuasion" is a promising approach in the same direction. It combines ambient

intelligence (AmI), which refers to the pervasion of the everyday world with information technology [3] and persuasive technology, which describes a system designed to change people's attitudes or behaviors [4]. Ambient Persuasion allows surrounding the users with persuasive technology in their work environment. This approach enables operators to intervene at the right time and in the right place. We have already shown that ambient persuasion in the factory context is a fruitful scientific approach [6].

In this paper we present the Operator Guide (OG), an ambient persuasive system that aims to improve operator work efficiency, and simultaneously reduce error response times of operators by providing information on the next best working steps. This paper provides a detailed description of the user-centered design approach for the OG, and covers important design decisions. Additionally we provide an overview of the system architecture and implementation.

II. BACKGROUND

In this chapter we present a short overview on the research field of human-computer interaction and the related user-centered design approach. Based on that we will explain the interaction paradigms ambient intelligence, persuasive technologies, and ambient persuasion.

A. Human-Computer-Interaction and User-centered Design

Human-Computer Interaction (HCI) is an interdisciplinary research field, which combines computer science with aspects, such as human factors, ergonomics, industrial engineering, and cognitive psychology. The SIGCHI, the Association for Computing Machinery Special Interest Group on Computer-Human Interaction, describe HCI as "A discipline concerned with the design, evaluation and implementation of interactive computing systems for human use and with the study of major phenomena surrounding them." [17]. Another definition is provided John Carroll, who defines HCI as "The study and the practice of usability. It is about understanding and creating software and other technology that people will want to use, will be able to use, and will find effective when used" [16].

The essential concepts of HCI are: the user, the computer, and the interaction between those two. The user can be an individual, a group of users who work together or even a sequence of users within an organization, who are using a computing system. The term computer does not only include PCs, but any kind of technology, from desktop computers to large-scale computer system, a process control system in a manufacturing facility, or an embedded system, such as a washing machine. These two components—the user and the computer—are linked by the interaction, which refers to any communication between the user and the computer. The communication between the user and the computer is mediated by an interface, which provided feedback to the user and can be actively used by him/her. Thus, the goal of HCI is to optimize the interface and subsequently the interaction with it. A common approach in HCI to do so is user-centered design.

User-centered design (UCD) is a design philosophy, which focuses on the needs and wants of potential users in the process of system development. The goal of this approach is, that users do not need to change their habits or working routines after the novel system is integrated in their working environment. The International Standard Organization (ISO) defines UCD as process consisting of four activities [18]:

1. Specify the context of use

2. Specify requirements: Requirements on the user, business and interface side need to be considered

3. Create design solutions

4. Evaluate designs: The quality testing of the system.

The cycle (process) ends, when the system meets all specified requirements. For the development of the operator guide we adapted this UCD approach, according to the requirements of the clean room setting (see Section III) with the goal to create a system, the operators are willing to use.

B. Ambient Persuasion

Ambient persuasion is a concept in HCI that is a combination of various paradigms and philosophies, namely awareness, distributed cognition, ambient intelligence, and persuasive technology.

Firstly, awareness is regarded as a main influence factor for successful cooperation between co-workers who are mediated by computing systems. Rittenbruch and McEwan define awareness as a "fundamental quality of collaborative work, the ability of co-workers to perceive each others' activities and expressions and relate them to a joint context" [2].

Secondly, the theory of distributed cognition aims at understanding the relationship between people and technologies, emphasizing the whole environment beyond the individual. This includes interaction between people, but also between people and artifacts. Thereby, cognitive processes are not limited to the individual, but can be distributed across a team and stretch across materials and time [5].

Thirdly, ambient intelligence (AmI) refers to the pervasion of the world with information technology [3]. Users in an AmI environment are every time in contact with multiple systems, which are mostly invisible. The flow of information is supported by linked up and embedded systems as sensors, processors, and actuators, which communicate with each other and interact with users. AmI research developed constantly to an innovative und sustainable concept, which includes natural interaction, ambient displays, and implicit interaction. Natural interactions indicate the usage of gestures, speech, gaze and movement to communicate with the system. Ambient displays integrate digital information into the physical environment, manifesting themselves as subtle changes in form, movement, sound, color, smell, temperature, or light [15]. In contrary to traditional displays, ambient displays are perceived and not used actively [11]. Implicit interaction relates to input, which is not explicitly entered by the user, but is implicitly dependent of the user's context [12].

Fourthly, persuasive technology is defined as „any interactive computing system designed to change people's attitudes or behaviors" [4]. One of the key prerequisites for a successful persuasive strategy is to interact with the user at the right time and place. Important strategies for persuasive technologies are social facilitation, persistence and simplicity. Among others, persuasion has been studied regarding the improvement of health behavior [7], promoting physical exercise [9] and mobile marketing [8].

However, little work has been done in the factory context with respect to persuasive technology until now. As mentioned by Reitberger [14] ambient persuasion connects the field of ambient intelligence and persuasion technology. This research field provides the possibility to enclose persuasive technologies into the users daily life, giving the opportunity for persuasively intervention independent of time or place.

III. USER-CENTERED DESIGN PROCESS

The implementation of the OG followed a user-centered design process (see figure 1).

Figure 1. The four user-centered design phases: (1) requirements analysis, (2) design prototype, (3) implementation, (4) user study.

Within the user requirement analysis phase (1) we conducted a series of qualitative studies within the semiconductor factory Infineon Technologies Austria AG in Villach, directly in the clean room area. These studies provided us with a holistic understanding of the clean room environment in general, basic operator work tasks and already deployed interfaces. Thereafter we used these insights to inform the

978-1-61284-408-4/11 $26.00 © 2011 IEEE

design of a prototype of the OG (2). We decided to develop three design solutions, which reflect different depths of information granularity. Subsequently, the OG was implemented and deployed within a specific semi-conductor factory (3). In a future step we will conduct a user study (4) and evaluate its usability and related user experience factors.

This paper focuses on the first three steps in the UCD process. In the next sections the results from the requirement analysis phase, three different designs, and the implementation of the OG will be described.

IV. CONTEXT SEMICONDUCTOR FACTORY

Within a semiconductor factory several entities interact with each other resulting in a rather complex structure. The overall aim of the semiconductor factory is to as many error free integrated circuits as possible. Thereby sets of silicon crystal wafers—typically groups of 25 or 50 pieces combined into "lot boxes"—follow a complex path through the factory. During this process they pass through different process steps (e.g. etching, exposure, etc.) performed on different equipment. Due to different manufacturers and types the interfaces of the equipment are very heterogeneous. Additionally signal lights are attached to each equipment, which have the purpose to inform about the equipment status. A red light indicates an error. A blinking yellow light shows that the processing of a lot is finished. A constant yellow light illustrates that the equipment is idle. A green light signals that a lot is being processed on the specific equipment.

Typically to each type of equipment highly specialized and well-trained operators are assigned. Whenever lot boxes need a particular processing step on their well-defined way through the factory, operators have to perform three basic tasks: Firstly, they take it from a delivery shelf. Secondly, they process them on the right equipment with the appropriate recipe. Thirdly, operators deliver the lot box to the next destination.

Based on the manufacturing algorithm wafers may spend days, mostly even weeks within the production line. Due to this long processing time, wafers are becoming rather expensive. Therefore a main goal company perspective is to avoid errors leading to a zero-defect-factory paradigm. Nevertheless a high productivity and efficiency is also desired. As a result, an overall aim of the company is to process lots as fast as possible with little idle time between processing steps. Thus, a maximized equipment load is desired. If all lots would have the same priority a first in, first out (FIFO) principle—meaning that lot boxes are assigned based on their arrival time—could be a good strategy to maximize operating grade. In reality different lots often have different priority levels. Lots with a higher priority (e.g., express lots, or so called "hand to hand" lots) need to be preferred by the operator compared to lower priority lots. More priority lots within the production cycle may have a negative effect on the rest of the production [13].

For the operators the balance between maximizing equipment load and preferring priority lots has consequences. At any time the operator has to assign lot boxes according to their priority and the equipment status (e.g., idle, operating) to one certain equipment. Therefore, for each lot box, which is delivered at the delivery shelf, the operator has to make at least two decisions:

- Firstly, the operator has to decide whether he processes the specific lot box now or later (depending on the priority of the lot).
- Secondly, the operator has to select the right equipment on which he processes the wafers of this lot box (spending on the status of the different equipment).

The fact that a specific lot box can be processed at different equipments for the same program step make this task challenging for operator. Our observations showed that operators often rely on the FIFO principle, unless a lot box with a higher priority is delivered. Nevertheless, a high throughput and therefore high equipment load was assumed to be most important by the operators. From an individual perspective, the operators' goal was to maximize their work outcome.

From an overall company perspective, the sum of these local maxima does not necessarily lead to a global optimum, when looking at the factory as a whole. Any locally optimized equipment load may lead to a bottleneck at the next process steps, causing undesired effects for the entire factory along the remaining path of the lot box. Due to the high complexity of the processes within the factory individual operators cannot anticipate these after-effects.

To improve these shortcomings, dynamically generated "dispatch lists", which rank lot boxes accordingly to their priority, have been introduced. Dispatch lists provide visual information of the ranked list of lot boxes situated on displays in the vicinity of every equipment. From a global perspective, the uppermost lot box displayed on the dispatch list should be processed next. Since one and the same lot may be processed on various equipments, it is possible that one and the same lot may be displayed on various dispatch lists. Thus, the ultimate decision, which lot box to be chosen first is still in the responsibility of operators. The reason for this is the flexibility of the manufacturing process (e.g. malfunctions of equipment).

According to Fogg's definition [4], the described dispatch list clearly constitutes a persuasive technology. It aims to influence the operators' decisions locally, in order to act according to overall goals. Nevertheless, the dispatch list bears some shortcomings:

Firstly, the operator is encouraged to follow the ranking provided by the dispatch list—which in the following is referred to as "compliance". A high compliance means a strict adherence of the dispatch list, whereas a low compliance indicates that the operator does not adhere the dispatch list. The compliance may contradict the operator's goal to achieve a high throughput, which means to process as many lots as possible and therefore reduce equipment idle times, since a high compliance may lead to longer equipment idle times as the operator might be forced to wait for high priority lots. This indicates that the operator needs more information about the global status of the lot.

978-1-61284-408-4/11 $26.00 © 2011 IEEE

Secondly, the dispatch list is not situated on every place the operator makes a decision. In our study we observed that operators made the decision which lot to take next at the delivery shelf for incoming lots. At this site no dispatch list was available and therefore the operator had no ranking information allocated. Additionally the operator had no input available on which equipment the next lot should be processed. This shows the importance of contextually appropriate information at the right moment.

Thirdly, we noticed that operators worked with numerous interfaces that caused an information overload. Various interfaces including the dispatch list visualize not only relevant information for this moment, but provide a lot of information that is not needed by the operators for most of their tasks.

V. OPERATOR GUIDE DESIGN

Based on these findings, we decided to design a novel interface for the semiconductor factory supporting the operator in his work. It is not supposed to replace the dispatch list, but seen as an additional information system to help the operator to make better-informed decisions. Based on the requirement analysis described above the overall design goals for the OG are as follows:

- foster a higher compliance by guiding the operator through the optimal next working steps

- improve efficiency and reduce errors by simplifying the interface and therefore reducing the information overload

- site the interface on each place the operator has to make a decision

- visualize not only lot specific information, but also equipment specific processing status

In the following three different designs are presented. The three designs incorporate different information levels. Design A includes basic elements of the OG including information which equipment has to be handled in the next step. This design is the simplest one, providing only information about the equipment status and the ranking of the next steps. The second design (Design B) adds also a temporal aspect. It provides additional information to the user, which is not relevant for the very next step. Finally, design C integrates additional information about the equipment status and links this with information about lot boxes to be processed next. Each new designs loses simplicity but provides more information.

A. Operator Guide Design A

In order to improve learnability we took advantage of mental models of two already deployed interfaces. To visualize equipment status we decided to use the color scheme for the signal lights for the equipment described above. Therefore red symbols signal an error, yellow icons imply that the operator has to take action and a green color visualizes equipment, which is processing at the moment. On the other hand the ranking order of the dispatch list is utilized. Analogue to it the uppermost line shows the next tasks to be done by the operator.

Figure 2.   Operator Guide design A

Figure 2 visualizes design A for the operator guide. Basically, it provides information about the equipment's status of five different equipments including an identifier for the specific equipment. The ranking (from the top to the bottom) indicates a priority of next equipment to be handled. The color provides information about the next step.

- The uppermost line shows that there has been an error at equipment "F02:136". This is indicated by the red color. The operator is obliged to solve the error or to inform the maintenance.

- The next two items signal, that the specified equipment has finished processing and is idle (yellow color). Both equipment can be unloaded and the finished lot box should be delivered at the next process step. First equipment "F02: 263" should be worked on, thereafter equipment "F04: 131".

- The green item visualizes, that equipment "F04: 133" is processing at the moment. No action needs to be taken.

- The last item shows that equipment "F02: 213" is idle. Whenever a new lot is delivered at the delivery shelf this lot may be processed at this equipment.

We have chosen to use five items to provide information to the operator for the next five steps. This number is freely selectable. On the one hand a lower number would make the interface simpler, on the other hand the operator would have less choice for the next step. Providing the operator with only one item, would force the operator to handle this equipment next. Too many items would lead to a more complex OG with even more freedom for the operator. Additionally we have decided to add a window-close-button to the top right corner of the OG. This makes it possible for the operator to close the OG window.

B. Operator Guide Design B

The second design only slightly differs from the first. It adds some information to the equipment, which do not need immediate interaction.

Figure 3 shows that equipment "F04: 133" is processing for the next 17 minutes. After this time the item would change its color into yellow and could be unloaded. This information could also be visualized by adding a dynamic bar. In this case the exact timing information (17 minutes) would be lost.

Figure 3.   Operator Guide design B

C. Operator Guide Design C

Design C of the OG provides the most information to the operator. Additionally to the representation of the tasks a symbol indicates the type of the task, which should be performed. Furthermore (optionally) a lot number is given. Figure 4 demonstrates design C of the OG. Again each item represents a different task to be worked on by the operator.

Figure 4.   Operator Guide design C

- Anon the uppermost item shows that there has been an error at equipment "F02:136". This is indicated by the red color and the symbol "X".

- The next line shows a symbol with an up and down arrow, an equipment ID as well as a lot number. It signals, that the specified equipment ("F02:139") has finished processing and is idle (yellow color). The symbol on the left side indicates that the finished lot can be unloaded and a new one loaded. The lot number "RU 751293" in the blue field visualizes the optimal next lot to be loaded into the specified equipment

- The third line is similar to second one except the symbol on the left side. The down arrow means that the equipment is ready to be loaded. Equipment ID "F02:263" and lot number "RU 915518" have the same meaning as above.

- The fourth line shows that a lot can be unloaded (indicated by the up arrow) at the specified equipment

"F04:131". Since no lot is ready to be processed on equipment no lot number is shown.

- The right arrow symbol in the fifth line indicates that the operator can carry the specified lot "RU 801250" from the delivery shelf for incoming lots to the specified equipment "F04:133". There the lot has to wait until the equipment is idle.

- The circle arrow and the green color in the sixth line symbolize a processing equipment. This means that the operator does not have to take actions at this moment. If needed the remaining time could be added.

D. Operator Guide Design Discussion

The usage of the OG is simple: the operator should perform the uppermost displayed task next. The list of tasks visualized in the OG is generated dynamically, which means that after completion of each task the corresponding line vanishes. The presented design is the result of three iterations. After each iteration we evaluated the design using a set of focus groups as an expert evaluation method. Both HCI experts as well as experts from the semiconductor factory discussed the design on the basis of usage scenarios.

Fogg's [4] distinguishes three roles a computing technology, which are tool, media and social actor. A tool can be persuasive by making the target behavior easier to do or follow or by directing people through a process. The OG is designed as a persuasive tool, because it increases ability in both ways. The operator will be well informed and made faster and effective decisions by providing a visualization of the prioritized working steps in a ranked order.

This is achieved by following three major persuasive principles: (1) reduction, (2) tunneling, and (3) suggestion [4]. The OG follows the reduction rule by minimizing information overload. It displays only relevant information for the next tasks. Tunneling is achieved by providing the next best steps along with the tasks to do visualized in a simple way within the OG. Each line of it shows the relevant information for the working task. It supports compliance and makes it easier to work efficiently by simply following the order. Suggestion is achieved since the OG provides the right information at the right time and place. It presents contextually appropriate information as it shows relevant information wherever the operator has to make a decision.

Since the OG provides contextually appropriate information it is also a good example for an ambient display. It is implicitly interwoven with its environment providing information in an unobtrusive way. This and the reduction of information overload by displaying only relevant information for the tasks. The interplay of ambient intelligence paradigms and persuasive strategies results in the OG as a real world example of an ambient persuasive interface.

VI.   OPERATOR GUIDE IMPLEMENATATION

The OG has been implemented following a server/client architecture consuming real time messages from the MES (Manufacturing Execution System) and the manufacturing integrated equipments themselves. The information on equipment and job states in the relevant work area is presented to the operator on large screens at the delivery racks (see figure 5), mobile devices, and directly at the equipment. The close link to manufacturing data and events from job controller, integrated equipments, and the dispatching system allows to immediately alerting the operator in case of any assistance required while reducing their effort in terms of time and empty miles for screening healthy equipments. The detailed but intuitive to follow instructions, the support on unload/load decisions and the easy way for fast orientation was highly appreciated by the operators.

Figure 5. Productive OG usage

The IT solution was realized as a Windows application for .NET Framework using C#. A multicast Middleware is used for messaging.

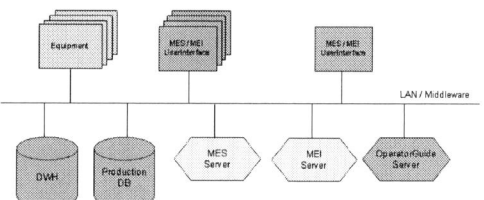

Figure 6. High-level architecture

The OG is a perfect example for a real world implementation of an ambient display. It is implicitly merged with its environment providing only relevant information based on its context.

VII. CONCLUSION AND FUTURE WORK

In this paper we have presented the Operator Guide as an ambient persuasive interface. We have described the first three phases of a user-centered-design process including three different designs with varying information granularity.

As described in the UCD process depicted in the chapter III follow-up studies about the implementation and usage of the OG are planned. Currently, we are therefore in the process of setting up a new series of experiments to find out how the OG is involved in the decision making process of the operator and how useful it is in the workflow.

ACKNOWLEDGMENT

The financial support by the Federal Ministry of Economy, Family and Youth and the National Foundation for Research, Technology and Development is gratefully acknowledged (Christian Doppler Laboratory for "Contextual Interfaces").

REFERENCES

[1] M. Bauer, L. Jendoubi, K. Rothermel, and E. Westkämper, "Grundlagen ubiquitärer Systeme und deren Anwendung in der "Smart Factory"," Industrie Management - Zeitschrift für industrielle Geschäftsprozesse, vol. 19 (6), 2003.

[2] M. Rittenbruch and G. McEwan, An Historical Reflection of Awareness in Collaboration, ch. An Histor- ical Reflection of Awareness in Collaboration, pp. 3–48. Springer London, 2009.

[3] E. Aarts and S. Marzano, The New Everyday: Views on Ambient Intel ligence. Rotterdam: 010 Publishers, 2003.

[4] B. J. Fogg, Persuasive Technology: Using Computers to Change What We Think and Do. Interactive Technologies, Morgan Kaufmann Publishers Inc., 2003.

[5] J. Hollan, E. Hutchins, and D. Kirsh, "Distributed cognition: Toward a new foundation for human- computer interaction research," vol. 7, no. 2, pp. 174–196, 2000.

[6] A. Meschtscherjakov, W. Reitberger, F. P̈ohr, and M. Tscheligi, "The operator guide: An ambient persuasive interface in the factory," in First International Joint Conference on Ambient Intelligence (Malaga, Spain, November 10-12, 2010), 2010.

[7] A. Grimes and R. Grinter, "Designing persuasion: Health technology for low-income african american communities," in PERSUASIVE '07, pp. 24–35, 2007.

[8] E. Holmen, "The four pillars of a successful mobile marketing vision," in Mobile Persuasion: 20 Perspec- tives on the Future of Behavior Change, (Stanford, California), Stanford Captology Media, 2007.

[9] A. Eyck, K. Geerlings, D. Karimova, B. Meerbeek, L. Wang, W. A. IJsselsteijn, Y. de Kort, M. Roersma, and J. H. D. M. Westerink, "E ect of a virtual coach on athletes' motivation.," in PERSUASIVE '07, pp. 158–161, 2007.

[10] T. McCalley, F. Kaiser, C. J. H. Midden, M. Keser, and M. Teunissen, "Persuasive appliances: Goal priming and behavioral response to product-integrated energy feedback," in PERSUASIVE '07, pp. 45-49, 2007.

[11] J. Mankoff, A.K., Dey, G., Hsieh, J., Kientzin, M., Ames, and S. Lederer, "Heuristic evaluation of ambient displays.," in CHI 2003, pp. 169-176, 2003.

[12] A. Schmidt, "Context-awareness, disappearing and distributed user interfaces: Experience, open issues and research questions." Presentation delivered at the Dagstuhl Seminar on Ubiquitous Computing, September 9-14 2001.

[13] C. D. DeJong and S. P. Wu, "Material handling: simulating the transport and scheduling of priority lots in semiconductor factories," in WSC '07, (San Diego, California), pp. 1387–1391, December 2002.

[14] W. Reitberger, M. Tscheligi, B. E. R. de Ruyter, and P. Markopoulos, "Surrounded by ambient persuasion," in CHI 2008: Extended abstracts on Human factors in computing systems (M. Czerwinski, A. M. Lund, and D. S. Tan, eds.), (New York, NY, USA), pp. 3989–3992, ACM, Apr. 2008.

[15] C. Wisneski, H. Ishii, A. Dahley, M. Gorbet, S.Brave, B. Ullmer, and P. Yarin, "Ambient Displays: Turning Architectural Space into an Interface between people and Digital Information", in CoBuild'98, 1998.

[16] J. Carroll, Human-computer interaction in the new millennium; Addison-Wesley Professional, 2001.

[17] ACM Special Interest Group on Computer-Human Interaction Curriculum Development Group, ACM, New York, 1992.

[18] ISO TR 18529, Ergonomics of human-system interaction - Human- centered lifecycle process description, 2000.

Looking Towards a Sustainable and Green Future

Sumita Basu, Intel Corporation
Technology and Manufacturing Group
Intel Corporation
Hillsboro, USA
Sumita.basu@intel.com

Sandra Viarengo, Intel Corporation
Technology and Manufacturing Group
Intel Corporation
Santa Clara, USA
sandra.j.viarengo@intel.com

For manufacturing-heavy industries such as the semiconductor world, one has to consider a lot of environmental issues. We need to consider the inputs to our operations and how to reduce and minimize consumption of energy, water and chemicals. We also need to think about the output of the operations and how to reduce and minimize emissions including air pollutants and global warming gases such as CO2 and PFC's from our manufacturing processes. In addition to the emissions, we need to look at limiting the creation of waste, both solid (metal/material) and liquid chemical), and reducing and recycling through treatment of wastewater and recycling programs.

Optimization of processes and logistics, efficiency in energy usage and transportation of products in a sustainable manner are the keys towards a cleaner environment. By doing these three, we can preserve our natural resources while putting money back in our pockets thereby literally making it a "green" future for all.

Overall Intel has a holistic approach to managing our operations responsibly. This paper gives a brief insight into the Environmental Social Governance (ESG) Program undertaken at Intel towards the 'green' initiative. The future of our value chain - from sand to consumer or industrial device depends on our ability to grow revenue and profit, allow the continuation of our licenses to operate and the ability to move forward as fast as we can innovate. This means we need to stay ahead of regulatory and community needs so that they do not become constraints.

Keywords-Environment, green, eco-friendly, Green House Gases (GHG), Supply Chain, sustainable

I. INTRODUCTION

If there's a singular topic that's on the minds of most manufacturing factories in the world today, it's environment. It is no longer just enough to meet environmental guidelines as set forth by a government or higher authority. The sustainability of our natural resources at the current pace of use and misuse can no longer be ignored. It is time that we collectively own the social responsibility and look at ideas and innovations to create a cleaner and greener future for the best home any one of us has ever known (planet Earth).

The complexity and urgency of the many factors that contribute to these problems before us requires us to think together, plan together and act together to combat challenges before us. Figure 1 shows the influencers that have the highest impact to this problem. While natural resources like water require preservation of quality and quantity, the emissions and energy consumption require diligent monitoring. Currently,

there appears to very little consistency in measuring or abating each of these factors. For example a consistent and standardized reporting approach to Green House Gas emissions is necessary to enable the whole value chain to communicate efficiently and effectively.

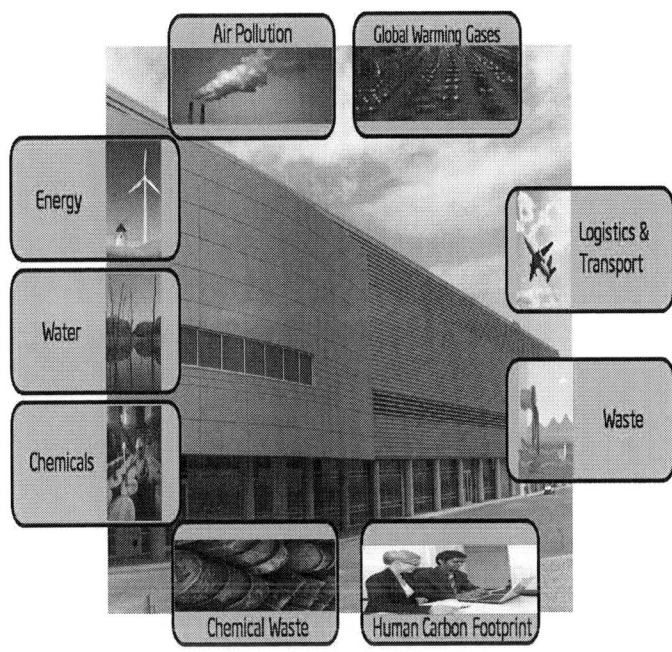

Figure 1. Factors with highest impact to the environment

This will also allow our government to understand us and provide us with the maximum opportunity to develop compelling roadmaps for sustained industry growth and innovation, while reducing cost to the environment.

(Note: All performance graphs and charts used in this paper have either been normalized or presented only in absolute values. References to "per chip" assume a chip size of 1 cm^2 although actual sizes vary depending on product.)

II. OUR APPROACH TO ENVIRONMENTAL SUSTAINABILITY

Figure 2 shows the typical life cycle of Intel products. Intel's approach towards achieving environmental sustainability starts during the research and development stage with building and designing some of the world's most sophisticated products with careful management of energy consumption, air emissions and conservation of natural resources. Intel is attempting to incorporate what's known as Design for the Environment principles throughout all phases of its product development process and well into manufacturing. Figure 3 shows the current state that is predominantly existent today where we combat environmental concerns with treating waste water and abating the gaseous emissions to meet strict regulations. It is desirable and to the best of our interest if we move away from this mindset and look towards a future where

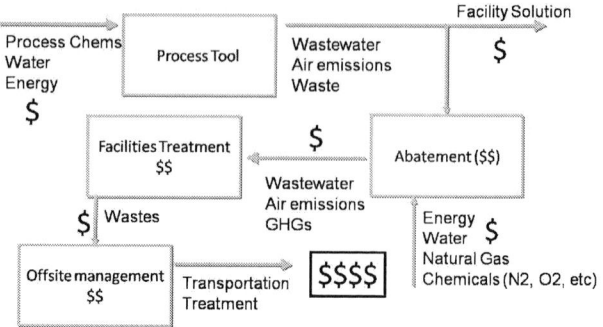

Figure 3. Current state: treat and abate (without Design for Environment)

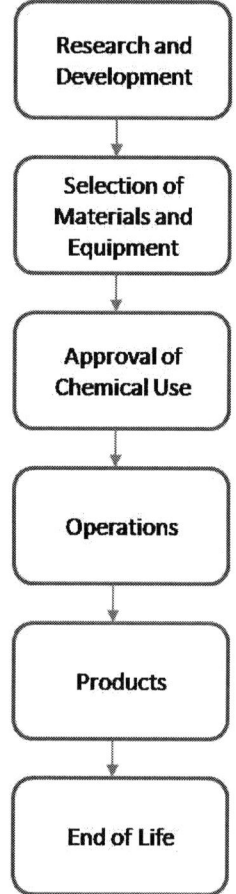

Figure 2. Life cycle of an Intel product

tools and processes are designed with environment in mid. Figure 4 depicts the ideal desirable state.

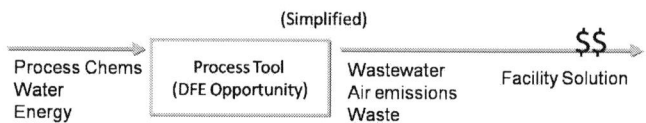

Figure 4. Ideal future state with Design for Environment

III. CLIMATE CHANGE AND ENERGY EFFICIENCY

Intel considers climate change and energy efficiency important environmental issues and for several years now the company has taken steps to reduce and manage its carbon footprint.

A. Reducing Green House Gas Emissions

More than a decade ago, Intel and several other semiconductor manufacturers entered into a voluntary agreement with the U.S. Environmental Protection Agency (EPA) to reduce emissions of poly-fluorocarbon compounds (PFCs). PFCs are materials used heavily in semiconductor manufacturing and are known to have a detrimental effect on the atmosphere leading to global warming or the 'green house effect'. By the end of 2009 [1] Intel had reduced its PFC emission by 50% on an absolute basis and 80% on a per chip basis as compared to the 1995 baseline. As we continue to strive towards reducing these emissions and making semiconductor manufacturing green, we realize that this cannot be achieved without collaboration with our suppliers and alignment with our competitors.

978-1-61284-408-4/11 $26.00 © 2011 IEEE   14

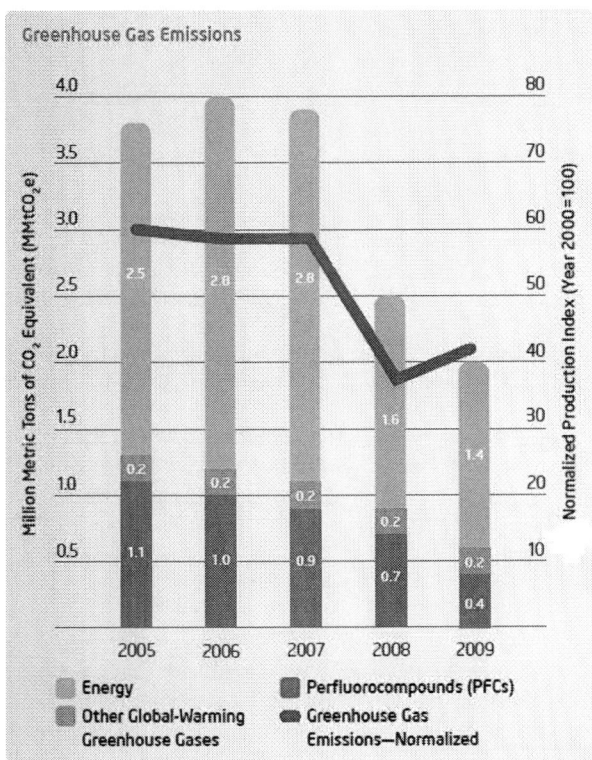

Figure 5. Overview of Green House Gas emissions over four years, normalized to 2000 baseline levels (100%) Source: CSR, 2009

In the past Intel has focused on Scope 1 and Scope 2 emissions. These are emissions that come directly from factories and purchased electricity respectively. Today, Intel is also focused on understanding with a goal towards reducing Scope 3 emissions, those that come indirectly from the result of operations or use of Intel products. The calculation of an overall CO_2 footprint is still full of limitations and has certain uncertainties [2]. Research has confirmed that approximately 60% of the CO_2 impact comes from semiconductor manufacturing with the balance attributable to supply chain, commute and business travel. Figure 5 shows a snapshot of Intel's Scope 1 and Scope 2 emissions over four years. It is evident from the graph that the emissions were down 20% in 2009 as compared to the previous year on an absolute basis but it may be noted that the emissions were actually up by 13% on a per chip basis. This increase was attributable to lower production in 2009 due to the global economy [1]. The decrease in absolute emissions was brought about by reduction of PFC emissions and Intel's undertaking of several energy efficiency projects. Intel's ongoing work on understanding the carbon footprint of the supply chain section is presented in detail in section IV.

B. Reducing Air Emissions

There are other volatile organic compounds (VOCs) and hazardous air pollutants (HAPs) that need to be monitored in addition to the direct emissions of CO_2. While some of these cannot be eliminated completely, Intel strives to invest in technology that can assist with reduction of these wastes. Installation of thermal oxidizers and wet scrubbers to neutralize

and absorb gases and vapors are a preferred approach. The oxidizers concentrate the VOCs and oxidize them into CO_2 and water vapor. Intel pays very close attention to the selection of such equipment to ensure efficient use of heat and fuel. Wet scrubbers that utilize water to neutralize acids and other contaminants are chosen keeping in mind their water recycle capability. Figure 6 shows an overview of other air emissions and their history over a four year span.

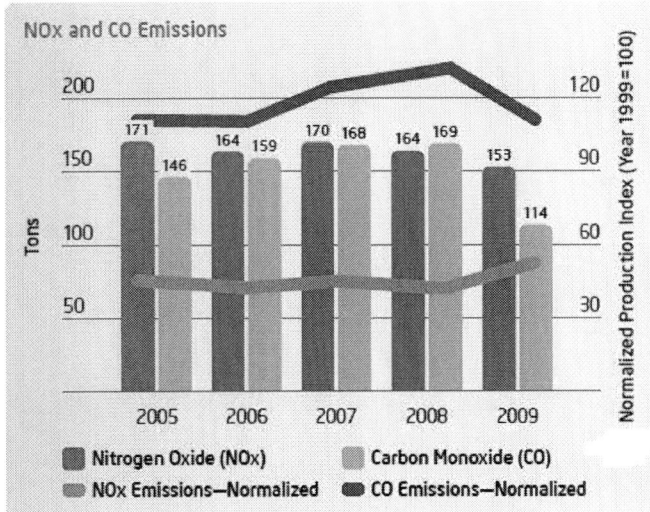

Figure 6. Overview of other emissions over four years, normalized to 1999 baseline levels (100%) Source: CSR, 2009

C. Energy Efficiency and Usage

The move from 200mm to 300mm wafers inherently improved the energy usage in semiconductor industry as more chips could be produced at a time. The trend towards smaller chip sizes and the advancements on Moore's Law help reduce energy per chip by allowing more chips per wafer. All these initiatives towards saving on energy stems from the fact that semiconductor manufacturing rely heavily on usage of electricity. Figure 7 shows Intel's energy usage over four years.

While much has been achieved in the realm of installing solar panels, improving product energy efficiency and promoting the green grid in terms of business computing ecosystems, a yet bigger task lies ahead of us. Manufacturing equipments are the largest users of energy in this industry and it is now time to turn our focus on improving their efficiency. Most of these equipments are left in idle state and not powered off so that they can return to production in a short span of time (few hours vs. few weeks). In their idle state, these tools consume 80% of the energy required for full use. Intel's focus and long term objective is on collaborating with its suppliers towards developing a solution where equipment can be idled such that they can return to active production in a very short period of time (a few hours) while cutting down power consumption by 80 to 90% as compared to a full shut down state. In the short term, the company is focused on making

978-1-61284-408-4/11 $26.00 © 2011 IEEE

incremental improvements on reducing energy consumption within its manufacturing factories by 5-10% per technology.

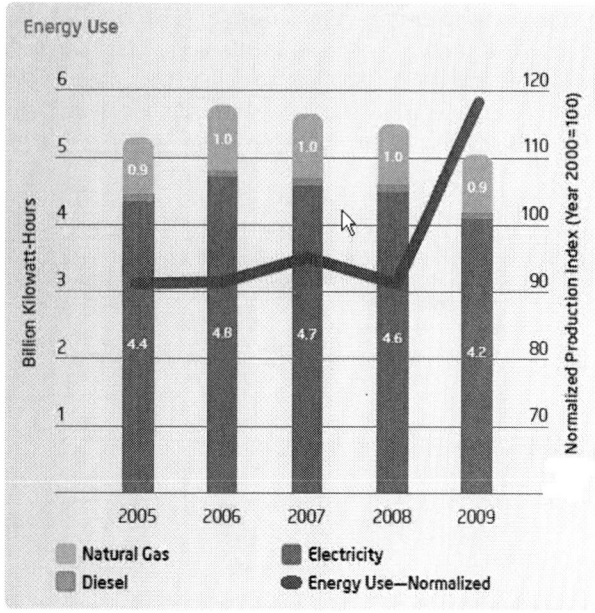

Figure 7. Overview of other energy use over four years, normalized to 2000 baseline levels (100%) Source: CSR, 2009

Three years ago, Intel became the largest voluntary purchaser of green power in the USA [3]. The purchase commitment includes a balanced portfolio of wind, solar, small hydroelectric and biomass sources. This commitment has the equivalency of taking 200,000 passenger vehicles off the road every year and hopes to stimulate the market to purchase green power and reduce energy costs.

D. Water Conservation

One of our key focus areas in the stride towards a green future is to manage our water usage at all our sites in a sustainable manner. This means reviewing access to sustainable water resources during site selection for a new factory as well as incorporating water conservation elements in our manufacturing processes. As an example: cleaning Silicon wafers require the use of ultra pure water (UPW). Our focus has been to increase the efficiency of the UPW process (going from using 2 gallons of water to make 1 gallon of UPW to using 1.25 gallons only). Historically, the same water used to clean the wafers can be treated minimally and reused to meet other industrial purposes as well as irrigation. Intel also has a reuse policy wherein we harvest as much water from our manufacturing processes as possible and direct it to equipment such as cooling towers and scrubbers. As a result of the various efforts, Intel reported reclaiming approximately 2 billion gallons of water in 2009 [1]. Figure 8 shows the water usage at Intel over the last couple of years.

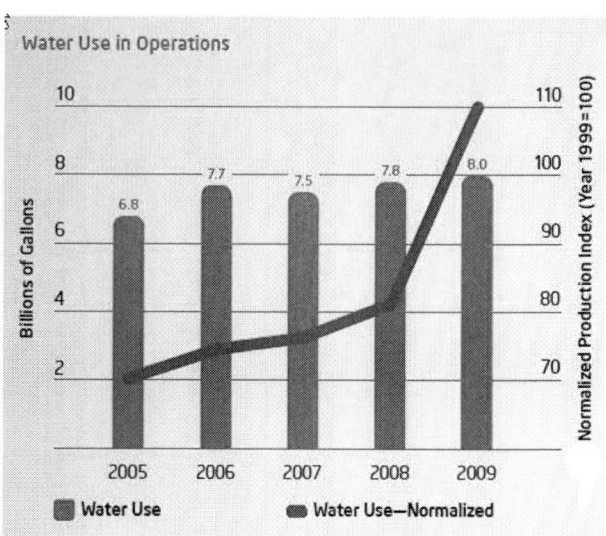

Figure 8. Overview of water usage at Intelover four years, normalized to 1999 baseline levels (100%) Source: CSR, 2009

Our ultimate vision for the semiconductor manufacturing industry must be to look at achieving a continuous reuse of water without having to send part of it to waste as is currently done in compliance with local permits. Similar to the scopes of GHG management, water conservation has three scope levels. Scope 1 calls for the direct impact within factories during Intel operations and is associated with use, recycle and reuse, Scope 2 lies with water management amongst our energy providers and finally a key piece (Scope 3) comes from our suppliers and the capability of their tools to handle water efficiently.

E. Waste (reduce, reuse or recycle)

In addition to all of the above, Intel strives to reduce its solid and chemical waste generation as well. Over the last 4 years the company has recycled about 80% of the solid waste generated in our operations each year. Solid waste reduction takes on an added challenge during the construction of our two new upcoming factories: D1X and Fab 42. Several programs are being worked on to ensure that we continue our work in the realm of recycling and reusing waste where possible.

With the complexity in our processes increasing with every technology we are seeing an unfortunate but unavoidable rise in chemical waste generation. This is attributed to added layers, more cleansing for the smaller features, etc. Attempts are on to reverse this trend and the goal is to reduce our chemical waste generation by 10% by 2012 (compared to 2007 baseline levels). Installation of systems to recover valuable dissolved metal waste (copper) thereby eliminating off site shipment of the waste promises to provide the much needed help.

IV. ENVIRONMENTAL SOCIAL GOVERNANCE

With all said and done, we must realize that to bring about a complete integrated understanding of our carbon and water footprints we need to engage with our supply chain world. We must collaborate with other companies in our industry to develop processes, standards and tools that will bring about lasting environmental improvements in the global electronic supply chain.

Intel interfaces with over 9000 suppliers in over 100 countries to bring in parts, equipments, materials and services to its factories. Since adopting the Electronic Industry Code of Conduct (EICC) in 2004 we expect our employees and suppliers to adhere to the code. In short the code encompasses human rights and social responsibility towards carrying out business with tools for monitoring ethical practices, safety of workers, etc. To help the company meet and exceed its goals of complying with the codes of the Environmental Protection Agency (EPA), an internal group has been formed. Its charter is to set the direction and strategy for all supply chain corporate responsibilities such as tool selection criteria, social ethics and looking into our supplier base for green initiatives and opportunities.

V.   OTHER ENVIRONMENTAL FACTORS

There are several other non-manufacturing factors that Intel also considers in its effort to create a sustainable future.

Intel uses its IT resources to reduce energy consumption and minimizing carbon footprints in a parallel effort. With its achievements in this field, Intel was recently ranked fifth amongst the top 12 green companies in the world [2].

VI.   ACKNOWLEDGMENT

The authors of this paper would like to acknowledge Cody Philpot and several key members of the Technology and Manufacturing Group at Intel for their valuable contributions towards this paper.

REFERENCES

[1]   Corporate Sustainability Report, Intel 2009.

[2]   "Developing on Overall CO_2 Footprint for Semiconductor Products", IEEE, 2009

[3]   National Top 25 Green Power Partnership and Fortune 500 Challenge lists

[4]   *Computerworld* and *Network World* published Oct 25th, 2010

Dynamic Management of Controls in Semiconductor Manufacturing

Justin Nduhura Munga [1,2], Stéphane Dauzère-Pérès [1], Philippe Vialletelle [2], Claude Yugma [1]

[1] Ecole des Mines de Saint-Etienne – Centre Microélectronique de Provence
880 avenue de Mimet, – F-13541 Gardanne – France
(nduhura@emse.fr, dauzere-peres@emse.fr, yugma@emse.fr)

[2] STMicroelectronics – 850 rue Jean Monnet – F-38926 Crolles Cedex – France
(justin.nduhura-munga@st.com, philippe.vialletelle@st.com)

Abstract – **In order to optimize the number of controls in semiconductor manufacturing, a Permanent Index per Context (IPC) has been developed to evaluate in real-time the risk on production tools. Depending on the context which can be defined at tool level, chamber level or recipe level, the IPC allows a very large amount of data to be managed and helps to compute global risk indicators on production. A prototype based on the IPC has been developed and deployed for the CMP workshop. Results show that various indicators can be determined in real time to control risks. The number of measurements without actual added value can be strongly reduced. Moreover, the dispatching of lots on production tools can be improved.**

Keywords – Control; risk; defectivity; dynamic sampling.

I. INTRODUCTION

In the context of worldwide competition, high technology products, transition from 200mm to 300mm wafers, semiconductor manufacturing companies have to pay great attention to guarantee high product quality and lower their production costs. An important consequence is the introduction of numerous measurement or control steps at different stages of product fabrication in order to detect as soon as possible potential "excursions". One of the explanations is that the increase of the wafer size, combined with a reduction of the transistor size, made the wafer more expensive [1]. Furthermore, the cost incurred because of these new types of controls has aroused a huge interest for control strategies.

Various works have been conducted on both static and dynamic sampling strategies in order to reduce the number of controls. Due to the drawbacks of static sampling, which consists of always selecting the same number of lots to control and thus missing critical information [2], dynamic sampling strategies have been developed. They are more efficient since sampling decisions are taken based on the actual production state. However, most of the time, authors do not specify the way the solution could be deployed in practice, especially when coming to high-mix semiconductor manufacturing [3] [4] [5]. The complexity of process flows, the variability of product flows, the heterogeneity of information systems and the large amount of data to be processed can strongly impact the solution. The integration of such solutions is often time consuming and may drag huge investments depending on the IT infrastructure of the semiconductor company.

In this paper, we propose a novel approach that aims at evaluating in real-time the risks on various production tools and selecting the best lot to schedule or sample to reduce risk. In this paper, the risk is defined as the number of wafers processed on a production tool since the last control performed on a wafer processed on the tool, or as the number of wafers processed on the same chamber with the same operation, or with the same technology and with the same operation, etc. To evaluate the risk, a global counter called *Permanent Index per Context* (*IPC*) was developed to allow both very quick and easy computations of the context. The context can be the tool, the chamber, the recipe, the technology, the resin, etc. Once the context is computed, we provide the information to help taking the best decision in order to reduce the risk on the production tools. The decision can be the validation of the best lot at the measurement step, the skipping of a lot that does not bring further information, the dispatching of lots on production tools, etc.

The algorithm has been embedded in a prototype and deployed for the CMP Workshop. The practical implementation helped to divide by two the number of lots processed on CMP production tools with a risk indicator larger than the one expected by the company. Capacity in metrology has been saved thanks to the skipping of lots that do not bring relevant information, and the dispatching of lots on the production tools has been strongly improved based on the real time evaluation of risk indicators on production tools.

The paper is organized as follows. Section II describes controls in semiconductor manufacturing. In Section III, we explain in more detail the IPC and formalize it in Section IV. In Section V, we describe the CMP workshop where the IPC has been deployed and present some results on actual experimentations. Section VI concludes the paper with recommendations for further research.

II. CONTROLS IN SEMICONDUCTOR MANUFACTURING

In semiconductor manufacturing, different types of controls exist depending on the level. Authors in [6] define four main levels of control plans:

(1) The Product level is concerned with electrical tests.

(2) The process level is concerned with metrology measurements.

(3) At tool level, tool variables are monitored and sometimes regulated.

(4) Manufacturing line audits can be assimilated to an organizational control plan.

However, there is a strong link, rarely highlighted between these different control levels. Let us focus on controls at the product level. Most of the time, controls performed at the product level will give us information on both processes and tools depending on the type and the step of the control. Critical Dimension (CD) measurement and Defectivity controls for example will be done at the product level. But the CD measurement [7] will mostly focus on the process whereas Defectivity controls [8] will directly give information on the state of production tools.

To be more general regarding controls at product level, we can define two main groups: intermediate and final controls. Intermediate controls will concern all controls which are done during the chip fabrication and between two process steps: CD Measurement, Defectivity, Overlay Measurement, Thickness, etc. Final controls will concern all controls which are done at the end of the process fabrication: Electrical Wafer Sorting (EWS), parametrical tests, etc.

In this paper, we focus on intermediate controls at the product level and on Defectivity controls in particular. The choice of Defectivity controls is motivated by the fact that:

- Controls are done on products in order to have information on production tools,
- The Defectivity area adresses all workshops and all production tools,
- The depth of controls, the capture rate ,and the case where a process operation can be validated by more than one control operation lead to an increasing complexity,
- And the large amount of data to treat can strongly impact the solution.

III. PROBLEM FORMULATION

A control without a real added-value is waste of time and money. Let us take a simplified example to illustrate the purpose:

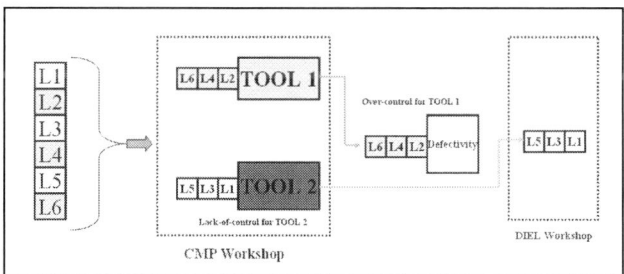

Figure 1. Problem of static sampling

Figure 1 shows the main drawback of static sampling. Let us suppose that we have six lots coming into the CMP workshop for different process operations. The control plan, designed by the engineering team at the start of the production is to control one lot each two. In this case, lots **L2**, **L4** and **L6** will be "flagged" for a Defectivity control after the CMP process operation.

A control on a lot is called to reduce the risk on the production tool regarding the number of wafers processed on the tool since the latest control performed i.e. if a control performed on a lot is validated, the risk is released on the wafers of all lots processed in the same production tool since the last control performed. In the case described in Figure 1, an optimal control plan would be to process at least one lot "flagged" on each production tool. But, because of the variability, the availability of production tools, and the complexity of process flows in modern high-mix semiconductor manufacturing, the situation such as the one described in Figure 1 is often frequent. **TOOL 1** process all "flagged" lots i.e. **L2**, **L4**, **L6** whereas **TOOL 2** does not process any. This results in an over-control for the production tool **TOOL 1** and lack-of-control for the production tool **TOOL 2**. As quality control is defined at product level, without taking into account tool information, all the information is biased to monitor tool drifts.

A simple solution could be to impose a limit set for each tool. This means to control **L3** or **L1** and **L5** for example. The problem of this solution is that the control capacity is limited. The better solution would be to release the control on **L4** and control **L3**. This implies to "flag" lot **L3** for a Defectivity control operation at the next step. The question here is: "what if potential defects generated by the current process operation can not be captured at the next control operation for the lot **L3**?

These kinds of situations have brought some designers to put in place additional controls often redundant [6] while a good repartition of different lots on production tools could allow an efficient sampling rate. This results in increasing the number of controls and, most of the time without real added value.

To overcome this problem, the risk on different production tools should be known in real time and took into account when dispatching lots on production tools. But, because of the large amount of data to treat, the complexity of process flows and control plans, most algorithms have been seen impracticable in modern high mix semiconductor manufacturing because of time and space consuming.

To issue this question, we developed an indicator called Permanent Index per Context (*IPC*) to allow very quick and fast computation of risk indicators. The risk depends on the context as described in the next section.

IV. THE *IPC* MECHANISM

The Permanent Index per Context (*IPC*) is a counter which is increased each time a context is verified. The context can be a tool, a chamber, a recipe, a technology, a resin, the combination of an operation and a technology, etc. This counter is never reset except when a special event occurs (Preventive Maintenance, intermediary qualification, etc.).

The *IPC* has been introduced to allow both very quick and easy computations for any given context. In our first implementation, the context has been defined at the equipment level in order to control the risk on production tools. In this paper, the risk is evaluated in term of number of wafers processed on a production tool since the last control performed. To each lot *l* and tool *m* is associated a Permanent Index per Context (*IPC*), which is equal to 0 if *l* is not processed on *m*. Let *M* be the number of production tools, and *NW(l)* be the number of wafers in lot *l*. The goal is to update in real time the following parameters:

- $LLM(m)$ Index of the Last Lot that has been

    Measured for production tool *m*

- IPC_l^m *IPC* of lot *l* for production tool *m*

Using these parameters, it is possible to calculate the following indicators:

- RI_m Risk indicator on production tool *m*

- NI_l^m The number of wafers potentially impacted

    on tool *m* if lot *l* was measured

- NI_l The number of wafers impacted on all

    production tools if lot *l* was measured

When lot *l* is processed on production tool *m*, an IPC IPC_l^m is associated to *l*. The *IPC* value is equal to the *IPC* value of the lot *l'* processed just before *l* on *m* plus the number of wafers in *l*:

$$IPC_l^m = IPC_{l'}^m + NW(l) \qquad (1)$$

The risk indicator on a production tool *m* is then given by:

$$RI_m = IPC_l^m - IPC_{LLM(m)}^m \qquad (2)$$

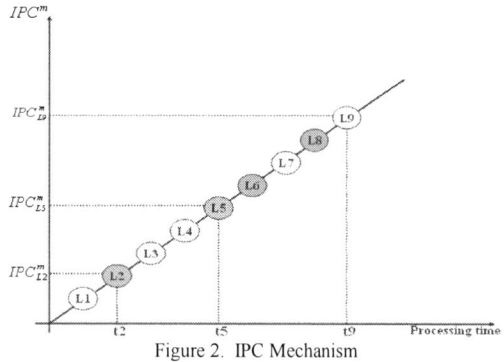

Figure 2. IPC Mechanism

The use of the *IPC* indicator simplifies the computations of risk indicators since they are reduced to calculating a simple difference between two integer values. This implies very low resource usages, the possibility to manage a very large amount of data and compute risk indicators for all production tools for a complex high mix semiconductor plant. Instead of computing each time the risk indicators with complex algorithms, we assign to each lot an index (*IPC* of the lot) when the context is verified.

Figure 2 represents a sequence of different lots processed on a production tool *m*. Lots **L1, L2, ..., L9** have been processed on tool *m* . **L2** and **L5** have been validated by a Defectivity control and, in this case, **L5** corresponds to the last lot that has been measured *LLM(m)*. According to (1) and (2), the risk indicator on tool *m* at time t9 is given by:

$$RI_m = IPC_{L9}^m - IPC_{L5}^m$$

Where

$$IPC_{L9}^m > IPC_{L5}^m$$

It is also possible to identify very quickly the best lot *l* to validate at the metrology step for a production tool *m*. This lot *l* is chosen such that its *IPC* verifies the following property:

$$IPC_l^m = \max\left\{0, \left\{IPC_{ll}^m \,|\, IPC_{ll}^m > IPC_{LLM}^m, \;\; ll \in LM\right\}\right\}$$
$$(3)$$

Where *LM* is the set of lots waiting at the metrology step.

In Figure 2, lots **L6** and **L8** have been processed on tool *m* and are waiting at the metrology step. According to (3), the best lot to select for *m* will be **L8** since $IPC_{L8}^m > IPC_{L6}^m$ and $IPC_{L8}^m > IPC_{L5}^m$.

A control is defined as a measurement plus an action [6]. It is then crucial to be able to evaluate in real time the number of lots potentially impacted whenever a problem occurs on a lot *l*. This number can be determined for a given production tool *m* (NI_l^m) and for all the production (NI_l):

$$NI_l^m = \max(0, IPC_l^m - IPC_{LLM}^m) \qquad (4)$$

And

$$NI_l = \sum_m NI_l^m \qquad (5)$$

In Figure 2, at time t9, NI_l^m will be given by $(IPC_{L9}^m - IPC_{L5}^m)$ corresponding to the sum of wafers contained in L6, L7, L8, and L9.

V. The CMP Workshop Prototype

A. The IPC and the CMP prototype

The algorithm as described in section IV has been embedded in a prototype and deployed for the CMP area in STMicroelectronics 300mm. The main objective was to ensure that the maximum risk expected by the company would not be exceeded and, when the situation is not under control, to provide the best decision to take regarding the production state.

The final user should be the operator and for that, information should be gathered and presented in way such as it could be understood by everybody. To do so, we defined three levels of alert associated to three different colors.

Green: the maximum risk indicator allowed by the company is not reached and the situation is under control regarding the production tool.

Orange: the risk indicator is very close to the maximum value specified by the company and so, actions should be taken to reduce risk.

Red: the maximum risk indicator allowed by the company is reached and the tool should be stopped or an action taken immediately.

RI represents the risk indicator on the production tool. It corresponds to the number of wafers processed on a production tool since the latest control performed. For confidential reasons, all data presented here were normalized.

Figure 3 illustrates the main concept of the prototype deployed for the CMP workshop in STMicroelectronics Crolles 300mm. We have three production tools: **CMP-T1, CMP-T2,** and **CMP-T4**. To each tool is associated a box that shows the situation of the tool risk indicator compared to the maximum risk indicator allowed by the company. This maximum risk indicator is denoted A in the sequel.

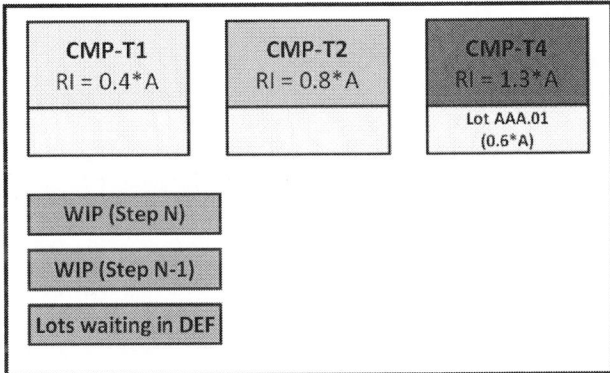

Figure 3 - CMP prototype illustration

The **CMP-T1** box is green because $RI_{CMP-T1} < A$. The **CMP-T2** box is orange because RI_{CMP-T2} is very close to A, and the **CMP-T4** box is red because $RI_{CMP-T4} > A$.

Below the **CMP-T4** box, we can see lot AAA.01 which corresponds to the best lot to validate in order to reduce the risk on this production tool. 0.6*A corresponds to the value of the

reduction of the risk indicator, i.e. if lot AAA.01 is validated at the measurement step, the new value of the risk indicator will be: (1.3-0.6)*A = 0.7*A for tool **CMP-T4**. This best lot is determined as discussed in previous section.

Tool **CMP-T2** is orange and there is not a lot associated as in **CMP-T4**. This means that, among all lots processed on the tool, there are no lots waiting at the metrology step or no lots which can be validated at the next metrology step. But, because the situation starts to become critical (orange status), we provide information on lots waiting to be processed in the CMP area (WIP-step N). If such a situation arises, the operator is called to check (by a simple click) among lots waiting for to be processed and, if there is no lot which can be assigned to tool **CMP-T2**, then the operator is called to check in WIP (step N – 1) corresponding to the list of lots actually being processed in the step before CMP. This allows anticipating and accelerating some lots in order to reduce the value of RI on production tools.

B. Results and discussions

By using the prototype only twice a day to support decisions, the global risk indicator has been strongly reduced in the CMP workshop. As shown in Figure 4, the number of lots processed on production tools with a risk indicator larger than 0.33 has been divided by two, where 0.33 represents the maximum risk allowed by the company.

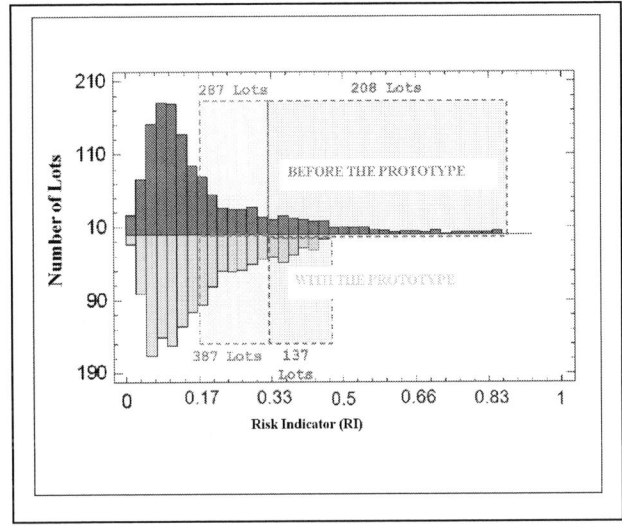

Figure 4 - Result on global risk reduction (after deploying the prototype)

Figure 5 illustrates the impact of the prototype on production. The evaluation has been realized on several weeks. What we can see is that, once the prototype was introduced, the risk was strongly reduced. During the holyday period, where there are less qualified operators, the risk has significantly increased until the prototype was used again.

We can underline the fact that the prototype is used only twice a day by operators. It shows the potential benefits of deploying such a solution in the entire fab

978-1-61284-408-4/11 $26.00 © 2011 IEEE

Moreover, thanks to the use of the prototype and various analyses, many cases of over and lack-of-control on production tools were highlighted, and the dispatching of lots based on the risks on tools was improved.

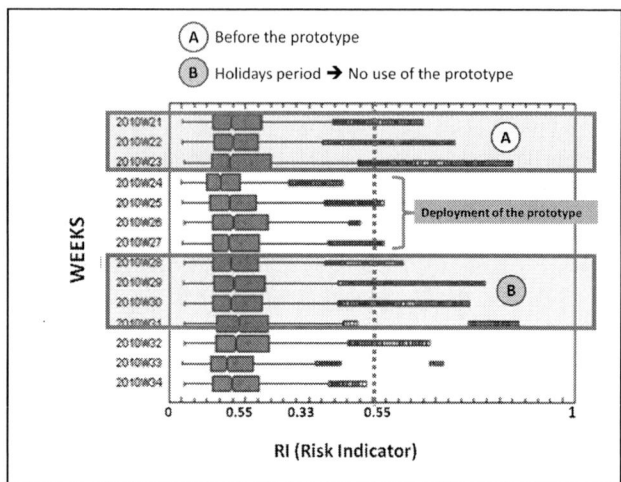

Figure 5 - Impact of the prototype on the overall risk

VI. CONCLUSION AND PERSPECTIVES

In this paper, an index has been introduced to support the reduction of the number of controls without added value. This index helps to evaluate in real time the production risk for most complex high-mix semiconductor plants, and correctly dispatching lots on production tools based on their states.

The Permanent Index per Context (*IPC*) allows very quick and easy computations of various types of information depending on how production evolves. It has been embedded in a prototype and deployed in the CMP workshop at the 300mm plant of STMicroelectronics in Crolles in order to evaluate in real-time the risk on production tools related to Defectivity controls.

The prototype helped operators to divide by two the number of lots processed on production tools with large risk indicators. Many cases of over and lack-of controls have been highlighted and the dispatching of lots on production tools has been improved.

However, in this first study, the impact on other workshops has not been analyzed. The amount of data can strongly increase and it could therefore be interesting to define new contexts such a global *IPC* which will include information for the lot on various production tools. This index could be used to compute the Global Sampling Indicator used in the smart sampling algorithm proposed in [9].

Future works will be dedicated to the deployment of similar solutions in the entire fab and on the optimized management of excursions based on IPC information.

ACKNOWLEDGMENT

This work has been done within the framework of STMicroelectronics. The scientific part of the work has been done within the framework of Centre Microélectronique de Provence – George Charpak at Ecole Nationale Supérieure des Mines de Saint-Etienne in Gardanne, France. The paper has been written as a part of the European Union project IMPROVE.

REFERENCES

[1] S. Joe Qin, G. Cherry, R. Good, J. Wang, and C. A. Harrison, Semiconductor manufacturing process control and monitoring: A fab-wide framework, Journal of Process Control, pp. 179-191, 2006.

[2] J. H. Lee, "Artificial intelligence-based sampling planning system for dynamic manufacturing process," Expert System with Applications, 22(2), pp. 117-133, 2002.

[3] A. Boussetta and A. Cross, Adaptive sampling methodology for In-line Defect Inspection, IEEE Advanced Semiconductor Manufacturing Conference and Workshop, Munich (Germany), pp. 25-31, April 2005.

[4] C. Mouli and M. J. Scott, Adaptive Metrology sampling techniques enabling higher precision in variability detection and control, IEEE Advanced Semiconductor Manufacturing Conference and Workshop, pp. 12-17, 2007.

[5] R. Good and M. Purdy, An MILP approach to wafer sampling and selection, IEEE Transactions on Semiconductor Manufacturing 20(4), pp. 400-407, 2007.

[6] S. Bassetto and A. Siadat, "Operational methods for improving manufacturing control plans: case study in a semiconductor industry," Journal of Intelligent Manufacturing, vol. 20, pp. 55-65, 2009.

[7] A. Tay, Real time estimation and control of CD uniformity in lithography, IEEE Advanced Semiconductor Manufacturing Conference, pp. 185-190, 2008.

[8] D. Pepper, O. Moreau, G. Hennion, Inline Automated defect Classification: a Novel Approach to Defect Managament, IEEE Advanced Semiconductor Manufacturing Conference and Workshop, pp. 43-48, 2005.

[9] S. Dauzère-Pérès, J.-L. Rouveyrol, C. Yugma, P. Vialletelle, A Smart Sampling Algorithm to Minimize Risk Dynamically, ASMC 2010 (IEEE/SEMI Advanced Semiconductor Manufacturing Conference), pp. 307-310, San Francisco, USA, 2010.

AUTHOR BIOGRAPHIES

Justin Nduhura Munga received his Engineering degree in Computer science, Microelectronics, and Automation from the Ecole Polytechnique de Lille in Lille, France in 2009. Currently he is a Ph.D. student in Industrial Engineering at the Ecole des Mines de Saint-Etienne in Gardanne, France, and works at STMicroelectronics in Crolles, France.

Stéphane Dauzère-Pérès is Professor at the Provence Microelectronics Center of the Ecole des Mines de Saint-Etienne, where he is heading the Manufacturing Sciences and Logistics Department. He received the Ph.D. degree from the Paul Sabatier University in Toulouse, France, in 1992; and his Habilitation à Diriger des Recherches from the Pierre and Marie Curie University, Paris, France, in 1998. He was a PostDoc Fellow at the Massachusetts Institute of Technology, U.S.A., in 1992 and 1993, and Research Scientist at Erasmus University Rotterdam, The Netherlands, in 1994. He has been Associate Professor and Professor from 1994 to 2004 at the Ecole des Mines de Nantes in France. He was invited Professor at the Norwegian School of Economics and

Business Administration, Bergen, Norway, in 1999. Since March 2004, he is Professor at the Ecole des Mines de Saint-Etienne. His research mostly focuses on optimization in production and logistics, with applications in planning, scheduling, distribution and transportation. He has published more than 35 papers in international journals and 100 communications in conferences.

Claude Yugma is an Associate Professor at the Ecole des Mines de Saint-Etienne, in the Manufacturing Sciences and Logistics Department. He received the Ph.D. degree from the Institut National Polytechnique de Grenoble, France, in 2003.

Philippe Vialletelle is manager of the Operations and Methods System group at STMicroelectronics. After receiving an Engineering degree in Physics, he entered the semiconductor industry working on ESD and physical characterization. His next experiences were Metrology and Process Control where he drove the deployment of methodologies and tools for a 200mm fab. He finally integrated Industrial Engineering and is now responsible for the development of advanced programs for the management of Crolles 300mm production line. At European level, Philippe is in charge of the definition and follow-up of collaborative programs in the field of Manufacturing Sciences such as HYMNE or IMPROVE.

Characterizing the Value of Technological Knowledge for Lean Manufacturing

Charles M. Weber

Department of Engineering and Technology Management
Portland State University
Portland, Oregon, USA
WeberCM@gmail.com

Abstract — **This paper argues that managing technological knowledge is the last frontier of lean manufacturing. An empirically grounded model of the structure of organizational technological knowledge allows manufacturing managers to determine the true value of lean manufacturing practices over lifecycle of their processes. The approach described in this paper also helps fab managers enhance profitability by choosing appropriate strategies organizational learning and problem-solving for three semiconductor businesses: VLSI, commodity components and specialty parts.**

Keywords - semiconductor, value, technological, knowledge, lean manufacturing

I. Introduction

Traditional approaches to lean manufacturing strive to increase factory productivity through reducing waste. To improve their bottom line, semiconductor manufacturers take measures that cut costs, improve their cost structure and take advantage of economies of scale. Unfortunately, these measures are reaching diminishing returns. New approaches, which integrate efforts that increase value with those that reduce cost, need to be investigated for lean manufacturing practices to make substantial contributions towards the top line and the bottom line once again.

Semiconductor manufacturers compete on the basis of trading off economies of scale and speed [1]. Economies of scale are required for reasons of capital productivity [2] – the manufacturer needs to recover large investments in research, development and plant equipment. Economies of speed [3] are required because leading-edge, Very Large-Scale Integrated (VLSI) circuits are produced in an urgent environment [4], [5] in which the unit sales price of the integrated circuits decays exponentially within a short period of time [3]. A significant portion of the profit potential of a VLSI product line can be lost before the first lot that realizes the product exits the fab. Economies of scale have traditionally been associated with increasing the number of goods to be produced or reducing the unit cost of the goods to be sold [6], [7]. Lean practices consequently tend to enhance economies of scale. Economies of speed are inherently linked to revenue generation as well -- they depend upon how the unit price of the good to be sold varies over time. They are generally achieved by accelerated yield learning [8] and the reduction of fab cycle time [3].

This paper is the first in a series of papers, which argue that managing technological knowledge is the last frontier of lean manufacturing. Technological knowledge is defined as "understanding the effects of the input variables (of a technological system) on the output. Mathematically, the process output, Y, is an unknown function $Y=f(\mathbf{x})$; $\mathbf{x}$ is always a vector (of indeterminate dimension)" ([9], p. 62). It is estimated that in semiconductor manufacturing the vector $\mathbf{x}$ consists of over 1000 control variables and at least as many exogenous environmental variables that affect the process output [10].

This paper looks primarily at the organizational aspects of managing technological knowledge and their impact on profitability, which can be characterized by analyzing how well known, subsystem-level input variables such as fault density, survival yield and throughput impact a wafer fabrication facility's (fab's) ability to generate bottom-line profit per unit time [8] . (Die-sort yield and cycle time are intermediate variables.) Subsequent papers will try to characterize how individual control variables and environmental variables affect profitability, and how the nature of the various kinds of knowledge for which these variables act as proxies evolve as the semiconductor manufacturing process matures.

Empirical evidence for this paper comes from real-time observation of practices in a wafer fabrication facility [11], from quantitative performance data from said fab, and from using well-known qualitative research methods [12]-[14] to analyze data gathered in over 100 interviews with practitioners that are employed by more than 25 firms in semiconductor industry. Models of lifecycles of three kinds of manufacturing process that operate in three distinctive economic environments and produce three different product classes – VLSI, commodity components and specialty parts with high value added – have emerged from the data. Inputs are varied, *ceteris paribus* (all other things being equal), to determine the sensitivity of fab performance to accumulating different kinds technological knowledge in different economic environments.

The primary contribution of this paper is a model of organizational technological knowledge that gives practicing manufacturing managers the ability to make investment decisions through scenario planning. The primary differentiators of this model are the incorporation of value of ownership [1], [11], [15] and the model's ability to span the complete investment horizon of a semiconductor process lifecycle. These features allow manufacturing managers to determine the true value of lean manufacturing practices over time. The approach described in this paper also helps fab managers enhance profitability by choosing appropriate organizational learning [8] and problem-solving [16] strategies. In particular, the model shows when to invest in accelerated learning as opposed to lean practices that reduce cost. Finally, this paper provides insight into the structure of technological knowledge in semiconductor manufacturing, which has historically been treated as a black box [9].

II. THEORETICAL FRAMEWORK

Figure 1 depicts the theoretical framework for technological knowledge that has emerged from the data. It recognizes three kinds of actors that accumulate technological knowledge through a variety of learning activities. Production personnel engage in 'learning by doing' (LbD), through which they reduce unit wafer cost at a decreasing rate as they gain production experience (e.g., [17]). The proxy measure for production experience is the cumulative number of wafers produced, which, for a particular product line 'j', is given by

$$q_j(t) = \int_{t_i}^{t} W_j(t)\, dt , \qquad (1)$$

where $W_j(t)$ denotes the wafer start rate during a particular time interval Δt, and $t\text{-}t_i$ represents the time that has elapsed since the fabrication of wafers that realize chips for the product line under consideration began. The organizational learning variable that measures the performance of learning-by-doing is the variable cost of producing a wafer, $c_{wj}(q)$.

Production personnel engage in two deliberate learning activities – Production Volume Learning (PnVL) and Production Quality Learning (PnQL). $W_j(t)$, is the proxy measure for PnVL, and PnQL is measured by the survival yield, $S_j(t)$, i.e. the number of wafers started during Δt that survive the production line. Process engineers engage in Process Quality Learning (PcQL), which essentially consists of reducing fault density, $F_j(t)$, over time. Design engineers engage in design learning (DL), which manifests itself in a smaller chip area, $A_j(t)$, and a greater number of die per wafer, $N_j(t)$, from design revision to design revision.

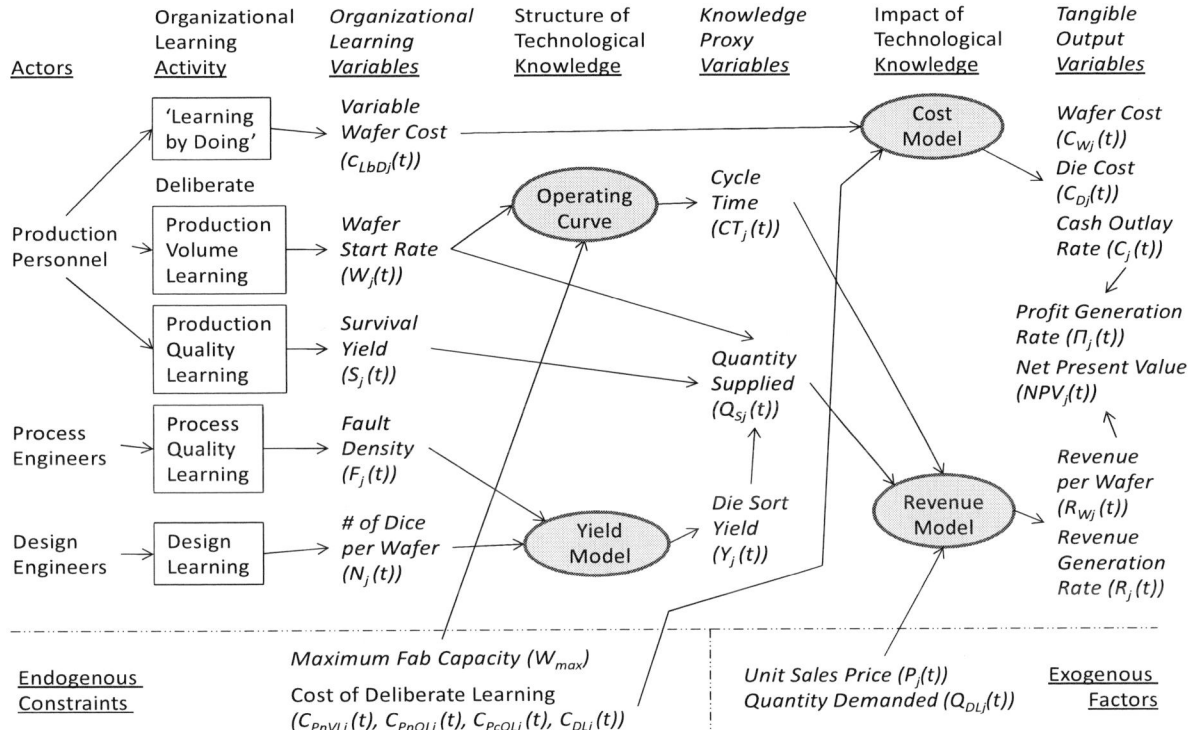

Figure 1. Structure of technological knowledge in semiconductor manufacturing for a particular product line 'j' and its impact on a fab's bottom line.

A. Learning Rates and Technological Knowledge

The learning rates associated with the various forms of organizational learning are given in Table I. As the theory of the learning curve (e.g., [17]) predicts, learning-by-doing causes $c_{wj}(q)$ to decay as a power function in which $c_{wj}(1)$ denotes the variable production costs of the first wafer; k_{wj} is a rate constant; and $c_{wj}*$ represents the theoretical minimum of $c_{wj}(q)$. Production Volume Learning is a linear function of time because knowledge is gained when an organization encounters new problems and challenges as it scales up linearly [18] (although the ramp to volume production is

978-1-61284-408-4/11 $26.00 © 2011 IEEE   25

frequently managed as a series of steps (e.g., [3]). Consistent with extant theory on continuous improvement and waste reduction [19], the rate equations for PnQL, PcQL and DL take on exponential form. For example, when a product line is released into production at t=0, it exhibits a performance gap in fault density of $F_j(0)-F_j^*$, where $F_j(0)$ denotes the fault density at t=0 and F_j^* is the target value of $F_j(t)$. Fault density decays exponentially at a rate given by the constant k_{Fj}. PnQL and DL proceed in an analogous manner, although DL can only be measured when a new design revision is released.

Two kinds of technological knowledge accumulate in every fab. The first kind is *operating knowledge*, which manifests itself in the operating curve. The operating curve characterizes a product's production cycle time, $CT_j(t)$, as a highly non-linear function of $W_j(t)$ and W_{tot}/W_{max}, the ratio between the sum total of all wafer start rates of all product lines and the maximum throughput rate of the fab [11], [20]. The second kind of technological knowledge is *yield knowledge*, which is manifests itself as a yield model in which die sort yield, $Y_j(t)$, is expressed as a highly non-linear function of $A_j(t)$ and $F_j(t)$ (see, for example, [21] for a review of the possibilities). According to [10], the quantity of functioning integrated circuits (ICs, chips) that a fab can supply can be estimated as

$$Q_{Sj}(t) = W_j(t)\,S_j(t)\,Y_j(t)\,N_j(t) \,. \qquad (2)$$

B. Profit, Revenue and Cost

"It is widely agreed upon among students of corporate finance that, for practical purposes, the most appropriate investment criterion for a corporate investment project is the net present value of that project" [22, p. 822], which represents the expected cumulative real profit of that investment project. Assuming continuous cash flows, this criterion can be given as

$$NPV_j(t) = \int_{t_i}^{t} [R_j(t) - C_j(t)]\, e^{-\rho t}\, dt \,, \qquad (3)$$

where $NPV_j(t)$ denotes the net present value of the project at a particular point in time t, discounted at the corporation's effective annual cost-of-capital rate, κ; the continuously compounded discount rate ρ is equal to $\ln(1+\kappa)$; $R_j(t)$ and $C_j(t)$ respectively denote the stream of cash revenues and cash outlays at time t [22]. The quantity $t - t_i$ represents the investment horizon. The profit generation rate (measured in $ per unit time) that is commensurate with (3) is given by

$$\Pi_j(t) = [R_j(t) - C_j(t)]\, e^{-\rho t} \,. \qquad (4)$$

TABLE I. SEMICONDUCTOR LEARNING RATES

	Learning by Doing (LbD)	Production Volume Learning (PnVL)	Production Quality Learning (PnQL)	Process Quality Learning (PcQL)	Design Learning (DL)
Actors	Production Personnel	Production Personnel	Production Personnel	Process Engineers	Design Engineers
Action	Engage in Production	Scale up Deliberately	Improve Production Deliberately	Improve Process Deliberately	Improve Designs Deliberately
Type of Improvement	(Continuous Improvement)	(Linear Improvement)	(Continuous Improvement)	(Continuous Improvement)	(Continuous Improvement)
Type of Knowledge	Tacit Production Knowledge	Production Volume Knowledge	Production Quality Knowledge	Process Quality Knowledge	Design Knowledge
Input Variable	Variable Production Cost	Wafer Start Rate	Survival Yield	Faults/Wafer	Products/Wafer
Rate Equation	$c_{wj}(q)=\{c_{wj}(1)^{-k_{wj}}\}+c_{wj}^*$	$W_j(t) = m_{wj}t + W_j(0)$	$S_j(t)=\{[S_j(0)-S_j^*]10^{-k_{Sj}t}\}+S_j^*$	$F_j(t)=\{[F_j(0)-F_j^*]10^{-k_{Fj}t}\}+F_j^*$	$N_j(t)=\{[N_j(0)-N_j^*]10^{-k_{Dj}t}\}+N_j^*$
Rate Constant	k_{wj} (unit cost reduction)	m_{wj} (scaleup)	k_{Sj} (improved survival)	k_{Fj} (fault reduction)	k_{Dj} (design shrink)

TABLE II. THE COST OF LEARNING

	Learning by Doing (LbD)	Production Volume Learning (PnVL)	Production Quality Learning (PnQL)	Process Quality Learning (PcQL)	Design Learning (DL)
Cost of Learning (Cash Outlay Rate)	$c_{LbD}(q)=c_{wj}(q)=\{c_{wj}(1)^{-k_{wj}}\}+c_{wj}^*$ (Power Function)	$c_{PnVLj}(t)=w_s(dW_j(t)/dt)=w_s m_s$ (Constant)	$c_{PnQLj}(t) = C_{PnQLj}(0)$ (Constant)	$c_{PcQLj}(t) = C_{PcQLj}(0)$ (Constant)	$c_{DLj}(t) = C_{DLj}(0)$ (Constant)

In figure 1, a revenue model and a cost model assess the impact of technological knowledge on profitability. The revenue model is a variant of Leachman and Ding's formula [3, eq. (18)]; it integrates cycle time and quantity supplied with two exogenous factors: unit sales price and quantity demanded, $Q_{Dj}(t)$. When $Q_{Dj}(t) > Q_{Sj}(t)$, then the fab under study operates under capacity constraint -- all chips that are produced are sold. When $Q_{Dj}(t) \leq Q_{Sj}(t)$, then the fab under study operates under market constraint – price is determined by supply and demand [23]. Let $P_j(t)$ denote the unit sales price of a chip of type j when a product wafer of type j is

started and $P_j(t+CT_j(t))$ denote the unit sales price of a product of class j when that wafer exits the fab. The revenue generation rate of product line j can consequently be given as

$$R_j(t) = P_j(t + CT_j(t))\, Q_{Sj}(t) \,. \qquad (5)$$

Table II illustrates the structure of the cost model, which integrates the cost of 'learning by doing' with the cost of the various forms of deliberate learning that go on in a fab. All cost variables are given in terms of cash outlay rates; they align with its rate equations of the learning efforts that they respectively characterize. The cost of learning-by-doing,

$c_{lbdj}(q)$, is essentially equivalent to $c_{wj}(q)$, and consequently decreases at a decreasing rate (e.g., [17]). By contrast, the cash outlay rates of PnQL, PcQL and DL are constant over time. They primarily consist of paying for technical personnel to improve production quality, process quality and designs, respectively, plus the overhead that these learning activities incur. If PnVL consists of scaling up production volume linearly, then the cost of Production Volume Learning, $c_{PnVL}(t)$, is also constant over the time span during which it transpires. However, $c_{PnVL}(t)$ can be much larger than the cost of other forms of deliberate learning because it is likely to involve purchasing capital equipment, as well as hiring and training new personnel, all at an approximately linear rate. The total cost (cash outlay rate) of developing and producing product line j is given by

$$C_j(t) = W_j(t)\, c_{LbDj}(q) + c_{PnVLj}(t) + c_{PnQLj}(t) + c_{PcQLj}(t) + c_{DLj}(t) , \quad (6)$$

where q is given by (1). The traditional measure for wafer cost can be obtained by dividing $C_j(t)$ by $W_j(t)$.

TABLE III-A. ASSUMPTIONS FOR VLSI CIRCUIT MANUFACTURING

	Learning by Doing (LbD)	Production Volume Learning (PnVL)	Production Quality Learning (PnQL)	Process Quality Learning (PnQL)	Design Learning (DL)
Initial Value of Performance Metric	$c_{w1}(1) = \$1k/wafer$	$W_1(0) = 10k\ wafer/qtr.$	$S_1(0) = 0.95$	$F_1(0) = 355\ F/wafer$	$N_1(0) = 200\ units/wafer$
Target Value of Performance Metric	$c_{w1}* = \$0/wafer$	$W_1* = 90k\ wafers/qtr.$	$S_1* = 1.00$	$F_1* = 0\ F/wafer$	$N_1* = 1000\ units/wafer$
Rate Constant	$k_{w1} = 0.2/year$	$m_{S1} = 10k\ wafers/qtr.^2$	$k_{S1} = 0.1/year$	$k_F = 0.5/year$	$k_{D1} = 0.1/year$
Cost Constant	$k_{w1} = 0.2/year$	$w_{s1}m_{s1} = \$200M/qtr.$	$c_{S1} = \$0.4M/year$	$c_{F1} = \$4M/year$	$c_{D1} = \$1M/year$

TABLE III-B. ASSUMPTIONS FOR MANUFACTURING COMMODITY COMPONENTS

	Learning by Doing (LbD)	Production Volume Learning (PnVL)	Production Quality Learning (PnQL)	Process Quality Learning (PnQL)	Design Learning (DL)
Initial Value of Performance Metric	$c_{w1}(1) = \$10k/wafer$	$W_1(0) = 0k\ wafer/qtr.$	$S_1(0) = 0.95$	$F_1(0) = 632\ F/wafer$	$N_1(0) = 2000\ units/wafer$
Target Value of Performance Metric	$c_{w1}* = \$0/wafer$	$W_1* = 90k\ wafers/qtr.$	$S_1* = 1.00$	$F_1* = 0\ F/wafer$	$N_1* = 10000\ units/wafer$
Rate Constant	$k_{w1} = 0.2/year$	$m_{S1} = 20k\ wafers/qtr.^2$	$k_{S1} = 0.1/year$	$k_F = 0.5/year$	$k_{D1} = 0.1/year$
Cost Constant	$k_{w1} = 0.2/year$	$w_{s1}m_{s1} = \$10M/qtr.$	$c_{S1} = \$0.4M/year$	$c_{F1} = \$4M/year$	$c_{D1} = \$0.4M/year$

TABLE III-C. ASSUMPTIONS FOR MANUFACTURING SPECIALTY PARTS

	Learning by Doing (LbD)	Production Volume Learning (PnVL)	Production Quality Learning (PnQL)	Process Quality Learning (PnQL)	Design Learning (DL)
Initial Value of Performance Metric	$c_{w1}(1) = \$10k/wafer$	$W_1(0) = 500\ wafer/qtr.$	$S_1(0) = 0.80$	$F_1(0) = 632\ F/wafer$	$N_1(0) = 400\ units/wafer$
Target Value of Performance Metric	$c_{w1}* = \$0/wafer$	$W_1* = 600\ wafers/qtr.$	$S_1* = 1.00$	$F_1* = 0\ F/wafer$	$N_1* = 1000\ units/wafer$
Rate Constant	$k_{w1} = 0.2/year$	$m_{S1} = 5\ wafers/qtr.^2$	$k_{S1} = 0.1/year$	$k_F = 0.5/year$	$k_{D1} = 0.1/year$
Cost Constant	$k_{w1} = 0.2/year$	$w_{s1}m_{s1} = \$100k/qtr.$	$c_{S1} = \$0.4M/year$	$c_{F1} = \$4M/year$	$c_{D1} = \$1M/year$

III. SIMULATING SCENARIOS

The model described in the previous section has been customized into three baseline scenarios so that the value of organizational technological knowledge can be characterized for three product lines: VLSI (j=1), commodity components (j=2) and specialty parts (j=3). Tables III-A, III-B and III-C respectively detail key assumptions pertaining to each baseline scenario, which differ greatly from each other because manufacturers of VLSI circuits, commodity components and specialty product lines tend to operate in diametrically different economic environments. All baseline scenarios assume a fab that produces wafers with a diameter of 200 millimeters; the maximum capacity of that fab (W_{max}) is 100,000 wafers per quarter. Consistent with prior observation [11], [20], the fabs that are stylized in these simulated scenarios all exhibit a very steep operating curve that is approximated in table IV. Consistent with prior studies (e.g., [8], [24]), the effective annual cost-of-capital rate, κ, of integrated circuit manufacturing is pegged at 15% or 0.15 for all three scenarios.

TABLE IV. OPERATING CURVE OF SIMULATED FABS

Fab Load (W_{tot}/W_{max})	0.1	0.2	0.3	0.4	0.5	0.6	0.7	0.8	0.9	0.99
Cycle Time (days)	28	31	34	37	40	44	48	53	64	>100

The analysis of all three scenarios proceeds in an analogous manner. First, it is assumed that all subsystem-level learning occurs concurrently and on a continuous basis. $R_j(t)$, $C_j(t)$ and $\Pi_j(t)$ are calculated for this set of assumptions over a significant portion of the lifecycle of a semiconductor product line. Then, the impact on $\Pi_j(t)$ and $NPV_j(t)$ of varying only one form of subsystem-level learning, *ceteris paribus*, is calculated and compared to the calculation that assumes all forms of subsystem-level learning occur concurrently.

A. VLSI Circuits (j=1)

The economic environment of VLSI circuit manufacturing is characterized prolonged periods of capacity constraint [23] and by rapidly eroding unit sales prices [3], [25]. The model of the VLSI product line assumes that every chip that is produced will be sold at the unit sales price of the day, which can be expressed as

$$P_j(t) = P_j(0)\, 10^{-k_{Pj}t}, \qquad (7)$$

where $P_j(0)$ denotes the unit sales price when the product line is released into production, and k_{Pj} determines the rate at which the unit sales price decays over time. Consistent with semiconductor pricing data for VLSI chips with a high value added (e.g., a microprocessor), $P_j(0)$=US\$ 500 per chip, and k_{P1}=-0.5 year^{-1}. This implies that a circuit, which sells for US\$ 500 when it is released into production, will sell for US\$ 50 two years later.

The model of the VLSI product line also assumes that extensive Process Quality Learning and Production Quality Learning, as well as some Production Volume Learning [26], [27] have transpired prior to product release. These forms of deliberate learning accrue substantial sunk costs by prior to product release. Finally, VLSI circuits are relatively large; it is consequently assumed that about 200 of them fit onto a 200-mm wafer when the product is released ($N_1(t=0)$ = 200 dice per wafer).

Figure 2a shows that in the early stages of the production are driven by economies of speed. The revenue generation rate significantly outpaces the cash outlay rate. The profit generation rate increases sharply if yield learning is accelerated [8], or if fab cycle time is decreased [3]. The situation reverses in the fourth year, when the real revenue generation rate, $R_1(t)e^{-\rho t}$ approaches the real cash outlay rate, $C_1(t)e^{-\rho t}$ in magnitude. Traditional lean manufacturing techniques, which are focused on cost reduction, become more valuable at that stage of the production lifecycle. The radical drop in $C_1(t)e^{-\rho t}$ results from a cessation of Production Volume Learning once the fab runs near capacity and no new capital equipment is added.

Figures 2b and 2c illustrate the synergy that results from concurrent learning. *Ceteris paribus*, PnVL, PnQL and DL have a negative value throughout the production lifecycle. PnVL by itself results in a particularly negative profit generation rate. It is associated with two expensive propositions: purchasing plant equipment and hiring new personnel. LbD and PcQL have a net positive value initially,

which erodes very rapidly as time progresses. However, if all learning activities occur concurrently and continuously, then the VLSI product line generates real profit for about four years.

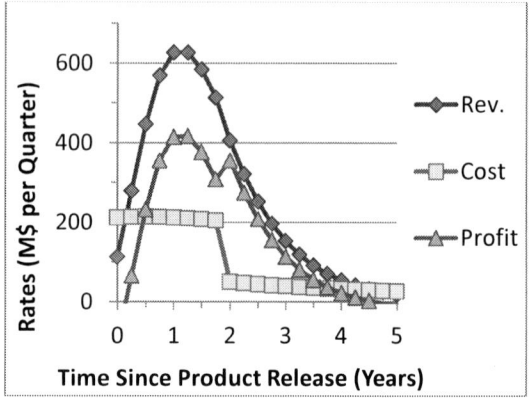

Figure 2a. $R_1(t)\,e^{-\rho t}$, $C_1(t)\,e^{-\rho t}$ and $\Pi_1(t)$ for VLSI Circuit Manufacturing

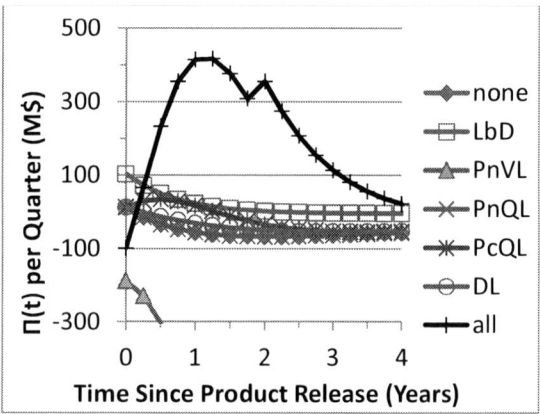

Figure 2b. $\Pi_1(t)$ as a function of learning mode (VLSI).

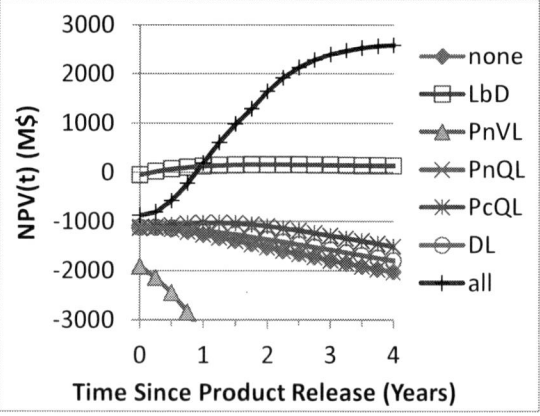

Figure 2c. $NPV_1(t)$ of VLSI circuit product line.

B. Commodity Components (j=2)

Producing commodity components (Com) is very different from VLSI manufacturing. The baseline scenario for commodities assumes capacity constraint, as well as unit sales prices that are stable and low ($P_2(t)=P_2(0)=$US\$ 0.25$). The model of the Com product line also assumes a mature process, and that the product line to be installed replaces another. Thus a substantial amount of PcQL has transpired and significant sunk costs have accrued prior to product release ($t=0$). However, little PnVL and PnQL have transpired before $t=0$. Finally, commodity components are relatively small; it is consequently assumed that about 2000 of them fit onto a 200-mm wafer when the product is released into manufacturing ($N_2(t=0)=2000$ dice per wafer).

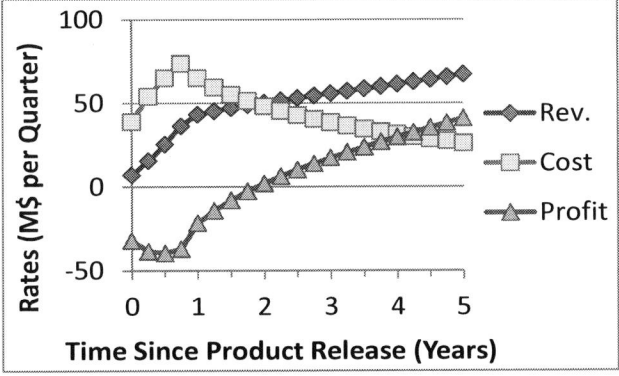

Figure 3a. $R_2(t)\,e^{-\rho t}$, $C_2(t)\,e^{-\rho t}$ and $\Pi_2(t)$ (commodity components)

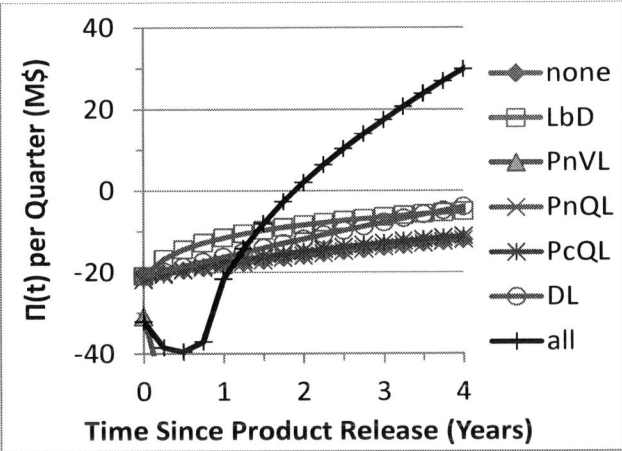

Figure 3b. $\Pi_2(t)$ as a function of learning mode (commodity components).

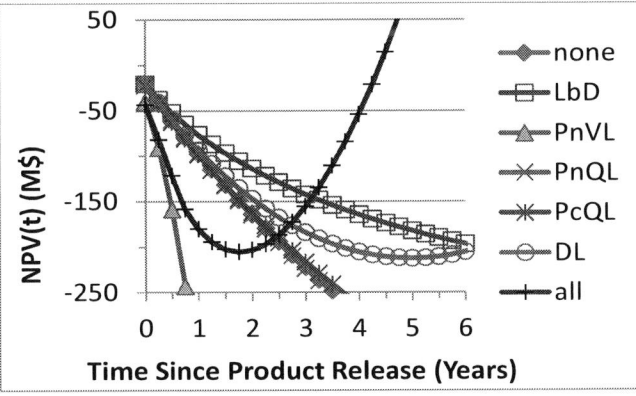

Figure3c. $NPV_2(t)$ as a function of learning mode (Com).

The commodities business is driven by economies of scale; economies of speed are less important because the unit price of the good to be sold does not erode. Figure 3a shows that the real cash outlay rate exceeds the real revenue generation rate for the first two years; the product line loses money during that time period. The product line generates positive cash flow thereafter, but profit margins are slim. The magnitude of the real revenue generation rate is barely above that of the real cash outlay rate. Therefore, traditional lean manufacturing practices, which are focused on cost reduction, are the best approach to increasing the margins in this economic environment.

Once again, the synergy that results from concurrent learning is paramount. Figures 3b and 3c illustrate that, *ceteris paribus*, PnVL, PnQL and DL have a negative value throughout the production lifecycle. PnVL by itself results in a particularly negative profit generation rate. It is still associated with purchasing some plant equipment and hiring some new personnel. All other forms of learning have a net value that is negative initially but grows gradually as time progresses. Design learning even generates positive cash flow after five years. However, if all learning activities occur concurrently and continuously, then the Com product line generates real positive cash flow after two years; breakeven occurs in the fifth year. It should be noted, however, that commodity businesses are notoriously volatile and that the model is highly sensitive to changes in unit price fluctuations. The conclusions of the model may consequently change if major fluctuations in unit sales price do occur.

C. Specialty Parts (j=3)

The economics of making specialty parts (Spec) differ significantly from those of the Com or the VLSI business. Specialty parts are usually built to order; both the unit price and the quantity demanded are fixed in advance. The unit sales price is generally high; the quantity demanded is low and does not change significantly over time. The baseline scenario for Spec assumes $P_3(t) = $US\$ 100$, and $Q_{D3}(t) = Q_{S3}(t) = $ chips 10,000 per month for the time interval $0 \leq t \leq 5$ years; $P_3(t) = Q_{D3}(t) = Q_{S3}(t) = 0$ before and thereafter. The model of the Spec product line assumes a relatively immature process, and that the product line to be installed replaces another. Thus very little process quality learning, production quality learning and production volume learning have transpired, and few sunk costs have accrued prior to product release. However, specialty parts are relatively large; it is consequently assumed that about 400 of them fit onto a 200-mm wafer when the product is released into manufacturing ($N_3(t=0)=400$ dice per wafer).

The specialty parts business is driven by economies of scale; economies of speed are less important because the unit price of the good to be sold does not erode. Figure 4a shows that the real cash outlay rate exceeds the real revenue generation rate over the whole production period, but barely. Therefore, traditional lean manufacturing practices, which are focused on cost reduction, are the best approach to increasing the margins in this economic environment.

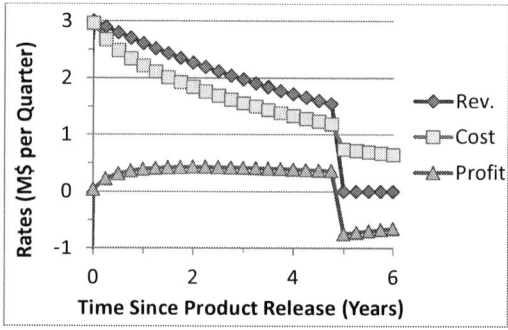

Figure 4a. $R_3(t) e^{-\rho t}$, $C_3(t) e^{-\rho t}$ and $\Pi_3(t)$ (specialty parts)

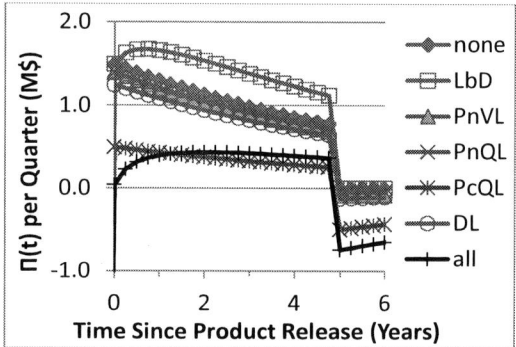

Figure 4b. $\Pi_3(t)$ as a function of learning mode (specialty parts).

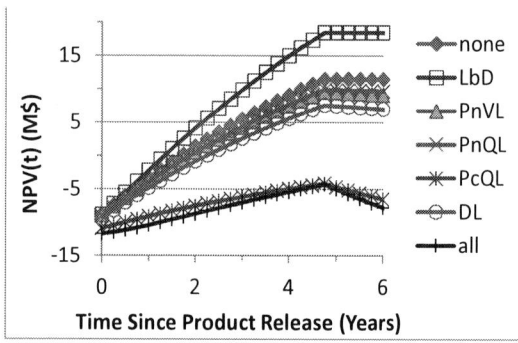

Figure 4c. $NPV_3(t)$ as a function of learning mode (specialty parts).

Figures 4b and 4c illustrate that the value of all forms of learning other than LbD is less than not learning at all. In addition, the synergy that results from concurrent learning is negative. The value of conducting all forms of learning is lower than that of every individual form of learning with the possible exception of Process Quality Learning. This suggests that the most valuable skill in Spec may be the ability to bring in another contract to produce specialty parts.

IV. SUMMARY AND DISCUSSION

This paper has described a study, which has attempted to characterize the value of organizational technological knowledge in semiconductor manufacturing. The study was based on real-time observation of semiconductor factories [11] and interviews with experts in various areas of practice related to semiconductor manufacturing [24]. The following findings, which can most likely be generalized beyond semiconductor manufacturing, have emerged from the data.

- The external economic environment strongly influences the value of different kinds of technological knowledge.
- The value of knowledge varies over time.
- The relative value of different kinds of technological knowledge varies over time.
- The value of knowledge can be positive or negative.
- The synergy between different kinds of knowledge can be positive or negative.
- Useful theories of technological knowledge must be value-driven and time sensitive.

The findings of this study have significant implications for semiconductor manufacturing practice. First and foremost, the paper clearly establishes the importance of developing, acquiring and learning how to use technological knowledge as the key to profitability. For example, the value of purchasing and installing plant equipment is highly negative unless training and prolonged experimentation takes place on said equipment. The value of production volume learning, which involves both training and experimentation [18], [27] on capital equipment, can be negative for prolonged periods of time. Making a fab truly profitable evidently requires conducting a variety of highly diverse learning activities in parallel.

It should be noted that the model described in this paper can separate revenue from cost. This capability gives practitioners insight into whether their product lines are driven by economies of speed [3] or economies of scale [6], [7]. In former case, measures that accelerate yield learning [9] or shorten factory cycle time [3], [20] should be given preference. In the latter case, traditional lean manufacturing practices that reduce the cost of production are appropriate.

The model allows managers to develop scenarios for specific kinds of product lines that can characterize the structure of technological knowledge and its impact on profitability *a priori*. This capability enables managers to design choose appropriate strategies and design practices that are well-suited for managing their operations. However, the paper has not described the most common scenario in modern semiconductor manufacturing – a wafer fab that runs a mix of all three kinds of product lines and perhaps more. Models of multi-product-line manufacturing that are based on the principles presented in this paper have been developed [1], but still require broadly based quantitative validation in multiple fabs.

This paper has focused on the impact of organizational technological knowledge on profitability. It has used departmental or subsystem-level metrics as inputs and financial metrics as outputs. It has not looked at the how knowledge that resides within the mind of the individual person contributes to the fab's bottom line. Research on that subject is ongoing. Its goal is to characterize the granularity of technological knowledge to the point where practices at each work station can be designed and executed in a manner that maximizes profitability for the whole fab.

ACKNOWLEDGMENT

The material presented in this is based upon work supported by the National Science Foundation under Grant No. 0822062 (Enabling Timely Revolutions in Organizational Performance).

REFERENCES

[1] C. M. Weber and A. Fayed, "Scale, scope and speed: Managing the challenges of semiconductor manufacturing," *IEEE Transactions on Semiconductor Manufacturing*, Vol. 23, No. 1, February 2010, pp. 30-38.

[2] P. Silverman, "Capital productivity: Major challenge for the semiconductor industry," *Solid State Technology*, Vol. 37, No. 3, 1994, p. 104.

[3] R. C. Leachman and S. Ding, "Integration of speed economics into decision-making for manufacturing management," *International Journal of Production Economics*, 2007, pp. 39-55.

[4] C. Terwiesch and R. E. Bohn, "Learning and process improvement during production ramp-up," *International Journal of Production Economics*, Vol. 70, No 1, 2001, pp. 1-19.

[5] C. J. G. Gersick, "Time and transition in work teams: Toward a new model of group development," *Academy of Management Journal*, Vol. 31, No. 1, 1988, pp. 9-41.

[6] M. Porter, "Note on the structural analysis of industries," Harvard Business School Case Study 9-376-054, 1983.

[7] R. S. Pindyck & D. L. Rubinfeld, *Microeconomics*, Upper Saddle River, NJ: Prentice Hall, 1998.

[8] C. Weber, "Yield Learning and the Sources of Profitability in Semiconductor Manufacturing and Process Development," *IEEE Transactions on Semiconductor Manufacturing*, Vol. 17, No. 4, November 2004, pp. 590-596.

[9] R. E. Bohn, "Measuring and Managing Technological Knowledge," *Sloan Management Review* Vol. 36, No. 1, Fall 1994, pp. 61-73.

[10] R. E. Bohn, "Noise and learning in semiconductor manufacturing," *Management Science*, Vol. 41, No. 1, 1995, pp. 31-42.

[11] C. M. Weber and A. Fayed, "Optimizing your position on the operating curve: How can a fab truly maximize its performance?" *IEEE Transactions on Semiconductor Manufacturing*, Vol. 23, No. 1, February 2010, pp. 21-29.

[12] Yin, R. K., *Case Study Research*, Sage Publishing, Newbury Park, CA, 1994.

[13] K. M. Eisenhardt, "Building theories from case study research," *Academy of Management Review*, Vol. 16 (1989), pp. 620-627.

[14] Miles, M. B., and Huberman, A. M. *Qualitative Data Analysis*, Sage, Beverly Hills, CA, 1984.

[15] C. Weber, V. Sankaran, K. Tobin, and G. Scher, "Quantifying the value of ownership of yield analysis technologies," *IEEE Transactions of Semiconductor Manufacturing*, Vol. 15(4) Nov. 2002, pp. 411-419.

[16] C. Weber, " Knowledge Transfer and the Limits to Profitability: An Empirical Study of Problem-Solving Practices in the Semiconductor Industry," *IEEE Transactions of Semiconductor Manufacturing*, Vol. 15, No. 4, November. 2002. pp. 420-426.

[17] L. Argote and D. Epple, "Learning curves in manufacturing," *Science*, Vol. 247, 1990, pp. 920-924.

[18] K. Mishina, "Learning by new experiences: Revisiting the Flying Fortress learning curve," In Lamoreaux *et al.* (eds.) *Learning by doing in markets, firms and countries*, A National Bureau of Economic Research Conference Report. The University of Chicago Press, Chicago, IL. 1999.

[19] W. I. Zangwill and P.B. Kantor, "Toward a theory of continuous improvement," *Management Science* Vol. 44, No. 7, 1998, pp. 910-920.

[20] S. S. Aurand and P. J. Miller, "The operating curve: A method to measure and benchmark manufacturing line productivity," *IEEE/SEMI-ASMC*, Sept. 10-12, 1997, pp. 391 – 397.

[21] J. A. Cunningham, "The use and evaluation of yield models in integrated circuit manufacturing," *IEEE Transactions on Semiconductor Manufacturing*, Vol. 3, No. 2, May 1990, pp. 60-71.

[22] U. E. Reinhardt, "Break-even analysis for Lockheed's Tri Star: An application of financial theory," *The Journal of Finance*, September 1973, Vol. 28, No. 4, pp. 821-838.

[23] R. E. Bohn and C. Terwiesch, "The economics of yield-driven processes," *Journal of Operations Management*, vol. 18, no. 1, 1999, pp. 41-59.

[24] C. M. Weber, "Do learning organizations have strokes of genius?" *Proceedings of PICMET '06*, Istanbul, Turkey, July 8-13, 2006.

[25] W. Trybula, "Presentation of International SEMATECH's Global Economic Model," Portland, Oregon, May 20, 2003.

[26] G. P. Pisano, "Learning before doing in the development of new process technology," *Research Policy*, Vol. 25, 1996, pp. 1097-1119.

[27] J. E. Carillo and C. Gaimon, "Improving manufacturing performance through process change and knowledge creation," *Management Science*, Vol. 46, No. 2, 2000, pp. 265-288.

Charles Weber holds (among other degrees) a B.S. degree in engineering physics from the University of Colorado, Boulder; an M.S. degree in electrical engineering from the University of California, Davis; and a Ph.D. in management from MIT's Sloan School of Management. He joined Hewlett-Packard Company as a process engineer in an IC manufacturing facility. He subsequently transferred to HP's IC process development center, working in electron beam lithography, parametric testing, microelectronic test structures, clean room layout, and yield management. From 1996 to 1998, Charles managed the defect detection project at SEMATECH as an HP assignee. In December 2002, he joined the faculty of Portland State University, where he is an associate professor of engineering and technology management. Charles Weber's research interests are in organizational learning, problem solving, knowledge management, innovation and entrepreneurship.

In-line Metrology of High Aspect Ratio Structures with MBIR Technique

Delphine Le Cunff
STMicroelectronics
Crolles, France

Leif Jonny Höglund and Nicolas Laurent
SEMILAB AMS
Billerica, MA, USA

Abstract — **This paper presents recent results obtained applying the Model Based Infrared Reflectometry (MBIR) technique as an in-line monitoring technique for high aspect ratio structures. The paper will focus on the description of the metrology development and its implementation in a manufacturing environment for the particular case of few microns Deep Trench Isolation (DTI) structures. The technique is demonstrated to be robust method for the in-line geometry control of etched structures. Experimental data relating to high aspect ratio Through Silicon Via (TSV) structures will be presented as well. The results confirm that the technique is very sensitive to the geometry even for small Via of diameter down to 3 microns.**

Metrology, high Aspect Ratio, deeep trench, TSV

I. INTRODUCTION

The introduction of microns scale deep structures for integrated circuit devices required an in-line automated metrology to control the geometry of those highly vertical structures. The scatterometry technique is the classical manufacturing method that one could think of to address this need. Nevertheless, depending on the exact optical configuration, multi wavelength- single angle, this technique can be inappropriate due to the lack of signal coming from the bottom of the structures. Therefore, the control of the etched geometries versus the process specifications is performed through high cost SEM cross section for the depth and through CD-SEM for the top surface critical dimension (CD). The MBIR technique offers an attractive alternative as being a versatile, non-contact and non-destructive measurement of vertical structures [1-3]. The recent development of small spot systems allows in addition this technique to be fully integrated in a metrology production environment by offering the capability to measure within the scribe lines of production wafers themselves. We have implemented this technique with the IR3100S tool from SEMILAB for the case of DTI structures and are exploring the possibility to extend its applications to TSV structures.

II. MODEL BASED INFRARED REFLECTOMETRY

A. Optical technique

The MBIR technique uses an infrared light with a wavelength range of 0.9 – 20 µm, for which silicon is transparent. The instruments proprietary optical configuration blocks the reflection from the backside of the wafer, allowing for photometrically accurate measurements of the top layers. A glowbar and a halogen lamp are used as light sources and are combined as input to a Fourier Transform InfraRed spectrometer. A stirling cooled Mercury Cadmium Telluride detector is used to detect the reflected light [3]. The small spotsize of 80x40 microns and pattern recognition system allow measurement on small targets of 100x52 microns.

B. Effective Medium Approach

One of the advantages of the long infrared wavelengths employed by the MBIR instrument is the ability to approximate arrays of trenches and vias as effective mediums. The dielectric function of the effective medium is calculated as a fractional mixture of the optical properties from the host material and the air inclusions [4]. In the case of visible light, the shorter wavelength gets comparable to the pitch values and diffraction of the light occurs, and the effective medium approximation is not accurate anymore [5]. One approach to model diffraction effects, which is commonly applied in scatterometry measurements, is to use Rigorous Coupled Wave Analysis (RCWA) [6]. The longer computational time of the RCWA diffraction efficiencies prevents the calculations to be used in on-the-fly fitting routines. At runtime when a measured spectrum has been acquired it is typically compared with the spectra in a pre-calculated library of spectra. The structure parameters corresponding with the best fit from the search are reported as the measurement result. In order to achieve accurate measurements, it is therefore needed to have a large library with small steps in the variable parameters. The step for library generation can indeed be very time consuming and following process changes that may occurs can even lead to a need for the whole library to be regenerated.

In the case of MBIR technique, the short computational time to obtain a reflectance spectrum when applying the effective medium approach allows calculations to be performed in a fitting routine as the measured reflectance spectra are acquired by the MBIR instrument. The robustness of the effective medium approach also eliminates the need for a larger calibration wafer set and three DOE wafers are typically sufficient to develop a calibrated model, making the recipe setup time short.

978-1-61284-408-4/11 $26.00 © 2011 IEEE

An example for the effective medium approach applied to a trench structure is provided in figure 1, where linear trenches etched in silicon are modeled as a mixture of silicon and air. The trench thickness is varied together with the top and bottom air fractions during the fitting process in order to calculate the geometry of the trenches.

One of the strengths of the MBIR technique applied to effective medium trench structures is the simple behavior of the spectrum shape vs. trench geometry and the lack of correlation between the fitted parameters. For simplicity the mechanics of the reflectance spectrum model fit can be described as trench depth being extracted by fitting the period of the regular interference fringes, top CD by fitting the average reflectance of the spectrum and bottom CD by fitting the amplitude of the regular fringes [1]. For scatterometry measured spectra, typically composed of multiple diffraction peaks that sometimes overlap each other, the correlation between structure parameters becomes more challenging to assess [6].

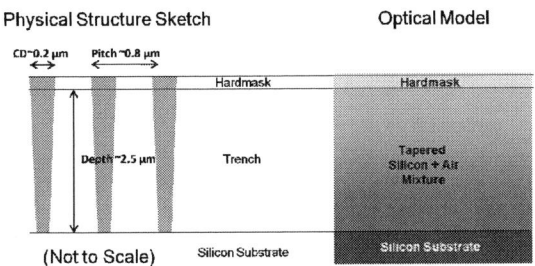

Figure 1 – DTI on EPI substrate structure description. Tapered silicon + air mixture optical properties describe the trench etch profile.

III. APPLICATIONS FOR DEEP TRENCH ISOLATION

Deep trench isolation is a widely used approach for dielectric isolation. Typical dimensions are 0.18 to 1μm in width and 2 to 5 μm in depth. Trench isolation puts extreme demands on the etch process. Not only the trench must have smooth walls but the slope of the walls relative to the plane of the wafer and the taper are also critical parameters. Considering the density of voids in the trench array is high enough, the etched geometry of such structures can be controlled with the MBIR technique through an in-line monitoring of top CD, bottom CD and trench depth parameters.

A. Modeling of the DTI Structures

Figure 2 is showing a cross section of the DTI structure from the 100x52 μm test area located in the scribeline of the production wafers. For this case, it has a trench depth of ~2.5 μm, pitch of ~0.8 μm and CD of ~0.2 μm with an etch profile narrowing towards the bottom of the trenches.

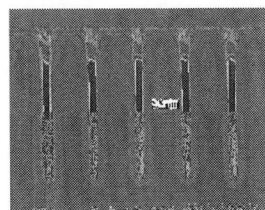

Figure 2 – Example x-SEM image showing the DTI structure

An optical model as described in the figure 1 is used to perform a model fit to the measured spectrum in order to extract the measured trench depth, top CD and bottom CD. From figure 3 we see that the optical model describes the measured structure well in the wavenumber range 1500-3500 cm-1, where there are regular interference fringes present and the effective medium approximation allows for fast calculations that can be performed online as the measured spectra are acquired.

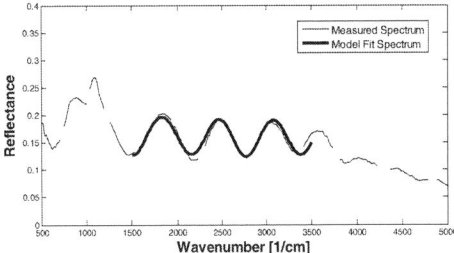

Figure 3 – Example of measured DTI reflectance spectrum (dashed line). DTI model-fit spectrum (solid line).

Excellent correlation results (figure 4) have been achieved with cross section SEM as reference for trench depth and bottom CD, and CD-SEM as reference for top CD. Repeatability and reproducibility (R&R) studies were also performed. %R&R values of 3.7% were found for depth and 8.5% for top CD measurement respectively. This assessed that the levels of measurement error are acceptable compared to the process variation and then validate the technique for an in-line process control.

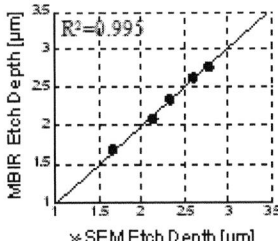

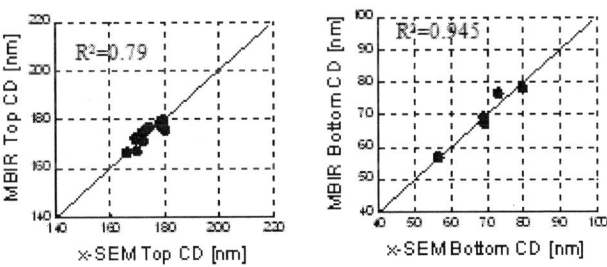

Figure 4 – DTI structure MBIR vs. x-SEM correlation.

B. Monitoring of DTI on Silicon substrate

The benefit of the MBIR in-line technique is to allow intra-wafer measurement in order to fine tune etch process recipe. 2D mapping (figure 5) of the depth is quick and easy to obtain.

978-1-61284-408-4/11 $26.00 © 2011 IEEE

This is obviously a key asset for etch process monitoring and etch tool requalification in a production environment.

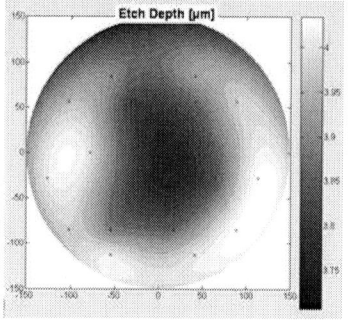

Figure 5 – 2D mapping of depth with mean value of 3.9μm and intra wafer dispersion of 250nm.

The MBIR technique has also been providing valuable trench geometry data while setting up the etching process. Figure 6 presents the depth values for several production wafers depending on the top CD and on the process recipe. Recipe A only differs from Recipe B by a slightly different number of etching cycles. As expected, it is found that larger CD leads to deeper DTI structures and it is also clear that the technique is very sensitive to the trench depth.

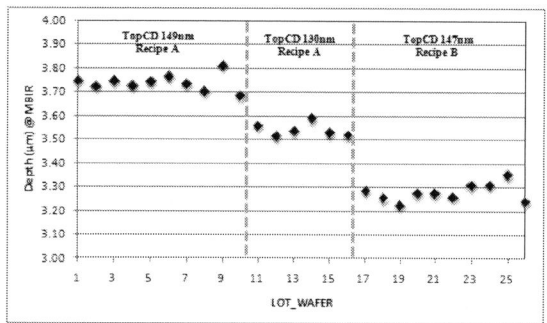

Figure 6 – Change in Depth versus Top CD and process recipe

The tool stability was monitored over time on the same DTI control wafer. Figure 7 reported the variation of trench depth top and bottom CD over a period of three months. The measurement results have been very stable over the evaluation period although tool service occurs during that period.

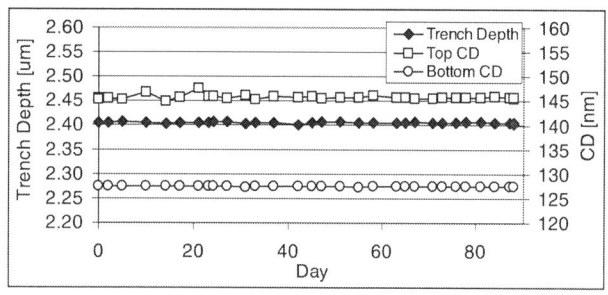

Figure 7 – Tool stability measurement over 3 months

The standard deviation during the evaluation period was 0.05% for trench depth, 0.3% for top CD and 0.04% for bottom CD.

The top CD of a large number of production wafers have been measured in-line using both the CD-SEM and MBIR

techniques. Note that the CD-SEM and the MBIR measurement target structures used for in-line measurements have different design. The correlation results, included in figure 8, are showing that the two techniques are in good agreement with a slight constant offset due to the difference in target structures.

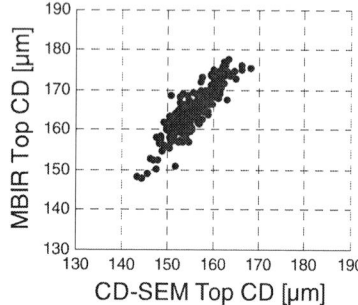

Figure 8 – Top CD MBIR vs. CD-SEM correlation.

Since both the trench depth and CD's are measured by the MBIR system the CD-SEM step could be eliminated, saving time in the production process flow.

C. Monitoring of DTI on SOI substrates

For DTI structures etched in SOI layers it is critical that the trenches are etched all the way down to the buried oxide layer without any remaining silicon at the bottom of the trenches. A test wafer has been provided for which the DTI etch was stopped approximately 300nm short before reaching the buried oxide. The MBIR technique offers non-destructive process monitoring to ensure there is no remaining silicon between the bottom of the trenches and the buried oxide.

MBIR and X-SEM measurements have been performed on sites with variation in the thickness of the unetched silicon layer between the bottom of the DTI structures and the buried oxide. The correlation results, which are included in figure 9, are excellent.

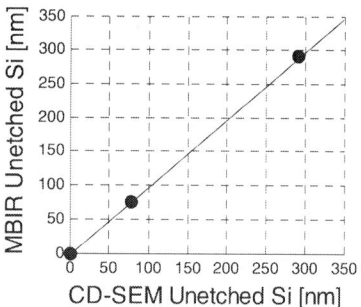

Figure 9 – Unetched silicon in SOI DTI MBIR vs. X-SEM correlation

Figure 10 is showing the 17 point MBIR unetched silicon data for a wafer where the DTI structures were not etched all the way to the buried oxide. A concentric pattern is present with a thicker unetched silicon layer near the center of the wafer. The within wafer variation is large with a standard deviation of 90 nm. This is coherent with the etch process tool signature with etch rate lower at the center of the wafer.

This is again very valuable data for process etch monitoring as the metrology can identify quickly if the DTI openings reach the buried oxide layer of the SOI on all locations on the wafer.

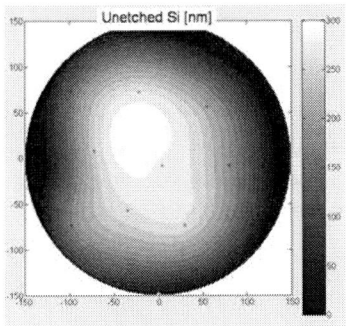

Figure 10 – 17 point MBIR measurement of unetched silicon on a DOE wafer

IV. APPLICATIONS FOR THROUGH SILICON VIAS

New developments in the electronic integration look increasingly to the third dimension. There are a wide variety of technologies that can be used to realize 3D interconnect but of particular interest is the through silicon via (TSV) technologies used for 3D-Wafer Level Packaging and 3D-Stacked Integrated Circuit for which the 3D TSV process is typically integrated in the Si-wafer fabrication line. One of the main challenges for 3D metrology involves measuring of such high aspect ratio via of (10:1) approaching (20:1) in future roadmaps.

The capability of the MBIR technique was explored on such challenging applications. Specific test wafers with various combinations of CD (1 to 4µm) and pitch (6, 8 and 10µm) were processed to assess the capability of the technique. Periodic TSV arrays with depths in the range 22.5-25.5 µm and CD's in the range 2-4 µm have been measured using the MBIR instrument. A tilted SEM view of the TSV array is provided in figure 11.

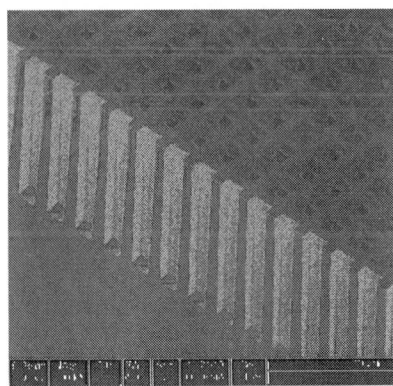

Figure 11 – Tilted SEM view of TSV array

A. Modeling of TSV Structures

The same modeling technique as described for DTI structures etched in silicon is applied to the TSV structure. We note that the effective medium approximation is valid when the pitch of the TSV array is small compared to the wavelength of light. The MBIR technique is therefore mainly suitable for measurements of high density TSV arrays and will benefit from the shrinking TSV design rules, as specified in semiconductor roadmaps. As expected it was found that the 6 µm pitch with 3 and 4 µm CD provided the better results due to the higher density of vias.

B. Measurement of TSV Structures

An MBIR diemap measurement of the TSV structure, presented in figure 12, reveal a concentric etch depth profile with deeper vias towards the edge of the wafers. The vias are consistently tapered throughout the wafer, with smaller CD at the bottom compared to the top. Variation in depth inside the wafer is found to be about 1µm.

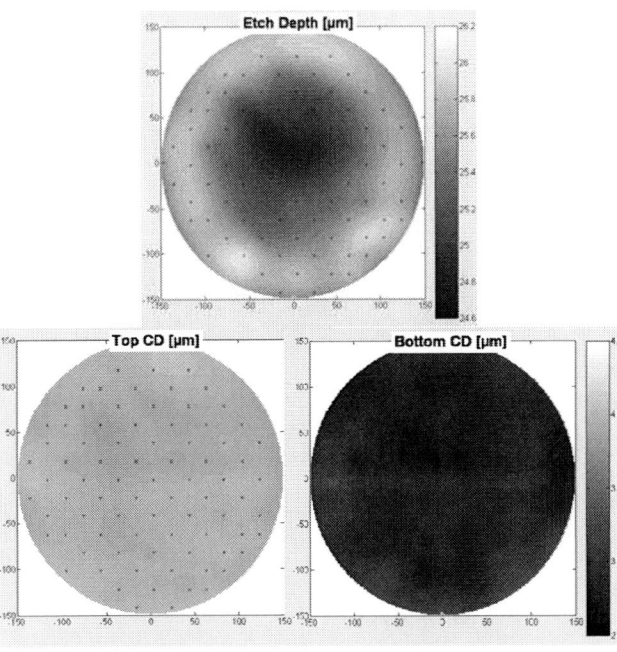

Figure 12 – Depth, Top and Bottom CD wafer maps measured by MBIR for 6 µm pitch and 4 µm CD.

Excellent results ($R^2 > 0.99$) have been achieved when correlating the MBIR TSV depth and top CD measurement with cross section SEM measurements, as seen in figure 13.

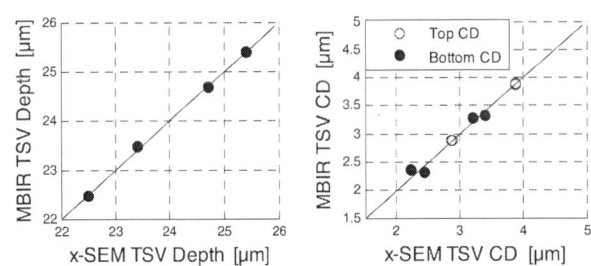

Figure 13 – TSV structure MBIR vs. x-SEM correlation

For the same pitch configuration, the variation of the depth is plotted (figure 14) versus the Top CD and clearly shows the general trend of larger CD's yielding deeper TSV's during the etch process.

The dynamic repeatability was evaluated by measuring 5 sites and 30 repeats, using load/unload of the wafer between the measurements. The results obtained were found to be good with a standard deviation of 0.32% for via depth, 0.86% for top CD and 2.8% for bottom CD.

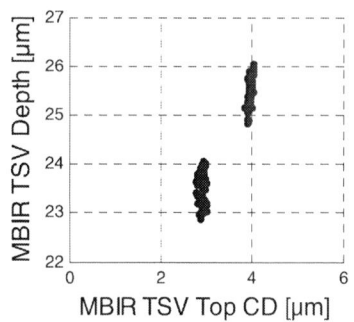

Figure 14 – Variation of TSV depth vs. top CD

V. CONCLUSION

We have presented recent results using the MBIR technique for geometry measurements of high aspect ratio structures. The MBIR technique has proven to be a reliable and versatile measurement technique for non-destructive process in-line monitoring of few microns deep DTI structures of about 200nm CD. The technique could also be adopted for measurement of DTI on SOI substrate with the benefit to measure the remaining Silicon layer on top of the buried oxide layer.

Moving to the field of few microns CD such as 3D-TSVs, the MBIR is also found to be a promising technique with excellent correlation with cross section SEM.

ACKNOWLEDGMENT

The authors acknowledge Arnaud Tournier and François Leverd for their support in the introduction of the MBIR technique for DTI applications. They also would like to thank Laurent-Luc Chapelon for providing wafers for the TSV applications and Fabienne Pico for the SEM analysis. Han Chang and John Byrnes are acknowledged for valuable technical discussions.

REFERENCES

[1] M. Gostein, P.A. Rosenthal, A. Maznev, A. Kasic, P Weidner, and P.-Y. Guittet, "Measuring Deep-Trench Structures with Model-Based IR", Solid State Technology, March 2006, pp. 38-42.
[2] P.-Y. Guittet, "Model-Based Infrared Reflectometry: In-Line Applications for DRAM Manufacturing," Future Fab, 2006.
[3] P. A. Rosenthal, C. Duran, J. Tower, A. Mazurenko, U. Mantz, P. Weidner and A. Kasic, "Model-Based Infrared Metrology for Advanced Technology Nodes and 300 mm Wafer Processing", in Characterization and Metrology for ULSI Technology 2005, D.G. Seiler, Ed. (AIP Conference Proceedings, 2005), pp. 620-624.
[4] G.E. Jellison, "Physics of optical metrology of siliconbased semiconductor devices", in *Handbook of Silicon Semiconductor Metrology*, A.C. Diebold, Ed. New York: Marcel-Dekker, 2001, pp 723-760
[5] C.A. Durán, A.A. Maznev, G.T. Merklin, A. Mazurenko and M. Gostein, "Infrared Reflectometry for Metrology of Trenches in Power Devices", 2007 ASMC conference.
[6] C.J. Raymond, "Scatterometry for semiconductor metrology", in *Handbook of Silicon Semiconductor Metrology*, A.C. Diebold, Ed. New York: Marcel-Dekker, 2001, pp 477-513

Full automatic on the fly optical macro wafer edge inspection system

Denis Kirmizigül, Dresden University of Technology, Institute of Artificial Intelligence
Heiko Fröhlich, Infineon Technologies Dresden GmbH

Abstract - **We present a low cost optical macro inspection system for the wafer edge. The system is able to inspect the full wafer edge (front-, backside and apex) and provides a short feedback loop to the unit processes. Furthermore the use of image processing methods enables inspection without additional time loss. This inspection system can be installed on all wafer rotating tools with free space for hardware installation.**

I. INTRODUCTION

In recent years, wafer inspection technology has been extended beyond the patterned area of the wafer to the edge. The goal is to understand defect mechanisms and prevent defect migration from the edge that might contaminate devices across the wafer surface and process tools. These defects can result from wafer handling, all dry and wet processes and can cost yield. Particularly thermal processes, like anneal and bake processes, lead to manifold defect mechanism at the wafer edge. In many cases macro defects at the wafer edge have the biggest influencing factor regarding yield critical defect generation. Therefore an optical macro inspection system is required.

Goals of all defect inspections are high wafer and lot sampling rate, as far as possible short inspection time, automatic defect classification and very short automatic feed back loops to control of critical processes. The new idea is the implementation of wafer edge defect scan at different required process steps. First step on this way is the installation of inspection equipment within a full automatic controlled and robot provided wafer sorter tool (FAROS). So it is possible to scan 100% of lots and wafers with no additional process time.

II. EXPERIMENTS SETUPS AND DETAILS

As a demonstrator we implemented the system in a fully automated wafer sorter cluster for detecting and measuring of process edges (e.g. copper plating edge, litho resist EBR edge and backside and frontside wafer edge cracks. With the integrated active feed back loops to process tools and the wafer transport system, the further handling of affected wafers and stop of bad tools is triggered. Staggering in rotation of not aligned wafers is completely compensated with real time anti staggering system of HSEB [2] inspection module. The camera system needs only one complete turn of the wafer and gives three 360° view images of all wafer edge zones by putting line scan images side by side. Fig.1 shows the different

zones of the wafer edge and which zones are viewed by which camera.

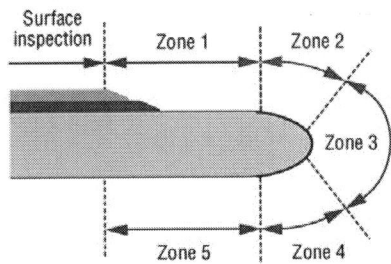

Fig. 1. Schema of all scanned wafer zones. zones 1,2 – camera 1, zone 3 – camera 2, zones 4,5 – camera 3

The camera systems are installed on eight full automatic controlled and robot provided wafer sorter tools (FAROS). Two sorters each work in a robot provided sorter cluster. The four clusters operate in two FAB modules (Fig.2).

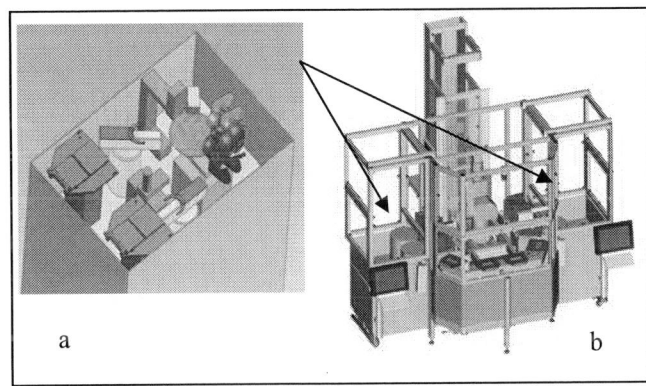

Fig. 2. Two wafer sorters with camera modules (a) installed in a robot provided sorter cluster (b)

Depending on production area the camera systems work with white- or yellow light bright field illumination sources. Geometrical resolution of optical systems is assured for defects of interest greater than 50µm. All of the defect results are available in real time in FAB defect data base "Knights" for further evaluations and triggering of feed back loops and bad wafer handling. The anti staggering unit of camera

978-1-61284-408-4/11 $26.00 © 2011 IEEE

systems made staggering of wafers up to 4mm during rotation possible without loss of best focus.

III. EXAMPLE APPLICATION

As an example application for the inspection system we demonstrate its use for the detection and measurement of copper removal distances from the wafer edge. This is done by means of image processing and pattern recognition methods.

For the example application the images the zones 1 and 2 are used. These are acquired by camera 1 (Fig. 1). In these images we measure the distance of the copper plating relative to the wafer apex in pixels. The conversion from pixels to micrometers is simply done by linear mapping. We identify three regions in the image: Coppered area, de-coppered area and background area (Fig. 3)

The background area in the image comprises real world background and upper bevel (zone 2). For convenience we will denote the transition in the image from background area to de-coppered area as wafer edge curve and the transition from de-coppered area to coppered area as copper curve. Therefore the quantity we are interested in is the distance of the copper curve relative to the wafer edge curve in the observed image. As can be seen in Fig. 3 a simple image gradient based approach to find the transitions between the different regions in the image will give poor results. This is because there are plenty of different image gradients which do not belong to the region transitions of interest. On the other hand a method solely based on color values is also not applicable because the different regions have a lot of colors in common.

In the following we propose a probabilistic model for the joint observation of an image with a wafer edge curve and a copper curve. Based on this model we formulate the detection of the two curves as seeking the maximum a-posteriori curves when an image is observed.

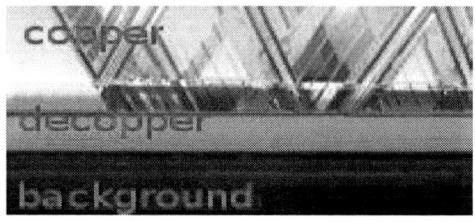

Fig. 3. Detail of an image obtained with the HSEB wafer inspection system. Light upper area is coated with copper. Dark lower area is background and in the middle is the region where copper plating is removed.

IV. MODEL FORMULATION FOR THE EXAMPLE APPLICATION

Let I and J denote a set of row and column indices. An image x can then be written as a mapping $x : I \times J \to F$ from the Cartesian product $I \times J$ to a set of color values F. According to this $x(i,j) = x_{i,j} = f$ denotes the color value of the pixel at position $(i,j) \in I \times J$ with color $f \in F$. Here the usual coordinate system used by the image processing

community with the topmost left image pixel as the origin and row (column) numbers increasing downwards (rightwards) is used.

The sequence $k = (k_1, ..., k_{|J|})$ denotes the wafer edge and the sequence $d = (d_1, ..., d_{|J|})$ denotes the copper curve. Thereby $k_j, d_j \in I$ give the image row through which the wafer edge curve and the copper curve pass in column $j \in J$. Since it is not possible for the copper curve to fall below the wafer edge curve in the image, it holds that $k_j > d_j$ for all $j = 1, ..., |J|$. To a wafer curve k the probability

$$P(k; \varphi_k) = \frac{1}{Z_k} e^{-\frac{1}{2} \sum_{j \in J} \left(\frac{[k_j - \rho(j; m, a, z)]}{\sigma_k} \right)^2} \quad (1)$$

is assigned, in which

$$\rho(j; m, a, z) = m + a \sin(\omega(j - z)) \quad (2)$$

forms an ideal sinusoidal curve with mean height m, amplitude a and zero-crossing at z. The probability is parameterized with $\varphi_k = \{m, a, z, \sigma_k\}$. Z_k is a normalizing constant to form a proper probability distribution. This distribution assigns wafer edge curves higher probabilities if they have a shape similar to the sinusoidal curve $\rho(j; m, a, z)$. With this we take into account that the wafer edge curve might not be a straight horizontal line in the image because of a miss-alignment between the wafer center and the rotation center during the image capturing process. Next we state a probability distribution for copper curves:

$$P(d; \varphi_d) = \frac{1}{Z_d} e^{-\frac{1}{2} \frac{H(d, \mu)}{\sigma_d^2}} \quad (3)$$

The function H is defined as follows:

$$H(d, \mu) = \sum_{j \in J} (d_j - \mu)^2 + \sum_{(j,j') \in J \times J; |j-j'|=1} (d_j - d_{j'})^2 \quad (4)$$

$\varphi_d = \{\mu, \sigma_d\}$ is again the set of parameters for the distribution and the parameter μ represents the expectancy about the height of the copper curve in the image. So in the first summation in H copper curves are more penalized, the more they are away from the expectancy about copper curves in images. The second sum in H penalizes non-smooth copper curves. Z_d is normalizing constant to render $P(d; \varphi_d)$ into a proper probability distribution. Therefore the probability distribution (3) assigns a high probability to curves

d which are smooth and located near to the expected height μ of copper curves.

Furthermore a relationship between an observed image x and given curves k and d is established. This is done via a conditional probability distribution for observing an image x given a wafer edge curve k and a copper curve d:

$$P(x|k,d) = \prod_{j \in J} P(x_{.,j}|k_j, d_j). \tag{5}$$

That means the image columns $x_{.,j}$ are independent given a wafer edge curve k and copper curve d. We define the probability of observing image column $x_{.,j}$ given that the curves pass through row k_j and d_j to be

$$P(x_{.,j}|k_j, d_j) = e^{Q(x_{.,j}, k_j, d_j)} \tag{6}$$

wherein

$$Q(x_{.,j}, k_j, d_j) = \sum_{i=0}^{d_j} q_c(x_{.,j}) + \sum_{i=d_j+1}^{k_j} q_d(x_{.,j}) + \sum_{i=k_j+1}^{|J|} q_b(x_{.,j}) \tag{7}$$

is a summation over the image rows for column j summing up the different image color values $x_{i,j}$ weighted according to what region they fall in. The weighting is done via the functions q_c, q_d and q_b, which are the logarithms of the normalized color histograms for the different regions (copper area, decoppered area and background area) in the image. In the experiments we approximated the histograms for the different regions by coarsely taking the values from the different regions.

Based on the previously defined probability distributions we get the joint probability of observing image x, wafer curve k and copper curve d as

$$P(x, k, d) = P(k)P(d)P(x|k,d). \tag{8}$$

IV. RECOGNITION

Observing an image x, we seek the wafer curve k^* and copper curve d^* which maximize the joint probability (8). Therefore we have to solve

$$
\begin{aligned}
k^*, d^* &= \mathrm{argmax}_{k,d} P(k, d|x) \\
&= \mathrm{argmax}_{k,d} P(x, k, d) \\
&= \mathrm{argmax}_{k,d} \left[P(k)P(d)P(x|k,d) \right]
\end{aligned}
\tag{9}
$$

This can be done efficiently by dynamic programming [1] because of the special structure of the copper curve distribution (3).

The quantity we are interested in, namely the distance in pixels between the wafer edge curve and the copper curve, can then be calculated for each image column j by subtracting d_j^* from k_j^*.

Although the formulated model allows the unsupervised learning of the parameter sets φ_d and φ_k, we set them in our experiments for performance reasons by hand. Fig. 4 shows the result of solving (9). It can be seen that the copper curve follows the transition from the copper plating area to the uncoppered area although the patterned wafer area superimposes the transition zone. The wafer edge curve is located at the transition from zone 1 to zone 2.

Fig. 4. Recognition results: wafer edge curve k^* (red) and copper curve d^* (green)

IV. Results and Conclusion

First results of installed inspection system within a full automatic controlled and robot provided wafer sorter tool (FAROS) meet customer's expectations. Complete scan of wafer edge side by side with the normal sorter process without loss of process time is possible. The functionality of tested anti wafer staggering system is very good. The compensation of the wafer staggering is better than the limited field of depth of optical system. Other applications like edge crack detection and litho resist/ARC EBR monitoring are successful tested (actually in high volume test). So with all four sorter cluster we are able to check all wafers in FAB regarding critical wafer edge cracks.

REFERENCES

[1] Schlesinger, Michail I. and Hlaváč, Václav, Ten lectures on statistical and structural pattern recognition, Kluwer Academic Publishers, Dordrecht, The Netherlands, 2002.
[2] HSEB Dresden GmbH, http://www.hseb-dresden.de/

Application of CGS Stress Metrology to Advanced Process Control & Monitoring

David M. Owen and Jeff Hebb
Ultratech, Inc.
San Jose, CA 95134, USA
dowen@ultratech.com

Christian Otten
Texas Instruments Inc.
Dallas, TX 75243

Abstract— The improvement of device performance associated with the intentional manipulation of stresses on the transistor scale is an integral part of device fabrication at advanced technology nodes. However, comparatively little attention is given to stress management at within-die and within-wafer length scales. Process variations that occur on these longer length scales can induce significant within-wafer stress variations. In turn, the die-to-die, wafer-to-wafer and lot-to-lot stress variations may have significant impact on device performance and yield. This paper describes the use of the CGS (Coherent Gradient Sensing) stress metrology to characterize the detailed within-wafer and wafer-to-wafer stress variations. The CGS stress maps consist of more than 700,000 data points, enabling new potential applications for stress metrology. Case studies are presented that summarize the use of the CGS data to reveal correlations between stress variations and device performance, lithographic overlay and tool matching.

Keywords-stress metrology; process control; device yield

I. INTRODUCTION

The development of advanced semiconductor devices at the 45nm device node and beyond routinely involves the use of strain engineering to manipulate mobility in the transistor channel. In this case, the scale of the relevant stress variations is on the order of 10's of nanometers. For example, embedded silicon germanium is often used to impose strain on the channel in the development of advanced PMOS devices [1]. The characterization of channel-scale stresses is challenging and necessarily involves the measurement or inference of lattice strains as directly as possible.

On the other hand, thin film stress characterization in the context of manufacturing, has traditionally been used to monitor thin film deposition processes. Various stress metrology systems, based on a variety of techniques, have featured low point-density line scans or maps of the wafer shape comprised of typically no more than a few hundred points. The differences in wafer shape measurements made before and after the deposition process are used to compute the average film stress, σ_f from the change in substrate curvature, $\Delta\kappa$ using the well-known Stoney equation [2]:

$$\sigma_f = \frac{M_s h_s^2}{6 h_f} \Delta\kappa \qquad (1)$$

where M is the biaxial modulus, h is thickness and subscripts s and f refer to the substrate and film, respectively.

Those measurements of stress (or strain) on the device scale and on the wafer scale are separated by approximately 6 to 7 orders of magnitude in their in-plane dimension. Consequently, there is an entire range of in-plane length scales, from within-die to within-wafer, across which stress variations often go unevaluated. Many real manufacturing issues may exist across those exact length scales; for example, yield of die in the center of the wafer may be greater than those at the edge. As a result, there is an opportunity to explore whether characterizing apparent stress variations occurring within-die and die-to-die may be a useful way to identify and resolve manufacturing issues. If so, advanced stress metrology could be used for process control to monitor critical processes.

In the following section, the physical basis for the use of stress for process control metric is described, where the concept of 'direct' and 'indirect' stress effects is introduced. Subsequently the specific attributes of the CGS technology that enable detailed within-wafer stress characterization are discussed. Finally, a series of case studies that demonstrate the relationship between stress variations and a) device performance variation, b) lithographic overlay and c) tool matching are summarized.

II. STRESS FOR PROCESS CONTROL

Equation 1 indicates that the change in wafer shape, expressed as $\Delta\kappa$ can be used to compute stress in a thin film of thickness h_f. Implicit in equation 1 is the notion that the curvature change is represented by a single value (i.e. stress is constant for a uniform curvature change). Curvature is defined as the second derivative of the surface shape, such that uniform curvature implies a spherical shape. For an arbitrary, non-spherical shape, the curvature is more accurately represented by three curvature components; two direct curvatures and a twist or shear curvature (e.g κ_{xx}, κ_{yy} and κ_{xy} in Cartesian coordinates). The simple recognition that real wafers may not remain perfectly spherical as processing proceeds reveals that stresses will likely vary from point-to-point and at a particular location, curvature may be different in different orientations. In other words, the deviations of the wafer shape from spherical imply much more complex stress state across the wafer.

A. Sources of Curvature Variation

As noted above, stress metrology often relies on the measurement of curvature change, and variations of $\Delta\kappa$ may come from many sources. In order to explore these sources, equation 1 can be re-arranged to express $\Delta\kappa$ (i.e. what is measured) in terms of the other variables:

$$\Delta\kappa = \frac{6\sigma_f h_f}{M_s h_s^2} \qquad (2)$$

Furthermore, the film stress can be alternately expressed in terms of the mismatch strain, ε_m between the film and substrate:

$$\sigma_f = M_f \varepsilon_m \qquad (3)$$

Mismatch strains may arise due to differences in lattice constants, a between the film and substrate or can be thermally induced, as a result of differences in coefficients of thermal expansion, α:

$$\varepsilon_m = (a_s - a_f)/a_f \qquad (4a)$$

$$\varepsilon_m = (\alpha_f - \alpha_s)\Delta T \qquad (4b)$$

Inspection of equations 2 – 4 reveals that the measurement of $\Delta\kappa$ actually combines contributions due to numerous possible sources: stress, temperature, thickness or geometry (h), material properties (a, α, M). Therefore, the stress variations computed from a process-induced curvature change may include several if not all of the factors listed above. As a result, stress represents a very sensitive metric for evaluating process non-uniformity because the local stress combines several sources of process non-uniformity simultaneously.

B. Direct and Indirect Stress Effects

The case studies that will be presented in section IV provide examples of the relationship between stress variations and semiconductor fabrications issues such as device performance variation and overlay errors. It should be noted that the correlation of stress to device performance, for example, does not necessarily imply that 'stress' is the root cause of the problem per se. Consequently, when considering the physical mechanisms associated with any particular issue it should be recognized that there are those for which the effect of stress directly causes the problem and those for which stress is an indirect indicator of some other root cause.

Direct stress effect may include issues such as dislocation formation, strain relaxation, stress migration, film delamination and cracking. In these cases, the mechanical response of the device structure and substrate to processing causes potential failure and yield loss. Indirect effects are those where process non-uniformities such as temperature variation, material property & composition variation and geometric variation are manifested as variations in substrate curvature in a manner consistent with the above equations. An example of an indirect effect is a measured stress non-uniformity that is due to the temperature non-uniformity of a thermal process, where the temperature non-uniformity leads to variations in dopant activation. Stress does not cause variation in dopant activation, but stress variations still may be related to variations in dopant activation since both quantities are affected by temperature variations. Stress non-uniformity also causes topography variations that may lead to focus errors and variations in critical dimension in lithography. Again, stress variations do not necessarily 'cause' the variation in critical dimension. Instead stress non-uniformity caused by any arbitrary upstream process may create issues when the wafers reach a lithography step.

By recognizing that a detailed stress measurement reveals many aspects of process non-uniformity, it is possible that a variety of process monitoring challenges may be well-suited for stress metrology beyond traditional application of a thin film deposition monitor. More broadly, the evolution of wafer shape during semiconductor manufacturing provides a general-purpose indicator of both incremental and cumulative process uniformity.

C. Stress Calculation Beyond the Stoney Formula

The Stoney equation was derived for the case of a continuous thin film of uniform stress on a substrate, with both the film and substrate having uniform thickness. There are further assumptions, but nevertheless it is clear that real device structures, featuring a wide variety of materials and within-die variations in geometry, do not conform to the Stoney assumptions.

Extensions to the Stoney formula that consider more complex geometries and more general deformation states have been developed [e.g. 3, 4]. It is beyond the scope of this paper to review such formulations, but even with additional levels of complexity considered by these analyses of idealized cases, they do not reasonably resemble the conditions relevant to practical device geometries and patterned wafers. Indeed, the development of the detailed relationship between local stress and wafer-level curvature for a specific device combining unique materials, geometry and process history is currently well beyond the reach of current analytical tools.

Nevertheless, there are critical insights to be gained from evaluating the various extensions to the Stoney formula. The most notable is that the Stoney formula is recovered for the case of small, uniform curvature; i.e. the simpler Stoney solution is embedded in the more complicated solutions. The second observation is that the additional terms that account for the various complexities (e.g. material property variations in a structure of conducting lines separated by dielectric) are distinct from the 'Stoney' term, such that the specific details only modify the Stoney result by a constant. Therefore, the key insight is that the computation of stress from curvature change using the over-simplified Stoney (or similar) equation still provides stress values whose *relative* values have integrity. From a process control standpoint, this realization is adequate to enable reasonable comparison of apparent stress values from die-to-die and wafer-to-wafer in the absence of a device-specific stress / curvature relationship.

This section can be summarized as follows. First, measured variations in curvature change (and calculated stress) are a result of the combination of several factors including material properties, device geometry and process history. Second, the sensitivity of apparent stress to many factors makes stress metrology uniquely suited to characterize process non-uniformity and can be related process issues that are both directly (e.g. dislocation formation) and indirectly (e.g. temperature non-uniformity) related to stress. Third, even though typical stress calculations feature simplifying assumptions, they can still be used effectively to monitor stress variations in patterned wafers because the relative values have integrity.

III. CGS OVERVIEW & STRESS CONSIDERATIONS

A. CGS Overview

Details regarding the CGS method and the attributes of the CGS-300 system can be found elsewhere [5, 6]. The CGS is a type of lateral shearing interferometer that uses two parallel diffraction gratings to generate the so-called 'shear' or lateral offset. The fringes of the CGS interference patterns represent contours of constant slope or tilt on the wafer surface. The surface topography can be computed by numerically integrating two orthogonal slope components (i.e. slope x & slope y). Surface curvatures can be computed by numerical differentiation of the slope information.

The CGS-300 system used as part of the case studies described below features a 310mm diameter collimated laser beam that images the entire wafer instantaneously. Phase-shifted interference patterns are generated using the beam reflected off the wafer surface; the reflected beam is imaged on a CCD array. The in-plane resolution of the CGS technique is simply a function of the area imaged (in this case, ~300mm x 300mm) and the pixel dimensions of the imaging array (1024x1024). As a result, the data maps generated by the system are ~700,000 points (~300µm / point).

All the data used in the case studies described below are from patterned device wafers. The attributes of the CGS technique and system are such that accurate measurements can be obtained on patterned wafers. The two significant attributes that enable patterned wafer measurement are:

- Self-referencing: The fringe patterns of the CGS are generated from the interference of two 'copies' of the reflected wavefront (as opposed to an independent reference). This ensures that fringe contrast may remain high even for wafer surfaces of low reflectivity.

- Phase-shifting: The CGS system acquires multiple interference patterns at discrete offsets in relative phase. The analysis of phase shifting interference patterns is a common approach in interferometry [7] and enables the isolation of fringe contrast from systematic or background intensity variations that may be due to the pattern or beam non-uniformity.

Both the inherent self-referencing of CGS and the implementation of phase shifting increases significantly the signal-to-noise of the fringes.

B. Curvature Measurement & Stress Calculation

The detail of the surface topography, if measured by CGS or any other technique contains information from different physical sources. For example, there are contributions due to surface roughness, stress-induced curvature and substrate thickness variations. These various sources can have different magnitudes and occur at different in-plane spatial frequencies. Since the stress or stress change associated with a process is determined from the *change* in substrate curvature (i.e. post-process minus pre-process curvature), many of the other contributors to surface shape are eliminated by subtraction.

It is important to note that stress-induced curvature changes will occur at longer spatial wavelengths. These curvature changes are physically limited by the substrate geometry. Specifically, a substrate cannot sustain curvature changes whose radius is less than twice the substrate thickness (i.e. the deformation is limited to locally 'folding' the wafer onto itself). The limit of $\Delta\kappa < 1/(2*h_s)$ has several important consequences. First, based on the Nyquist sampling theorem, no additional curvature information is obtained from decreasing the spacing of measurement sites to below the dimension of the substrate thickness. In the context of the CGS measurement described above, the in-plane data resolution of ~300µm / point represents an oversampling of the stress-induced curvature. Second, with curvature-based stress measurement, it is possible to evaluate stress differences within-die, but only on the scale of a few millimeters (assuming h_s~750µm). It is not possible to resolve stress differences on the scale of the structure of typical device geometries; the substrate acts as a 'mechanical filter' that constrains and homogenizes the deformation over the millimeter scale.

IV. CASE STUDIES

The data associated with the case studies below provide examples for which the high-point density obtained using the CGS system enables new applications for stress metrology.

A. Correlation of Stress Variation to Device Performance

The effect of die-scale stress variations on device performance was studied by 'fingerprinting' the stress using the CGS system across multiple process steps in the front-end. The device used for these measurements was a 65nm DSP device. CGS measurements were made on an entire lot (24 wafers) at 16 distinct points (15 sets of differential or delta data) in the process flow between wafer start and source-drain anneal. Note that for many of the post-process minus pre-process data sets, multiple processes were included. Generally the process steps are identified numerically; the names of specific process steps are provided only where potentially proprietary information would not be compromised.

Electrical parameters measured after the completion of manufacturing were evaluated for 21 distinct die locations. Details of the process flow and the methodology for correlating device performance to stress are described in prior reports [8, 9]. The three main goals of this effort were: 1) Establish the correlation, if any, between die-to-die stress variations and device performance variations, 2) Identify the process steps for which stress variation has the greatest impact on performance and 3) evaluate whether within-wafer or wafer-to-wafer stress variations contribute more significantly to device performance variations.

978-1-61284-408-4/11 $26.00 © 2011 IEEE

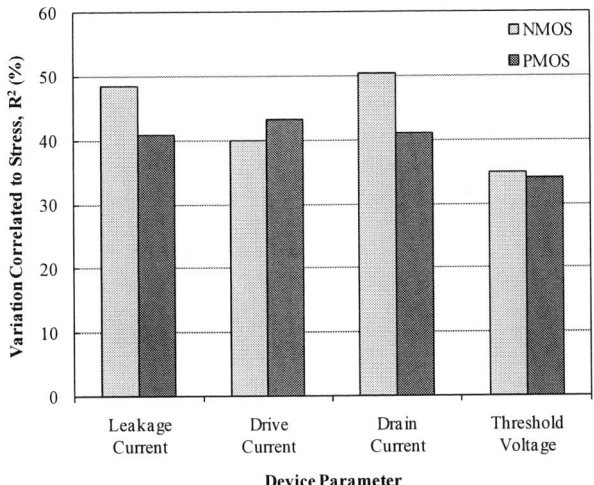

Figure 1. Chart of correlation coefficients between multi-step stress and leakage current, drive current, drain current and threshold voltage for both NMOS and PMOS.

Fig. 1 shows a chart that summarizes the correlation between device performance and stress for NMOS and PMOS drive current (I_{ON}), leakage current (I_{OFF}), drain current (I_D) and threshold voltage (V_T). The value shown in Fig. 1 is the correlation coefficient squared, R^2 for each of the eight correlations (i.e. the percentage of variance in device performance that can be attributed to multi-step stress). Inspection of Fig. 1 reveals that measured stress across multiple process steps accounts for between 40 and 50% of the device performance variation for I_{ON}, I_{OFF}, and I_D and approximately 35% for V_T.

The correlations indicate that active stress management in front-end processing may provide an effective approach to reduce device performance variation and improve yield. The significance of the contribution of each of the 15 process segments to the overall correlation can be evaluated as shown in Fig. 2 for NMOS and PMOS device parameters. The data shown in Fig. 2 shows the relative contribution of the stress variations at each process to the device performance. Note that the sum of all the NMOS or PMOS values in the figure is 100% and the respective data sets represent an average from

the data for I_{ON}, I_{OFF}, I_D and V_T. Using the data in Fig. 2, it is the critical process steps that may benefit from a stress-based process monitor can be identified.

For NMOS, four of the fifteen process segments account for ~60% of the device performance variation correlated to stress. These are steps 1 (liner oxide), step 4 (damage anneal & high voltage gate oxidation) and steps 10 and 12. For PMOS, steps 1 and 4 account for >50% of the device performance variation correlated to stress. None of the other steps contributes more than 8% to the PMOS stress / device performance correlation. From a manufacturing perspective, these four steps (1, 4, 10 & 12) are candidates for both further characterization and the implementation of stress as a process monitor. It is interesting to note that step 1 (liner oxide) and step 4 (damage anneal & high voltage gate oxidation), represent two of the higher thermal budget processes in the flow.

The role of within-wafer and wafer-to-wafer stress variations can be evaluated by simply averaging the stress / device performance data across all wafers for a given die location and across all die locations for each wafer, respectively. Fig. 3a shows a plot of the NMOS drain current versus the multi-step stress when the data is averaged across all wafers for each die location. In this plot, the wafer-to-wafer variation is averaged out and the remaining 21 data points indicate the average within-wafer drain current versus stress (i.e. one data point for each die location). Fig. 3b shows the

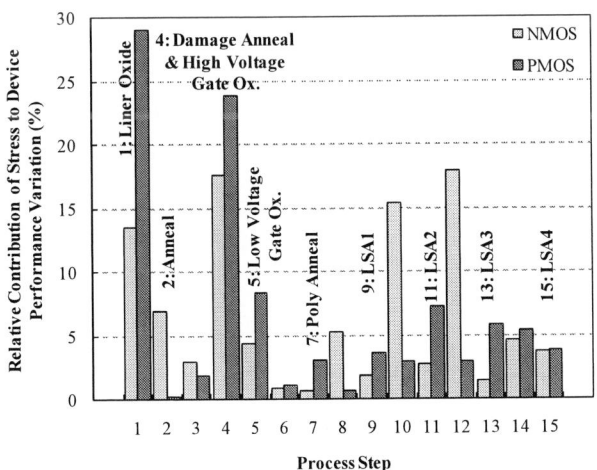

Figure 2. Relative contribution of the various process steps to stress / device performance correlation for both NMOS and PMOS.

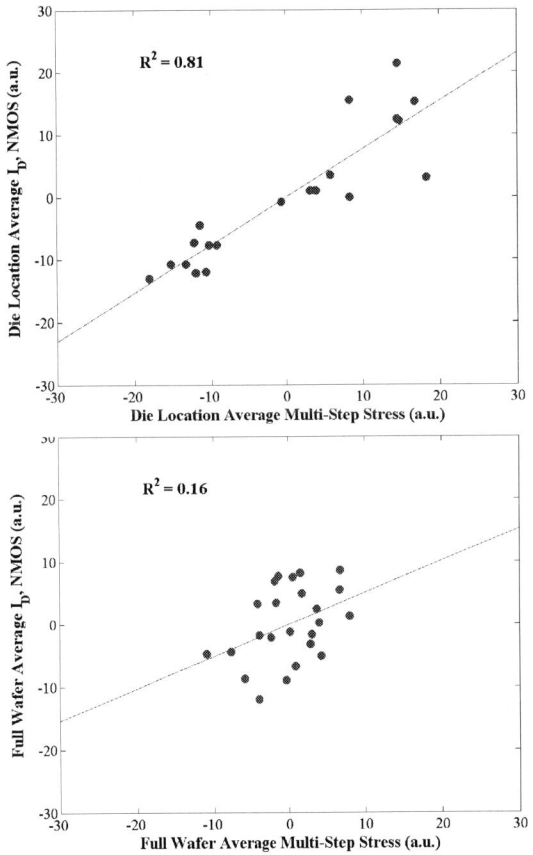

Figure 3. Role of (a) average within-wafer stress variations and (b) average wafer-to-wafer stress variations on NMOS drain current.

978-1-61284-408-4/11 $26.00 © 2011 IEEE

analogous data to evaluate the effect of wafer-to-wafer variations on the NMOS drain current. The 24 data points in Fig. 3b are the full wafer average of stress and drain current for each wafer in the lot. The annotation in each figure indicates the value of R^2 for each case. The average within-wafer stress variations correlate strongly to the average drain current variations with a value of $R^2=0.81$, whereas the average wafer-to-wafer stress variations do not with $R^2=0.16$. Note that for all 504 points evaluated independently of die-location or wafer, the value of R^2 shown in Fig. 1 is ~0.50. The apparent difference between the correlation of drain current to wafer-to-wafer and within-wafer stress variations implies perhaps that the variation s are caused by distinct attributes of the process.

Using CGS data obtained across a series of process steps, the relationship between die-to-die stress variations and device performance variations has been explored. The data set has been used to identify critical process steps that may benefit from stress-based process control and monitoring to reduce device performance variation.

B. Correlation of Stress to Overlay

It is commonly understood that it may be difficult to expose extremely warped wafers within specification during lithography. However, the warpage and stress-induced lithography issues could be managed more effectively if critical processes upstream of lithography were monitored for the stress variations. The ability to monitor for potential issues significantly upstream of lithography would enable early identification of process excursions and reduce material-at-risk. In order to implement such an approach, a detailed understanding of the relationship between wafer deformation, stress and lithography overlay errors is required.

Overlay errors resulting from pattern slip can be related directly to stress gradients [6, 10, 11]. Specifically, uniform stresses cause uniform dilatation of the patterned structure, such that the pattern displacements are linear and can be compensated for during lithography. Non-uniform stresses, on the other hand, cause non-uniform distortion of the pattern structure, and induce shear stresses along the horizontal interfaces between the pattern and underlying structure. Using straightforward mechanical equilibrium concepts, the horizontal shear stresses, σ_{xz} & σ_{yz} can be inferred from the in-plane stress gradients:

$$\sigma_{xz} \propto Q_x = \frac{\partial \sigma_{xx}}{\partial x} + \frac{\partial \sigma_{xy}}{\partial y} \qquad (5a)$$

$$\sigma_{yz} \propto Q_y = \frac{\partial \sigma_{xy}}{\partial x} + \frac{\partial \sigma_{yy}}{\partial y} \qquad (5b)$$

where the quantities Q_x and Q_y define the relative magnitude of the shear stresses and can be compared directly to measured overlay errors in the x and y directions. Note that the high point density of the CGS data maps facilitates the computation of gradient quantities described by equation 5.

A case study using this approach for characterizing stress-induced overlay and subsequently optimizing a millisecond anneal process has been published previously [10]. Millisecond annealing conditions of varying temperature and dwell time

were explored in order to simultaneously improve device performance and minimize overlay error.

Fig. 4 shows the variation of the wafer-average stress gradient magnitudes, either Q_x or Q_y versus the wafer-average overlay residuals for five wafers. Four wafers have been processed by millisecond anneal using annealing times of 200 or 400µs and temperatures of 1235 or 1270°C. The fifth wafer had no millisecond anneal. Inspection of Fig. 4 reveals that all the data fall on a straight line with a value of $R^2=0.93$. Note that the overlay residuals in the x and y directions correlate to the x and y stress gradient magnitudes independently.

In addition, the overlay residuals for the 'no anneal' wafer indicates the contributions from the processes other than millisecond anneal. It should be noted the relationships between stress, deformation and lithography performance cited above can be used to quantitatively evaluate any individual step or series of steps in a process flow. In other words, data from a process fingerprinting study can be used to identify the processes that influence overlay in a manner similar to that described in the previous section on device performance.

C. Stress to Characterize Tool Matching

As discussed in section 2, stress is inherently and simultaneously sensitive to several different types of process non-uniformities. Consequently, stress metrology may provide a versatile method to evaluate tool matching and characterize subtle, but significant differences between tools or individual chambers on device wafers. In this case study, device wafers from numerous lots were measured at a consistent point in the process flow. At least three wafers were measured for each lot, and in a few cases, all the wafers from the lot were measured. In all more than 1500 wafer from ~120 lots were measured. The goal of the case study was to quantify within-lot and lot-to-lot stress variations observed in manufacturing.

The stress maps across all wafers and lots generally exhibited one of four distinct stress patterns. Fig. 5 shows a representative von Mises stress map from each of the four typical patterns. Each map in Fig. 5 is shown on a unique scale so as to best show the features of the stress map. The average stress and stress non-uniformity (1-sigma) is annotated at the

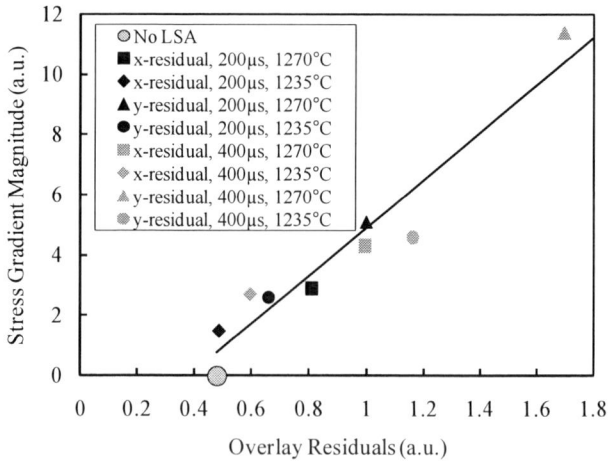

Figure 4. Correlation of stress gradient magnitude to overlay residuals, after [10].

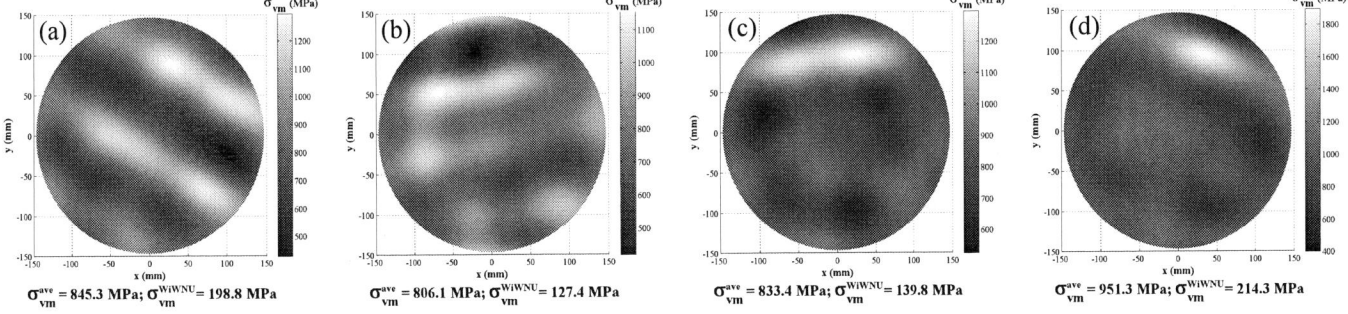

σ_{vm}^{ave} = 845.3 MPa; σ_{vm}^{WiWNU} = 198.8 MPa σ_{vm}^{ave} = 806.1 MPa; σ_{vm}^{WiWNU} = 127.4 MPa σ_{vm}^{ave} = 833.4 MPa; σ_{vm}^{WiWNU} = 139.8 MPa σ_{vm}^{ave} = 951.3 MPa; σ_{vm}^{WiWNU} = 214.3 MPa

Figure 5. Representative stress maps for four distinct stress signatures observed: (a) type 1, (b) type 2, (c) type 3, and (d) type 4.

bottom of each map. Comparing the values below each map, the average stresses range from 800 to 950 MPa and the within-wafer non-uniformity values are between 125 and 215 MPa. The maps shown in Fig. 5 are all of the same device type and were subjected to nominally identical process histories.

The within-lot stress distributions were evaluated for two lots for which all wafers in the lot were measured. Fig. 6 shows the cumulative probability distribution of the wafer average stress for two different lots: one that had a 'type 1' stress signature and a second that had a 'type 2' stress signature. Inspection of the two distributions shows that very different behavior. The 'type 1' distribution is relatively narrow with average stress values between ~700 and 750 MPa (i.e. ±3.5%). On the other hand, the 'type 2' distribution appears to be bimodal with about half the wafers exhibiting average stresses of ~750 MPa and the other half of ~810 MPa.

The data indicates that the lot-to-lot variation is ~15 to 20% based on average stress and almost a factor of two with respect to stress uniformity. The within-lot data shows distinctly different probability distributions of average stress that appeared to be related to the specific stress signature. The data shows that nominally identical processing can lead to distinctly different stress response.

V.   SUMMARY

Full-wafer stress mapping is a potentially powerful approach for process monitoring and control due to the inherent sensitivity of stress and wafer deformation to process non-uniformities. CGS stress metrology can provide detailed data maps that enable the resolution of within-die and die-to-die stress variations. These stress variations have been demonstrated to correlation to device performance and overlay, while also revealing details lot-to-lot and within-lot stress variations from semiconductor manufacturing.

ACKNOWLEDGMENT

The authors would like to thank Dr. Amitabh Jain of Texas Instruments for his contributions to the overlay study.

REFERENCES

[1]   K.J. Kuhn, A. Murthy, R. Kotlyar and M. Kuhn, "Past, present and future: SiGe and CMOS transistor scaling," ECS Transactions, 33(6), pp. 3-17, 2010.

[2]   G.G. Stoney, "The tension of metallic films deposited by electrolysis," Proc. R. Soc. London, A82, pp. 172-175, 1909.

[3]   T.S. Park and S. Suresh, "Effect of line and passivation geometry on curvature evolution during processing and thermal cycling in copper interconnect lines," Acta Materialia, 48, pp. 3169-3175, 2000.

[4]   D. Ngo, X. Feng, Y. Huang, A.J. Rosakis and M.A. Brown, "Thin film/substrate systems featuring arbitrary film thickness and misfit strain distributions. Part I: Analysis for obtaining film stress from non-local curvature information," Inter. J. Solids Struct., 44, pp. 1745–1754, 2007.

[5]   A.J. Rosakis, R.P. Singh, Y. Tsuji, E. Kolawa and N.R. Moore, Jr., "Full Field Measurements of Curvature Using Coherent Gradient Sensity: Application to Thin Film Characterization," Thin Solid Films, 325, pp. 42-54, 1998.

[6]   D.M. Owen, Y. Wang, A. Hawryluk, S. Zhou and J. Hebb, "Characterization of deformation induced by micro-second laser anneal using CGS interferometry," in 16th IEEE Inter. Conf. On Adv. Thermal Processing of Semiconductors – RTP2008, pp. 127-133, 2008.

[7]   D.C. Ghiglia and M.D. Pritt, Two-Dimensional Phase Unwrapping. New York, NY. John Wiley & Sons, Inc., 1998.

[8]   D.M. Owen, C. Otten, H. Bu, Y. Wang, S. Shetty and J. Hebb, "Correlation of Device Performance to Die-Level Stress Variations," in Proc. Inter. Workshop Junction Tech., pp. 15-18, 2009.

[9]   D.M. Owen, C. Otten, H. Bu, Y. Wang, S. Shetty and J. Hebb, "The Effect of Die-Level Stress Variations on Device Performance," in 17th IEEE Inter. Conf. On Adv. Thermal Processing of Semiconductors – RTP2009, pp. 127-133, 2009.

[10]   S. Shetty, A. Jain, D.M. Owen, J. Mileham, J. Hebb, and Y. Wang, "Impact of Laser Spike Annealing Dwell Time on Wafer Stress and Photolithography Overlay Errors," in Proc. Inter. Workshop Junction Tech., pp. 119-122, 2009.

[11]   D.M. Owen and J. Hebb, "Application of Die-scale Stress Management to Advanced Annealing Optimization," Solid State Technology, , http://www.solid-state.com/display_article/368397/5/none/none/TCHNE/Die-scale-stress-management-to-advanced-annealing-optimization.html, 2009.

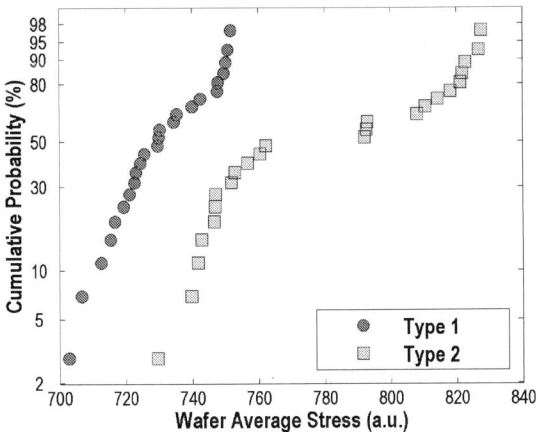

Figure 6. Within lot stress distributions of average stress for two lots having different stress signatures.

Inline Control of an Ultra Low-k ILD layer using Broadband Spectroscopic Ellipsometry

Ronny Haupt, Jiang Zhiming, Leander Haensel
KLA-Tencor Corporation
One Technology Drive, Milpitas 95035, CA

Ulf Peter Mueller, Ulrich Mayer
GLOBALFOUNDRIES Module One Limited Liability
Company & Co. KG
Wilschdorfer Landstrasse 101, 01109 Dresden

Abstract— **As chip dimensions are scaling down new challenges develop in the back-end-of-line. In order to keep the capacitance small while decreasing the volume of the inter-layer dielectric (ILD), new materials and processes have been introduced over the past years to lower the dielectric constant of the ILD layers. For design rules of 45nm and below porous Ultra low-k materials are widely used in today's semiconductor process flows. Beside process challenges this introduces stringent requirements for metrology not only to monitor film thickness but other properties of the material as well. This paper discusses the development and implementation of a Broadband Spectroscopic Ellipsometer for inline process control of a SiCOH based porous ultra low-k film. After deposition the material is cured with UV light to introduce the porosity. The challenge for the metrology is to measure both the thickness and an adequate metric for the chemical properties which do no longer correlate with optical properties. In addition the chemical properties vary as a gradient from top to bottom of the film. We discuss the methodology to develop a metrology recipe resulting in the thickness of a metric layer and the percentaged thickness shrink being the best parameters to sense and track the process adequately. Results demonstrate the sensitivity of the technique to process variations. Short term precision, long term stability and tool-to-tool matching results prove that the technique enables routine process monitoring in a high volume automated semiconductor fab.**

Inter-layer dielectric, ILD, ultra low-k, ULK, Metrology, Spectroscopic Ellipsometry, UV cure

I. INTRODUCTION

Semiconductor manufacturers are driven to an aggressive roadmap to improve device performance in order to remain competitive and provide technology that enables complex computations to be performed in reasonable time with high reliability. In the back-end-of-line (BEOL) of semiconductor processing one of the keys to stay competitive is the inter-layer dielectric (ILD). In order to avoid unwanted effects like capacitive charging or cross-talk between layers the capacitance of the ILD has to be kept small while its volume is decreasing due to shrinking chip dimensions. This led to the introduction of several new materials and processes to lower the dielectric constant of the ILD layers [1]. One of the major

innovations in the recent past has been the introduction of porous ultra low-k (ULK) materials in the BEOL process flow. These materials with a typical dielectric constant of 2.6 and below are SiCOH-based and cured with UV light after deposition to generate the pores [2]. Besides the challenge to handle the process itself these new ILD films generate new metrology requirements in order to appropriately monitor and control the process conditions. This paper discusses the capabilities of optical film metrology to measure both the thickness and an adequate metric for the chemical properties which do no longer correlate with optical properties of the ILD material necessarily and vary as a gradient in the film. The existing metrology regime on another spectroscopic ellipsometer (SE) system only monitors the thickness after SiCOH deposition and secondly thickness and refractive index of the total film after UV cure.

II. ELLIPSOMETRY METHODOLOGY

This study uses the Aleris 8350 Broadband Spectroscopic Ellipsometer (BBSE) from KLA-Tencor. The signal captured by the ellipsometer ($\tan\Psi$ and $\cos\Delta$) is sensitive to the refractive index and thickness of the film being measured. The chemical composition of the film is correlated with the optical properties i.e. the refractive index of the film. This is the basis for sensitivity of the ellipsometer to composition changes in the film. The Aleris 8350 is a broadband SE with effective wavelength range from 220nm to 800nm. The system has an effective measurement box size of 50 x 50 μm which enables in-line measurements on product wafers. The tool is further equipped with software for model based analysis of collected spectra. A homogenous model representing the film or stack being measured is used as a starting point with reasonable space where the thickness and refractive index can vary. A mathematical regression is performed to match the measured spectra to the modeled spectra resulting in an estimation of the thickness and refractive index. Advanced models such as Harmonic Oscillator (HO), Bruggeman Effective Medium Approximation (BEMA) or parametric gradient models can be implemented. In the case of the SiCOH-based ULK film a two layer approach with HO models consisting of two oscillators each was used as described in the next chapter.

978-1-61284-408-4/11 $26.00 © 2011 IEEE

III. BBSE Measurement Setup

In order to understand the key challenges and parameters of interest for the film metrology one needs to understand some characteristics and effects of the process to generate the SiCOH-based ULK film. A low-k SiCOH material is deposited together with an organic additive using a PECVD process. The UV-light cure process removes the organic additives which generates the pores in the material and leads to shrinkage of the ILD film. In order to enhance the mechanical properties esp. the cross-linking to the layer underneath the UV cure is extended further. In summary the cure process induces four effects in the ILD film: densification of the material, enhancement of mechanical parameters [4], loss of carbon and shrinkage [3]. Because the film is exposed with the UV light from above, these four effects fall off from top to bottom. In order to control the process adequately, this gradient in chemical properties of the film needs to be measured.

An adequate process control metric would be the carbon concentration of the ULK film. This could be measured using Secondary ion mass spectrometry (SIMS). However, because the bottom of the film is of main interest in order to determine the condition of the interface to the layer underneath, SIMS cannot measure the gradient in carbon concentration. In addition SIMS analysis cannot be implemented as in-line metrology regime. That is why the carbon concentration response was not studied for this paper. The shrinkage of the film can be used as an overall monitor for the cure process and easily obtained by SE measurements after SiCOH deposition and after UV cure. However, this does not provide a measure for the gradient of the chemical properties nor the film characteristics at the bottom of the ILD film. In the past years composition measurements developed as a new field of application for optical thin film metrology (e.g. SiGe, Nitrided Gate, High-k Gate). The composition of the film is derived from some parameters of its optical properties, e.g. the refractive index with the correlation to reference metrology. In the case of the ULK film discussed herein this approach is challenged by the gap in sensitivity of reference metrology to the carbon concentration at the bottom of the layer and the nonmonotonic trend of the refractive index across the curing process duration as shown in "Fig. 1" for the wavelength of 633nm. Ideally, a measured or derived parameter can be found, which correlates with the chemical properties i.e. the carbon concentration at the very bottom of the film.

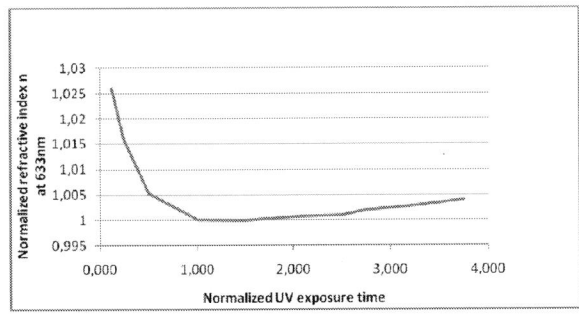

Figure 1. The refractive index n at 633nm over UV exposure time taken from a reference SE system. n has a nonmonotonic trend across the UV curing duration. Both axes are normalized to the value of the process target.

A. BBSE Sensitivity Study (DOE#1)

In order to extract the film dispersions, gain an understanding which modeling approach could be used and study the sensitivity of BBSE to the parameters of interest, a first set of wafers was designed. The SiCOH films were deposited with approximately 25% of the target thickness and cured with different UV exposure durations. In addition, one uncured wafer was used to extract the dispersion of unexposed SiCOH. "Fig. 2" illustrates the raw BBSE spectra from wafers with different exposure level. The spectral difference proves that the sensitivity of the Aleris 8350 BBSE signal to the variation in the process. After the dispersions have been extracted and compared for all exposure levels, different dispersion models and modeling approaches were investigated in order to figure out which gives the best sensitivity to the process variation and especially reflects the chemical properties at the bottom of the film. It was found that a two layer approach provided the best results. The model consists of a thin layer of unexposed SiCOH (T1) and a thick layer of exposed SiCOH (T2) on top. The thickness of the bottom layer is used as a metric for the exposure level at the very bottom of the film, which is of interest in order to control the film characteristics at the interface to the layer underneath. For the top layer, floating of the dispersion model is enabled in the regression to reflect the changes in optical properties of the film. The overall thickness of the ULK is represented by the sum of both thicknesses (SumT). As shown in "Fig. 3, 4 and 5," all measured parameters correlate well to the variation in exposure time and match to the process expectation. In summary, the two-layer model approach is providing a metric for the film characteristics at the bottom of the film.

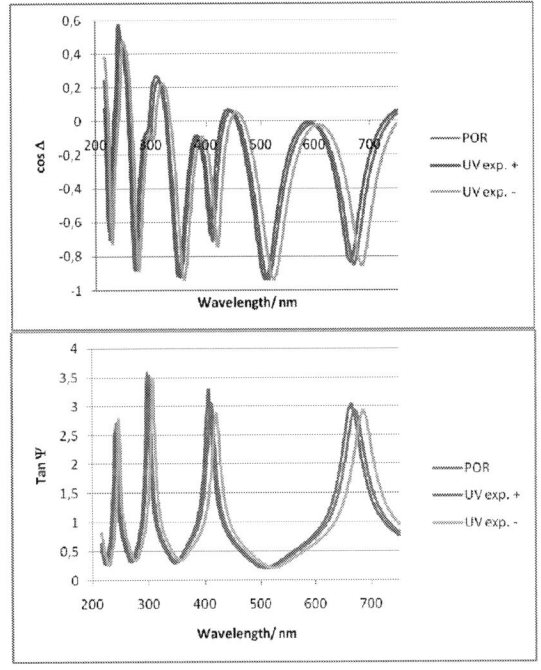

Figure 2. Raw spectra (tanΨ and cosΔ as functions of wavelength in nanometers) from three silicon wafers with ULK film of similar deposited

978-1-61284-408-4/11 $26.00 © 2011 IEEE

thickness but different UV curing duration illustrating the optical sensitivity to the process variation.

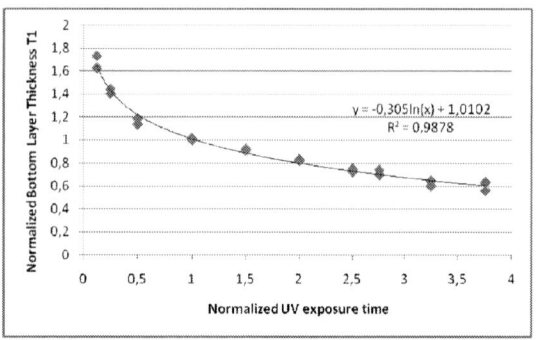

Figure 3. Correlation of the measured thickness of the bottom layer T1 to the UV exposure time. T1 is sensitive to the process variation and provides a metric for the film characteristics at the interface to the underneath film. The fraction of unexposed SiCOH in the stack decreases with extended UV curing, following a logarithmic dependency. All values are normalized to the process target.

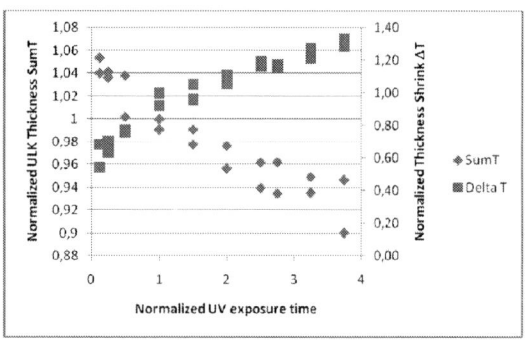

Figure 4. Measured total thickness SumT and the thickness shrink vs. the UV exposure time. SumT decreases with extended UV curing causing the material shrink to increases. This behavior matches with the process expectations. All values are normalized to the process target.

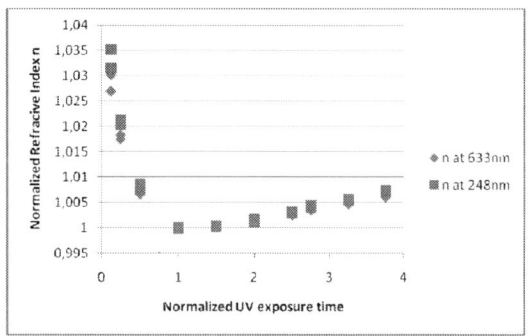

Figure 5. Refractive index n at two wavelength vs. UV exposure time. N shows the characteristic behavior over extended UV curing. All values are normalized to the process target.

B. Study of Sensitivity to Process Parameters (DOE#2)

In order to study the influence of all potential process parameters and verify the recipe modeling approach, a second DOE wafer set was designed as described in "Table I". The UV curing time is the major process parameter. Other influencing factors are the initial thickness of the SiCOH film before curing and the intensity of the UV lamp used in the curing process tool. The variation was limited to a reasonable range around the process target. The DOE was designed with two groups of wafers. In the first group (sample 2 to 11) only one parameter was varied and the other two fixed. In the second group (sample 12 to 19) all parameters were varied within the process range. Sample 1 was generated with all parameters at process target that represents the process of record (POR).

TABLE I.   DOE#2 OVERVIEW, ALL VALUES ARE NORMALIZED TO THE PROCESS TARGET

Sample#	Nominal deposited thickness	UV exposure time	lamp power	Measured deposited thickness
(POR) 1	100.0%	1.00	100%	100.0%
2	100.0%	1.00	115%	99.5%
3	100.0%	1.00	85%	100.5%
4	103.2%	1.00	100%	103.2%
5	101.6%	1.00	100%	101.8%
6	98.4%	1.00	100%	98.3%
7	96.8%	1.00	100%	96.5%
8	100.0%	1.27	100%	100.4%
9	100.0%	1.13	100%	100.7%
10	100.0%	0.87	100%	100.7%
11	100.0%	0.73	100%	100.1%
12	101.6%	1.13	115%	102.0%
13	101.6%	1.13	85%	101.6%
14	101.6%	0.87	115%	102.2%
15	101.6%	0.87	85%	101.8%
16	98.4%	1.13	115%	98.4%
17	98.4%	1.13	85%	98.8%
18	98.4%	0.87	115%	98.8%
19	98.4%	0.87	85%	99.0%

Analyzing the measured spectra using the recipe and dispersion model from the first study, it was found that the metric bottom layer thickness did no longer correlate to any of the process parameters, esp. the UV exposure time. This is mainly an effect of the larger film thickness which causes only a small portion of the BBSE signal to interact with the very bottom of the film. The measured value of the overall film total thickness and refractive index were still reasonable and correlated very well with the process parameters. In order to make the measurement sensitive to the more important metric for the bottom of the film, the model had to be revised. The final solution was developed using the same two layer approach dispersion models as extracted from the first DOE but allowing the dispersion parameters of the bottom layer to float in the model regression as well. Because both dispersions are similar, this increased the effect of correlation between both layers in the model. In order to keep the correlation between both layers at a minimum, the regression of the bottom layer parameters had to be limited within narrow ranges.

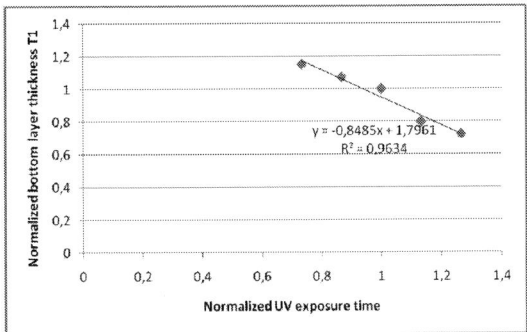

Figure 6. Correlation of the bottom layer thickness T1 to the process parameter UV exposure time. T1 shows a linear correlation trend and can be used as a metric for the film characteristics at the bottom of the layer. All values are normalized to the process target.

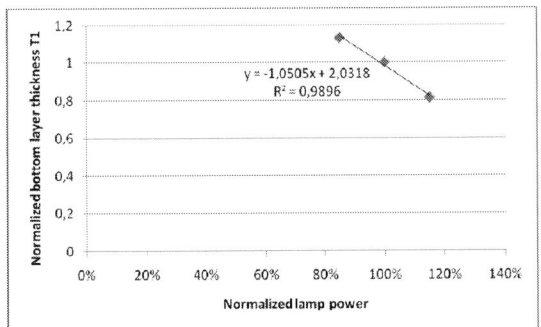

Figure 7. Correlation of the bottom layer thickness T1 to the process parameter lamp power. T1 shows a linear correlation trend and can be used as a metric for the film characteristics at the bottom of the layer. All values are normalized to the process target.

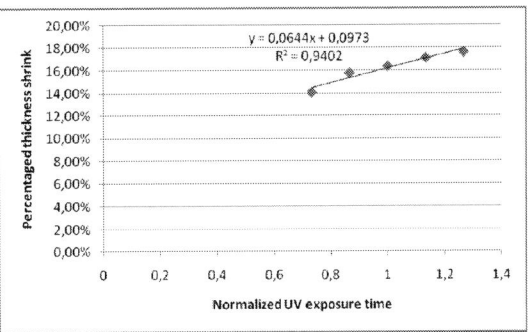

Figure 8. Correlation of percentaged thickness shrink to the process parameter UV exposure time. The relative shrink shows a linear correlation trend and can be used as a process control parameter for the overall film. UV exposure time values are normalized to the process target.

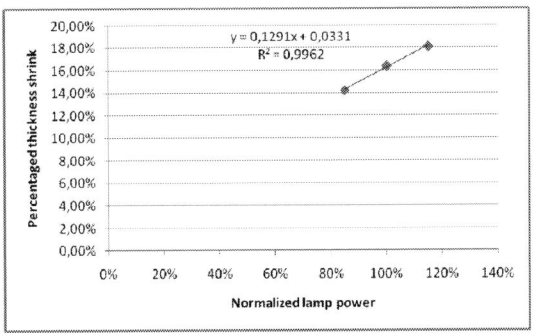

Figure 9. Correlation of percentaged thickness shrink to the process parameter lamp power. The relative shrink shows a linear correlation trend and can be used as a process control parameter for the overall film. Lamp power values are normalized to the process target.

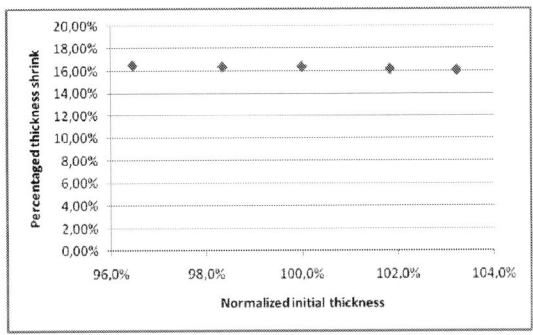

Figure 10. Percentaged thickness shrink vs. process influence initial thickness. The relative shrink is constant and independent from the deposited SiCOH thickness. This behavior meets the process expectations. Thickness values are normalized to the process target.

Using the adjusted model the results for the metric layer thickness T1 improve significantly. As illustrated in "Fig. 6 and 7," T1 provides a reliable metric for the film characteristics at bottom of the ULK film. Only the variation in initial thickness is not tracked by the measurement results. This is explainable with the correlation effects within the model, where a change in initial thickness is mainly compensated in the model's top layer. This gap in sensitivity is not an issue going forward because the SiCOH thickness after deposition is monitored independently in the process flow anyway.

Furthermore, the measured percentaged shrink of the thickness due to the curing process was found to be an excellent measure for the effect to the overall film. The percentaged shrink provides linear correlation to all process parameters independent from the initial thickness before the curing process. "Fig. 8, 9 and 10" illustrate that the measured parameter correlates well to the process parameters and meet the process expectations.

In summary the developed recipe provides good sensitivity and meets the process expectations for all parameters of interest. The unique approach compared to standard SE measurements is the thin bottom layer which is just used as a metric layer with its thickness correlating to the chemical properties at the very bottom of the ILD film.

IV. VERIFICATION OF MEASUREMENT RESULTS

Usually reference metrology is used in order to verify results of indirect measurements like SE. As already described, in the case of ULK, this reference is not available for the major parameter of interest – the chemical properties i.e. the carbon concentration at the bottom of the film. That is why the Aleris results have to be correlated to the process parameters and compared with process expectations. As described in the previous chapter, this was successful for all major parameters of interest. Furthermore the measured overall thickness and refractive index were compared to the existing metrology step on a different SE system. An excellent linear correlation with a slope close to 1 as illustrated in "Fig. 11" confirms the measurement results of Aleris. Additionally the measured wafer uniformity was compared before and after the curing process. As shown in "Fig. 12," the uniformity of the thickness remains in the same range whereas the uniformity of the refractive index becomes worse. This matches the process expectations. The huge variation in refractive index across the wafer and even locally within small areas after UV curing was confirmed for another project using a variable angle SE lab tool. A third measure for the reasonability of indirect measurements is to perform a line scan across the wafer, plot the results over the wafer diameter and compare the uniformity and shape of the plots between different wafers of a DOE set. The shape should be smooth, without flyers and show the same trend for all wafers as illustrated in "Fig 13" for the major measured parameter – bottom layer thickness T1. For all other parameters, the diameter scan shows consistent results across the entire DOE as well.

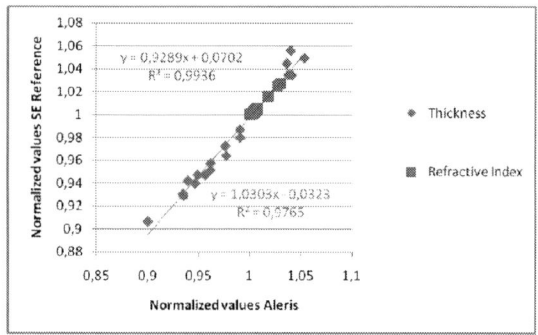

Figure 11. Thickness and refractive index of reference SE system vs. Aleris results. The excellent correlation confirms that the Aleris results are matching to the baseline. All values are normalized to the results of the POR wafer.

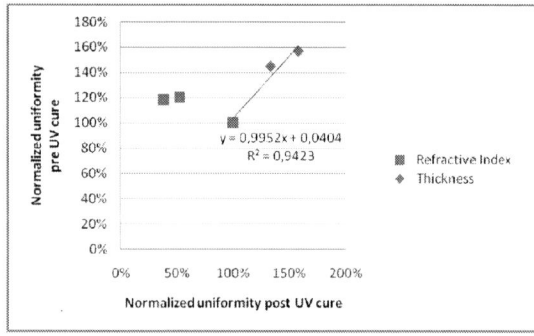

Figure 12. Wafer uniformity of thickness and refractive index before vs. after UV curing. The excellent correlation for thickness confirms the Aleris results. The missing correlation for the refractive index is explainable with a proven induced non-uniformity for optical parameters. All values are normalized to the results of the POR wafer.

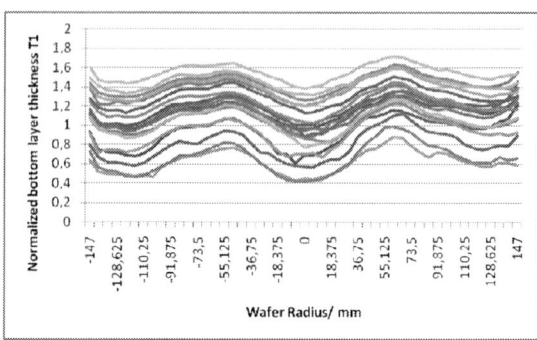

Figure 13. Bottom layer thickness T1 over the wafer diameter for all wafers of DOE#2. The consistent shape across the entire process range confirms the Aleris results. Thickness values are normalized to the result in the center of the POR wafer.

V. THROUGHPUT, MEASUREMENT PRECISION, STABILITY & MATCHING

The tool throughput was recorded using the same measurement setup to achieve the performance discussed in this chapter. The throughput is based on 5- or 49-site measurement per wafer and measuring all 25 wafers of a lot. The results are listed in "Table II".

TABLE II. THROUGHPUT COMPARISON OF A 5-SITE- AND 49-SITE RECIPE USING THE MEASUREMENT SETUP BKM.

Recipe	MAM time/ s	Throughput/ wph
5-site	5.3	85.2
49-site	5.3	11.3

In order to prove the in-line capability and production worthiness of the Aleris measurement, the standard performance measures have been recorded with the developed recipe on a subset of 3 wafers, which were cured with exposure levels at the process target, slightly above and below. Precision was recorded on one Aleris tool measuring the same measurement site 10 times without unloading the wafer (static repeatability). The 3sigma of the 10 single results was calculated. Long-term stability was recorded on one Aleris system measuring a 49 point wafer map once per day over 5 days (dynamic repeatability). The 3sigma of the 5 wafer means was calculated. Tool-to-tool matching was recorded by measuring the 49 point wafer map once on 3 different Aleris systems. The range of wafer means was calculated.

As illustrated in "Fig. 14 and 15," the measurement performance of the Aleris 8350 is production worthy. Precision and matching is excellent for the measured parameters SumT and n. In comparison the performance for the metric layer T1 is poorer because the SE signal has much lower sensitivity to the very bottom of the layer. However the measurement is still repeatable. The result from the long-term test could not be used because the wafers where changing over time and the long-term results overlaid by changing wafer conditions.

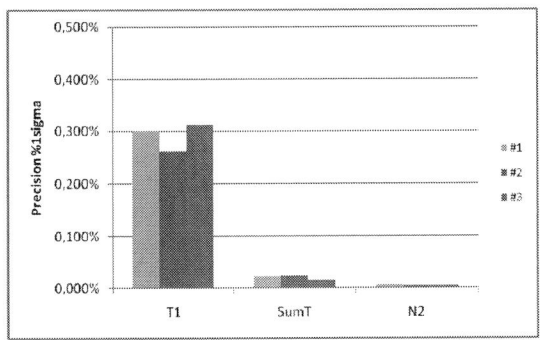

Figure 14. Relative static precision results for T1, SumT and n on a subset of 3 wafers. The precision for SumT and n is excellent. The precision for T1 is higher due to the low sensitivity of SE to the bottom of the film.

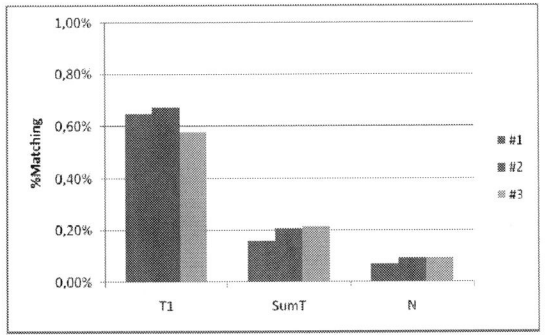

Figure 15. Relative tool-to-tool matching results for T1, SumT and n on a subset of 3 wafers. The matching for SumT and n is excellent. The matching for T1 is higher due to the low sensitivity of SE to the bottom of the film.

VI. SUMMARY AND OUTLOOK

This paper discussed the development of an optical measurement approach to control and monitor the thickness and chemical properties of a UV cured, graded ultra low-k ILD film. The measurement is performed on KLA-Tencor's Aleris 8350 broadband spectroscopic ellipsometer tool.

The approach to study the sensitivity of BBSE and the development of the model-based recipe for this indirect measurement was described. As outcome a recipe using a two-layer model was used where thickness and refractive index of both layers are floated in the model regression and reported as results. It was found that the thickness of the thin metric bottom layer can be used as a monitor for the chemical properties at the bottom of the ILD film which is the main parameter of interest. The percentaged thickness shrink can be used as a process control parameter for the overall film. All measured parameters correlate well with the process parameters and match the process expectations. The collected measurement precision, matching and throughput results confirm that the Aleris 8350 provides a production worthy inline metrology solution for this application.

For a final confirmation of the measurement approach, a correlation of the measured bottom layer thickness to the real carbon concentration at the interface to the film underneath

would be desired, but could not be performed due to missing analysis capabilities.

REFERENCES

[1] H. Geng, "Semiconductor Manufacturing handbook," McGraw-Hill, 2005

[2] H. Ruelke, "Integration of low-k and ULK interconnects" TechArena SEMICON Europa 2009, October 2009

[3] U. Mayer, M. Hecker, H. Geisler, "High volume ULK production," TechArena SEMICON Europa 2010, October 2010

[4] H. Geisler, U. Mayer, M.U. Lehr, P. Hofmann and H.-J. Engelmann, "Profiling of the Mechanical Properties of Ultralow-k Films Using Nanoindentation Techniques," 2010 MRS Fall Meeting Proceedings, January 2011

Advanced Overlay Control in Volume Manufacturing

Timothy Wiltshire, Christopher Ausschnitt,
Nelson Felix, Emily Hwang, Michael Pike,
Allen Gabor
IBM Corporation
Hopewell Junction, New York, USA
wiltsh@us.ibm.com

Moshe Preil[1], Vincent Couraudon, James
Schreiber, Timon Fliervoet, Geert Simons
ASML
Veldhoven, Netherlands

[1] Current affiliation GLOBALFOUNDRIES,
Sunnyvale, California, USA

Erica Rottenkolber
ASML
Fishkill, New York, USA

Abstract—**The work reviewed will describe a particular effort in the area of overlay matching based on the BaseLiner™ package marketed by ASML. BaseLiner relies on the concept of Scanner Baseline Constants (SBCs). Traditionally, optical lithography systems are controlled by many numerical parameters known as Machine Constants (MCs). MCs are typically generated by a lithography system during in situ or other system based tests that generate and optimize the MCs for a very specific test condition set. The concept of SBCs introduces a "middle layer" of offsets that forces tools to be closely matched to one another under general lithography conditions, not just the specific test conditions used for MC generation. The methodology is designed to handle specific product layouts.**

overlay; baseline; lithography controls; lithography; advanced process control

I. INTRODUCTION

State-of-the art semiconductor lithography systems are typically water immersion systems with maximum numerical aperture (NA) values of 1.35. The overlay performance requirements for a 22 nm product design is typically on the order of 10 nm or less for the most critical levels of a high performance product design. The tool focus capability requirements for such a product would be on the order of 100 nm total included range (TIR). During installation and setup of the systems, very detailed calibrations are performed in order to assure that the systems meet the stringent overlay and focus requirements. These calibrations can take many hours and are typically by self-metrology of some sort on the lithographic systems. These calibration procedures set the values of machine

constants (MCs) and a typical system can have thousands of machine constants indicating the set points of various subsystems.

State-of-the-art semiconductor manufacturers typically use an Advanced Process Control (APC) system to feed back overlay, dose, and even focus corrections for various product flows. Such APC systems [3] are able to estimate ideal overlay or exposure values after-the-fact, and using hierarchal feedback, apply those updated operating conditions for product on an ongoing basis. The corrections generated by a simple APC system, however, may not differentiate between corrections driven by random changes in the manufacturing process and corrections driven by instabilities of the lithography system itself. These APC systems compensate for non-centered processes, without regard to the actual sources of the variations. The focus of the work described in this paper is the reduction in overlay variability of the lithography equipment itself apart from product or process induced effects.

Manufacturers frequently will employ use of some sort of monitor artifact wafer set to track overlay variation. The data can be tracked in some type of statistical process control (SPC) system. The SPC system will indicate when a given parameter is outside acceptable performance limits.

As indicated in Fig. 1, a system can be allowed to drift until the data violate some kind of critical value limit. At that point, the lithographic system is typically taken out of production and one or more of base calibration procedures performed. The step function in performance caused by the repair and calibration actions can be problematic for simple APC systems. The APC system must be capable of compensating for such step perturbations or allow for other actions to be taken, such as invalidating the use of historical data.

™ BaseLiner is a trademark of ASML.

A preferred approach for compensating for system drifts (minor or significant) is to introduce a new set of corrections that can be used to compensate for such changes without extensive calibrations. The idea behind Scanner Baseline Constants (SBCs) is depicted graphically in Fig. 2.

II. CONCEPT IMPLEMENTATION

A. Specific embodiment

The SBC approach described in this paper is comprised of key components depicted in Fig. 3 and includes:

1. Reference monitor wafers

2. Metrology systems

3. Host lot and information control systems

4. Lithography specific feedback generation software and systems

Monitor wafers are processed through the litho tool and measured. The manufacturing execution system sends logistic information for both the expose and metrology steps to the BaseLiner server. The BaseLiner server in turn retrieves data files from the litho and metrology tools, and then uses that combined information to generate SBCs. The SBCs are then uploaded to the manufacturing execution system along with a report file containing key performance indicator (KPI) metrics describing the quality of raw and SBC corrected metrology data. The manufacturing control system loads the updated SBCs to the litho tool and the quality metrics to the host SPC system. The KPIs are used by the SPC system to flag out of control conditions or unexpected host system program termination conditions, and consist of both raw and corrected data summary parameters. Since the goal of this application to keep the litho tool overlay optimized

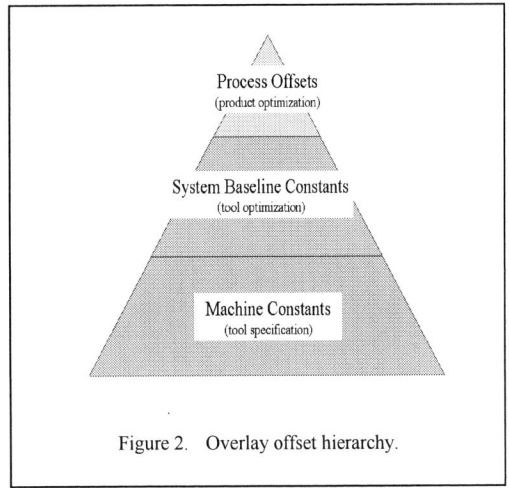

Figure 2. Overlay offset hierarchy.

against a reference wafer set, the corrected KPI values are expected to be stable over time.

B. Overlay Specifics

For overlay, an IBM internal test reticle set has been used. The first level patterns are exposed in single chuck mode on an XT1400E non-immersion lithography system after optimization of the system to minimize stepping and exposure induced overlay errors. The wafers are subsequently etched and measured to verify the quality and consistency of the first level prints. The second level exposures are made on the various immersion lithography systems and processed as indicated above.

III. TECHNICAL DETAILS

A. Overlay model

The overlay model used includes both interfield

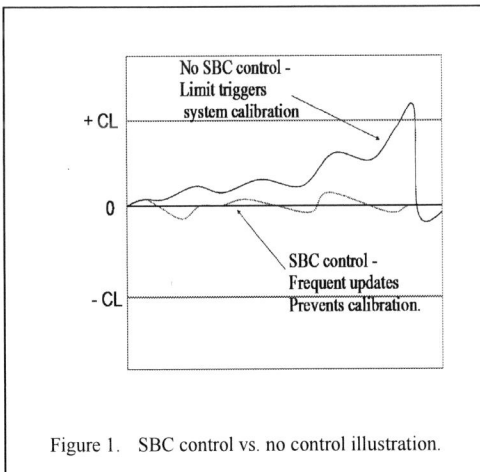

Figure 1. SBC control vs. no control illustration.

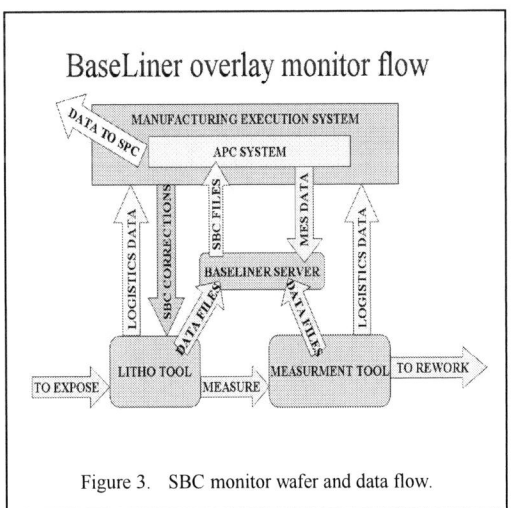

Figure 3. SBC monitor wafer and data flow.

978-1-61284-408-4/11 $26.00 © 2011 IEEE

and intrafield error terms indicated in (1) and (2).

$$dx(x,y,x_f,y_f) = k_1(x,y) + k_3(x,y)*x_f + k_5(x,y)*y_f + g(x_f,y_f) \qquad (1)$$

$$dy(x,y, x_f,y_f) = k_2(x,y) + k_4(x,y)* y_f + k_6(x,y)* x_f + h(x_f,y_f) \qquad (2)$$

Where:

k_1 through k_6 are actually 5th order polynomials;

x,y are center of field coordinates relative to center of the wafer;

x_f,y_f are field coordinates relative to center of the exposure field;

$g(x_f,y_f)$ and $h(x_f,y_f)$ are linear polynomials including various 2nd and 3rd order terms.

The overlay modeling uses different approaches for low order and high order terms. The six low order terms k_1 through k_6 are fitted to 5th order polynomials as a function of wafer position. The higher order intrafield terms (second and third order) are calculated as single best fit values from the entire data set. The actual minimization scheme used is a proprietary iterative procedure, and the details are not given here. The overlay model is executed independently for all unique combinations of chuck and scan direction.

B. Correction and control options

BaseLiner supports control options "control to zero" (C2Z) and "control to reference" (C2R). In C2Z mode, overlay corrections are generated to minimize measured overlay errors. In C2R mode, overlay corrections are generated to minimize deviation from the initial overlay reference as C2R mode controls to the initial state. C2Z mode is used for the IBM monitor and product data reported below, but C2R mode has also been exercised and the capability verified using monitor runs during the initial BaseLiner evaluation period. In addition to the C2Z and C2R control options, BaseLiner supports a filter based historical weighting of SBCs in "Control-to-baseline" (C2B) mode or use of only the most recent data set in "Recover-to-baseline" (R2B) mode.

IV. RESULTS

A. Monitor data

The SBC corrections are generated from overlay monitors exposed with no SBC corrections applied.

After SBC corrections have been generated from the overlay monitor wafers, additional overlay monitors are exposed with the updated SBC corrections applied. The use of these 2 types of monitors allows a user to monitor the native and SBC corrected states in parallel.

Monitor lot data from 1.35NA immersion lithography systems are plotted in Fig. 4 and summarized in Table I. For purposes of this analysis, we assume:

$$OVres = \sqrt{(OVwres^2 + OVfres^2)} \qquad (3)$$

Where:

OVres is the overall overlay residual,

OVwres is the wafer grid overlay residual and

OVfres is the expose field overlay residual.

The residuals indicated in Table I. are based on a 10 term linear overlay model analysis. The data for Tool B show significant improvements in both residual overlay and in overlay mean control when SBC corrections are applied. The data for Tool A also show measurable improvements in overall residual overlay when SBC corrections are applied but most of that residual overlay improvement is from the wafer residual. The field residual shows very little improvement with SBC application. This is believed to be the case because Tool A had a very good "native" set up and not because of any failure or inherent deficiency in the SBC calculation procedure.

The wafer-to-wafer mean control data for Tool A show significant improvement in the X direction when using SBC corrections (3.3 to 1.5 nm), but actually shows a degradation in the Y direction (3.8 to 4.3 nm). There are two suspected causes for this

TABLE I.   1.35NA SYSTEM MONITOR OVERLAY DATA.

Tool / SBC Correction	Axis	A / No	A / Yes	B / No	B / Yes
OVres (nm 3σ)	X	6.1	5.1	8.2	5.3
	Y	5.3	4.8	5.1	4.1
OVwes (nm 3σ)	X	4.8	3.6	5.3	3.6
	Y	3.6	2.1	3.7	2.2
OVfres (nm 3σ)	X	3.9	3.8	6.6	4.1
	Y	4.1	4.3	3.6	3.5
OVmvar (nm 3σ)	X	3.3	1.5	2.1	1.7
	Y	3.8	4.3	3.4	2.9

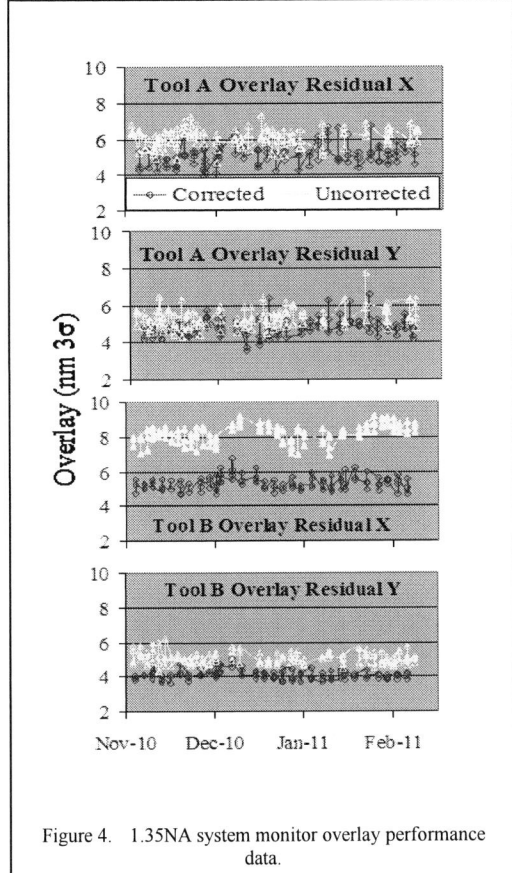

Figure 4. 1.35NA system monitor overlay performance data.

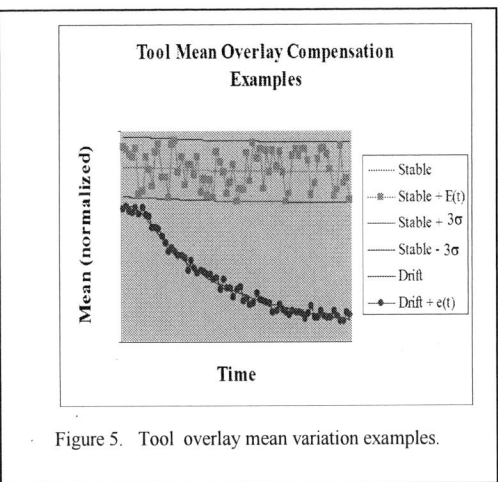

Figure 5. Tool overlay mean variation examples.

degradation being investigated: short term Y overlay mean instability, and sampling population differences between SBC generation and overlay monitoring data sets.

In the case of short term Y overlay mean instability the determined mean from each overlay run may have a very large confidence interval about it. Fig. 5 features a hypothetical depiction of this extreme case. The behavior of a litho system with stable long term mean control but highly variable short term overlay behavior is indicated by curves "Mean Stable" and "Mean Stable + E(t)". For illustration purposes the approximation that the maximum value of abs(E(t)) is equal to 3*standard deviation is made and +/- 3σ bounding curves portrayed. In this case, it is clear that compensating for the apparent mean overlay errors based on limited sampling would increase product overlay variation, not improve it.

By contrast, hypothetical curves of a litho system having long term smoothly drifting mean and low short term variation are also indicated in Fig. 5 by curves "Mean Drifting" and "Mean Drifting + e(t)". It is clear by inspection for this case that the uncertainty of the mean overlay error based on a limited sample size is small compared to the size of the actual mean overlay drifts. For this case, SBC driven corrections would reduce the realized product overlay errors, not increase them.

To help illustrate the case of increased apparent Y overlay mean instability due to population sampling differences Fig. 6 depicts the BaseLiner monitor overlay sampling (full field, full wafer out to a radius of 145 mm) and Fig. 7 depicts the overlay monitoring sampling (17 fields per wafer). The SBC calculations fit coefficients that optimize overlay for the entire wafer. Experience is indicating that the chuck edge locations will be more time variant for overlay for various reasons than interior locations. As the SBC generating code attempts to minimize errors, it is possible to even likely that the high order coefficients introduce mean shifts for subpopulations of the wafer internal fields.

B. Product data

1) Background

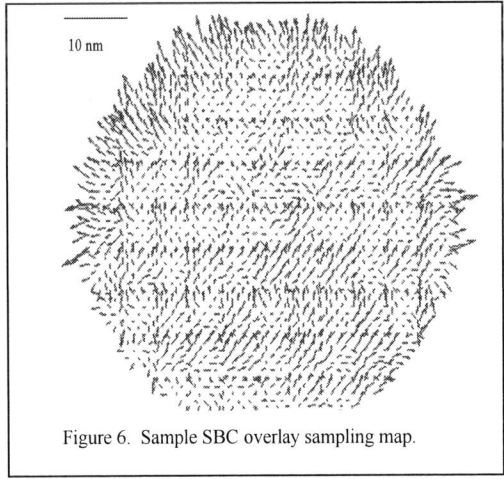

Figure 6. Sample SBC overlay sampling map.

Prior to using an SBC correction strategy, IBM had developed an entire infrastructure to implement the GridMapper™ package to force product specific overlay matching across lithography systems [4]. This system is effective for product overlay performance, but the monitor wafer overhead impacts equipment production. A single set of SBC corrections by contrast - because of the polynomial fitting [5] approach - is universally applicable to multiple product layouts and only one set of monitor wafers is required for use across all lithography tools. When implementing SBC corrections, a strategy was devised whereby new products are processed with SBC control only, while legacy products are processed with SBC and GridMapper corrections to ensure continuity in overlay signatures.

2) Results

Because product disposition overlay sampling is sparse (4 to 8 sites per field) compared to the SBC overlay sampling (see Fig. 7) it is expected that the product disposition measurements will show very little improvement for field overlay residuals, even though the actual within field overlay should improve significantly (see Fig. 8). Furthermore, since legacy products operate using GridMapper controls the observed wafer overlay residuals are not expected to show significant change when implementing SBC control.

Table II and Table III summarize the overlay performance for a high volume technology "Tech1" product over a multiple month period before and after SBC control implementation. The data analysis is separated into periods "Pre SBC", "SBC (debug)" and "SBC". The overlay results from these periods are inline with expectations. During these periods various tool and APC system issues occurred and no attempt has been made to account for these excursions in the data. Of particular note,

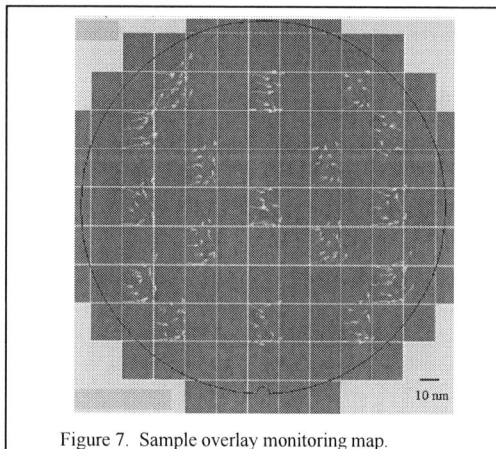

Figure 7. Sample overlay monitoring map.

™ GridMapper is a trademark of ASML.

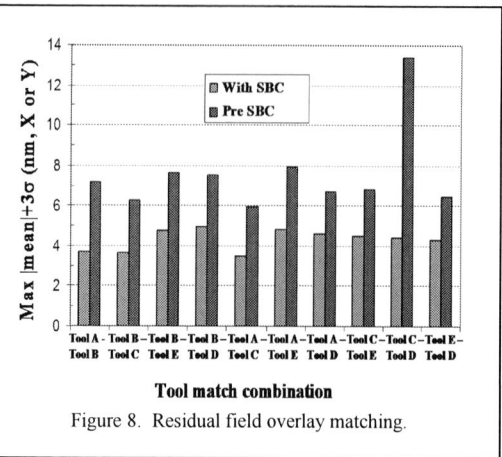

Figure 8. Residual field overlay matching.

however, is the increase in lot-to-lot mean overlay control during the end of 2010 and into 2011. This is not fully understood, but is likely related to the edge-of-chuck issues discussed in the monitor results section above.

Table IV and Table V summarize the overlay performance for two technology "Tech2" products being exposed on 1.35NA immersion lithography systems. Detailed plots of the data are also shown in Fig. 9. Product "Prod1" is older and processes in manufacturing using GridMapper and SBC control as described above. Product "Prod2" was released in a later timeframe and processes using SBC control only.

The overlay performance of the products are statistically similar and inline with expectations. During the time frame there have been both tool and APC feedback system issues – particularly with Tool A - but no attempt has been made to remove impacted points or adjust data values.

TABLE II. TECH1 LAYER L1 OVERLAY.				
Control Method		**Pre SBC**	**SBC (debug)**	**SBC**
Average Overlay (nm 3σ)	X	7.4	6.5	7.0
	Y	7.4	6.8	7.3
Lot-to-lot Mean (nm 3σ)	X	4.0	4.4	3.3
	Y	3.5	3.5	4.2
N Lots		370	731	544

TABLE III. TECH1 LAYER L2 OVERLAY.				
Control Method		**No SBC**	**SBC (debug)**	**SBC**
Average Overlay (nm 3σ)	X	7.4	6.5	7.0
	Y	7.4	6.8	7.3
Lot-to-lot Mean (nm 3σ)	X	4.0	4.4	3.3
	Y	3.5	3.5	4.2
N Lots		370	731	544

TABLE IV. TECH2 LAYER L1 OVERLAY.

Product / Control Method		Prod1 / GM & SBC	Prod2 / SBC Only
Average Overlay (nm 3σ)	X	7.5	8.3
	Y	8.0	8.4
Lot-to-lot Mean (nm 3σ)	X	3.2	4.0
	Y	3.9	3.7
N Lots		375	64

TABLE V. TECH2 LAYER L2 OVERLAY.

Product / Control Method		Prod1 / GM & SBC	Prod2 / SBC Only
Average Overlay (nm 3σ)	X	6.6	5.9
	Y	7.1	6.9
Lot-to-lot Mean (nm 3σ)	X	4.3	4.8
	Y	5.9	4.9
N Lots		189	70

V. POTENTIAL SYSTEMS IMPROVEMENTS

Based on the results observed to date some potential modifications that could improve utility or system robustness are:

1. Incorporation of the ability to specify terms for SBC equation fitting, in particular high order grid terms.

2. Modification of the model to treat scan up and down, and similar terms as parameters in a single model.

3. Evolution of the overlay model to optimize corrections by wafer regions [6] to isolate variable regions and prevent impact to stable wafer regions.

VI. SUMMARY

An automated method of monitoring and correcting the overlay performance of multiple immersion lithography systems in a volume 300 mm manufacturing facility has been implemented. Monitoring data with SBC corrections applied show significant improvements in overlay control over extended periods. Overlay metrology results for products of 2 different technologies indicate at least equivalent performance with reduced monitor wafer levels using SBC control compared to the previous product specific forced matching approach.

ACKNOWLEDGMENT

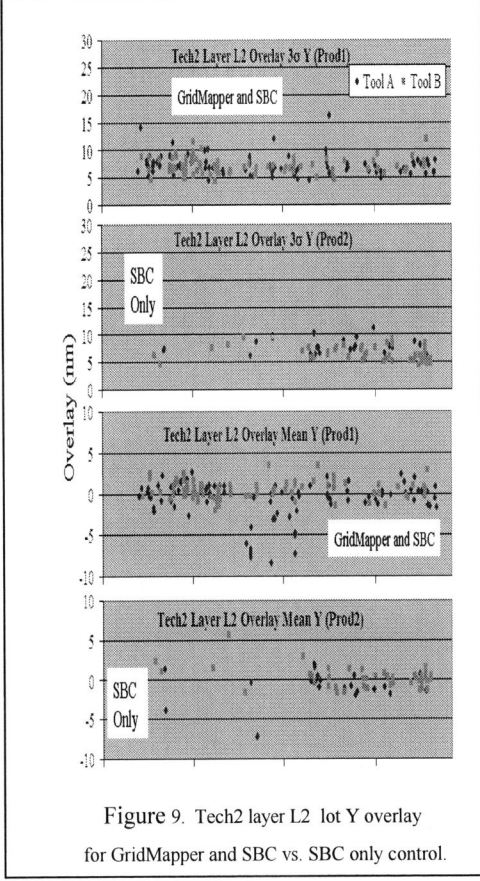

Figure 9. Tech2 layer L2 lot Y overlay for GridMapper and SBC vs. SBC only control.

The authors would like to thank the combined management teams at both the IBM Corporation and ASML for their support in bringing this work to a productive result over an extended period of development and implementation.

REFERENCES

[1] I. Bouchoms et al., "Advanced imaging with 1.35 NA immersion systems for volume production", Proc. SPIE 7640, 76401R (2010)

[2] S. Sivakumar, "Lithography of the Future: A Technical and Economic Challenge", SPIE Advanced Lithography 2010, Plenary Session Presentation

[3] A. Topac, Y. Wang, "Advanced method for run-to-run control of photolithography overlay registration in high-mix semiconductor production", Proc. SPIE, vol. 5755, pp. 1 – 8, May 2005

[4] A. Gabor et al., "The GridMapper challenge: how to integrate into manufacturing for reduced overlay error", Proc. SPIE 7650, 764015 (2010)

[5] H.M. Lin et al., "Improve overlay control and scanner utilization through high order corrections", Proc. SPIE 6922, 69222R (2008)

[6] C. T. Hung et al., "Advance overlay correction beyond 32nm DRAM process", Proc. SPIE 7140, 71400M (2008)

On the Technology and Ecosystem of 3D / TSV Manufacturing

Klaus Hummler, Larry Smith, Raymond Caramto, Robert Edgeworth, Stephen Olson, Daniel Pascual, Jamal Qureshi, Andy Rudack, Roger Quon, Sitaram Arkalgud

3D Interconnect Division
SEMATECH
Albany, New York

Abstract—Three-dimensional (3D) die stacking using through silicon viae (TSV) promises significant improvements in performance, power consumption and size over traditional edge - connected die stacking (e.g. wire bonds) or package -on-package (PoP) based approaches. 3D integrat ed circuits (3D -IC) using TSV will enable new system in package (SiP) applications, especially where ultra -high memory bandwidth at moderate power consumption is needed. This paper describes the technology elements for a succ essful implementation of TSV in high volume manufacturing (HVM) with special focus on the so - called "TSV mid" integration flow   being developed at SEMATECH. The maturity and readiness of each process module for HVM is assessed.   In addition to technological feasibility and manufacturing readiness, 3D-IC adoption requires an environment of agreed upon specifications, standards, and tools (3D ecosystem). Progress toward a well defined 3D ecosystem by SEMATECH and many other organizations is described in part III of this paper.

Keywords-3D; TSV; TSV mid; manufacturing readiness; standards

I. INTRODUCTION

For nearly two decades   through silicon viae (TSV) have been a solution looking for a problem [1,   2]. Recently, it appears that the problem   – or rather the opportun ity – has finally presented itself. In the semiconductor industry, for the mainstream adoption of a revolutionary technology to occur, the inertia of existing technologies, familiar concepts, capital invested, and installed manufacturing infrastructures has to be overcome. Three-dimensional (3D) stacking of dice using TSV is gathering momentum for mainstream adoption through a combination of driving forces. Application requirements for power, performance and size are emerging which cannot be fulfilled by existing technologies at a reasonable price. Most of these applications can be found in the mobile, high performance computing, and networking areas, and they are characterized by a common need for high density, ultra -high bandwidth, low power memory access by a logic device. The evolutionary extension of existing technologies, such as two - dimensional (2D) systems on chip (SOC) and lithographic scaling, is progressively unable to meet these requirements or becomes prohibitively expensive (wire delays, power consumption, cost of advanced lithography nodes, etc.)[3-6].

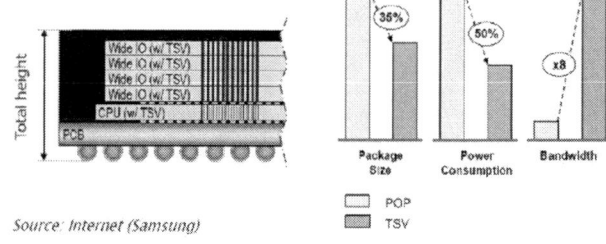

Figure 1.   Benefit of 3D integration of wide -IO DRAM vs. POP

3D integration using TSV has the potential to significantly reduce wiring   lengths between   different parts of the same circuit (e.g., CMOS processor) and between multiple circuits with varying func tions (e.g., CMOS, memory, RF, analog, etc.). Die stacking opens up completely new ways of partitioning an IC or an SOC in the third dimension and therefore alleviates concerns with heterogeneous integration, yield, and cost associated with large die. Due   to the shorter wiring distances and higher wiring densities of TSV based 3D integrated circuits (3D-IC) compared to 2D or edge -connected 3D systems, significant improvements in performance, power, and size can be realized (Fig. 1).

While TSV are being used in packaging of light emitting diodes (LED), CMOS image sensors, and power devices, no true 3D-IC products are in high volume manufacturing (HVM) as of today.   3D stacking of heterogeneous die still presents significant challenges for thermal and stress man agement, test, reliability and design. Not surprisingly, recent TSV   -based product announcements have in one way or another circumvented most of these issues. Silicon interposers with TSV and side -by-side placement of dice (also known as 2.5D integration) gain momentum as a way to reap some of the 3D - IC benefits without high demands for   novel co-design or thermal management methods (e.g. Xilinx Virtex -7 FPGA) . Similarly, announcements of 3D die products have been limited so far to homogeneous memory   -on-memory stacking (Samsung, Micron, Elpida) . True, heterogeneous 3D   -IC stacking as a revolutionary new approach still faces implementation hurdles related to TSV and packaging technology as well as the creation of an ecosystem in which all parts of the supply ch ain work together smoothly and provide economical benefit to all parties. SEMATECH is committed to accelerating the adoption of 3D stacking by contributing to the development of TSV technology as well as to shaping the ecosystem.

II. TSV Technology

A. TSV Integration

As Table I shows, there are a multitude of options of integrating the TSV into device and packaging process flows. The TSV can be formed before device processing (TSV first), after device processing and before back -end-of-line (BEOL) metallization (TSV mid), or after completion of BEOL metallization (TSV last). Furthermore, the TSV processing can be preformed from the front-side of the wafer or from the back-side of the wafer, typically after wafer thinning. If wafer -to-wafer (W2W) stacking is target ed, TSV can be manufactured before or after wafer stacking. These schemes have their individual merits depending on the target application and depending on the effect of device processing on the TSV (via first) or vice versa (via mid and last). For example , TSV first processing enables the use of high temperature TSV dielectrics, but prohibits using copper as a TSV fill, because of the subsequent high temperature device processing. Via last processing might play a major role in W2W stacking and is already widely used in packaging of light emitting diodes LED, CMOS image sensors, power devices and silicon interposers. For heterogeneous die -to-wafer (D2W) and die -to-die (D2D) stacking, industry consensus has emerged that via mid processing at various stages before or during BEOL formation will likely be the mainstream technology[7].

TABLE I.   COMPARISON OF VARIOUS TSV INTEGRATION SEQUENCES

TSV first	TSV mid	TSV last (front-side)	TSV last (back-side)
TSV patterning, liner, metallization, CMP			
Device processing	Device processing	Device processing	Device processing
	TSV patterning, liner, metallization, CMP		
BEOL die wiring	BEOL die wiring	BEOL die wiring	BEOL die wiring
		TSV patterning, liner, metallization, CMP	
Carrier wafer bond	Carrier wafer bond	Carrier wafer bond	Carrier wafer bond (optional)
Back-side thinning	Back-side thinning	Back-side thinning	Back-side thinning
TSV reveal	TSV reveal	TSV reveal	TSV patterning, liner, **contact etch**, metallization, CMP
Back-side metallization	Back-side metallization	Back-side metallization	Back-side metallization
Carrier wafer de-bond	Carrier wafer de-bond	Carrier wafer de-bond	Carrier wafer de-bond
Die or wafer stacking	Die or wafer stacking	Die or wafer stacking	Die or wafer stacking

Therefore, SEMATECH has focused on TSV mid processing before BEOL formation for its technology development and ecosystem activities . The following paragraphs describe the individual process modules for the SEMATECH via mid flow, compare with process alternatives and assess their readiness for HVM.

B. TSV etch, metallization and planarization

In the via mid flow, TSV patterning and etch take place after device formation and before BEOL processing . For our baseline process at SEMATECH, we target TSV with 5 um diameter and 50 um depth (Fig. 2) . Etching 50 um deep viae

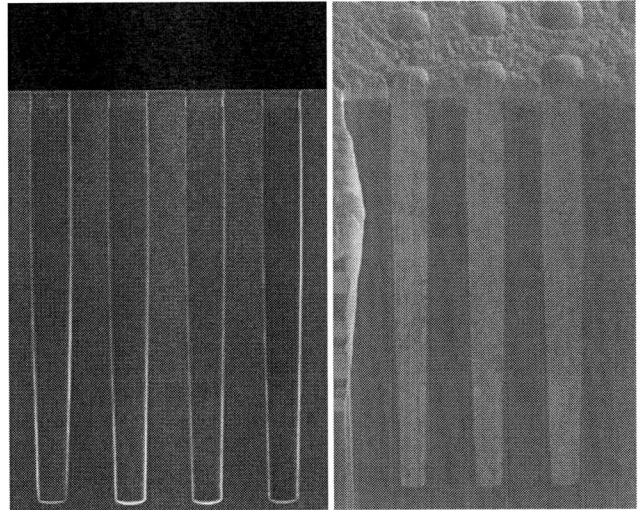

Figure 2.   5 um x 50 um TSV after RIE (left) and bottom -up ECD (right)

into silicon requires either thick resist or a hard mask. Thick resist may present a problem for some front -end fabs, because mid UV tools may have been phased out. For the same reason we have chosen to etch TSV with an SiO2 hard mask. We use a non-Bosch process with an SF6 / SiF4 / O2 chemistry, because it shows better selecti vity to the hard mask than SF6 / O2 alone. The etch times are approximately 12 min per wafer. Due to the heavy use of deep silicon reactive ion etching (DRIE) in other industries (e.g. MEMS, trench DRAM), DRIE processes and tools are relatively mature and ready for HVM.

To isolate the TSV electrically from the silicon substrate, a dielectric liner must be deposited before metallization. Several material and process choices (e.g. CVD SiO2, vapor deposited polymers, wet deposited organic liner, etc.) are bei ng pursued, and the industry has not converged on one specific method yet. Ozone TEOS based CVD SiO2 (HARP) is a common choice, and is used at SEMATECH as well. Compared to other methods it shows good isolation properties, but may have worse conformality. Apart from isolation properties and dielectric constants, the stress behavior of the dielectric liner in the presence of copper metallization is an important selection criterion.

To prevent copper diffusion into the dielectric liner and the surrounding sil icon, a diffusion barrier material has to be deposited followed by a copper seed layer. Mostly conventional BEOL barrier materials (e.g. TiN, TaN) are used, but the high aspect ratio of the TSV presents a challenge for the formation of c ontinuous barrier and seed layers . PVD, CVD, ALD and wet deposition methods are known. The SEMATECH process utilizes PVD Ta/ TaN with a CVD Ru adhesion promoter, followed by PVD copper seed. This process combination performs very robustly at 10:1 aspect ratio. While a variety of material, process and tool choices exists, barrier and seed processing will need further improvement to reduce cost of ownership and extend to higher aspect ratios. When assessing the cost of ownership of any barrier and seed process choice one must ke ep in mind that all

978-1-61284-408-4/11 $26.00 © 2011 IEEE

these layers will have to be removed later during the planarization step.

For the TSV metal fill, initially, conformal electrochemical deposition (ECD) of copper was applied. A conformal process will inevitably lead to center fill voids and a high amount of copper overburden on the wafer surface, which leads to stress issues and high planarization cost. The current SEMATECH process uses so -called "bottom up" plating chemistries . Through an intricate mix of additives, those chemistries are able to fill high aspect ratio TSV from the bottom without seam voids and with minimal overburden. We have achieved less than 1.2 um overburden for 50 um deep TSV (Fig. 2). Plating times are approximately 30min per wafer which leaves room for improvement. For HVM copper TSV ECD, the analytics and dosing of additives in the plating chemistries will have to become more stable and automated.

The last step in the front-side TSV process sequence is the removal of metallization layers from the wafer surface , thus isolating the individual TSV. This is typically accomplished by means of chemical mechanical polishing (CMP). CMP equipment and consumables for this task are widely available and perform well with respect to removal rates, selectiv ies and uniformi ty. Low overburdens, as achieved by advanced bottom up plating chemistries and thin seed layers, eliminate the need for a separate CMP step with high copper removal rates. As mentioned above, some liner or barrier materials may present a special challenge for th e CMP step, and may negatively affect overall processing cost.

This concludes the front-side processing of the TSV. In the via mid flow, the device wafers are now ready to receive standard BEOL processing and then proceed to back -side processing and packaging.

C. Wafer bonding, thinning and TSV reveal

At a target post -thinning thickness of 50 um or less, the silicon wafer is not mechanically stable enough to be handled with conventional robots. Therefore, it must be supported by a handling substrate whic h is compatible with conventional wafer handling and processing. The choice of support media depends on the selection of temporary adhesive materials, in particular the de -bonding method. Glass substrates are compatible with laser de -bonding, while silicon substrates are most widely used for thermal release of the temporary adhesive. Substrates are also available with holes for chemical de-bonding. At SEMATECH we chose a standard silicon substrate with alignment features patterned in silicon dioxide (SiO2) as the carrier wafer. The adhesive material is spun onto the carrier wafer and partially cured. The device wafer is then aligned to the carrier wafer and bonded in a vacuum chamber. Better alignment than mere wafer edge alignment is necessary for successful coarse alignment of the back -side lithography steps and for die yield close to the wafer edge exclusion zone. In our experience, better bonded wafer pair flatness is achieved by maintaining pressure while reaching the cure temperature and cooling down the wafer pair. Adhesive bonding should be performed in a dedicated vacuum chamber, separate from any metal-to-metal bond operations to avoid organic contamination. Temporary adhesive bonding is prese ntly operating on the

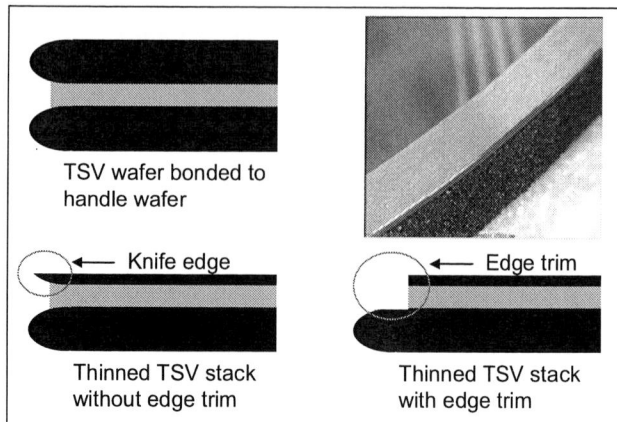

Figure 3.   Schematic and photo of edge -trimmed bonded wafer pair

order of one wafer pair per hour. Fo r a viable HVM solution the throughput would have to be at least an order of magnitude higher. While a variety of temporary adhesive materials is available, new materials need to be developed which support high processing temperatures (>300 C) for direct m etal-to-metal bonding and allow low temperature de -bonding for compatibility with packaging materials such as lead free solders. After joining with the carrier wafer, the device wafer is now ready for back-side thinning.

The removal of silicon is performe d using wafer grinding equipment similar to that used in conventional packaging operations. Multiple equipment vendors are available. The target for the backgrind process is to remove the majority of silicon, stopping just before the TSV. For 50 μm deep TSV on a standard 300 mm wafer, the grind must remove approximately 730 μm of silicon. Good grind processes finish with less than 2 μm total thickness variation across the wafer. These tools are on the edge between ultra high precision machine tools and s emiconductor processing equipment. Grinding is usually performed using at least three steps. A course grind step removes most of the Si at a high removal rate. This leaves the surface of the wafer rough. A fine grind step is used to remove the roughness caused by the course grind. Finally, a polish step or other method is applied to remove a thin layer of silicon together with any remaining crystal defects. When a bonded wafer pair is thinned the edge bevel results in a sharp corner (known as a "knife ed ge") which can lead to chipping and breakage of the wafer(Fig. 3). To mitigate this problem, the edge bevel region is removed from the top wafer (edge trim).

After the backgrind step, approximately 1 -5 μm of silicon remains above the TSV. The goal of t he TSV reveal process step is to removes silicon from above and around the TSV and to leave the TSV protruding from the surface of the wafer. Any silicon etching technology with good selectivity to the via dielectric liner material can be used. The minim um required via height depends on the passivation and back -side metallization (MB) process following the TSV reveal. The SEMTAECH process deposits less than 1 um SiO2 and

978-1-61284-408-4/11 $26.00 © 2011 IEEE

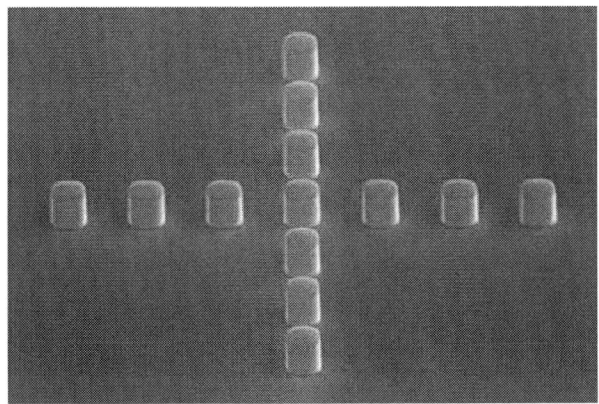

Figure 4.   Tilt SEM pictograph of wet -revealed TSV

Figure 5.   300 mm TSV wafer populated with Cu -Cu bonded die

palanrizes using CMP to form the inter level dielectric (ILD) below a copper single damascene MB layer. TSV protrusion heights of 0.5 to 1 µm are required to ensure a robust contact between MB and TSV. Achieving well controlled TSV reveal heights can be challenging, because they are a function of several earlier processing metrics, such   as TSV RIE depth, temporary adhesive and carrier wafer thickness uniformity, grind and etch profiles, etc.

Process choices for TSV reveal are wet etching, dry etching, and CMP, and SEMATECH has practiced all of these options. Dry etch processes are the most common, and can be tuned for a wide range of selectivity and etch rates. Etch rates of 1   -3 µm/min can be achieved. A loss of liner selectivity can significantly affect the etch rate due to copper contamination of the etch chamber.

Wet etch processe   s utilizing high pH chemistries have excellent selectivity to SiO2 liners   (Fig. 4)   . When implemented in a single wafer spin tool, the etching uniformity is as good as or better than dry etch. Etching rates typically increase with temperature, but achieving greater than 0.5µm / min is difficult. Acid type chemistries utilizing an oxidizing and HF chemistry can achieve higher etch rates but show poorer selectivity to the SiO2 liner.

CMP methods can be used for   TSV reveal as well. These processes typica lly planarize the TSV during the reveal and are therefore able to equalize in   -coming process variations (see above). Other CMP processes are designed to leave the TSV protruding above the silicon surface and may be able to serve as a one-step TSV reveal process.

The TSV reveal step presents a significant hurdle to HVM readiness. All proposed methods are single wafer processes with resulting throughput limitations. These tools are usually found in a front -end fab operation, and are not typical for a packaging environment. The need for front -end type precision tools in this step is essentially driven by the process window issues mentioned above.

D.   Wafer / die stacking and package integration

While there is some consensus on a via mid integration flow for wafer processing, there is much less consensus about

processes and sequences in the area of die or wafer stacking and final packaging of 3D systems.   To support discussions about standards and specifications, some of the activities outlined in Part III of this   paper aim at defining a reference flow for 3D stacking and packaging. At SEMATECH we currently focus mostly on direct copper   -to-copper (Cu-Cu) bonding of dice to wafer . Die-to-wafer (DtW) bonding was chosen because it achieves higher yield by joining only   pre - tested good die, and permits stacking of differently sized die. However, compared to wafer -to-wafer (WtW) bonding, DtW bonding achieves much lower throughputs when completing the bond process for each die individually. At SEMATECH, we developed a collective bonding process where we align and place die individually with the help of a thermo -decomposable adhesive, followed by the application of temperature and pressure to all die collectively (Fig. 5). Excellent electrical and mechanical interface charac teristics have been achieved with this approach (Fig. 6) [8]. While collective DtW bonding improves productivity significantly, die placement speeds (with sufficient accuracy) and Cu -Cu bonding times, pressures and temperatures need to be improved   significantly before these

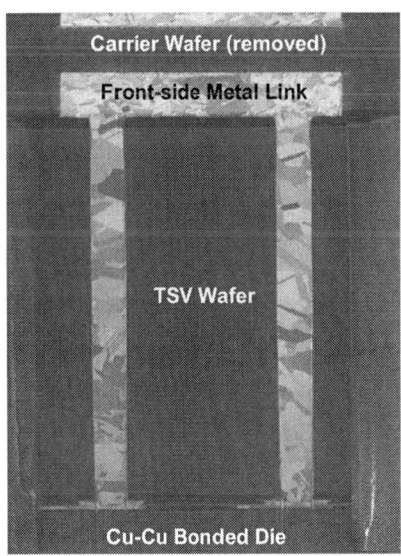

Figure 6.   TSV daisy chain link, completed by C u-Cu bonded die

978-1-61284-408-4/11 $26.00 © 2011 IEEE   61

steps can be considered ready for HVM. Since the underlying Cu-Cu diffusion bond mechanism is the same for DtW and WtW bonding, future work will explore opportunities to reduce processing times, and lower process temperatures and pressures, while maintaining interface properties superior to conventional solder-based connections. Another aspect of 3D-IC HVM readiness is the handling of thin dice or wafers. Most 3D package integration schemes require de-bonding, singulation, solder bumping, and pick-and-place operations performed on dice or wafers with thicknesses of 50 um or less.

III. 3D ECOSYSTEM

Successful HVM of TSV-based 3D-ICs will require the coordination of multiple segments of the semiconductor industry supply chain, including design houses, EDA suppliers, foundries, memory manufacturers, and OSATs. Industry-wide standards and specifications are needed to enable this coordination. New standards, specifications and business models will have to be defined, so that 3D integration can take place smoothly and cost effectively. As seen in Part II of this paper, some parts of TSV manufacturing are firmly rooted in a traditional CMOS fab environment, while others are situated somewhere between the CMOS fab and a traditional packaging environment (e.g. wafer thinning, TSV reveal and back-side metallization). In what we call "supply chain partitioning" the industry will have to decide which conventional manufacturing entities perform which part of the TSV processing, testing and packaging, or if to form new, hybrid manufacturing entities for the processing aspects between the conventional front-end fab and packaging houses.

Fig. 7 [9] shows the results of a SEMATECH survey of companies in the supply chain representing Memory and Logic IDMs, foundries, fabless companies, and OSATs. These companies were asked to prioritize gaps in the Via-mid ecosystem in the areas of missing Standards and Specifications, and regarding tools, processes, and materials needing significant process development or cost reduction. Heterogeneous stacking of wide-IO DRAM on logic was identified as a key driver. Silicon interposers incorporating TSVs was also of high interest. The priorities in Figure 1 represent the gaps that were identified as having the highest priority for the case of heterogeneous integration. There are several standards bodies who are currently engaged in developing standards to support this technology, including JEDEC, SEMI, IEEE, Si2, and others. Fig. 8 attempts to summarize the 3D standards landscape based on a presentation given by Nicholas Yu from Qualcomm[10]. SEMI has formed a new 3D Stacked IC committee with task forces addressing bonded wafer stacks, thin wafer handling, and inspection/metrology. New standards activities have been approved for parameters associated with bonded wafer stacks (including thickness, edge bevel, notch, mass, bow/warp and diameters, etc.), wafer identification for bonded wafer stacks, and shipping containers for thinned wafers; and proposals for additional activities are being developed. For another example, standards supporting Wide IO DRAM are being developed in JEDEC.

SEMATECH, with the support of the SIA and SRC, has formed a new program, the 3D Enablement Center, whose mission is to enable industry-wide ecosystem readiness for cost effective TSV-based 3D stacked IC solutions. Membership is open to all companies, and includes, in addition to SEMATECH's core members, fabless companies and OSATs. Objectives include assembling a summary of ongoing standards activity in order to identify needs that are not currently being addressed or risk being later than needed by industry; accelerating the development of new SEMI standards,

SEMATECH survey: gaps in via-mid ecosystem

Gaps in Standards and Specifications

- EDA exchange formats
  - Partitioning and floorplanning; Logic verification; Power/Signal integrity analysis; Thermal analysis flow; Stress analysis flow; Physical verification; Timing analysis
- Reliability
  - Reliability test methods
- Test
  - DFT test access architecture
- Inspection/metrology
  - TSV voids, defect mapping, microbump inspection and coplanarity
- Chip interface
  - Stackable memory pin assignment; Stackable memory physical pinout
- TSV
  - Keep out area, fill materials, dimensions
- Thin wafer handling
  - Universal thin wafer carrier

Technology Development and Cost Reduction

- Reliability
  - Criteria; Test methods; ESD
- Temporary bond/debond cost reduction
  - Materials and release mechanisms cost reduction; Equipment cost reduction
- TSV
  - Keep out distance/area
- Microbumping and bonding
  - Pad metallurgy and layer thickness; Bump metallurgy
- Inspection/metrology
  - Microbump inspection and coplanarity; TSV voids; BWP voids
- Test
  - Probing microbumps cost reduction

Figure 7. Gaps in TSV mid ecosystem (SEMATECH survey results)

Category	Technical Area	Driver	Standards Body
Die layout compatibility	Bump layout	Die Supplier	JEDEC
	Bump Array Layout		
	Bump Assignment		
Modeling Compatibility	Electrical model format	SEMATECH & GSA	Si2
	Thermal model format		
	Stress model format	SEMATECH	
Design database Compatibility	Partitioning and Floorplanning Exchange Formats	IMEC	Si2
	Stress Exchange Formats	EDA	
	Temperature Exchange Formats		
	Power Distribution Exchange Formats		
	Signal Integrity Exchange Formats		
	Design for Test Exchange Formats	IMEC	IEEE
Material compatibility	Metallurgy pairs	SEMATECH	TBD
	Maximum dT Safe Operating Area		
	Reliability SoA		
QA Incoming specs	Metrology (e.g., warpage)	SEMATECH	JEDEC
	Die/wafer QA metrics	SEMI	
Handling Specs	Shipping Carrier Specs - Bonded wafer stacks	SEMATECH	SEMI
	In-assembly ESD	OSAT	ANSI/ESDA
Test	KGD / pre-bond test	IMEC	IEEE
	probe cards		

Figure 8. 3D standards landscape proposed by Nicholas Yu, Qualcomm [10]

development of silicon -validated inspection and metrology specifications for TSV voids, bond voids, and wafer thickness and warpage, and the development of a microbumping and bonding materials compatibility table to evaluate the compatibility for 3D stacking of die that are manufactured by different suppliers.

IV. CONCLUSION

After almost t wo decades of virtual dormancy, 3D integration using TSV finally has the momentum to develop into the next technology revolution of the semiconductor industry. Much work remains to be done in maturing processes,

materials, tools and standards for HVM readi ness, especially with respect to wafer back -side processing and package integration. SEMATECH continues to accelerate the advent of 3D-IC by its 300 mm pilot technology development and by forging industry consensus within the framework of the newly formed 3D Enablement Center and other industry bodies.

ACKNOWLEDGMENT

The authors would like to acknowledge valuable contributions by Tawfeeq Alzaben, Steve Bennett, Jose Colon, Doug Coolbaugh, Jeremiah Hebding, John H udnall, Brian Ji, Brian Martinick, Jerry Ma se, Megha Rao, Pratibha Singh, Travis Smith, Chris Taylor, Weng Hong Teh, and Brian Thomas.

REFERENCES

[1] P. Ramm et al, "Three dimensional metallization for vertically integrated circuits," Microelectronic Engineering 37/38, ed S. Namba, J. Kelly, M. van Ros sum, Elsevier Science (1997) p. 39; P. Ramm, R. Buchner, "Method of making a vertically integrated circuit," US Patent 5,766,984, priority Sep. 22, 1994 [DE].

[2] Bertin et al., "Three dimensional multichip packages and methods of fabrication", US Patent 5,202,704.

[3] International Technology Roadmap For Semiconductors http://www.itrs.net/

[4] M. Bohr, "Interconnect scaling - the real limiter to high performance ULSI"; Proceedings of the 1995, IEEE International Electr on Devices Meeting; pp241-242

[5] W. Haensch, " 3DI the next best thing f or MPU design?", IEEE International Solid-State Circuits Conference, 2007

[6] H. Hedler, "Status, opportunities and trends of 3D integration by thru - silicon-via stacking", IEEE International S olid-State Circuits Conference, 2007

[7] 3D panel discussion, IMAPS annual meeting, Raleigh, 2010

[8] D. Pascual, J. Hudnall, A. Gracias, K. Hummler, J. Castracane, " Novel die-to-wafer interconnect p rocess for 3DIC utilizing a thermo-decomposable adhesive and Cu -Cu thermo -compression b onding", IMAPS International Conference and Exhibition on Device Packaging, Scottsdale, 2011

[9] Larry Smith et. al., "TSV Manufacturing Assessment", 3D Architectures for Semiconductor Integration and Packaging Conference, December, 2010

[10] Nicholas Yu, "Interface Standardization for 3D Product Realization", SEMICON Taiwan 2010, September, 2010

Parasitics Extraction, Wideband Modeling and Sensitivity Analysis of Through-Strata-Via (TSV) in 3D Integration/Packaging

Zheng Xu*, Xiaoxiong Gu†, and Jian-Qiang Lu*

* Department of Electrical, Computer and Systems Engineering, Rensselaer Polytechnic Institute, Troy, NY 12180, USA.
† IBM T.J. Watson Research Center, Yorktown Heights, NY 10598, USA.

Abstract—**This paper reports on a number of extraction techniques to investigate the TSV parasitics in 3D integration/packaging, including 3D fullwave electromagnetic (EM) simulator, 3D quasi-static EM simulator, static SPICE simulator, and empirical calculations. All the TSV RLGC values extracted from the fullwave simulation are in good agreement among different approaches over the entire frequency range of interest. Empirical calculations indicate close results to fullwave extractions, while the quasi-static simulation underestimates TSV parasitics. A wideband SPICE model is generated from TSV EM solution with good agreement for both magnitudes and phases of return loss and insertion loss. Further sensitivity analysis results indicate the isolation layer thickness weighs most in the signal gain at 20 GHz. This work provides some insight to TSV electrical characteristics and helps TSV physical design to maximize the benefits of 3D systems.**

Index Terms—**3D integration, modeling, packaging, RLGC extraction, sensitivity analysis, TSV**

I. INTRODUCTION

Since traditional CMOS scaling governed by Moore's Law gradually slows down its pace today, three-dimensional (3D) integration is paving a potential path towards diversified technologies in the "More than Moore" era. 3D integration [1], by means of stacking function blocks and connecting them vertically, helps overcome some physical/technological/economic limits encountered in planar ICs at present. Through-strata-via (TSV) has been well regarded as a key component in 3D integration technology. A massive number of TSVs can be used to electrically connect multiple strata of ICs and/or devices vertically, enabling high performance, high functionality, compact 3D heterogeneous systems with high data bandwidth and speed, and low power consumption and cost.

Since a TSV is an extra structure unlike other circuit parts on the silicon substrate, a via can be fabricated in either front-end-of-the-line (FEOL) or back-end-of-the-line (BEOL) processing. In the FEOL process, TSVs (e.g., filled with doped poly-silicon) are formed at the very beginning of the fabrication process, and then transistors and metal distribution layers are built, followed by 3D processes (such as wafer alignment, bonding and thinning). In the BEOL process, TSVs are made through the substrate after the active devices (even interconnects) are fabricated. There is another mode called the "post-BEOL" TSV process, in which TSVs are not fabricated until all IC components are completed. Here TSVs can be

formed before bonding or after bonding, the so-called process of "via first" or "via last". The post-BEOL TSV process can be used for heterogeneous integration with different foundry technologies. TSV technologies are currently being developed by foundries and outsourced semiconductor assembly and test (OSAT) vendors, towards volume manufacturing.

In the TSV process, typically a deep via hole is etched in a substrate by wet or dry methods, followed by lining an insulating layer to block the leakage between the TSV and the conductive silicon substrate. Following depositions of a barrier layer and a seeding layer, the TSV conducting materials are deposited, e.g., using electroplating for copper TSV fill. Finally excess conductive TSV materials are polished away and another thick layer of dielectric is deposited. Since these material properties and dimension parameters in the manufacturing process determine TSV performance, designers should have early awareness in the planning stage.

With the research and development of TSV technologies, a thorough understanding of TSV properties is vitally important to aid 3D system design and implementation. To electrically characterize this most important component in 3D integration/packaging, we have employed a 3D fullwave EM simulator (Ansoft's HFSS [2]), a 3D quasi-static EM simulator (Ansoft's Q3D [3]), a SPICE simulator (Agilent's ADS [4]), and analytical methods. In this paper, we present and discuss various methods to extract frequency dependent passive elements (RLGC) of TSVs. A SPICE wideband modeling approach is applied to fit the fullwave scattering matrix obtained from the EM simulator. Furthermore, a sensitivity analysis is conducted to unveil the factors most affecting the TSV physical design.

II. TSV PARASITICS EXTRACTION

Considering current 3D TSV technology and near future 3D applications, we use a pair of identical signal and ground copper (Cu) TSVs, where the ground TSV assists as the returning current path. As shown in Fig. 1, the TSV diameter is 10 μm and the height 30 μm, while a thin layer of 0.1 μm SiO_2 serves as the isolation layer between the silicon (Si) substrate ($\rho = 10$ Ω-cm) and the Cu TSV. The pitch of signal and ground TSVs is 20 μm. Two terminated ports (50 Ω) are assigned at the two ends of the signal TSV during the simulations. This is the baseline TSV structure throughout this

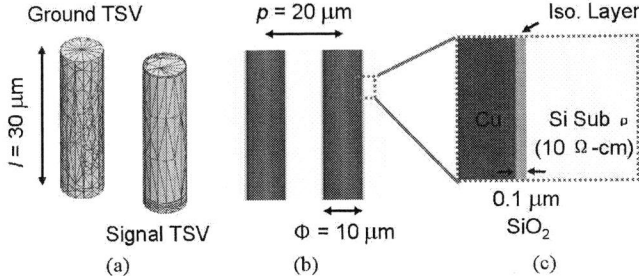

Fig. 1. Configuration of a signal-ground Cu TSV pair, (a) 3D mesh, (b) side view, and (c) enlarged view of one TSV. This signal-ground TSV pair serves as the baseline structure for TSV performance simulations in this work.

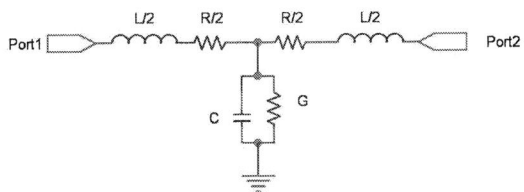

Fig. 3. "T" model of the signal-ground TSV pair.

work. TSV frequency-dependent parasitics are extracted based on the following methods.

A. "Π" Model

Ansoft's HFSS is employed as a 3D EM field solver, which is based on the finite element method (FEM) to differentially solve Maxwell's equations for arbitrary structures. The FEM meshes the TSV pair into many small tetrahedra and represents the field in an individual region with a local function. For typical passive RLC networks, Fig. 2 has a "Π" topology with one series subcircuit (resistance R and inductance L) and two shunt subcircuits (conductance G and capacitance C). Transforming the solved 2-port scattering (S) matrix to the 2-port admittance (Y) matrix,

$$\begin{bmatrix} I_1 \\ I_2 \end{bmatrix} = \begin{bmatrix} Y_{11} & Y_{12} \\ Y_{21} & Y_{22} \end{bmatrix} \begin{bmatrix} V_1 \\ V_2 \end{bmatrix},$$

where I and V are the port current and voltage, Y_{ii} is the input admittance seen looking into port i when the other port is short-circuited, and Y_{ij} ($i \neq j$) the transfer admittance between ports i and j.

Out of this, we can calculate the parameter values of the circuit elements as

$$R = Real\left(\frac{-2}{Y_{12} + Y_{21}}\right),$$

$$L = \frac{Imag\left(\frac{-2}{Y_{12} + Y_{21}}\right)}{\omega},$$

$$G = Real\left(2Y_{11} + (Y_{12} + Y_{21})\right),$$

$$C = \frac{Imag(2Y_{11} + (Y_{12} + Y_{21}))}{\omega},$$

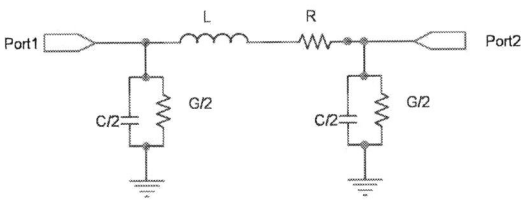

Fig. 2. "Π" model of the signal-ground TSV pair.

and the frequency is

$$f = \frac{\omega}{2\pi}.$$

B. "T" Model

TSV parasitics extraction can also be conducted according to a "T" model structure, as shown in Fig. 3. Unlike "Π" model, "T" model is implemented with one shunt subcircuit and two series subcircuits. An impedance (Z) matrix has the form,

$$\begin{bmatrix} V_1 \\ V_2 \end{bmatrix} = \begin{bmatrix} Z_{11} & Z_{12} \\ Z_{21} & Z_{22} \end{bmatrix} \begin{bmatrix} I_1 \\ I_2 \end{bmatrix}.$$

Similar to the admittance (Y) matrix, the impedance (Z) matrix also relates the voltage and current at the ports, and they are the inverses of each other,

$$[Z] = [Y]^{-1}.$$

The lumped components in Fig. 3 can be obtained,

$$R = Real\left(2Z_{11} - (Z_{12} + Z_{21})\right),$$

$$L = \frac{Imag(2Z_{11} - (Z_{12} + Z_{21}))}{\omega},$$

$$G = Real\left(\frac{2}{Z_{12} + Z_{21}}\right),$$

$$C = \frac{Imag\left(\frac{2}{Z_{12} + Z_{21}}\right)}{\omega}.$$

C. Transmission Line Model

In the fast-switching circuits produced by deep-submicron technologies today, the baseline signal-ground TSV pair had better be treated as a lossy transmission line. More precisely, this is the case when the rise and fall times of signals are comparable to the time of flight of a signal traveling along the TSV. As can be seen in Fig. 4, the model cascades a lot of identical RLGC units.

After a conversion from the scattering (S) matrix of the 3D EM field solution, the transmission (ABCD) matrix is good at characterizing such a component. Unlike in the definition used for Y and Z matrices, notice that I_2 is directed away from the port 2 in the ABCD matrix,

$$\begin{bmatrix} V_1 \\ I_1 \end{bmatrix} = \begin{bmatrix} A & B \\ C & D \end{bmatrix} \begin{bmatrix} V_2 \\ I_2 \end{bmatrix}.$$

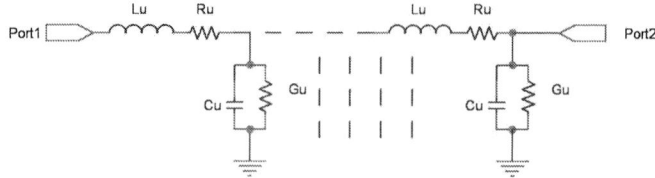

Fig. 4. Transmission line model of the signal-ground TSV pair.

Further, we get for the propagation constant γ of signal waveform and the characteristic impedance Z_0 of the TSV pair directly from the ABCD matrix,

$$\left[\begin{array}{cc} A & B \\ C & D \end{array} \right] = \left[\begin{array}{cc} cosh(\gamma l) & Z_0 sinh(\gamma l) \\ \frac{1}{Z_0} sinh(\gamma l) & cosh(\gamma l) \end{array} \right],$$

$$\gamma = \frac{cosh^{-1}(A)}{l},$$

$$Z_0 = \sqrt{\frac{B}{C}},$$

where l is the TSV length. According to the definitions of γ and Z_0, the equations below are derived,

$$R_u + j\omega L_u = \gamma Z_0,$$

$$G_u + j\omega C_u = \frac{\gamma}{Z_0}.$$

RLGC elements per unit length are given as,

$$R_u = Real(\gamma Z_0),$$

$$L_u = \frac{Imag(\gamma Z_0)}{\omega},$$

$$G_u = Real\left(\frac{\gamma}{Z_0}\right),$$

$$C_u = \frac{Imag(\frac{\gamma}{Z_0})}{\omega}.$$

D. Quasi-static Extractions

Ignoring the time-derivative coupling terms in Maxwell's equations, the electrical field and magnetic field can be solved separately, which significantly saves computing time and resources. Ansoft's Q3D is such a quasi-static tool implemented with both the finite element method (FEM) and the method of moments (MoM). FEM is principally the same as that in Ansoft's HFSS, while the MoM divides the surface (or volumes) of conductors and dielectrics into many triangular (or tetrahedral) elements to represent the charges and currents [3]. Q3D simulates the baseline signal-ground TSV pair (Fig. 1) and outputs the RLGC matrices as well.

E. AC Extractions

For a small signal simulation, AC analysis is introduced as a means of TSV parasitics extraction. First, a 1-Amp AC current source excites port 1 of the 2-port TSV fullwave solution, hence its impedance $Z = V/I = V$. Then a voltmeter detects the potential difference between port 1 and port 2 (V_{1-2}) when port 2 is grounded and between port 1 and reference (V_{1-0}) when port 2 is open-circuited, respectively. When frequency is swept, RLGC are obtained from,

$$R = Real(Z_{1-2}),$$

$$L = \frac{Imag(Z_{1-2})}{\omega},$$

$$G = Real\left(\frac{1}{Z_{1-0}}\right),$$

$$C = \frac{Imag\left(\frac{1}{Z_{1-0}}\right)}{\omega}.$$

F. Harmonic Balance Extractions

In general, harmonic balance (HB) is a frequency-domain analysis technique for studying distortion in nonlinear circuits and for RF and microwave applications. It offers several benefits over the time-domain transient analysis since it directly captures the steady-state spectral response. In addition, many linear models are best represented in the frequency domain at high frequencies.

We set up a testbench for HB simulation, put a voltmeter and an ammeter at port 1 of the 2-port TSV EM output file, and use a single-tone power source which outputs a cosine waveform with 1 mW power. After determining the spectral contents of voltage and current, at the fundamental harmonic we get R and L from the complex impedance ($Z = V_F/I_F$) by grounding the port 2 and get G and C from the complex admittance ($Y = I_F/V_F$) by grounding the reference port, respectively,

$$R = Real(Z),$$

$$L = \frac{Imag(Z)}{\omega},$$

$$G = Real(Y),$$

$$C = \frac{Imag(Y)}{\omega}.$$

G. Two-Port Input Z_{ii} and Y_{ii} Extractions

Scattering (S) parameter simulation is another type of AC simulation, and it is commonly used to characterize the small-signal properties of a passive RF component. We apply two terminated ports (50 Ω), thus the resulting scattering matrix becomes a 2×2 array. Through careful manipulations of the input impedance Z_{ii} and admittance Y_{ii}, RLGC are derived as,

$$R = Real\left(\frac{1}{Y_{11}}\right),$$

$$L = \frac{Imag(\frac{1}{Y_{11}})}{\omega},$$

$$G = Real\left(\frac{1}{Z_{11}}\right),$$

$$C = \frac{Imag(\frac{1}{Z_{11}})}{\omega}.$$

H. Two-Port Transfer Z_{ij} and Y_{ij} Extractions

Within the same 2×2 scattering matrix above, if we use the elements of the transfer impedance Z_{ij} and admittance Y_{ij}, the expressions of TSV passives are revised as below,

$$R = -Real\left(\frac{1}{Y_{21}}\right),$$

$$L = \frac{-Imag(\frac{1}{Y_{21}})}{\omega},$$

$$G = Real\left(\frac{1}{Z_{21}}\right),$$

$$C = \frac{Imag(\frac{1}{Z_{21}})}{\omega}.$$

I. Single-Port Input Z_{11} and Y_{11} Extractions

Another approach in extractions from S parameters is that we only introduce a single terminated port with a 50 Ω resistor, and conduct a 2-step testing procedure. First, ports 1 and 2 of the signal-ground TSV pair are connected to the two ends of this terminated port (50 Ω), leading to R and L when converting S parameters to Z parameters. Second, port 1 and the reference port are linked to the two ends of the terminated port (50 Ω), resulting in G and C when converting S parameters to Y parameters. The formulas are similar to those in the HB simulations above,

$$R = Real(Z_{11}),$$

$$L = \frac{Imag(Z_{11})}{\omega},$$

$$G = Real(Y_{11}),$$

$$C = \frac{Imag(Y_{11})}{\omega}.$$

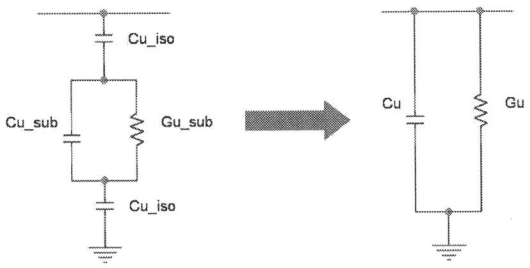

Fig. 5. Topology transformation of the shunt branch of the signal-ground TSV pair.

J. Empirical Calculations

Besides TSV passive element extractions using simulators, a set of analytical formulas were developed for quick hand calculations.

For the series resistance per unit length (R_u), we have $R_{u_{DC}}$ and $R_{u_{AC}}$. By Ohm's law,

$$R_{u_{DC}} = \frac{\rho}{\pi r^2}.$$

As frequency goes up. skin effect reduces the TSV conductor effective cross sectional area and increases resistance, which is proportional to the square root of frequency. A good approximation [5] is,

$$R_{u_{AC}} = 2\pi f\delta\frac{\mu}{\pi}\frac{p}{2r\sqrt{p^2 - 4r^2}},$$

$$\delta = \sqrt{\frac{\rho}{\pi\mu f}},$$

where ρ is the TSV conductor resistivity, δ the skin depth, μ the permeability, r the TSV radius, and p the pitch of signal and ground TSVs.

For the series inductance per unit length (L_u), it is the normalized number of Webers of field line loops around TSV per Ampere of current through it. We employ an expression for the two-wire transmission-line [5],

$$L_{u_0} = \frac{\mu}{\pi}cosh^{-1}(\frac{p}{2r}).$$

Enforcing causality from the Kramers-Kronig relations [6], inductance and resistance per unit length are interrelated with

$$L_u = L_{u_0} + \frac{R_u}{2\pi f}.$$

For the conductance and capacitance per unit length (G_u and C_u) along the shunt path, a little modification is applied in Fig. 5 to include the substrate conductance ($G_{u_{sub}}$), substrate capacitance ($C_{u_{sub}}$), and isolation layer capacitance ($C_{u_{iso}}$) per unit length as indicated in the real TSV structure.

Some empirical equations [7] are presented,

$$G_{u_{sub}} = \frac{\pi\sigma_{sub}}{ln\left(\frac{p}{2r} + \sqrt{(\frac{p}{2r})^2 - 1}\right)},$$

$$C_{u_{sub}} = G_{u_{sub}}\frac{\epsilon_{sub}}{\sigma_{sub}},$$

$$C_{u_{iso}} = \frac{2\pi\epsilon_{iso}}{ln\left(\frac{r+t_{iso}}{r}\right)},$$

where σ_{sub} and ϵ_{sub} are the conductivity and permittivity of the silicon substrate, ϵ_{iso} and t_{iso} the permittivity and thickness of the isolation layer.

Finally, the overall G_u and C_u illustrated in Fig. 5 are expressed as

$$G_u = \frac{-2\omega^2 C_{u_{iso}}C_{u_{sub}}G_{u_{sub}} + \omega^2 C_{u_{iso}}G_{u_{sub}}(2C_{u_{sub}} + C_{u_{iso}})}{4G_{u_{sub}}^2 + \omega^2(2C_{u_{sub}} + C_{u_{iso}})^2},$$

$$C_u = \frac{2C_{u_{iso}}G_{u_{sub}}^2 + \omega^2 C_{u_{iso}}C_{u_{sub}}(2C_{u_{sub}} + C_{u_{iso}})}{4G_{u_{sub}}^2 + \omega^2(2C_{u_{sub}} + C_{u_{iso}})^2},$$

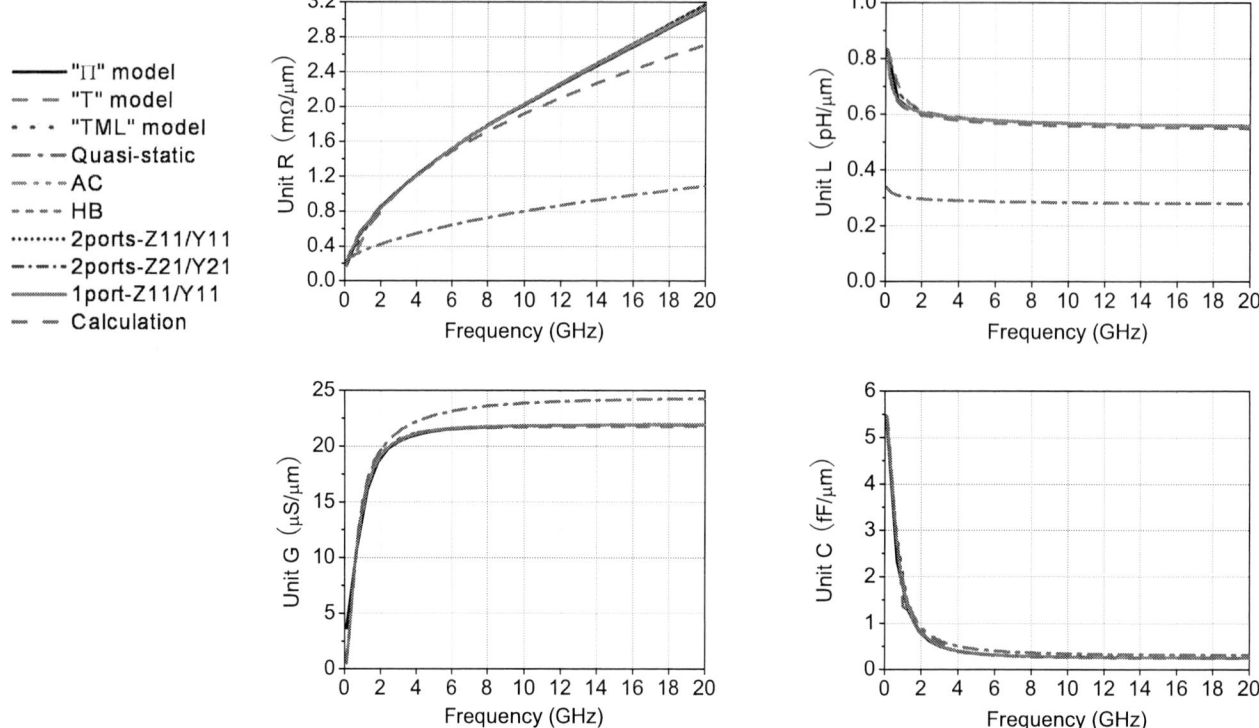

Fig. 6. Extracted results of TSV RLGC per unit length up to 20 GHz. A wide variety of methods based on the 3D fullwave EM simulation (HFSS) suggest very close results. The "calculation" results are the empirical calculations discussed in section II-J, they bring about similar results to the fullwave ones as well. while the quasi-static approach (Q3D) reports some differences.

both of which become functions of frequency. We import all the design parameters into the equations above and plot the analytical RLGC curves.

K. Extraction Results

Fig. 6 summarizes all the normalized RLGC extractions from the methods discussed above. The fullwave simulation results agree well among the various extracting methods over the entire frequency range. The empirical calculations accurately approximate the fullwave results. However, the quasi-static simulation engenders some discrepancies as it underestimates TSV parasitics. In Fig. 6, it shows good matchings for conductance and capacitance per unit length. However, the accuracy lack of the quasi-static simulation cause differences in TSV inductance and resistance per unit length.

III. TSV SPICE WIDEBAND MODELING

Instead of using passive elements (RLGC) to build an equivalent circuit explicitly as in [8]-[9], we use a SPICE netlist with voltage/current controlled sources to accurately model each port of the TSV.

For a signal-ground TSV pair as shown in Fig. 1, we use a SPICE tool [4] to convert frequency domain network parameters into equivalent circuit models. The input EM data for the SPICE tool is used to generate a rational expression,

which has the form of a series expansion in partial fractions:

$$\sum_{n}^{N} \frac{r_n}{s - p_n},$$

where r_n is the residue, s the Laplace variable and p_n the pole. In total, there are N poles, depending on the dynamics of the input EM data. After a successful convergence, N poles must be in the left half plane in order to enforce the passivity of the model. The netlist generated by the tool in this way describes the passive model, equivalent to the rational polynomial representation within the fitting error, using controlled sources (i.e., CCVS, VCVS, CCCS, and VCCS) at each port.

As shown in Fig. 7, the solutions of EM model and SPICE model are well matched in magnitudes and phases of return loss (S_{11}) and insertion loss (S_{21}). The resulting TSV SPICE netlist provides an accurate broadband model of scattering parameters, and can be further input to other commercially available simulators. This TSV SPICE netlist greatly saves time and resources to characterize arbitrary TSVs using field solvers, facilitating and expediting the design cycle of 3D systems.

IV. TSV SENSITIVITY ANALYSIS

To aid decision-making of several uncertainties in TSV physical design, sensitivity analysis helps rank the variables as to which variables affect the performance most and in

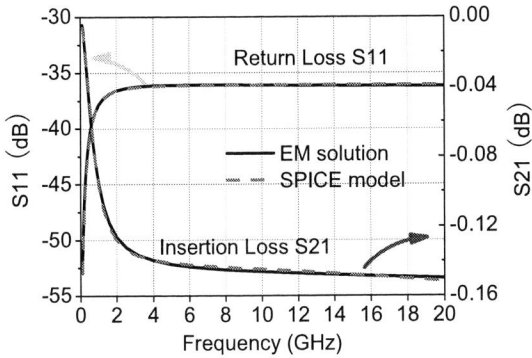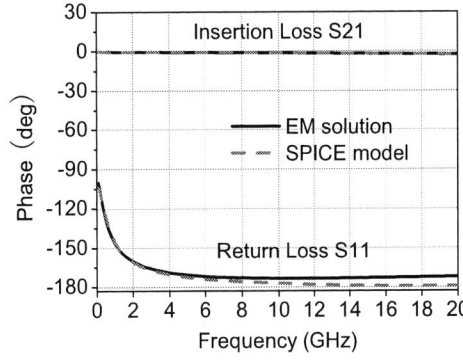

Fig. 7. Results of TSV EM solution and broadband SPICE model in magnitudes and phases of S_{11} and S_{21} agree well over the entire frequency range.

TABLE I
SENSITIVITY ANALYSIS OF THE PROPOSED SIGNAL-GROUND TSV PAIR AT 20 GHz.

Variables	Substrate ρ	TSV height (l)	TSV radius (r)	TSV pitch (p)	Isolation layer thickness (t_{iso})	Isolation layer ϵ_{iso}
1st derivative	-1.42E-02	-4.92E+03	-2.40E+04	2.28E-03	1.21E+05	-3.85E-03
2nd derivative	4.27E-05	-2.30E+06	-5.70E+09	-1.6E+03	-1.15E+11	1.72E-03

what way they affect the solution. Sensitivity analysis usually consists in computing derivatives of one or more outputs with respect to one or several independent inputs. In this work, we calculate the first derivative and second derivative of the signal gain (i.e., insertion loss S_{21}) with respect to each TSV design element, such as dimension/material variables, and then disclose which factors most affect the TSV physical design.

The absolute-sensitivity, i.e., the first derivative, of the function S_{21} to variations in the input parameter x is

$$S_x^{S_{21}} = \left. \frac{\partial S_{21}}{\partial x} \right|_{NOP},$$

and it should be evaluated at the certain normal operating point (NOP). The second derivative is for the interactions when changing x twice at the same time,

$$S_{x^2}^{S_{21}} = \left. \frac{\partial^2 S_{21}}{\partial x^2} \right|_{NOP}.$$

As an example, Table. I summarizes the results of the first derivatives and second derivatives of S_{21} with respect to 6 TSV design variables at 20 GHz, indicating that the isolation layer thickness is the most important in the proposed TSV physical configuration at this frequency point. Similarly, sensitivity results at other frequencies of interest can also be attained.

V. CONCLUSION

Since the electrical characteristics of TSV in 3D integration/packaging are still obscure, this paper utilizes ten extraction techniques to investigate the passives of a signal-ground TSV pair, including a 3D fullwave EM simulator, 3D quasi-static EM simulator, static SPICE simulator, and analytical methods. All the RLGC values extracted from the fullwave simulation agree well among the different methods over the entire frequency range, and hand calculations also show very close results. However, the quasi-static simulation

underestimates TSV parasitics. The scattering parameters of the signal-ground TSV pair simulated using the 3D EM field solver can be closely modeled by a wideband SPICE netlist. Furthermore, sensitivity analysis allows the designer to get hold of a problem with a great number of design variables, for instance, the isolation layer thickness weighs most in the signal gain at 20 GHz. In the early planning stage, this work can predict TSV electrical performance and optimize the TSV physical design.

ACKNOWLEDGMENT

We greatly appreciate K. Rose at Rensselaer, F. Baez, K. Han, and M. McAllister at IBM and S. Chickamenahalli, E. Acar and L. Polka at Intel for useful discussions. This work is supported by SRC GRC Interconnect & Packaging Sciences (IPS) program.

REFERENCES

[1] J.-Q. Lu,"3-D hyperintegration and packaging technologies for micro-nano systems," in Proc. IEEE, vol. 97, no. 1, Jan. 2009, pp. 18-30.
[2] Manual of Ansoft HFSS v12 2010.
[3] Manual of Ansoft Q3D v9 2010.
[4] Manual of Agilent ADS 2009.
[5] H.A. Wheeler, "Formulas for the skin effect," Proc. the Institute of Radio Engineers, vol.30, pp.412-424, Sept. 1942.
[6] T. Liang, S. Hall, H. Heck, and G. Brist, "A practical method for modeling PCB transmission lines with conductor surface roughness and wideband dielectric properties,", IEEE MTT-S International Microwave Symposium Digest, San Francisco, 2006, pp1780-1783.
[7] J.S. Pak, J. Cho, J. Kim, J. Lee, H. Lee, K. Park, and J. Kim, "Slow wave and dielectric quasi-TEM modes of metal-insulator-semiconductor (MIS) structure through silicon via (TSV) in signal propagation and power delivery in 3D chip package," Electronic Components and Technology Conference, Las Vegas, NV, June 2010, pp 667-672.
[8] Z. Xu, A. Beece, T. Zhang, K. Rose, and J.-Q. Lu, "Modeling and evaluation for electrical characteristics of through-strata-vias (TSVs) in three-dimensional integration," IEEE International 3D System Integration Conference, San Francisco, CA, Sept. 2009.
[9] G. Katti, M. Stucchi, K.D. Meyer and W. Dehaene, "Electrical modeling and characterization of through silicon via for three-dimensional ICs", IEEE Tran. on Electron Devices, vol. 57, no. 1, Jan. 2010, pp. 256-261.

978-1-61284-408-4/11 $26.00 © 2011 IEEE

Scaling of Copper Seed Layer Thickness Using Plasma-Enhanced ALD and an Optimized Precursor

Jiajun Mao, Eric Eisenbraun
College of Nanoscale Science and Engineering
The University at Albany-SUNY
Albany, New-York, USA

Vincent Omarjee, Andrey Korolev, Christian Dussarrat
Delaware Research and Technology Center
American Air Liquide
Newark, Delaware, USA
Vincent.omarjee@airliquide.com

A recently developed precursor, AbaCus, has been evaluated for use in ultra-low temperature copper deposition by PEALD. Film adhesion, platability and process window evaluation demonstrate a strong capability of this precursor to overcome current metallization challenges.

Keywords: AbaCus, Copper, PE-ALD,, BEOL

I. INTRODUCTION

Advanced metallization schemes require the use of low resistivity and high electromigration-resistant metal. Aluminum has been one of the first metals to be used, however with the scaling of semiconductor devices, copper became the only metal giving viable electrical performance. In order to fill a via with copper by electrochemical deposition, a seed layer is needed to initiate the growth. Up until now, a thin layer of copper has been used, this thin layer typically being deposited by iPVD. A major drawback to this approach is the poor conformality of PVD, causing overhanging and compromising void-free filling of the via. New barrier and metallization approaches are currently being used and can somewhat alleviate this issue. However, no solution is currently known for how to deposit a thin, continuous and conformal copper seed layer. Vapor depositions processes such as atomic layer deposition (ALD) are known to allow very high quality thin film deposition [1], however copper deposition has its own challenges; one major burden being that copper needs to be deposited at less than 100°C to allow thin copper film continuity. At such low temperatures, it is difficult to initiate the ALD growth process. Therefore, a plasma-enhanced processing option might be required to overcome this issue.

Plasma-enhanced processes, such as PEALD and PECVD, are promising techniques [2] which can allow producing high purity and high-density metal thin films at low growth temperatures, but require the development of advanced chemistries. Commercially available copper precursors do not satisfy the requirements for stability, volatility, low temperature deposition and high reactivity while allowing good film quality. To address those needs, a new copper precursor, "AbaCus", has been identified has a promising solution [3].

II. ALD REGIME EVALUATION

PEALD process development was carried out using a Tokyo Electron Limited (TEL) Phoenix™ 200-mm wafer capable CVD cluster tool modified for use with ALD processing. The PEALD process employed pulses of the Cu precursor and plasma-activated hydrogen for process simplicity, non-plasma hydrogen flow was used as the purge ambient (at reaction temperature, these chemistries are stable in the presence of non-plasma hydrogen). Thermal characteristics of the precursor were evaluated using thermal gravimetric analysis (TGA). In the TGA, the samples were heated at a ramp rate of 10°C/min in open cup configuration under atmospheric or vacuum condition – N_2 ambient. Film thicknesses were measured using a Carl-Zeiss LEO 1550 scanning electron microscope (SEM) for cross-section and confirmed by Rutherford backscattering spectrometry (RBS). Film purity was measured using a PHI 660 Auger electron spectroscopy (AES) system. Resistivity was assessed using a Signatone QuadProS-A8 four-point resistance probe.

A. Precursor thermal characterization

Figure 1 below shows the precursor TGA weight loss under vacuum and atmospheric conditions. A smooth single step mass loss is observed and no inflexion point is seen. While residue level is significant at 23% at high temperature, >250°C, under atmospheric condition, vacuum test shows complete evaporation without any residues, indicating that AbaCus can be easily delivered into a chamber at standard delivery temperatures, <200°C.

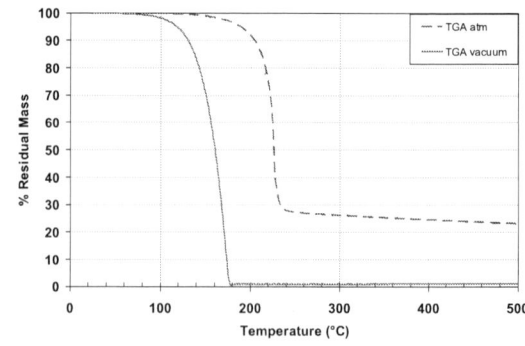

Figure 1. TGA plots of AbaCus under various conditions

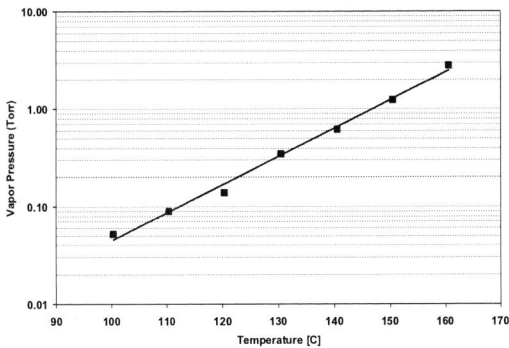

Figure 2. Vapor pressure plot of AbaCus

Precursor vapor pressure is an important factor for ALD processes. Figure 2 shows that AbaCus has a high vapor pressure, ~1 Torr at 145°C, making it suitable for use in vapor deposition processes. Being a non-viscous liquid at room temperature, it also allows flexibility in the type of delivery system (such as bubbler or vaporizer) that can be used.

Thermal stability is a critical parameter for manufacturing capabilities. Long term thermal stability is required to ensure process stability and avoid delivery issues. The thermal stability of AbaCus was assessed by monitoring its evaporation at constant temperature of 120°C under atmospheric pressure TGA conditions. Figure 3 shows steady linear mass loss over a ~4 hour period with no residue, indicating complete evaporation of precursor without noticeable decomposition. Thus it is concluded that AbaCus is stable up to 120°C. Higher temperatures were also successfully tested.

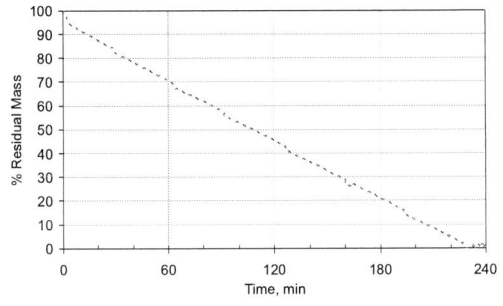

Figure 3. Isothermal TGA plot of AbaCus at 120 °C and atmospheric pressure. A complete evaporation without decomposition is demonstrated

B. ALD assessment

For PEALD process development using AbaCus, the process conditions were adjusted as follow: Wafer temperature 30-100°C, plasma power 60-160 Watts, and the pulse and purge times were varied from 3 to 15 seconds. Process pressure was fixed at ~2 torr using 300sccm H_2. The bubbler temperature was held at 100°C, and 40 sccm He carrier gas was employed. For half-reaction saturation experiments, all process conditions were held constant except for the specific pulse or purge time being evaluated in that particular experiment.

PEALD Cu growth rate was measured as a function of precursor pulse time. As shown in Figure 4, where the number of ALD cycles was fixed at 750, a 3 seconds long precursor pulse was sufficient to achieve saturation in the PEALD process at a wafer temperature of 60°C. All depositions used a 10 second, 120 Watt hydrogen plasma pulse. This indicates a fast transition to a saturated half-reaction.

Similarly, a plasma exposure saturation time was checked (Figure 6). Five seconds was determined to be sufficient to ensure full surface saturation.

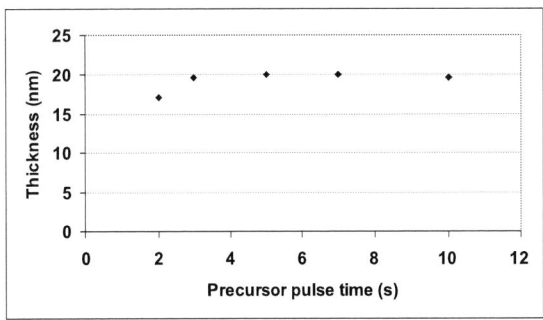

Figure 5. Cu thickness vs. precursor pulse time [750 ALD cycles]

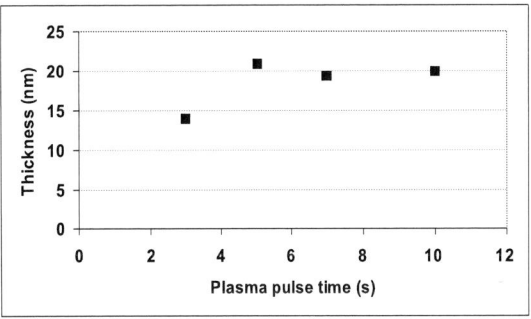

Figure 6. Cu thickness vs. plasma exposure time [750 ALD cycles]

Thin copper film quality is strongly dependent on the deposition temperature. AbaCus was tested over a wide temperature range. Figure 7 shows that AbaCus is very reactive and deposition at room temperature was possible with a stable growth rate of ~0.3Å/cycle. This capability of depositing a film at such a low temperature is expected to bring significant process advantages, especially with respect to film morphology and continuity.

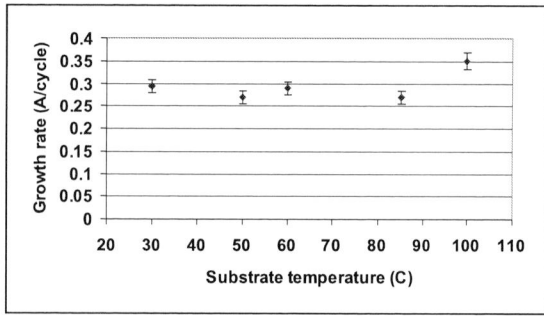

Figure 7. Process window showing stable ALD regime at a 30°C substrate temperature

C. Filling capability

For an application as a seed layer for electroplating in advanced interconnects, the conformality of the seed layer, both in terms of thickness and microstructure, is key in enabling complete, void-free via and trench filling. The use of plasma processes is usually a concern when considering high aspect ratio structures such as 3D through silicon vias (TSVs). If the film thickness decreases at the bottom of high aspect ratio structure because of rapid recombination of the plasma species, the film might not maintain continuity or it will have a higher resistivity, and therefore the plating may fail. In order to verify that a conformal Cu seed film can be deposited, growth over patterned samples was investigated. Figure 8 is a cross-section SEM indicating typical conformality of PEALD Cu deposited from AbaCus on 180nm (aspect ratio ~8) trench structures. The film appears uniform throughout the structure sidewalls and floor, although the thickness on the field is larger. As opposed to a PVD process, no "breadloafing", or pinching off of the structure, is noted. This indicates that plasma reduction is effective, even when using high aspect ratio structures.

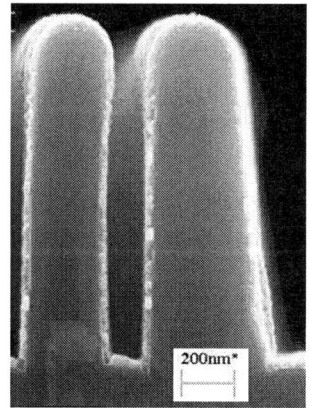

Figure 8. Cross-section SEM showing conformality of a ~40nm thick PEALD Cu film deposited with AbaCus. Step coverage is measured to be >70%, and the film thickness across the sidewall and floor of the structure is uniform

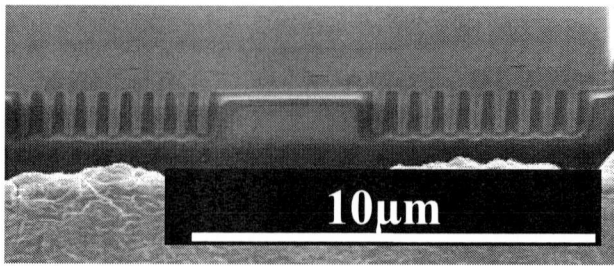

Figure 9. SEM of ECD Cu on 13nm Cu seed/ ALD TaN barrier. Cu-seed was deposited at 30C wafer temperature.

To determine the applicability as a seed layer for subsequent plating, PEALD-seeded structures were plated employing a standard bath consisting of 0.8M H_2SO_4, 0.8M $CuSO_4 \cdot 5H_2O$, 1.4mM HCl and the Viaform™ additive package from Enthone, Inc. The plating time was 40 seconds at a current density of 10mA/cm². High aspect ratio trench structures were coated with ALD TaN and 13nm PEALD Cu seed from AbaCus using a 30°C wafer temperature, then plated. As depicted in Figure 9, the resulting plates yield complete Cu filling, indicating that this combination of Cu precursor and PEALD method appears to be very promising for use in advanced interconnects.

III. FILM ANALYSIS: MORPHOLOGY, RESISTIVITY AND ADHESION TESTS

A. Grain morphology

As the process window indicated, an extremely low temperature deposition was achievable with this precursor. Copper deposition at room temperature by PEALD was demonstrated. No condensation effects were seen and the growth rate was stable. As expected, the suppression of copper agglomeration effects caused by substrate temperature was confirmed by atomic force microscopy analyses revealing that the smoothest film is obtained for the lowest deposition temperature (Figure 10). The surface roughness is correlated with film continuity. The film deposited at 30°C is very smooth with RMS roughness of 0.62 nm.

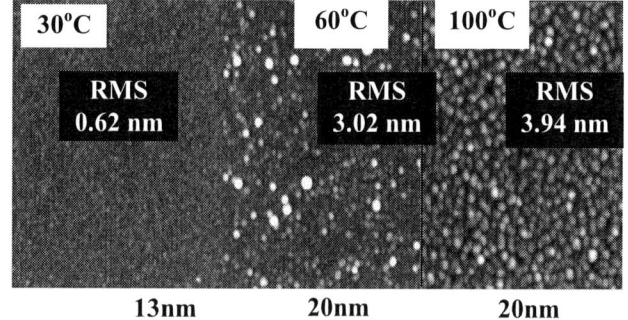

Figure 10. AFM analysis of PEALD Cu surface roughness as it relates to deposition temperature

B. Film resistivity

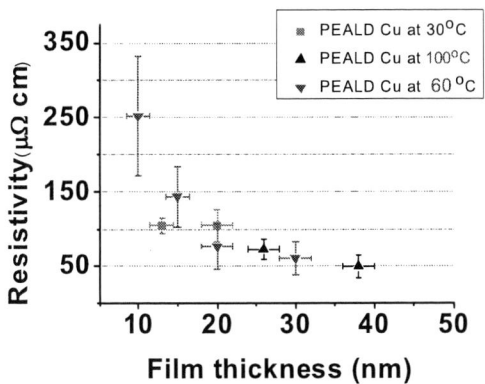

Figure 11. PEALD Cu resistivity for various film thicknesses for films deposited at different temperatures

Deposition temperature and film thickness were correlated with the film resistivity, as seen in Figure 11. To avoid substrate effects when measuring very thin film resistance, the measured films were deposited directly on SiO_2 substrates. As shown in Figure 11, the film thickness directly impacts the film resistivity. The thinner the film is, the higher the resistivity. This is even more pronounced for films below 20 nm. Within this overall trend, a lower resistivity is observed as the deposition temperature is decreased. By extrapolating the data obtained for a very thin film (13 nm) deposited at 30°C using AbaCus, it is predicted that a continuous film with a resistance low enough to allow void-free ECD can be achieved at a thickness in the 6 to 8 nm range. This is currently being confirmed.

C. Adhesion tests

Adhesion of the seed layer on the barrier is an important parameter to ensure successful integration. Usually, metallic tantalum or titanium is deposited on top of the standard TaN or TiN barriers to enhance copper adhesion. Adhesion of copper film deposited by PEALD is compared in Table I to the adhesion of copper deposited by PVD, on various representative substrates including silicon, TaN and Ruthenium.

TABLE I. Copper Adhesion Comparison: ALD vs PVD

	ALD Cu on TaN	ALD Cu on Ru	ALD Cu on Ta	PVD Cu on TaN	PVD Cu on Ru	PVD Cu on Ta
Tension (Mpa)	19.19	10.25	34.07	14.21	5.79	25.93
Fract. section	Cu/TaN	Ru / SiO_2	epoxy	Cu / TaN	Ru / SiO_2	epoxy
Tension (Mpa)	19.29	11.2	20.97	12.97	2.37	33.75
Fract. section	Cu / TaN	Inside SiO_2	Inside SiO_2	Cu / TaN	Ru / SiO_2	epoxy

N.B. fracture inside SiO_2 or epoxy indicates that Cu adhesion to barrier is higher than reported value

Cu films deposited at 30°C by PEALD using AbaCus have better adhesion than PVD Cu in all test condition.

IV. CONCLUSION

In this work AbaCus was investigated in PEALD of copper using hydrogen plasma reduction. Copper film was deposited at room temperature, and the obtained results indicate that this ultra-low temperature plays a key role in achieving smooth, continuous and uniform films. Plating using a seed layer as thin as 13 nm was found to be successful for void-free filling, showing that AbaCus can be a suitable precursor for advanced interconnect manufacturing processes.

REFERENCES

[1] L.Wu and E.Eisenbraun, "Hydrogen Plasma-enhanced Atomic Layer Deposition of Copper Thin Films," J. Vac. Sci. Technol. B(25), 2007, p2581-2585.

[2] J.Mao, E.Eisenbraun, V.Omarjee, A.Korolev, C.Lansalot, C.Dussarrat, "Ultra-Low Temperature Deposition of Copper Seed Layers by PEALD," ECS Trans., Volume 33, issue 12, 2010, p125-135.

[3] J.Mao, E.Eisenbraun, V.Omarjee, A.Korolev, C.Lansalot, C.Dussarrat, "Plasma-Enhanced Atomic Layer Deposition of Copper using New Precursors for Next Generation of Interconnections," unpublished ALD 2010 conference, Seoul, Korea, 2010.

Investigation of the structural and electrical characterization on ZrO$_2$ addition for ALD HfO$_2$ with La$_2$O$_3$ capping layer integrated metal-oxide semiconductor capacitors

C. K. Chiang[1,2,*], J. C. Chang[1], W. H. Liu[1], C. C. Liu[1], J. F. Lin[1], C. L. Yang[1], J. Y. Wu[1] and S. J. Wang[2]

[1]United Microelectronics Corporation, Science-Based Industrial Park, Hsinchu, Taiwan, ROC
[2]Institute of Microelectronics, Dept. of Electrical Eng., National Cheng Kung Univ., Tainan, Taiwan, ROC
*E-mail: ebony_chiang@umc.com

Abstract — In this work, we report on ZrO$_2$ position effect of ALD HfZrO$_x$ gate dielectric with a La$_2$O$_3$ capping layer for gate-first flow. The basic electrical characteristics of devices were compared with different ZrO$_2$ position in HfZrO$_x$ dielectric. Experimental results show : (1) Under top La$_2$O$_3$ capping layer for n-type Metal-Oxide-Silicon capacitor (nMOSCAP) device, ZrO$_2$ position on both of top and bottom in HfZrO$_x$ shows higher leakage (>x5) current and V$_{fb}$ shift (-0.18V) to band edge than HfO$_2$ dielectric. (2) For the top La$_2$O$_3$ cap device, ZrO$_2$ addition into ALD HfO$_2$ can have significant shift on J$_g$ and V$_{fb}$. Bottom La$_2$O$_3$ capping position stack has higher J$_g$ (>x4) and larger V$_{fb}$ shift (-0.15V) than the top La$_2$O$_3$ cap position for nMOSCAP device.

I. INTRODUCTION

With the continuous scaling of the complementary metal oxide semiconductor ⌐ (CMOS) technology, high-k gate dielectrics will be needed to replace conventional SiO$_2$ gate dielectrics for addressing the excessive high leakage concern. Hf based high-κ dielectrics have been thoroughly investigated to replace the conventional SiO$_2$ to reduce gate leakage current in MOS structures.. However, one of the key issues in high-κ gate stack is the high density of bulk traps, which degrade device mobility and result in poor reliability [1-2]. For HfO$_2$ based dielectrics, device thresholds are generally altered from the expected position due to charged defects [3-4]. Based upon earlier work showing shifted flatband (or threshold) voltages with the use of highly ionic oxides such as La$_2$O$_3$ [5], one approach to control V$_t$ is to introduce a threshold modifying material [6-9]. Many deposition techniques have been explored for high-κ dielectrics including molecular beam epitaxy (MBE), pulsed laser deposition (PLD), electron beam evaporation, metal organic chemical vapor deposition (MOCVD), and atomic layer deposition (ALD) [10-14]. In particular, ALD has been proven to be a very attractive technique for the fabrication of advanced gate dielectrics and DRAM insulators due to precise control of film thickness and excellent conformality. ZrO$_2$ and HfO$_2$ are considered twin oxides owing to their similarities in structural modifications, lattice constants,

chemical, and physical properties [15]. Among all the binary materials HfZrO$_x$ films were shown to present higher reliability and mobility than HfO$_2$ thin films [16-17]. But the HfZrO$_x$ shows higher leakage current due to the smaller band gap than HfO$_2$ device. In this paper, we have systematically investigated HfZrO$_x$/SiO$_2$ MOSCAP device characteristics by changing the ZrO$_2$ position in ALD HfO$_2$ gate stack with different La$_2$O$_3$ capping layer position for gate-first process.

II. EXPERIMENTAL

MOS capacitors were fabricated on p-Si ⌐ (100) 300mm wafers using conventional gate-first flow in this letter. After surface clean, the dielectric layer was grown on a 0.7-nm thermal oxide starting surface and 2-nm Hf$_{0.5}$Zr$_{0.5}$O$_2$ bi-layer film were deposited, respectively. HfO$_2$ and Hf$_{0.5}$Zr$_{0.5}$O$_2$ films were fabricated by ALD using HfCl$_4$, ZrCl$_4$, and H$_2$O precursors at a deposition temperature of 300 °C . All films were annealed at 1050 °C for 5 s in a nitrogen ambient. In order to study the MOSCAP characteristic of HfO$_2$ and HfZrO$_x$, different types of gate stack have been used,

Sample A: HfO$_2$ 20Å/SiO$_2$ 7Å (as a control wafer)
Sample B: ZrO$_2$ 10Å/HfO$_2$10Å/SiO$_2$ 7Å
Sample C: HfO$_2$ 5Å/ZrO$_2$ 10Å/HfO$_2$ 5Å/SiO$_2$ 7Å
Sample D: HfO$_2$ 10Å/ZrO$_2$ 10Å/SiO$_2$ 7Å
Sample E: ZrO$_2$ 5Å/HfO$_2$ 10Å/ZrO$_2$ 5Å/SiO$_2$ 7Å.

Finally, the top gate electrode comprising a 10-nm-thick TiN layer was formed using PVD and lithographic patterning. Samples with an area of 0.01×0.01 cm^2 were used for leakage-current and capacitance measurements. All samples are process on gate-first flow with 5Å top ALD La$_2$O$_3$ capping layer for effective work function (EWF) tuning. Finally, the ZrO$_2$ position effect of Hf$_{0.5}$Zr$_{0.5}$O$_2$ for nMOSCAP with top or bottom La$_2$O$_3$ capping layer device characteristics were also compared. Capacitance-voltage (C-V) measurements were performed using a HP4284 LCR meter at 100 kHz to obtain the EOT and flatband voltage. The leakage current density-voltage (J$_g$-V) characteristics were measured using a Keithley SCS

978-1-61284-408-4/11 $26.00 © 2011 IEEE

4200. The microstructures of $HfZrO_x/SiO_2$ gate stacks were observed by cross-sectional high resolution transmission electron microscopy (HR-TEM), and the depth profile of the high-κ/IL gate stack prepared samples was characterized by angle resolved x-ray photoelectron spectroscopy (AR-XPS) analysis on a Theta 300 XPS system from Thermo Fisher by using monochromatic Al Kα radiation (1486.6 eV) in an angle-resolved mode.

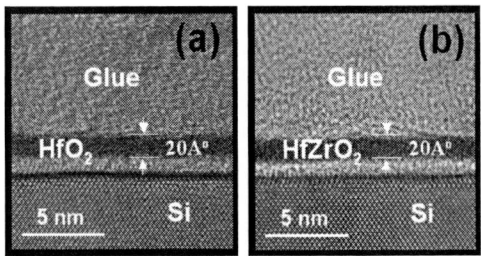

Figure 1. Cross section HR-TEM image of (a) HfO_2/SiO_2 (b) $HfZrO_x/SiO_2$ after annealing at 1050°C. A uniform contrast indicates a compositional uniformity

III. RESULTS AND DISCUSSION

To examine the impact of ZrO_2 addition into HfO_2, the HR-TEM micrographs of HfO_2 and $Hf_{0.5}Zr_{0.5}O_2$ are shown in **Figure 1(a) and 1(b)**, respectively. Both layers are amorphous after annealing at 1050°C for 5 s in a nitrogen ambient. Both dielectric are polycrystalline with similar bulk dielectric thickness (20Å) and interface oxide thickness (~8Å). **Figure 2 (a)** shows band gap (E_g) analysis by spectroscopic ellipsometry,

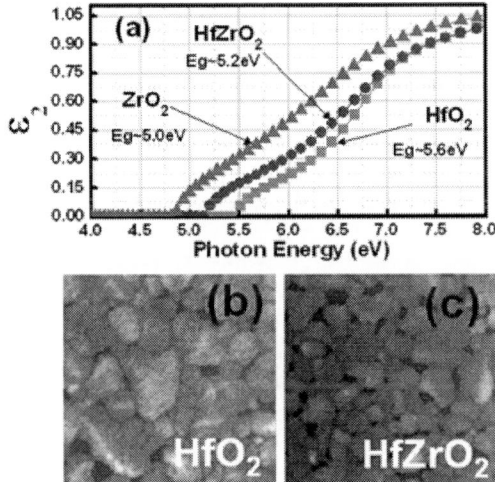

Figure 2. (a) Band gap (E_g) analysis by spectroscopic ellipsometry, the band gap of decreased when ALD ZrO_2 addition with HfO_2. Preliminary result of Plan-view TEM for (b) HfO_2 (c) $HfZrO_x$, the $HfZrO_x$ grain size shows smaller than HfO_2.

the band gap of $HfZrO_x$ decreased when ZrO_2 addition with HfO_2. Preliminary result of Plan-view TEM for HfO_2 and $HfZrO_x$, respectively [**Figure 2(b)**] and $HfZrO_x$

[**Figure 2(c)**]. The $HfZrO_x$ grain size shows smaller than HfO_2. We believe that the change in grain-size plays a role in reducing charge trapping in HfO_2.

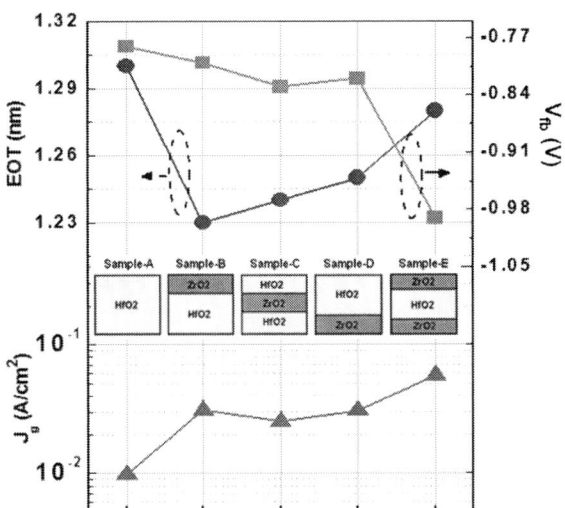

Figure 3. nMOSCAP device characteristics comparison between different ZrO_2 position in HfO_2 gate stack on gate-first process flow with top La_2O_3 capping layer.

Figure 3 shows samples A to E nMOSCAP device characteristics comparison between different ZrO_2 position in HfO_2 gate stack with top La_2O_3 capping layer. Compared to HfO_2 we found the higher J_g was observed for $HfZrO_x$ samples (B to E). This is due to smaller band gap and lower conduction band offset for $HfZrO_x$ than that of HfO_2. Sample B shows better EOT scaling , Sample E shows the worse J_g and larger V_{fb} shift to band edge for effective work function (EWF) tuning.

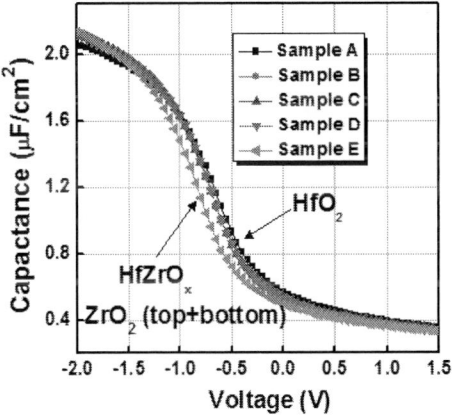

Figure 4. ZrO_2 position effect for $HfZrO_x$ gate stack nMOSCAP device (*C-V* plot) under *gate-first* process with top La_2O_3 capping layer. Sample E [ZrO_2 position on (top + bottom) of $HfZrO_x$] shows larger V_{fb} shift than sample A

Figure 4 shows the C-V plots for $HfZrO_x$ gate stack MOSCAP. ZrO_2 position on (top + bottom) of HfO_2 layer (Sample E) shows C-V shift to the negative direction. The

978-1-61284-408-4/11 $26.00 © 2011 IEEE

ZrO_2-HfO_2-ZrO_2 gate stack may enhance higher oxygen vacancy in bulk $HfZrO_x$ and exist more positive charge than HfO_2.

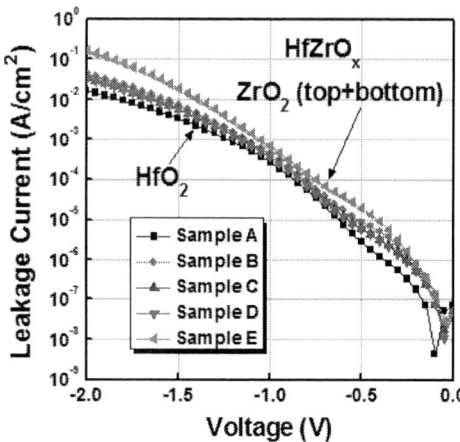

Figure 5. ZrO_2 position effect for HfZrOx gate stack nMOSCAP device (J_g -V plot) under gate-first process with top La_2O_3 capping layer. Sample E [ZrO_2 position on (top + bottom) of $HfZrO_x$] shows higher Jg behavior than sample A is observed.

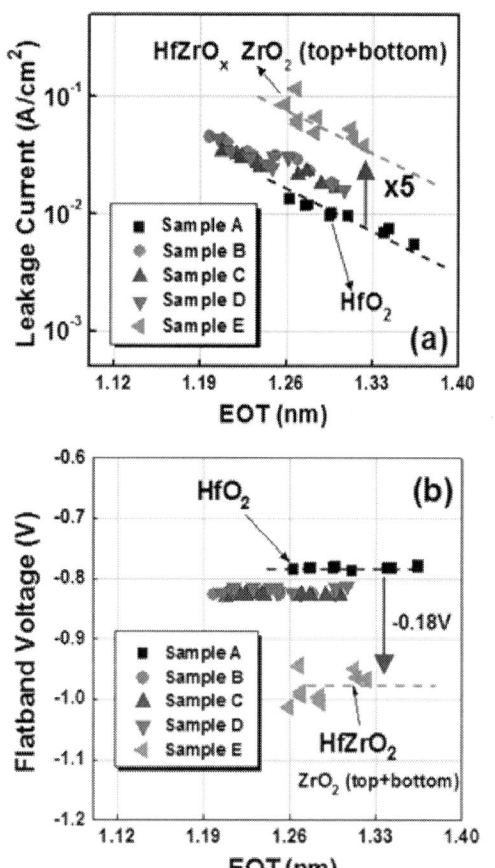

Figure 6. nMOSCAP with top La_2O_3 capping layer device characteristics for different ZrO_2 position in $HfZrO_x$/SiO_2 gate stack (a) EOT-J_g (b) EOT-V_{fb} plot.

In the **Figure 5**, it shows the J_g-V plots for $HfZrO_x$ gate stack MOSCAP. ZrO_2 position on (top + bottom) of HfO_2

layer (Sample E) shows higher leakage current (J_g) than other samples. We suspect it was due to higher multi-layer interface trapping when ZrO_2 contact with top La_2O_3 capping layer and bottom SiO_2 interfacial layer .

From the **Figure 6** shows the samples for nMOSCAP EOT-J_g [Figure 6(a)] and EOT-V_{fb} [Figure 6(b)] plots. Sample E stack has higher J_g (>x5) and larger V_{fb} shift (-0.18V) than sample A. We suspect it is due to higher bulk charge and multi-layer interface trapping.

In **Figure 7**, Gate stack depth profile for different ZrO_2 position of $HfZrO_x$/SiO_2 gate stack was obtained from angle resolved x-ray photoelectron spectroscopy (AR-XPS). The position of the Zr3d peak in the profiles was found to be accurately reproduced. Greater accuracy in this region would be expected if a larger number of data points are used. **Figure 7(a)~7(c)** show ZrO_2 on top position, center position and bottom position in $HfZrO_x$, respectively.

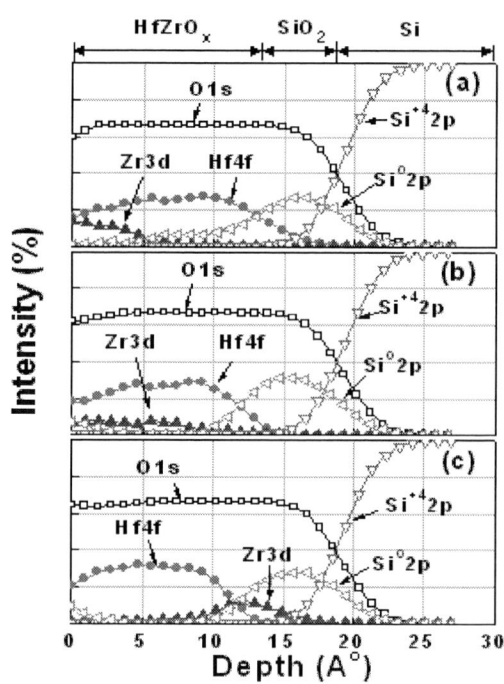

Figure 7. Maximum Entropy entropy depth profile for different ZrO_2 position of $HfZrO_x$/SiO_2 Gate Stack obtained from AR-XPS (a) Top ZrO_2 position (b) Center ZrO_2 position (c) Bottom ZrO_2 position

Sample A, B and D with different La_2O_3 capping positions were also used for nMOSCAP device comparison. **Figure 8** shows the EOT-J_g plot. For the bottom La_2O_3 cap device, the ZrO_2 addition effect can show the J_g different from HfO_2 due to the band gap decrease. For the top La_2O_3 cap device, ZrO_2 addition with ALD HfO_2 can have significant shift on J_g . Bottom La_2O_3 capping position stack has higher J_g (>x4) than the top La_2O_3 cap nMOSCAP device.

978-1-61284-408-4/11 $26.00 © 2011 IEEE

Figure 9 shows the EOT-V_{fb} plot. For the bottom La$_2$O$_3$ cap device, there is no clear impact on the V_{fb} shift. For the top La$_2$O$_3$ cap device, ZrO$_2$ addition with ALD HfO$_2$ can have significant shift on V_{fb}. Bottom La$_2$O$_3$ cap position stack has larger V_{fb} shift (-0.15V) than the top La$_2$O$_3$ cap nMOSCAP device.

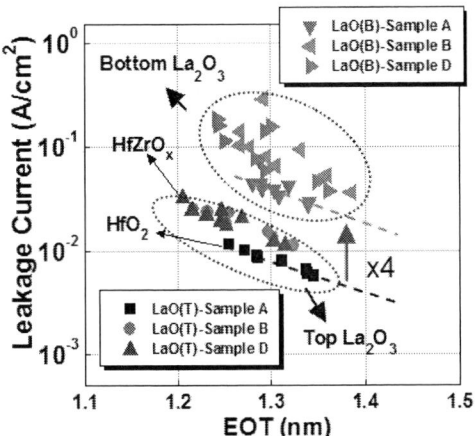

Figure 8. ZrO$_2$ position effect for HfZrO$_x$ gate stack nMOSCAP device under gate-first process with different La$_2$O$_3$ capping layer position (top and bottom). nMOSCAP device characteristics EOT-J$_g$ plot

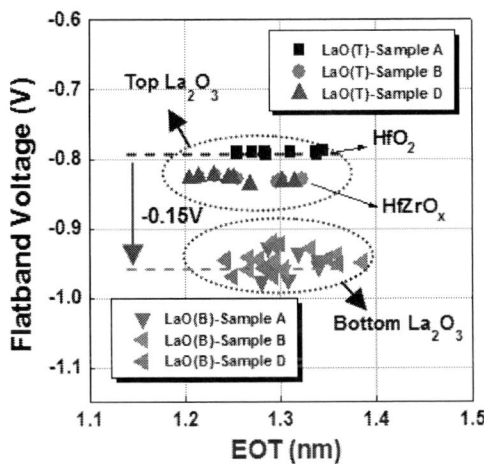

Figure 9. ZrO$_2$ position effect for HfZrO$_x$ gate stack nMOSCAP device under gate-first process with different La$_2$O$_3$ capping layer position (top and bottom). nMOSCAP device characteristics EOT-V$_{fb}$ plot.

IV. CONCLUSIONS

The experimental results provide the information of the structural and electrical characterization on ZrO$_2$ addition for ALD HfO$_2$ with La$_2$O$_3$ capping layer integrated MOSCAP for gate-first process flow. ZrO$_2$-HfO$_2$-ZrO$_2$ gate stack structure shows worse J$_g$ and larger V$_{fb}$ shift to band edge. Under top La$_2$O$_3$ capping device, ZrO$_2$ addition effects for ALD HfO$_2$ have significant shift on MOSCAP

characteristics than bottom La$_2$O$_3$ capping device.

ACKNOWLEDGMENT

This work was supported by the National Science Council (NSC) of Taiwan, Republic of China, under contract No NSC 95-2215-E-006-014, NSC 96-2221-E-006-081-MY2, and NSC 98-2218-E-216-002.

REFERENCES

[1] P. Sivasubramani, et al, VLSI Symp. P.68, 2007.

[2] I. P. Studenyak, et al, Thin Solid Films 476 p.137, 2005.

[3] J. K. Schaeffer, L. R. C. Fonseca, S. B. Samavedam, Y. Liang, P. J. Tobin, and B. E. White, Appl. Phys. Lett. 85, 1826 2004.

[4] E. Cartier, F. R. McFeely, V. Narayanan, P. Jamison, B. P. inder, M. Copel, V. K. Paruchuri, V. Basker, R. Haight, D. Lim, R. Carruthers, T. Shaw, M. Steen, J. Sleight, J. Rubino, H. Deligianni, S. Guha, R. Jammy, and G. Shahidi, Tech. Dig. VLSI Symp. p. 230, 2005.

[5] S. Guha, E. Carter, M. A. Gribelyuk, N. Bojarczuk, and M. Copel, Appl. Phys. Lett. 77, 2710 ┌ 2000.

[6] V. Narayanan, V. Paruchuri, N. Bojarczuk, B. Linder, B. Doris, Y. Kim, S. Zafar, J. Stathis, S. Brown, J. Arnold, M. Copel, M. Steen, E. Cartier, A. Callegari, P. Jamison, J.-P. Locquet, D. Lacey, Y. Wang, P. Batson, P. Ronsheim, R. Jammy, and M. Chudzik, Tech. Dig. VLSI Symp. P.178, 2006.

[7] X. P. Wang, M. F. Li, C. Ren, X. F. Yu, C. Shen, H. H. Ma, A. Chin, C. X. Zhu, J. Ning, M. B. Yu, and D. L. Kwong, IEEE Electron Device Lett. 27, 31 2006.

[8] H. N. Alshareef, M. Quevedo-Lopez, H. C. Wen, R. Harris, P. Kirsch, P. Majhi, B. H. Lee, R. Jammy, D. J. Lichtenwalner, J. S. Jur, and A. I. Kingon, Appl. Phys. Lett. 89, 232103 2006.

[9] S. Guha, V. K. Paruchuri, M. Copel, V. Narayanan, Y. Y. Wang, P. E. Batson, N. A. Bojarczuk, B. Linder, and B. Doris, Appl. Phys. Lett. 90, 092902 2007.

[10] M. Suzuki, T. Yamaguchi, N. Fukushima, M. Koyama, J. Appl. Phys. 103 34118 2008.

[11] S. Guha, E. Cartier, M.A. Gribelyuk, N.A. Bojarczuk, M.C. Copel, Appl. Phys. Lett. 77, 2710, 2000./

[12] E. Miranda, J. Molina, Y. Kim, H. Iwai, Appl. Phys. Lett. 86 232104 2005.

[13] H. Yamada, T. Shimizu, A. Kurokawa, K. Ishii, E. Suzuki, J. Electrochem. Soc. 150 G429 2003.

[14] K. Kukli, M. Ritala, V. Pore, M. Leskelä, T. Sajavaara, R.I. Hedge, D.C. Gilmer, P.J. Tobin, A.C. Jones, H.C. Aspinall, Chem. Vap. Dep. 12, p.158, 2006.

[15] D. H. Triyoso, et al., J. Vac. Sci. Technol. B 25(3) p.845, 2007.

[16] R. I. Hegde, et al, IEDM p.39, 2005.

[17] D. Y. Cho, et al., Phys. Rev. B 82. 094104 2010.

978-1-61284-408-4/11 $26.00 © 2011 IEEE

Low-k Etching Using CF₃I, A Path To Overcome Current BEOL Integration Issues

Adam J. Gildea, Justin C. Long, Eric Eisenbraun
College of Nanoscale Science and Engineering
The University at Albany-SUNY
Albany, New-York, USA

Vincent Omarjee, Nathan Stafford, François Doniat,
Christian Dussarrat
Delaware Research and Technology Center
American Air Liquide
Newark, Delaware, USA
Vincent.omarjee@airliquide.com

CF₃I, a low greenhouse warming potential gas, has been used for low-k etching using an ICP reactor. Key parameters such as reactor pressure, bias power, ICP power and total gas flow rate were investigated to develop an optimized etch process. A comparison with standard fluorocarbons such as CF₄, C₄F₈ or CF₃H has been made to illustrate the performances of this low environmental impact chemistry.

Keywords: CF₃I, Low-k, Etching, Plasma, BEOL

I. Introduction

BEOL manufacturing is facing major integration issues. The introduction of low-k materials (typically porous-SiCOH based with k=2.5 and below) is bringing new processing challenges. Compared to dense films, the introduction of porous low-k dielectrics with porosity varying from 15 up to 30% has resulted in more significant damage occurring during the etching process. Such damage can result in undesired film and structure modifications such as: pattern collapse, impurity penetration, and surface roughening, causing difficulties in the post metallization steps, moisture uptake that increases the effective k-value, and other issues. The sum of all those defects degrades the interconnect reliability and performance. Given the increasingly stringent constraints in manufacturing, these issues need to be mitigated.

CF₃I is a non-flammable gas that has been previously studied for low-k etching [1]. From the literature, it was shown that CF_3^+ is the main reactive species generated in a plasma environment. Samukawa et al [2] explained that with the weak C-I bond, CF_3^+ is preferentially generated and in high concentration compared to C₄F₈ and CF₄ under high plasma power conditions. In a later study [3,4], this low F radical generation (and low UV photon generation) was found to be beneficial for minimizing line edge roughness (LER) and low-k damage. The potential manufacturing advantages of CF₃I are the low LER and minimal low-k damage, as well as a low Global Warming Potential (GWP). Compared to standard gases such as C₄F₈ with GWP₁₀₀=8700years, CF₃I rapidly decomposed in the atmosphere and has therefore a GWP₁₀₀~1.

In this work, CF₃I was optimized to etch low-k dielectrics using a Design of Experiments (DOE) approach. A comparison to standard gases commonly used in the semiconductor industry namely CF₄, C₄F₈ and CF₃H, is also performed.

II. CF₃I Etching Behavior

All the experiments were conducted using a 200mm Inductive Coupled Plasma (ICP) reactor (PlasmaTherm Versalok). The planar low-k films used were porous low-k films deposited by Plasma Enhanced Chemical Vapor Deposition (PECVD) with k~2.4. Patterned wafers were comprised of a non-porous low-k film, k~2.9. The silicon etch rate was determined using SOI wafers. CF₃I specification was 99.99%.

All the films were characterized using Rutherford backscattering spectroscopy (SUNY 5 MeV Dynamitron), Auger electron spectroscopy (PHI-680) and SEM (Hitachi 4800). Film thickness measurements and etch rate calculations are based on SEM data.

A. Plasma Analysis

The CF₃I plasma has been studied by various groups. Jiao et al. [5] demonstrated the ionization cross section of CF₃I over a wide range of ionization energy. In the low energy range (10-20eV) that is representative of standard plasma etching conditions, CF₃I is mostly dissociated forming CF_3^+ species, and to a much lower extent CF₂I+. From this, it can be deduced that CF₃I generates very low amount of free fluorine.

This was later confirmed by S. Samukawa et al. [6] where they used CF₃I in an Ultra-High Frequency plasma and observed a dominant CF_3^+ generation.

The low free fluorine generation was found beneficial by Soda et al. [4] in lowering the low-k damage compared to other gases such as CF₄ that will by nature generate free fluorine when exposed to typical etch plasma conditions. The preferential dissociation of the C-I bond over the C-F bond can also be explained by a much weaker bonding energy between carbon and iodine compared to C-F.

B. CF₃I etching rate optimization

A Taguchi-based design of experiment approach was applied to optimize the etching rate of CF₃I and identify the key parameters of influence. Bias, ICP power, flow rate and pressure were varied. By controlling the bias and ICP power, we found that a high etching rate was achievable. Figure 2 shows that a very high etching rate was possible using high power settings. The etching rate was determined by measuring the remaining low-k film thickness by cs-SEM after a known etching time.

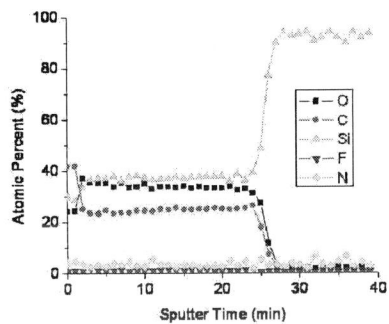

Figure 2. Low-k etching rate vs ICP power (bias=500W, pressure=2mTorr, flow rate=100sccm)

Fluorine incorporation is an important concern using fluorocarbons for low-k etching since it will affect the process integration, and likewise the use of CF_3I leads to concerns about possible iodine incorporation. In order to verify that no contamination by iodine or fluorine is seen, the etched films were analyzed, ex-situ, by AES and RBS. Figures 3 and 4 show that the amount of both fluorine (AES) and iodine (RBS) is below the detection limit of the analytical tools (~0.5 at.%).

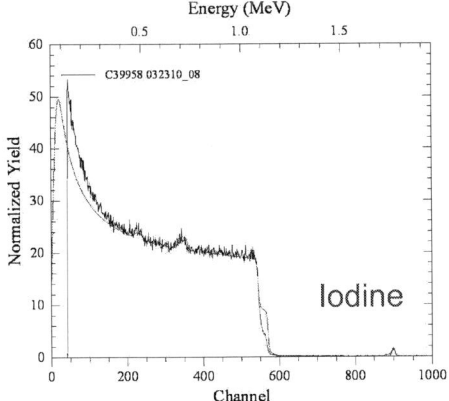

Figure 3. AES depth profile of an etched low-k film

Figure 4. RBS analysis of an etched low-k film

In all the experiments conducted using CF_3I, the detected amount of impurities is consistently very low. The low fluorine concentration can be explained by low amount of fluorine generated by CF_3I plasma; the low iodine, by the size of the iodine atom that makes it difficult to diffuse into the film.

C. CF_3I compared to CF_4, CF_3H and C_4F_8

Many fluorocarbons are being used to etch low-k dielectrics. Among the well known chemistries, CF_4, CF_3H and c-C_4F_8 are the one that were selected for comparison.

CF_3I etching rate was compared to other gases either as a pure gas or in mixture. Two different parameter sets were used for comparison. One "RIE mode" with ICP=0W – Bias = 400W and one "optimized mode" selected from previous etching rate optimization with ICP=700W – Bias = 500W. Those parameter sets and mixtures were used to measure the selectivity of the low-k etch over silicon and SiO_2.

Figure 5 shows the etching rate of CF_3I compared with the standard gases ("RIE mode"). Increasing the CF_3I ratio increased the etching rate when mixed with CF_3H or C_4F_8, however when mixed with CF_4 the etching rate decreased.

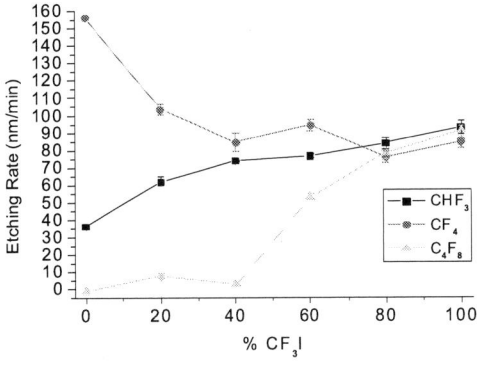

Figure 5. Low-k etching rate in RIE mode comparing CF_3I, CF4, C_4F_8, CF_3H and mixtures

As shown in Figure 6, the same behavior was observed in the optimized mode where the etching rates are considerably enhanced.

978-1-61284-408-4/11 $26.00 © 2011 IEEE

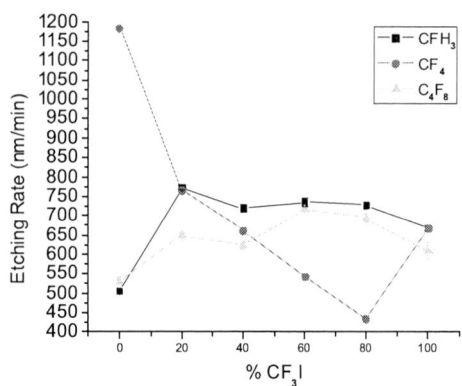

Figure 6. Low-k etching rate in optimized mode comparing CF_3I, CF_4, CF_3H and C_4F_8

The amount of fluorine was found to generally decrease while increasing the CF_3I flow rate in the mixtures. Only trace amounts of iodine were detected, ~0.3at.%.

D. Impact of CF_3I flow ratio on selectivity

The above described process was used to see the effect on selectivity of adding CF_3I to standard fluorocarbon etch gases. Increasing the amount of CF_3I in the $CF_3I:CF_4$ mixture was found to slightly increase the selectivity toward Si and SiO_2, up to ~2 independently of the power setting mode – at the exception of a 50:50% ratio in RIE mode that gave a surprisingly high selectivity for low-k / Si of ~10, as seen in Figure 7. This point is being reinvestigated.

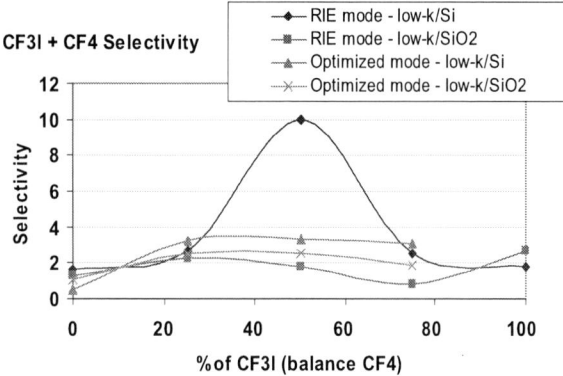

Figure 7. Low-k selectivity towards silicon and SiO_2 for CF_3I, CF_4 and their mixtures.

CF_3H is known to be selective to silicon [7]. In Figure 8, where CF_3I is added to CF_3H, a clear decrease of the selectivity low-k/Si was observed. When at least 50% of CF_3I was present in the mixture, the selectivity to silicon was stabilized at ~2. The selectivity to SiO_2 was constant at around 2.

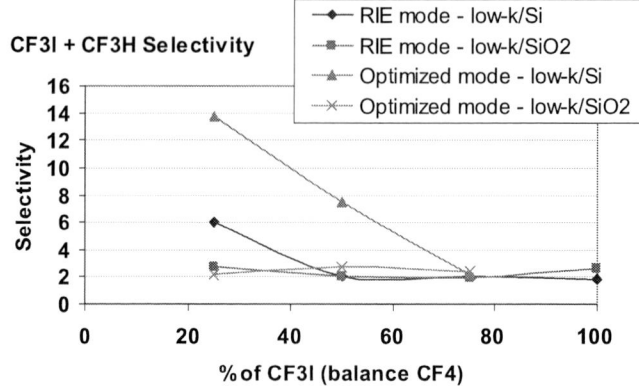

Figure 8. Low-k selectivity towards silicon and SiO_2 for CF_3I, CHF_3 and their mixtures

Mixing CF_3I and C_4F_8 resulted in a slight improvement in terms of selectivity, as seen in Figure 9. The best case is seen in the optimized process mode, where a low-k/Si etch selectivity of up to ~5 was achieved.

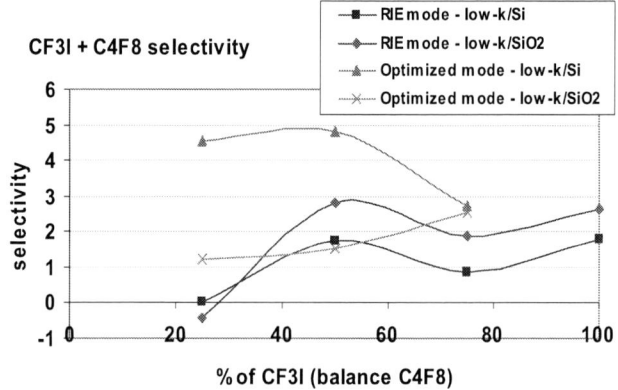

Figure 9. Low-k selectivity towards silicon and SiO_2 for CF_3I, C_4F_8 and their mixtures. 0% CF3I data is not presented, since at these conditions a polymer is deposited and no etching occurs. A thin polymer was also deposited using 25% of CF_3I in RIE mode explaining the negative selectivity value on SiO_2.

In order to quantify the damage generated on blanket films, the AES depth profiles were used. Knowing the sputtering rate of non plasma exposed low-k films and using the carbon depletion on top of the plasma-exposed (i.e damaged) low-k films, the thickness of the modified region was extrapolated. As depicted in Figure 10, which plots the damaged layer thickness as a function of the low-k etching rate (independently of any process parameter, the only constant being that approximately the same amount of low-k is being etched), it is very difficult to have a clear trend. For a given gas or mixture, it appears that high etching rate leads to lower surface damage. This can be attributed to the fact that usually the damage is proportional to the etching time therefore having a high etching rate can favor a low surface damage (8). Surprisingly, CF_3I+CF_3H mixture is showing very low surface damage almost independent of the etching rate.

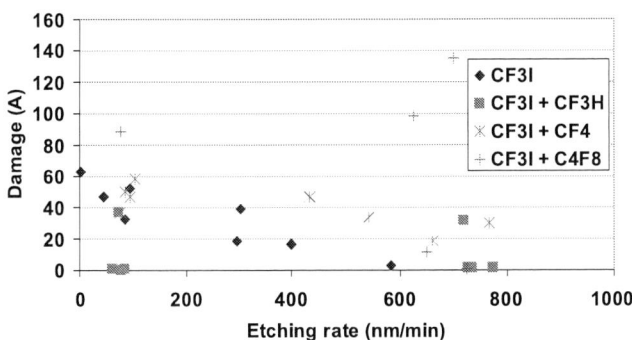

Figure 10. Etching rate correlation with low-k surface damage thickness

III. PATTERNED WAFER ETCH TESTING

A. CF₃I vs CF₄

A comparison of etching profile has been made between films etched using CF_3I and CF_4. Structures as presented in Figure 11 (non-porous low-k, k~2.9) having 120 and 140nm lines were etched under various conditions.

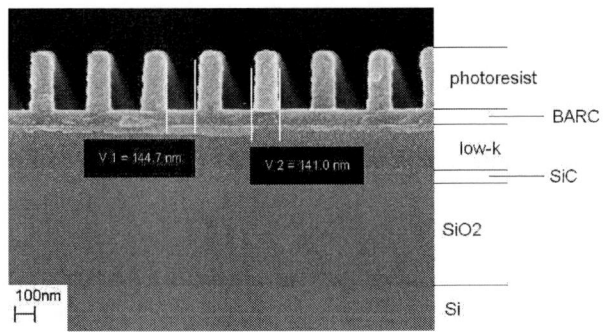

Figure 11. Structure used for patterning tests

In the following examples, only the 120nm lines will be presented for simplicity. However, the results are most of the time very similar between the 2 critical dimensions available. A DOE has been used to identify the optimized process regime, taking great care on optimizing the profile (maintaining critical dimension (CD) and having a tapering angle as close as possible to 90°).

As depicted in Figure 12, a very straight etch profile was achieved using CF_3I. In this experiment, the low-k etching rate was ~130nm/min and the etch selectivity to the photoresist ~3.1. In optimized conditions, it is also observed that the photoresist keeps a well defined structure with sharp edges.

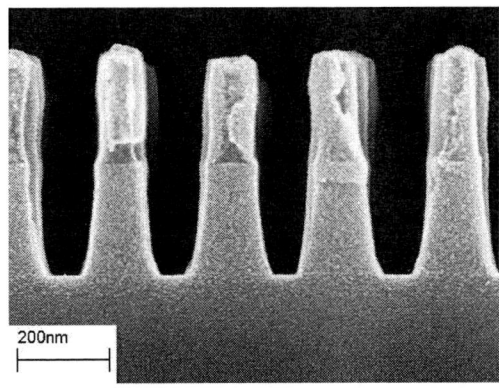

Figure 12. Completely etched low-k. (P=5mTorr, ICP=200W, CF_3I flow rate=45sccm, Bias=200W, time:120s).

By comparison, CF_4 usually gave higher etching rates. However, the profile of the final structure is not as vertical. As seen Figure 13, where only half of the total low-k film was etched, some microtrenching is appearing at the corners of the structure and the photoresist seems to be "rounded" on top. The low-k etching rate was very high, 677nm/min, but the selectivity to the photoresist was lower, ~1,36.

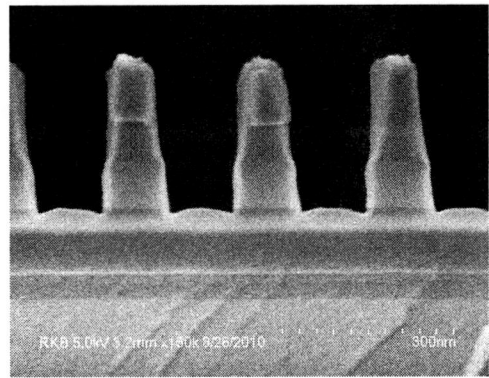

Figure 13. Structure etched with CF_4 plasma (P=5mTorr, ICP=700W, CF_4 flow rate=100sccm, Bias=500W).

B. CF₃I + CF₄ mixtures

Gas mixtures can be an effective way to improve etch performance and in some cases decrease the cost of ownership of a process. In this study, CF_3I was mixed at 50% with CF_4. The process has been also optimized using a DOE approach. While the etching profile obtained in optimized conditions was found to be similar to what was obtained using 100% CF_3I, the process variations were more pronounced using the mixture (50% CF_3I - 50% CF_4). The CDs were more difficult to maintain and this was emphasized at higher ICP power. On top of that, the micro-trenching that seems to be related to CF_4 is observed under some conditions.

Figure 14 shows an example of what etch performance is possible when mixing CF_3I+CF_4 at optimized conditions. A well defined structure was achievable at a high etch rate,

978-1-61284-408-4/11 $26.00 © 2011 IEEE

~160nm/min, but with a low selectivity to the photoresist (1.67).

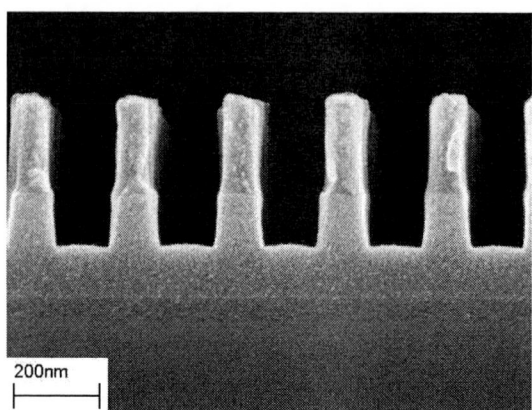

Figure 14. Structure etched with CF_3I+CF_4 mixture (pressure=5mTorr, ICP=400W, bias=250W, CF_3I=50sccm, CF_4=50sccm, t=90s)

IV. CONCLUSIONS

CF_3I has been studied for low-k etching and has demonstrated some promising features. Implementation wise, no process variation or throughput issues are anticipated. More importantly, this etch gas seems to provide a potential solution to some BEOL related etching issues. On blanket wafers it was confirmed that low damage is usually observed and that contamination by fluorine as well as by iodine is minimal and should not affect the integration process. Etching of fine patterned structures was successfully achieved and superior performance was seen using CF_3I compared to CF_4. Mixing those two gases did not bring process improvement. A next step will be to mix CF_3I with a polymerizing gas such as c-C_4F_8 that should help fine tuning the process allowing higher selectivity.

REFERENCES

[1] E. Soda et al., "Low-damage low-k etching with an environmentally friendly CF3I plasma," J. Vac. Sci. Technol. A 26(4), Jul/Aug 2008, p875-880.

[2] S. Samukawa, T. Mukai, K.Tsuda, "New radical control method for high-performance dielectric etching with nonperfluorocompound gas chemistries in ultrahigh-frequency plasma," J. Vac. Sci. Technolo. A 17(5), Sept/Oct 1999, p2551-2556.

[3] S. Samukawa, Y. Ichihashi, H. Ohtake, E. Soda and S.Saito, "Environmentally harmonized CF3I plasma for low-damage and highly selective low-k etching," Journal of Applied Physics 103, 053310, 2008.

[4] E. Soda, S. Kondo, S. Saito, K. Koyama, B. Jinnai, S. Samukawa, "Mechanism of reducing line edge roughness in ArF photoresist by using CF3I'", J. Vac. Sci. Technolo., B 27(5), Sept/Oct 2009, p2117-2123.

[5] C.Q. Jiao, B. Ganguly, C.A. DeJoseph, A. Garscadden, "Comparison of electron impact ionization and ion cheistries of CF3Br and CF3I'" International Jounral of Mass Spectrometry, 208, 2001, p127-133.

[6] S. Samukawa, T. Mukai, "high-performance silicon dioxide etching for less than 0.1um-high-aspect-ratio contact holes", J. Vac. Sci. Technol. B 18(1) Jan/Feb 2000, p166-171.

[7] M. Schaepkens et al., "Effect of capacitive coupling on inductively coupled fluorocarbon plasma processing," J. Vac. Sci. Techno., A 17(6), Nov/Dec 1999, p3272-3.

[8] Y.Iba et al., "Effects of Etch Rate on Plasma-Induced Damage to porous Low-k Films," Japanese Journal of Applied Physics, vol.47, 8, 2008, p6923-6930.

Advanced Elemental Analysis Methods for sub 30nm Defects in a Defect Review SEM

Dror Shemesh, Adi Boehm, Ofir Greenberg, Kfir Dotan
Process Diagnostics and control
Applied Materials
Rehovot, Israel

Abstract — as development phases became shorter, fast defect characterization is highly important. In addition, new chemicals involved in sub 28nm semiconductor processes introduce new types of defects. The variety of defect types complicates defect root cause analysis, so that image based defect analysis is limited, and defect material information becomes more important. Thus elemental analysis of defects in a defect review SEM gains importance in sub 28nm processes.

This work presents a new type of Energy Dispersive X ray Spectrometer (EDX), integrated together with a unique e-beam column and high collection solid angle. It allows x-ray count rates that are more than an order of magnitude higher compared to traditional detectors. The high count rate speeds up the analysis period to less than 2 seconds for sub 30nm defects. Short analysis time is critical for such small sized defects as they can disappear due to charging or evaporation. In addition, the new detector-column configuration becomes sensitive to very thin films (down to 1nm), something that was not reasonable for the traditional SiLi x-ray spectrometers.

Traditional EDX requires >5keV beam energies and >500pA beam current. New layers that are introduced to < 40nm processes, such as Ultra Low K materials, are sensitive to high electron doses. The new, fast EDX allows defect analysis on sensitive layers without damaging the analysis area

Keywords; EDX; silicon drift detector; elemental analysis; thin films; e-beam induced damage

I. INTRODUCTION

Defect review and defect characterization in semiconductor manufacturing environment has some unique requirements [1]. Beyond fundamental analysis requirements such as high sensitivity and high energy resolution, elemental analysis must be easy to operate and easy to understand. Analysis time should be fast enough in order to match the time that is budgeted for analysis [2]. Typical throughput of a defect review system is getting faster, and today is equal to more than 3000 defects per hour, or ~1 second per defect. EDX time should be targeted to be in the same range; otherwise its usefulness and production-worthiness is limited.

EDX has been an integral part of a defect review SEM for more than 10 years. Existing EDX technology, integrated in defect review SEMs allows taking spectra within 10 to 30 seconds and sometimes more. Main limiting factors are a) collection solid angle, b) EDX measurement speed and 3) column performance in high beam current. However, analysis speed of 10 to 30 seconds does not allow an in-line operation of EDX because the throughput loss is unacceptable. Defect review tool designers had to find ways to increase the collection solid angle, decrease EDX measurement time and increase analysis beam current.

Unique design of an e-beam column for better integration of EDX detector significantly improved the collection solid angle. In addition, the spot size at high beam current was reduced to a level of few nano-meters, as shows in Figure 1. Both effects caused the flux of x-rays hitting the EDX detector to be so high, that EDX measurement time became a bottleneck in the measurement cycle.

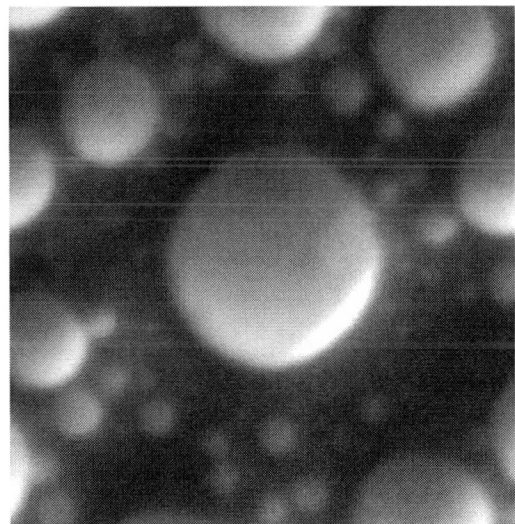

Figure 1: High resolution image recorded by the SEMVision defect review SEM, using beam energy of 1keV and beam current of 1nA. The image at 1 micron Field Of View, demonstrates a resolution better than 3nm.

II. FASTER AND MORE SENSITIVE EDX

Figure 2 describes the process of photon loss in EDX when the x-ray photon flux is high compared with the device measurement time. When the rate of x-ray photons hitting the detector is low enough, the EDX measures photon energy and it is ready for the next photon to arrive. However, when the rate of x-rays becomes higher, the probability of 2 photons arriving at the detector face during 1 measurement window increases. In case the detector detects 2 photons arriving at the detector face, it cannot accurately measures the photon energy and therefore neglects both photons. The time when the detector blocks 2 photons is usually called "dead time". Dead time increases with increasing beam current and collection solid angle.

If measurement time is shorter, as shown in the bottom part of Figure 2, the probability that 2 photons will arrive at the same measurement interval is smaller, resulting in a higher detection rate. With fast EDX detector, one can increase beam current and collection solid angle of x-rays, thus creating high signal to noise ratio (SNR) EDX spectra in a shorter time; that means – EDX which is sensitive to smaller particles.

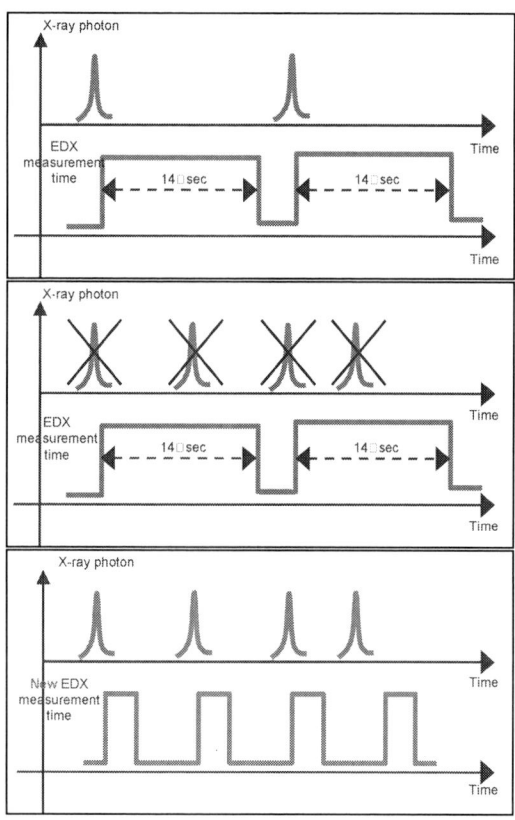

Figure 2: The influence of x ray flux on the detected signal by EDX detector. When the measurement time is long, EDX rejects incoming photons that arrive during the processing period. Short measurement time allows higher x-ray flux to be detected.

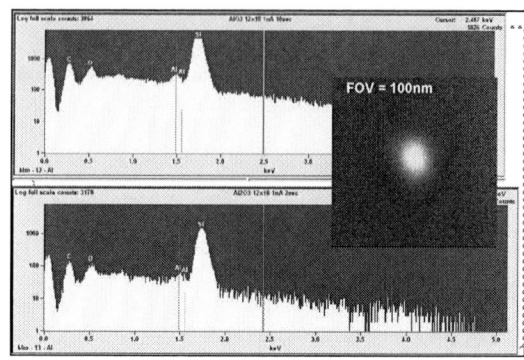

Figure 3: EDX spectra of sub 20nm particle taken using SEMVision defect review tool. Upper spectrum was taken in 10 seconds while lower spectrum was taken in 2 seconds. Both show clear detection of Al in the Aluminum Oxide particle.

A. Elemental analysis of sub-20nm particles

When analyzing particles or films thinner than 50nm, it is recommended to run the SEM with beam energies of 3kV and even lower. The interaction volume in Silicon is approximately 50nm in 3keV beam, and 800nm (!) in 10kV. Therefore, using low beam energies increases the contribution of the particle or the thin film to the resulted x-ray spectrum. The spectrum shown in Figure 3 demonstrates detection of Aluminum in an Aluminum-Oxide particle that is smaller than 20nm. Silicon Drift Detector (SDD) that is integrated so that the collection solid angle is maximized, together with a SEM with high performance in beam energies smaller than 3keV, enables the elemental analysis of this particle.

B. Elemental analysis of sub-10nm films

Another example demonstrating how powerful the SDD can be when integrated correctly is its sensitivity to analyze composition of thin films. Films with thicknesses < 10nm were previously analyzed only with Auger spectrometer, however now the SDD can be used as well [3], and produce elemental information even faster. Figure 4 describes a serious of measurements that were taken on Titanium Nitride (TiN) films deposited on a Silicon wafer. TiN film thicknesses were varied from 2.4 to 50nm. EDX signal were recorded with an e-beam energy of 3kV and a beam current of 1nA. It is shown that TiN peak is getting smaller for thinner thicknesses; however the signature is very clear even for the thinnest layers of 2.4nm. The TEM image in Figure 4 verifies that the thickness of the film is 2.4nm.

We did similar work on Hafnium Oxide that recently became useful in the high-k metal gate transistors. The thickness of HfOx gate is approximately 3nm, therefore sub 3nm thickness sensitivity is required in order to detect residue defects. Unlike the previous example of TiN detection, Hf detection is more challenging because the main Hf peak is 1644V, close to the big and significant peak of

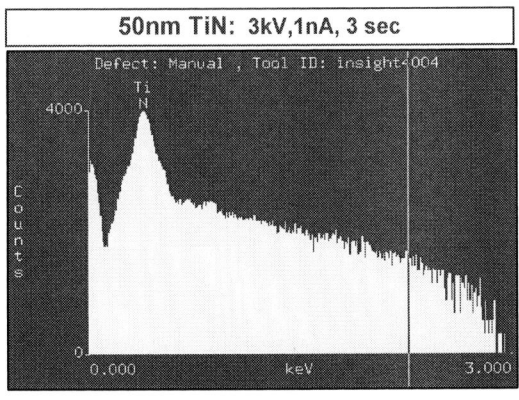

50nm TiN: 3kV,1nA, 3 sec

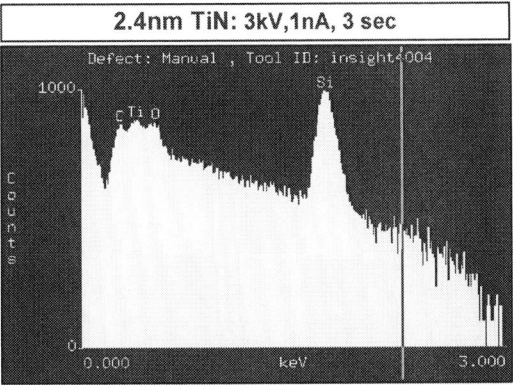

2.4nm TiN: 3kV,1nA, 3 sec

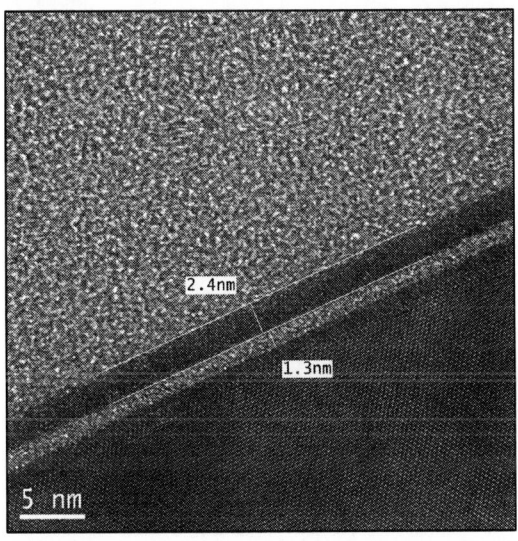

Figure 4: EDX spectra taken on 50nm and 2.4nm TiN films. 3kV e-beam does not penetrate through the 50nm film and Si peak does not exist. TiN is clearly identified on the 2.4nm film even in short acquisition time of 3 seconds. The image at the bottom is a TEM images of the TiN 2.4nm film.

Silicon at 1740V. In this case the challenge is to detect enough x-ray photons so that the shift in the tail of the Si peak caused by the Hf, is detectable.

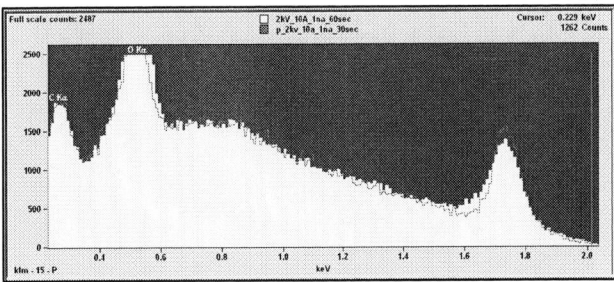

Figure 5: EDX spectrum of 2nm Hafnium Oxide film deposited on Silicon Oxide layer. HfOx spectrum is shown in yellow, while Si reference is in red. Hf is detected using beam energy of 2kV and beam current of 1nA.

Few wafers of Silicon Oxide covered by different thicknesses of HfOx were produced in Applied Materials process development lab. In order to quantify the sensitivity of the SEMVision SDD, we recorded EDX spectra of the wafers under test and compared to a spectrum of a bare Silicon wafer. An example of such spectrum is shown in figure 5, where the difference between Si peak and S1+Hf peaks is visible for a 2nm HfOx layer.

Signal to noise ratio (SNR) calculations were made in order to quantify the ability of spectrum analysis algorithm to separate Hf from Si. To do that, we composed new spectra by subtracting HfOx spectrum from the reference Si spectrum. The presence of Hf is more visible in the subtracted spectra, as shown in Figure 6, that presents spectra of 2nm HfOx layer took in 5, 10, 20 and 30 seconds.

978-1-61284-408-4/11 $26.00 © 2011 IEEE

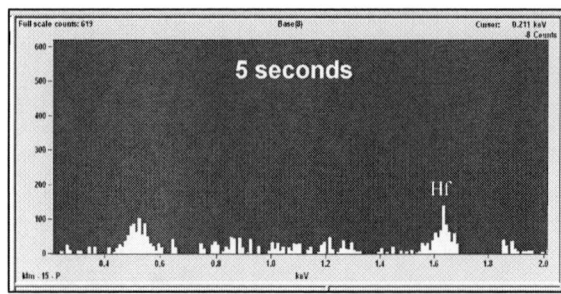

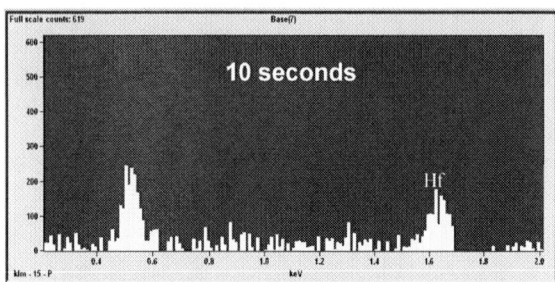

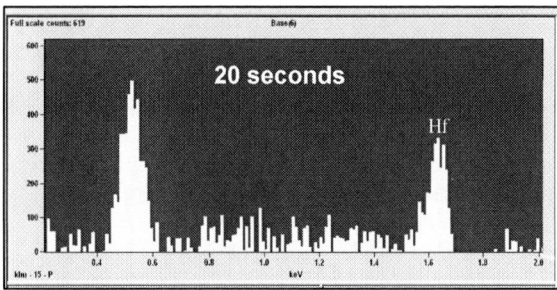

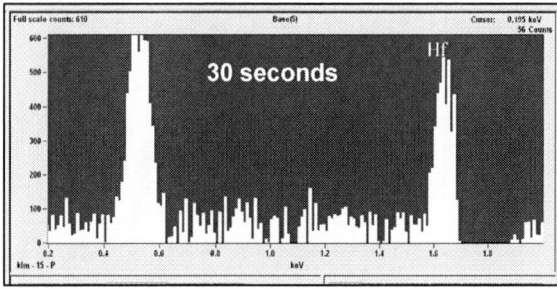

Figure 6: Subtracted spectra of 2nm HfOx film with different integration time of 5, 10, 20 and 30 seconds. Hf contribution increases with time

The signal to noise ratio (SNR) of a certain peak in the spectrum, is defined as

$$SNR = (P-B)/STD(B)$$

Where P is the peak value, B is the bottom value, and STD(B) is the standard deviation of the counts from the bottom of the peak. Usually SNR threshold of 5 or above is required to detect a certain peak in a spectrum. SNR calculations of Hf peaks for the above spectra of 2nm HfOx layer show that Hf detection is marginal for 5 seconds acquisition, while it is clearly detectable in 10 seconds or above, as shown in Figure 7.

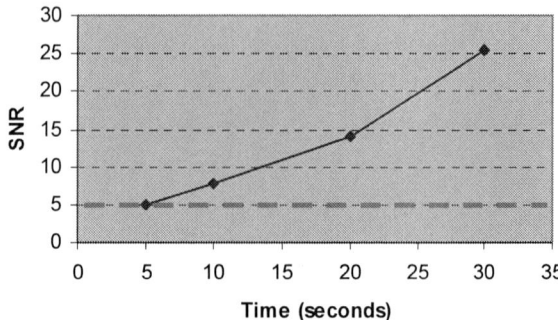

Figure 7: SNR calculations of Hf peaks shown in figure 6.

III. SUMMARY

The ability to analyze sub 30nm defects in less than 5 seconds, and the sensitivity to films as thinner as 2.5nm, brings powerful elemental analysis competency that can serve semiconductor manufacturing environment. The fast speed analysis allows elemental investigation of more defects, pointing faster to root cause analysis and short response time in finding defect source.

We demonstrated that the new EDX detector integrated in the SEMVision defect review SEM tool, together with new e-beam column that was designed for highest resolution at high beam currents, allow fast analysis 30nm defects and even smaller. In addition, films as thin as 2.5nm can be analyzed in less than 5 seconds. The fast speed, together with high sensitivity analysis, allows fast and high quality in-line elemental analysis of sub 30nm processes, contributing to faster and more accurate root cause analysis, and therefore higher yield.

REFERENCES

[1] Noam Dotan, "Method for enhancing topography and material contrast in automatic SEM review", Proc. SPIE 3677, 491, 1999

[2] Levin, Lior; Eilon, Michal; Porat, Ronnie; van der Sijs, Arjan; Stegen, Raf; van Brederode, Erik; , "In-line material analysis of 50nm defects by integration of Energy (EDX) and Wavelength (WDX) Dispersive X-ray analysis," *Semiconductor Manufacturing (ISSM), 2008 International Symposium on* , vol., no., pp.251-254, 27-29 Oct. 2008

[3] Fiorini, C.; Kemmer, J.; Lechner, P.; Kromer, K.; Rohde, M.; Schulein, T. "A new detection system for x-ray microanalysis based on a silicon drift detector with Peltier cooling," *Review of Scientific Instruments* , vol.68, no.6, pp.2461-2465, Jun 1997

Advanced excursion control and diagnostics for CMP process monitoring

Andrew Stamper
Microelectronics Division
IBM Corporation
Hopewell Junction, NY 12533
Stamper@us.ibm.com

Gangadharan Sivaraman, Ravi Sankar
KLA-Tencor
Gangadharan.Sivaraman@kla-tencor.com,
Ravi.Sankar@kla-tencor.com

Abstract—This CMP processes are generally well known and established critical steps in semiconductor manufacturing, but must be very closely monitored and controlled to maintain process uniformity and minimize process induced defects. CMP processes are generally monitored using a combination of blanket test wafers, short loop patterned wafers and product wafers, along with a variety of in-situ controls. This paper talks about the techniques demonstrated for defect excursion monitoring in CMP module using short loop test wafers and an advanced dark Field inspection tool. These techniques enable effective control and monitoring of the CMP tools used for manufacturing advanced semiconductor logic devices. We talk about three approaches that helped accomplish this objective: (i) Pattern wafer inspection at M1 Cu CMP level using Puma 9550 DF inspector (ii) use of In-line Defect Organizer (iDO) to effectively and automatically extract and classify scratches – the key defect of interest to the CMP process engineer; and (iii) use of quick recipe templates to diagnose and make required recipe changes for 32 nm and 22 nm devices using "Auto Derivative Recipe" feature on Puma 9550 to maximize engineering and tool efficiency

Keywords-iDO(inline defect organizer), auto derivative,

I. INTRODUCTION

This paper will discuss excursion control methodologies adopted at IBM using Puma 9550 DF wafer inspection system with Inline Defect Organizer for automatically classifying and monitoring scratches in the CMP module and Use of "Auto Derivative Recipe" feature to create multiple recipes to automatically monitor and diagnose issues due to small process changes on 32 nm and 22 nm product wafers. Having a functional, automated classifier will improve the Mean Time to Detect (MTTD) on the process tools and minimize product at risk to a potential tool or process issue. This paper will describe the methodology that enables fast and accurate defect detection and classification performance on Cu CMP short loop test wafers.

II. INLINE DEFECT ORGANIZER

iDO provides a way by which the user can use the classified defect data from inspection system and try to separate different defect types. We were able to use Puma 9550 + iDO to successfully bin scratches (polishing and handling

scratches) from other real defects (CMP residual slurry, Foreign Material, embedded contamination) and come up with a novel way to monitor scratches in CMP module and shut off CMP process tools based on results from iDO classification.

iDO helps to bin defects

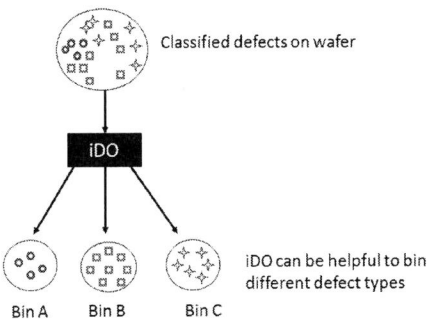

Figure 1: Concept of iDO for binning defects

Performance of iDO is measured by accuracy on purity of a bin. The accuracy and purity of a bin is defined based on Confusion matrix.

iDO Confusion Matrix

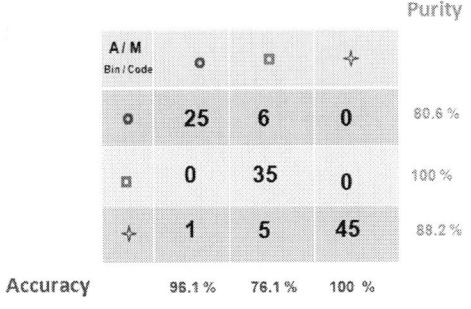

Figure 2: Confusion Matrix

Accuracy is the Percent of defects agreed by ADC and manual classification relative to Number of Manually Classified defects in that Class and purity is the percent of defects agreed by ADC and manual classification relative to number of Auto classified defects.

To represent them in confusion matrix

Accuracy = diagonal element /column total

Purity = diagonal element/row total

In a simple manner Accuracy is a measure of efficiency of the classifier in identifying a particular class (for ex: class x), while purity is the measure of how many the computer called bin x are really class x.

The goal for using iDO effectively in production is to have high purity and accuracy when detecting defects to shutdown tools without going to SEM review. In the current CMP scratch monitoring strategy 80% purity and 95% accuracy was demonstrated and this was just right for implementing IBM's automated tool monitoring strategy.

Bin/Code	DE	EC	FM	HM	HS	MP	NV	OP	OT	PD	PL	PR	PS	PU	UP	Purity
FM	7	8	23	450	72	44	17		10	7	6	534	31	4	4	2%
HS	20	2		100	1524	4	43	1	4	35	7	9	155		3	80%
PS	1			10	4	1	12	6			38	5				6%
Accuracy			100%		95%							3%				

Figure 3: Confusion matrix showing good accuracy and purity for HS (Heavy Scratch) bin

III. AUTOMATED TOOL MONITORING

The excursion monitoring is utilized on the CMP monitors and product in IBM's 323 Fab. This allows the tool to classify in real-time the defects from the wafer maps and directly uploads to the Yield management systems. An automated process sends a signal to the process tools to shut down. The SEM tool is not required or production to classify the defects which leads to a greater Mean Time to Detect (MTTD). This process saves hours, costs and less processing time for a bad tool.

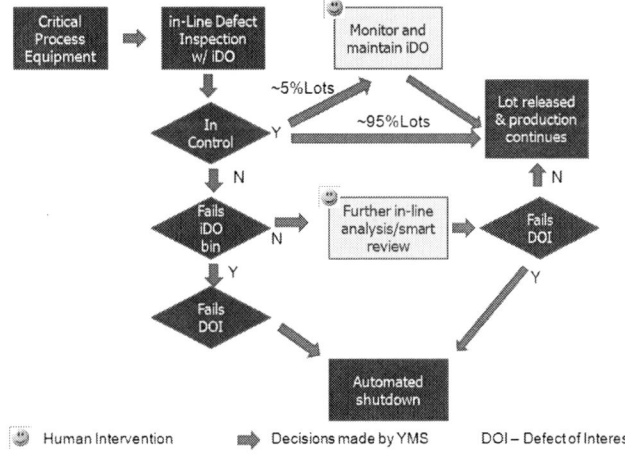

Figure 4: IBM methodology for automated tool monitoring

Automated tool monitoring strategy helped to flag bad CMP tools quickly

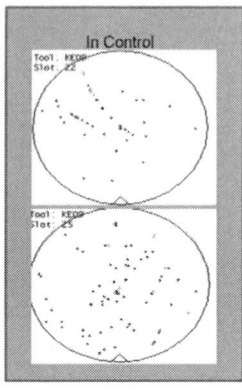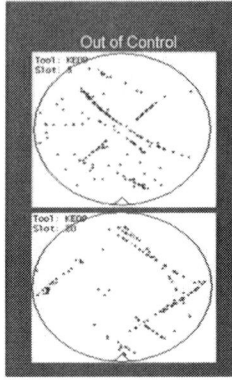

Figure 5: Wafer map indicated out of and in control lots flagged by the IBM automation system.

IV. AUTO DERIVATIVE RECIPE

Auto derivative recipe creation is a method by which wafer inspector creates recipes automatically with minimum user input done prior to the wafer arrival at the tool.

The benefits of auto derivative recipe are tremendous for several reasons.

1) Capability of 24x7 recipe build

2) Eliminates product going on hold for no recipe.

3) Improves cycle time and hence addresses some queue time issues on some layers.

4) Frees up tool user time for other responsibilities.

5) Fewer tool down time in manufacturing for recipe build.

6) Recipe parameters are set up on an offline station and by one person so accuracy has improved for recipe writing.

7) Provides the capability to create multiple recipes for different flavors of layers during process development.

V. PROCEDURE TO BUILD AUTO DERIVATIVE

Auto derivative recipe consists of two parts, namely on tool work and auto recipe run.

On tool Work comprises of creating a base recipe or a master recipe for every new device. Best known modes study need to be done for multiple layers to understand the sensitivity parameters and optics modes for inspection. With the knowledge of inspection modes, multiple recipes could be created offline without wafer on the tool.

During the auto recipe run the tool will relearn alignment sites and light level for inspection optics and perform an inspection scan and send the data. In the event of higher

978-1-61284-408-4/11 $26.00 © 2011 IEEE

nuisance rate on the recipe, offline tweak on the recipe is possible without having wafer on the tool.

Auto Derivative recipe

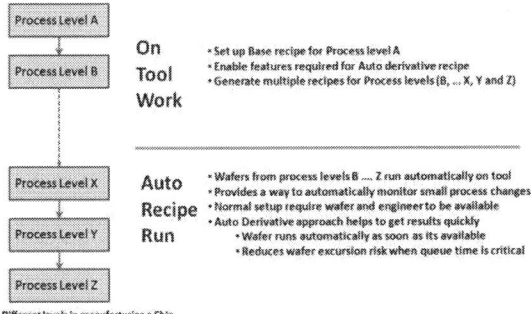

Figure 6: Concept of auto derivative recipe

As illustrated in Fig 6 Auto Derivative Recipe feature provides a way for the user to setup recipes for different process levels (Process level X, Y , Z) without the actual wafers using one base recipe (Process level A) and have the wafers from levels X, Y and Z to come on the inspection tool get trained automatically and run automatically without any user intervention.

At IBM We have demonstrated that this concept works well on multiple 32 nm wafers as a proof of concept and planning to use this on multiple 22 nm and 32 nm levels and provide a way to automatically quantify any small process changes in early development.

Figure 6 shows that when auto derivative recipe was scanned across three different lots, inspection tool was able to capture all the expected DOI population with non visual defects lower than 12%.

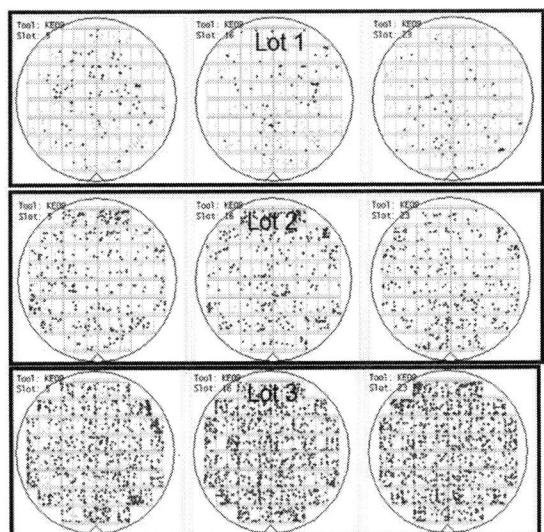

Figure 7: Auto Derivative recipe flagged DOI detection across three different lot flavors

VI. CONCLUSION

Implementation of real time iDO classification results in less SEM time and greater accuracy in classifying defects than production personal. As a result accurate iDO faster Feedback from inspection lots to shutdown tools results in direct cost savings to the factory.

Auto derivative recipe creates quicker recipes with pre existing information on similar process levels. With minimum to no tweaks on auto derivative recipes multiple recipes could be created for different process flavors which help to get faster MTTD in developing 32 and 22nm process technologies.

VII. ACKNOWLEDGEMENT:

IBM would like to recognize Robert Teagle, retired IBM, for his contribution to this work

Advanced Floating Gate CD Uniformity Control in the 75nm Node NOR Flash Memory

Sheng-Yuan Chang, Yu-Chung Chen, An Chyi Wei, Hong-Ji Lee, Nan-Tzu Lian, Tahone Yang, Kuang-Chao Chen
and Chih-Yuan Lu

Macronix International Co., Ltd., Technology Development Center,
Advanced Module Process Development Div.
No.16, Li-Hsin Road, Science-Based Industrial Park, Hsinchu 300, Taiwan ROC.

Abstract—**This paper describes the advanced control technology of critical dimension uniformity (CDU) by flash gate stack etch process. We have investigated the effective way of utilizing Tri-layer approach, which not only reduces the influence of topology step-height but also improves the range of ECD within die from 17.6nm to 4.9nm. Moreover, the influence of Etcher design on ECD variation becomes larger as the cell transistor size becomes smaller. The etch chamber effect is minimized by developing $CF_4/CHF_3/N_2$ plasma at 15mTorr pressure that provides better ECD uniformity within wafer.**

Keywords- Critical dimension uniformity (CDU); flash gate stack etch ; Tri-layer approach.

I. INTRODUCTION

The NOR Flash market has seen explosive growth in recent years, especially for code storage memory of portable device, i.e., a mobile phone and game consoles. High density NOR flash memory has posed increasing demand for faster read/write speed, which faced with more challenges as cell scaling down. [1, 2] In flash memory process technology, the flash gate stack etch is one of the most challenging process steps. Especially, the width uniformity of cell word line (WL) plays the most important role in programming and erasing performances.

In our 75nm NOR Flash memory product, we suffered from worse WL uniformity within die and within wafer, which results in worse programming and reliability performances. As a cell transistor area is scaled down to this generation, the influences of topology effect and Etcher design on WL uniformity becomes larger. [3] In this work, utilizing the nature of a novel material and optimizing etch plasma chemistry could improve critical dimension uniformity (CDU) during flash gate stack etch process.

II. EXPERIMENTAL

The film stacks of two different approaches studied in this work were shown in Fig. 1a and 2a. In an ashable hardmask approach, an amorphous carbon hardmask (a:C) deposited by PECVD is on the top of the floating gate structure, and is followed by anti-reflective coatings ARC (DARC and BARC). [4] The other film stack is that the tri-layer approach, which consists of a two-layer stack of organic underlayer layer (ODL) and Si-rich ARC (SHB) coated by spin method. [5] These two film stacks were pattern by conventional ArF dry lithography ASML XT1250B scanner, and then an etch consisting of various steps with chemical reactive ion etch (RIE) plasma in a Lam2300 STAR-t Poly chamber. The detail schematic flows of patterning 75nm cell array are as below (Fig. 1 and 2): The ARC or SHB etch is used to transfer the resist mask into ARC or SHB layer by a fluorine-based chemistry. Following organic layer (C-rich) is used for transferring ARC or SHB line pattern to underlying floating gate stack. The final poly etch uses a conventional HBr/Cl2/O2 poly-etch chemistry and fluorine-based chemistry.

The cross-sectional physical profiles were measured by scanning electron microscopy (SEM). The WAT test of cell word line was verified through the evaluation of threshold voltage distribution around boundary.

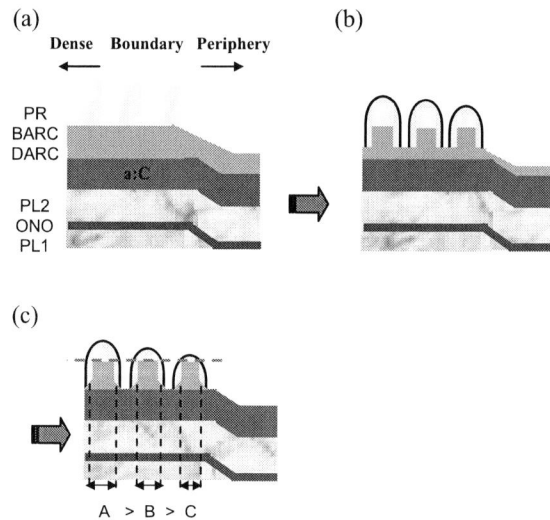

Figure 1. Possible mechanism of boundary effect during etching process in an ashable hardmask approach.

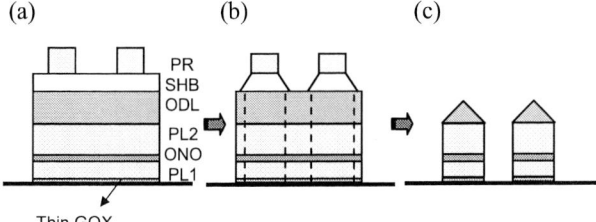

Figure 2. Process flow for floating gate formation with novel tri-layer approach: (a) Before Etching; (b) After PL2 SHB Etch; (c) After PL2 Etch without Dry/Wet strip.

III. RESULTS AND DISCUSSION

Due to the adoption of 193-nm ArF resist (PR) and cobalt silicide process [6], an ashable hardmask approach was evaluated to replace the traditional oxide hardmask approach for floating gate stack etch process. After full-run etching, the range of post-etch CDs (ECD) within die is up to 17.6nm, as shown in Table I. and Fig. 3. The cross-sectional SEM image indicated that the prior process-induced worse topology step-height around boundary area could not be flattened by a:C layer, where caused PR thickness difference between dense and boundary area. It is called "boundary effect", as described in Fig. 1 and 4(a). To improve ECD uniformity within die, a tri-layer approach that consists of a two-layer stack coated by spin method was studied instead of the ashable hardmask approach. As shown in Fig. 4(b), Table II and Fig. 5, the spin-coated ODL layer could provide excellent planarization to reduce the influence of topology step-height; after gate etching further improve the range of ECD within die from 17.6nm to 4.9nm.

TABLE I. SUMMARY OF IN-LINE ECD VARIATION FROM BOUNDARY AREA TO DENSE AREA WITHIN DIE IN AN ASHABLE HARDMASK APPROACH

Sector	L-Top	Middle	L-Bottom	Average
Max	Target+2.8	Target+0.2	Target+4.7	Target+2.6
Min	Target-10.3	Target-9.0	Target-12.9	Target-10.7
Variation	13.1	9.2	17.6	13.3

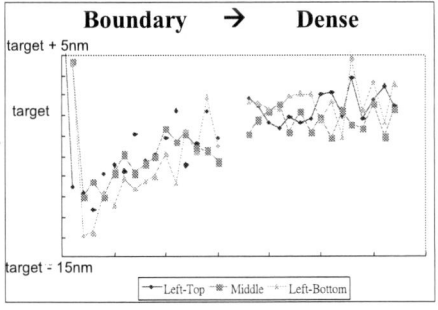

Figure 3. In-line ECD trend from boundary area to dense area within die in an ashable hardmask approach.

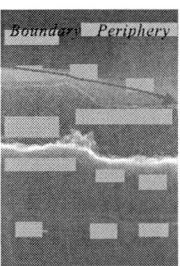 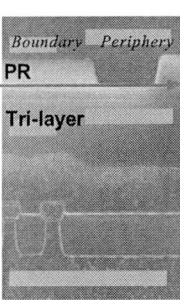

Figure 4. Cross-sectional SEM images of (a) an ashable hardmask approach; (b) a tri-layer approach.

TABLE II. SUMMARY OF IN-LINE ECD VARIATION FROM BOUNDARY AREA TO DENSE AREA WITHIN DIE IN A TRI-LAYER APPROACH

Sector	L-Top	Middle	L-Bottom	Average
Max	Target+0.5	Target+0.7	Target+1.0	Target+0.7
Min	Target-3.9	Target-3.9	Target-3.9	Target-3.9
Variation	4.4	4.6	4.9	4.6

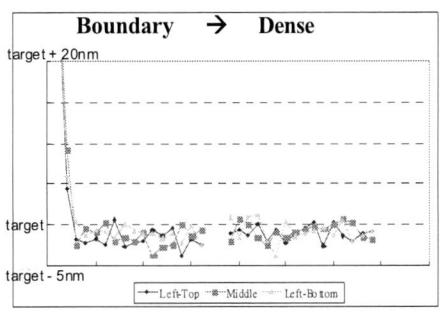

Figure 5. In-line ECD trend from boundary area to dense area within die in a tri-layer approach.

After resolving boundary effect within die, in-line ECD full map collection indicates another CDU issue within wafer, i.e., smaller ECD around the most outer edge of the wafer as shown in Fig. 6. Due to the limitations of lithography window, the polymer-rich CF_4/CH_2F_2 plasmas are used in SHB opening to gain the desired SHB ECD for approaching the final gate ECD target. However, the CF_4/CH_2F_2 plasmas resulting in smaller ECD around the most outer edge of the wafer is found strongly dependent on the specific chamber wall environments of the etcher (Lam2300 Star-t). Fig. 7 indicated that the most outer edge of the wafer is not located on ceramic electrostatic chuck, which results in poor thermal uniformity. Moreover, the material of chamber slit door different from chamber wall and asymmetric holder robot design also resulted in non-uniform polymer deposition during etch process. In order to reduce the chamber effect during plasma etching and keep the CD enlargement by using the heavy polymer chemistry in SHB etching, we developed the $CF_4/CHF_3/N_2$ plasmas at

978-1-61284-408-4/11 $26.00 © 2011 IEEE

15mTorr pressure instead of the CF_4/CH_2F_2 plasmas. The $CF_4/CHF_3/N_2$ plasmas could improve in-line (off-line) ECD variation between the wafer center and the outer edge from 3.8 ~ 7.3nm (14nm) to 0.7~ 2.1nm (7nm), as described in Fig. 6 and 8. WAT results also indicated that special fail die around wafer edge could be significantly improved by new etch chemistries, as shown in Fig. 9.

(a)

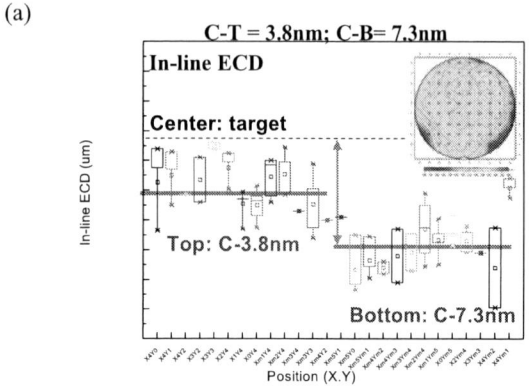

(b)

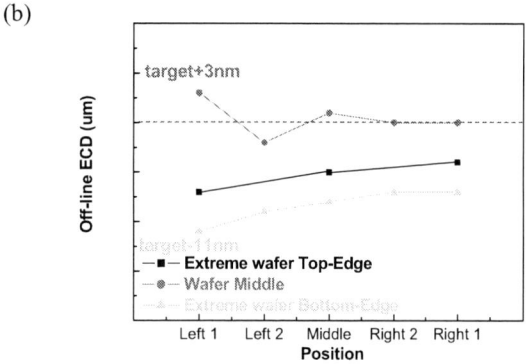

Figure 6. ECD map, in-line (a) and off-line (b) ECD trend within wafer of the most outer edge of the wafer as heavy polymer source CF_4/CH_2F_2 plasmas.

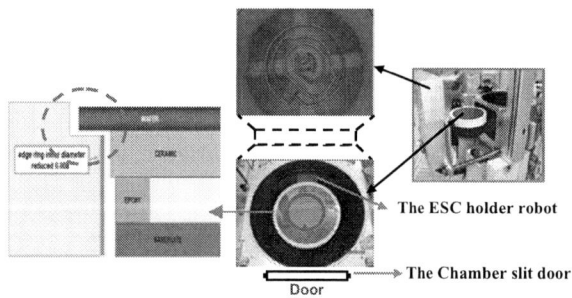

Figure 7. ESC holder, chamber slit door, etc. of Lam2300 STAR-t would result in non-uniform distribution of polymer deposition during etch process.

(a)

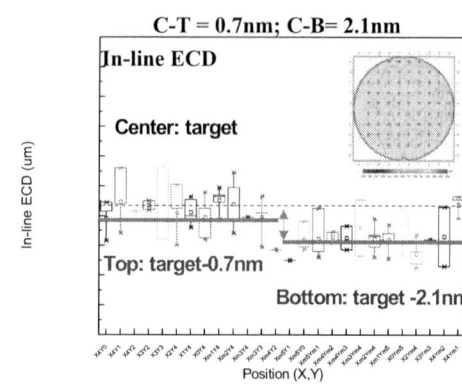

(b)

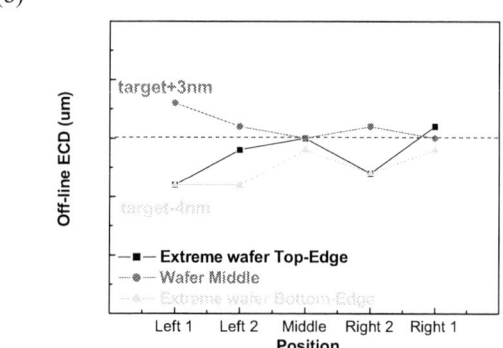

Figure 8. ECD map, in-line (a) and off-line (b) ECD trend within wafer of the most outer edge of the wafer as $CF_4/CHF_3/N_2$ plasmas.

(a)   (b)

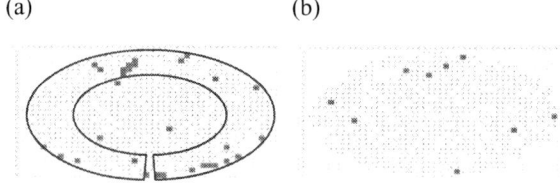

Figure 9. Flash cell of threshold voltage distribution around low boundary: (a) heavy polymer source CF_4/CH_2F_2 plasmas; (b) $CF_4/CHF_3/N_2$ plasmas. (Fail die: pink and red color)

IV. CONCLUSION

In this work, the boundary effect within die is successfully suppressed by excellent planarization of ODL in the tri-layer patterning approach. Furthermore, the optimized $CF_4/CHF_3/N_2$ plasmas at 15mTorr pressure could improve smaller ECD issue around the most outer edge of the wafer successfully.

ACKNOWLEDGMENT

The authors would like to thank MD420 and MD220 personnel, Macronix International Co., for their technical assistance and useful discussions during the course of this work.

REFERENCES

[1] C. Y. Lu, K. Y. Hsieh, and R. Liu, "Future challenges of flash memory technologies," Microelectronic Engineering, vol. 86, pp. 283–286, September 2008.

[2] Y. H. Song, J. Y. Lee, S. E. Lee, and J. H. Park,"Investigation for narrow cell threshold voltage distribution in NOR flash device," Jpn. J. Appl. Phys., vol. 46, pp. 5067–5070, August 2007.

[3] Y. Chiba, T. Matsumoto, M. Terahara, H. Kokura, and A. Hasegawa, "Chamber surface control for process stability," Dry Process International Symposium, pp. 21–22, 2009

[4] R. Ekwal Sah, J. Zhang, J. Deen, J. Yota, and A. Toriumi," PECVD bi-layer ARC for BARC-less immersion lithography," Electrochem. Soc. vol. 19, pp. 765–772, May 2009.

[5] D. C. Owe-Yang, T. Yano, T. Ueda, M. Iwabuchi, T. Ogihara, and S. Shirai,"Development of high-performance tri-layer material," Proc. SPIE, vol. 6923, 69232I, 2008.

[6] D. Codegoni, G. P. Carnevale, C. De Marco, I. Mica, and M. L. Polignano," Leakage current and deep levels in $CoSi_2$ silicided junctions", Materials Science and Engineering B, vol. 124–125, pp. 349–353, August 2005.

Automated Systematic Discovery for Development and Production

Brad Austin [1], Andrew Cross [2], Marcus Liesching [2]

(1) IBM Corporation, East Fishkill, NY, USA

(2) KLA-Tencor Corporation, Milpitas, CA, USA

austinb@us.ibm.com / andrew.cross@kla-tencor.com

Keywords: Design, Inspection, Design based binning, systematic defectivity

1. INTRODUCTION

In many semiconductor Fabs a combination of in-line photo inspection (PLY) and targeted physical failure analysis (PFA) is used to identify and monitor random defectivity, process excursions, and systematic defects. This analysis is fed back to process and design teams to create actions and fixes which drive yield ramps for integration and development. At the 45nm design node and below the number of process steps involved has greatly increased the time and cost from wafer start to testable product. To meet compressed time to market schedules, semiconductor companies must be more reliant on in-line wafer inspection and defect classification for yield learning, excursion flagging, and process split analysis.

Design based binning (DBB) of detectable defects is a relatively new capability of integrating design information with wafer inspection. This capability provides new opportunities for filtering and binning of defects based on both attributes of the defects and the design patterns at the precise locations where defects are detected. [1] Much value has been shown in the use of Design Based Classifiers (DBC) for more efficient SEM review sampling and for the historical analysis of known design patterns prone to systematic defects so that these patterns can be closely monitored. [2] Design based attributes are generated on tool concurrently with in-line wafer inspection.

A semi-automated method was needed to compare and monitor inspection results from multiple wafers for the identification of unknown systematic defects, process window deviations, and the comparison of process splits. In addition a baseline design clip pareto is required for overall qualitative design aware analysis. To achieve this a client initiated offline (off tool) design aware methodology was developed to identify potential systematic defects, identify excursion wafers based on user set statistical limits,

and allow for deep dive/drill down engineering analysis to find and resolve the source of the defects. The results of this analysis can be used to exclude excursion wafers from Design Based Group (DBG) analysis, and then identify potential systematic issues based on manually classified defect types.

2. METHOD

The new off tool design aware capability integrates KLA-Tencor's design based binning algorithms with a Defect Management System. This capability has been applied to IBM's advanced 3x and 2xnm design rule technologies. This has enabled the ability to run Design Based Grouping for multiple wafers offline to identify systematic patterns of interest. DBG (Figure 1) provides an unsupervised binning methodology for the identification of spatially random but structurally systematic defects.

This off tool design aware capability requires on-tool design aware analysis to provide defect data aligned to design. For off tool it is not possible to accurately translate the defect coordinate into design space.

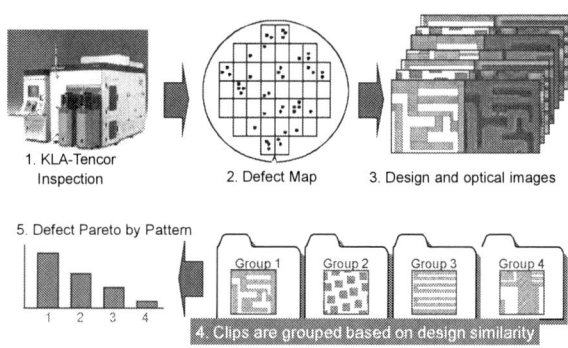

Figure 1: DBB – Design Background Grouping. Inline tool identifies defects and aligns each to design data. Similar design backgrounds are grouped into bins (DBGs)

This technique has provided the ability to identify systematic defect issues from nominal condition wafers and quantify differences between different process splits. In addition the methodology has been applied to inline defectivity monitor data to enable automated identification of excursions of defects in specific DBG Bins for large data sets.

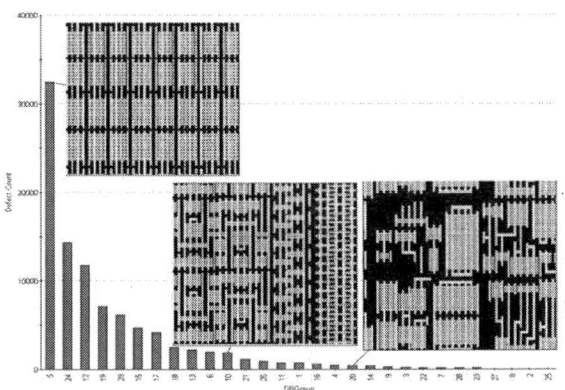

Figure 2: A Pareto of DBG bins for a set of scanned hardware. Design clips show similar pattern for selected bins.

Each DBG bin contains a particular design region or pattern of interest where defectivity has been detected, and on the resultant pareto design clips can be shown for each bin (Example shown in Figure 2). DBG bins generally provide very high accuracy and purity for the design features they group.

In addition Design Background Classification (DBC) can be utilized to investigate the historical impact of a particular systematic excursion. DBC allows supervised classification of known patterns of interest. When design background for a detected defect matches a previously defined design library, that defect is binned into the appropriate DBC bin (Figure 3).

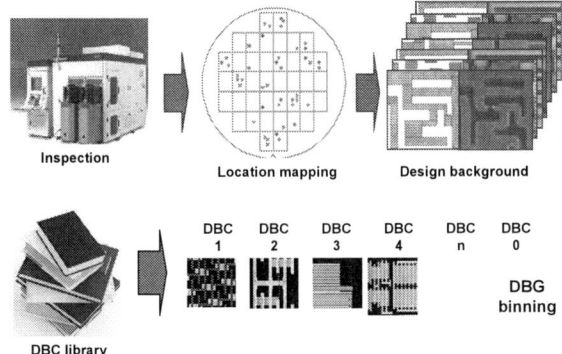

Figure 3: DBB - Design Background Classification (DBC)

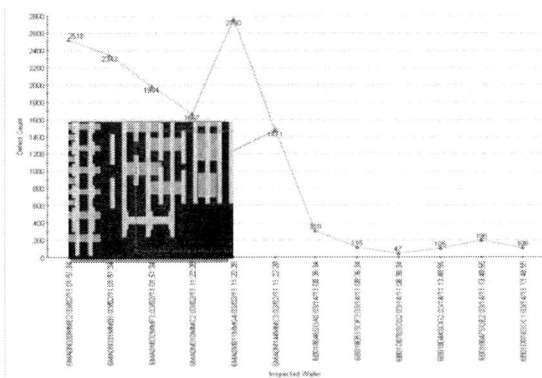

Figure 4: Historical tracking capability for DBC.

DBC has been shown to be effective for identifying both historical and current excursions of potential systematic defect types of interest. Figure 4 shows a monitoring example where the defect count for a specific pattern of interest was monitored before and after an improvement was put in place.

However given the comparative rarity of nominal wafer systematic excursions an automated excursion identification method was required. Figure 5 shows the excursion identification and control flow investigated. This methodology first allows the identification of systematic excursions on nominal wafers with the application of control limits on DBG bins. These control limits identify abnormal contributions to a specific bin from specific lots or wafers. In addition a control methodology for baseline identification of systematic issues related to DBG bins showing a difference to an expected value has been developed to identify baseline systematic issues. Because DBG bins are based on pattern matching and not area based, a nominal value based on defect density cannot be used. For this analysis the baseline for expected values is a random distribution of defects across the smallest repeating unit. In this case that is a single die. This is accomplished by populating an artificial die result with random distribution of defect coordinates. The DBB operation is executed in the same manner as a the true wafer results and thus a random defect pareto based on DBG bins is created and used as the basis for identification of systematic issues. DBG bins that vary greatly from the random distribution are considered systematic candidates.

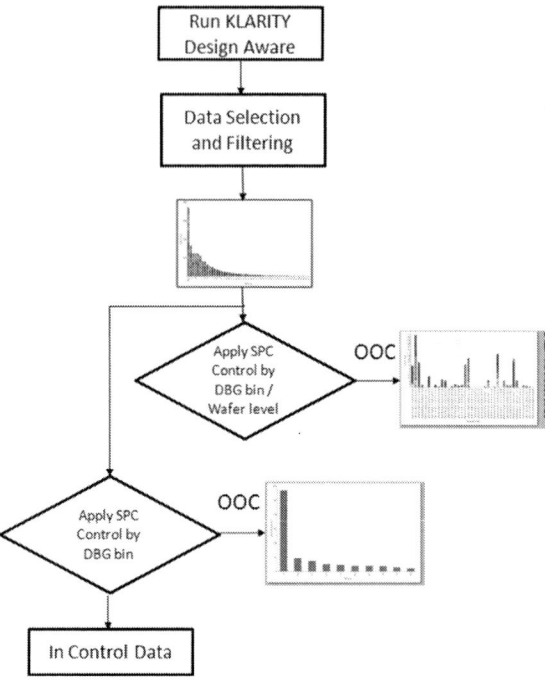

Figure 5: Excursion Control and Identification flow

The use of a simulated wafer approach allows consistent binning of design based groups in the database for comparison to or analysis of large temporal periods of data. Since nominal wafer excursions should be relatively rare the analysis of large datasets allows the potential identification of baseline deviations from the expected. Figure 6 shows an example of a simulated random defect distribution stacked on a single die.

Figure 6: Simulated random defect distribution only in areas included in tool scanning

The automated excursion flow first involves the running of multiple wafers using the offline DBB capability to generate consistent design based groups across all wafers. Once performed, DBG bins, and the background pattern represented, are consistent across the data set. In addition, by including the simulated / randomly distributed sample wafer, this can be used as a seed to enable consistent binning across different datasets with the resultant re-binned data saved back to a defect management system database.

The next step involves filtering of known systematic defect types identified by design based classification at runtime and spatially systematic but random defect events (in clusters).

The results dataset is then analyzed using the two automated control paths shown in Figure 5. The first path uses SPC control limits calculated by DBG group based on historical data. This path has the intention of identifying lots or wafers that show systematic differences to the typical wafer baseline for DBG bins. Various control metrics were investigated as well as investigating their typical distributions for the application of control limits. It was found that normal distribution statistics could be applied adequately for this control method. Since normal distribution control statistics are widely used within semiconductor process control this provides the advantage of a well understood method to be utilized.

Figure 7 shows an example of this excursion control path – identifying a potential excursion of a circular missing pattern affecting an SRAM cell.

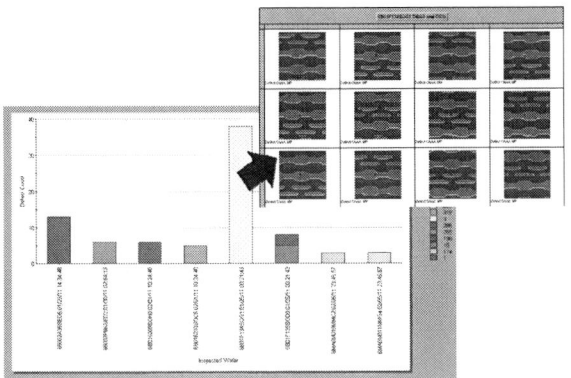

Figure 7: Wafer / lot level DBG excursion identification example. This DBG bin on this wafer was flagged as out of control (OOC)

The second control path makes use of the simulated random defectivity wafer to compare each DBG bin for the dataset under investigation to the expected defect percentage for each DBG bin. Figure 8 shows an example of such a comparison.

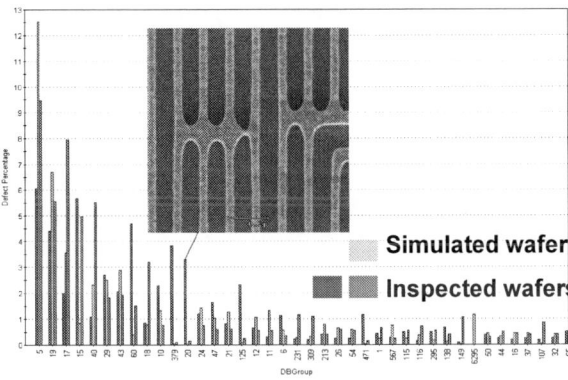

Figure 8: Comparison of actual DBG results to expected defect percentage. Show is a clip of a systematic missing pattern defect for an OCC DBG bin

Where a specific DBG bin has a much greater percentage in defectivity compared with the expected value we can consider this a potential bin of interest for further investigation. With control limits based on this delta between the expected and actual results baseline systematic defect types that potentially affect multiple wafers / lots within a dataset can be

automatically identified. An example of such excursion identification can be seen in Figure 9 – showing a systematic bridging defect at gate level.

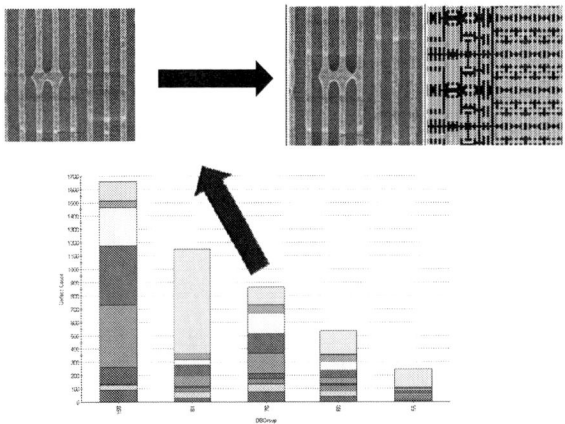

Figure 9: DBG type excursion identification

In this case a systematically weak structure is identified which can now be monitored with a DBC library. This allows for sensitized control of the pattern type of interest.

When a systematic pattern of interest is identified DBC can be run offline allowing an immediate quantification of the frequency of any pattern of interest occurring in historical data. This allows confirmation of whether a systematic event is a single temporal excursion or has been occurring at other intervals. Further, tool and process commonality or other deep dive analysis can be performed. Without this capability there would be a need to implement the design based classifier inline for data collection on future lots, requiring additional hardware be processed with this systematic issue before action could be taken.

When comparing a split lot, where wafers from a common start and process history are processed under different conditions at a particular step, this control methodology can be used to identify specific systematic structures which are sensitive to the differing conditions so that they can be identified apart from structures which are common to both sides of the process split. Figure 10 shows an example where two wafers were processed using different design masks. Although the overall random defectivity for the two wafers was similar, two DBG bins stood out. Investigation into these pattern types revealed a systematic bridging defect which was very dense on only one of the design masks. Because of the automated nature of the control method, this type of important analysis can be performed immediately

978-1-61284-408-4/11 $26.00 © 2011 IEEE

after wafer scan is completed. Reducing the time required to evaluate experimental process or design changes.

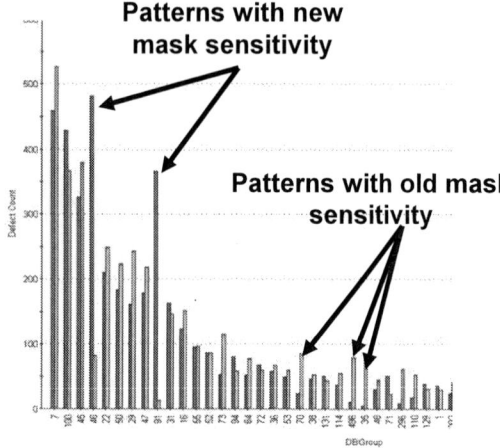

Figure 8: Lot split comparison where sensitive DBG bins are identified

This methodology is being investigated in the IBM East Fishkill facility to allow automated identification of potential systematic issues for critical devices.

3. CONCLUSIONS

The general approach for systematic defect discovery has been with the use of modulated wafers investigating the litho process window. This approach is seen as essential for initial hot spot identification and process window qualification. However the ability to process multiple wafers and lots with consistent design based binning provides a significant enabler for systematic defect discovery and monitoring from nominal wafers; to identify process interactions that cannot be fully simulated using a single wafer and to identify any process or device induced hot spots over time. In addition the ability to consistently bin multiple wafers based on design allows the speedy comparison of process splits where patterns of interest can be identified as potential responses to different process conditions on different wafers. Overall this automated methodology will allow for faster identification of problems in line and lessen the time to implement corrective actions.

References
[1] Park A and Yeh, J. "Novel Technique to Identify Systematic and Random Defects during 65 nm and 45nm Process Development for Faster Yield Learning". ASMC 2007.

[2] Jansen, S. Florence, G. Perry, A Fox, S. "Utilizing Design Layout Information to Improve Efficiency of SEM Defect Review Sampling" ASMC 2008.

Data Mining using PLS-Trees and other Projection Methods

Tamara Byrne

Umetrics, Inc.
MKS Instruments
San Jose, USA
tamara.byrne@umetrics.com

Svante Wold

Chemometrics Department
Umea University
Umea, Sweden
svante@nnsconsulting.com

Abstract—The amount of data measured during a typical manufacturing process is immense. To efficiently utilize these data without becoming overwhelmed with confusing and often conflicting information is difficult to impossible when using traditional univariate methods. Multivariate data mining methods can be used to examine large data sets by extracting relationships between variables to highlight variable correlations and deviations. Specifically, PLS-trees can be used to quickly identify significant clusters in large datasets and to highlight the differences within the groups.

Keywords- Cluster analysis; multivariate; PCA; PLS; time series data

I. INTRODUCTION

Data mining is an important task that should be prioritized in any manufacturing environment. Large percentages of the IT budget are spent in collecting and storing data, but these data are rarely utilized as thoroughly as they could be. Data in a semiconductor fab include logistics data, process time series data, metrology, electrical test, defect, and yield data. "Most measured quality control parameters (sheet resistance, line width, junction depth, etc.) are influenced by a number of variables in the process" [1]. These data may contain important information about how to improve the product quality and reduce scrap, but without the right analysis tools, we may never know what it is.

Measured variables are typically used for process control, but they are also useful for the overview of the process (process monitoring), for fault (upset) detection and classification, searching for differences in duplicate processing equipment to minimize chamber effects (equipment matching), predicting product or process properties like CD or thickness (soft sensors or virtual metrology), and for improved process understanding. In order to use the data in these ways efficiently, it is important to investigate properties of the data to determine how the analysis models should be created. Outliers, clusters, and trends can and should affect the way data are investigated. Advanced data mining techniques allow engineers to find and understand the trends and patterns existing in the vast amounts of data stored in data warehouses, and help them make well informed decisions.

II. MULTIVARIATE METHODS

Process data have several properties that make traditional statistical methods inappropriate for data mining. The data are almost always highly correlated, they are usually noisy, and there are often missing data from a failed sensor or IT issue. Projection methods like PCA and PLS [2] solve these problems nicely by relying on the correlation of the data to prevent noisy and missing data from having too big of an influence on the models. PCA is a good first step for getting an overview of the data [3]. It can be used to investigate outliers, trends, and groups, understand relationships between variables, and to get a feel for the underlying process. PCA creates a few summary variables, called scores (T), that explain a large percentage of the variation of the original variables. In this way, we can create plots and graphs in two dimensions that show as much information about systematic variation as possible. At the same time, non-systematic variation or noise is left out of the model and put into the residuals, or distance to the model (E).

$$X = 1 \cdot \bar{x} + TP' + E \qquad (1)$$

This way, the scores are not affected by the noise in the system, but we still have an indication if anything about the structure of the noise changes. These summary variables, the scores and distance to model, can be monitored in a control chart to indicate process deviations. PCA is an unsupervised method, so it explains the variation in the X data as much as possible [4]. PLS on the other hand is a supervised method that uses a Y variable (usually a quality measurement in manufacturing) to look for correlations between the data and the Y variable. It is a compromise between explaining systematic variation and correlation within quality. PLS also results in coefficients showing which data are predictive of the Y and in many cases can be interpreted as the critical quality parameters as shown in Fig. 1.

Although there are clearly many advantages to PCA and PLS, there are also some issues to be aware of. Both are sensitive to outliers and also to clusters in the data. Ideally, the data being modeled are homogeneous and fairly close to normal. Fortunately, PCA is good at identifying outliers as shown in Fig. 2, and multivariate cluster analysis can be used to determine clusters.

978-1-61284-408-4/11 $26.00 © 2011 IEEE

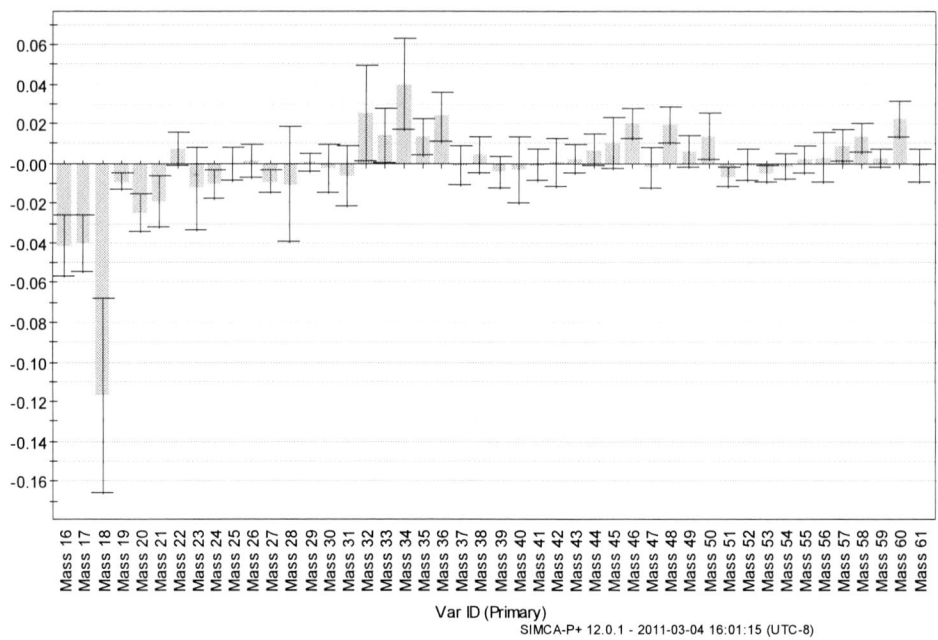

Figure 1. A coefficients plot from a PLS model shows how the X variables are correlated to the Y variables.

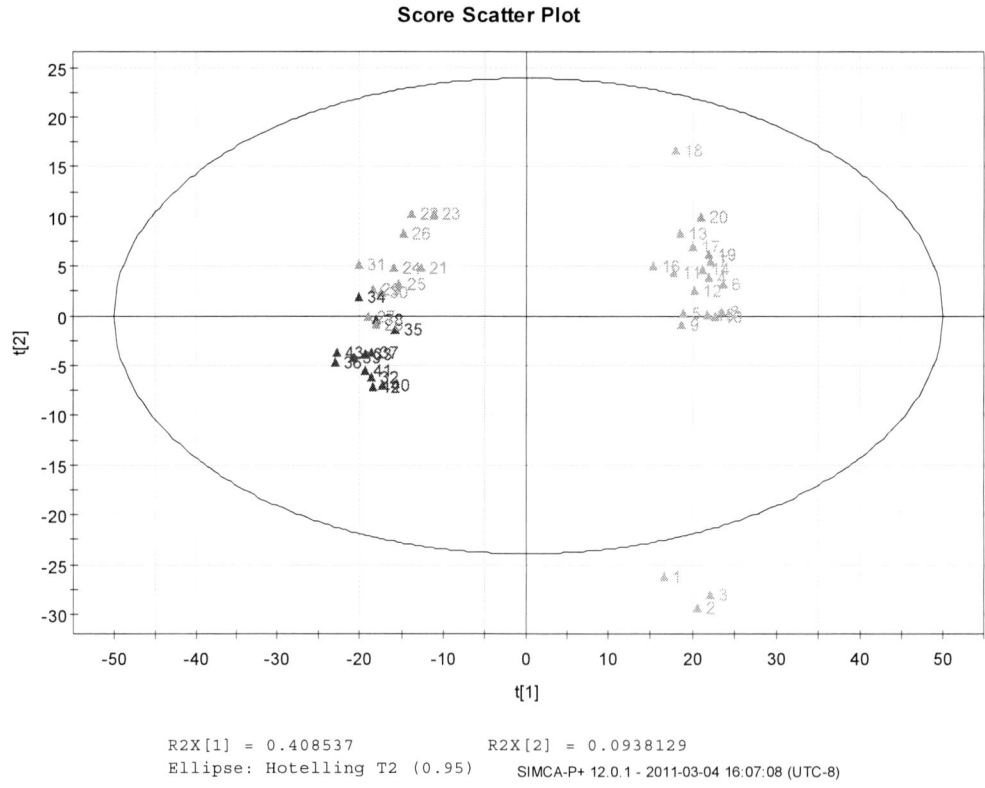

Figure 2. The ellipse in a PCA scores plot that indicates outliers corresponds to the 95% tolerance region, as specified by the T^2 statistic.

III. CLUSTER ANALYSIS

Cluster analysis methods can be used to identify data clusters so that local models can be built. These local models are very often more useful and have better predictive power than models built with heterogeneous data. Analogous methods like CART are based on regression and assume independent variables, which is not the case with process data. They also require more observations than variables, so there is a risk of discarding important information. An important point is that for matrices with N observations and K variables, multivariate projection methods work with long and lean (N >> K), short and fat (N << K) and almost square (N ~ K) matrices [5]. One possible multivariate method is Hierarchical Cluster Analysis (HCA). This method doesn't take into consideration the relationship of the data to quality attributes or other Y variables, so if no Y variable exists, it is a good choice. One

drawback is that, because HCA is a bottom up approach, it can be quite slow with very large datasets (Fig. 3). PLS-trees [6], on the other hand, work from a top down approach where all data are assumed homogenous, and then the data are split into two groups based on variation and correlation to the Y matrix, and so on until there is no longer a statistically significant difference between the groups as shown in Fig. 4. This makes the calculations much faster for large datasets. There are two settings available to adjust the clustering based on previous knowledge and preference. The first determines if it is more important to split the groups so there is an even number of observations on each side, or if the difference between observations should play the bigger role. The second setting makes either the X data or the Y data the most important for the splitting of the groups.

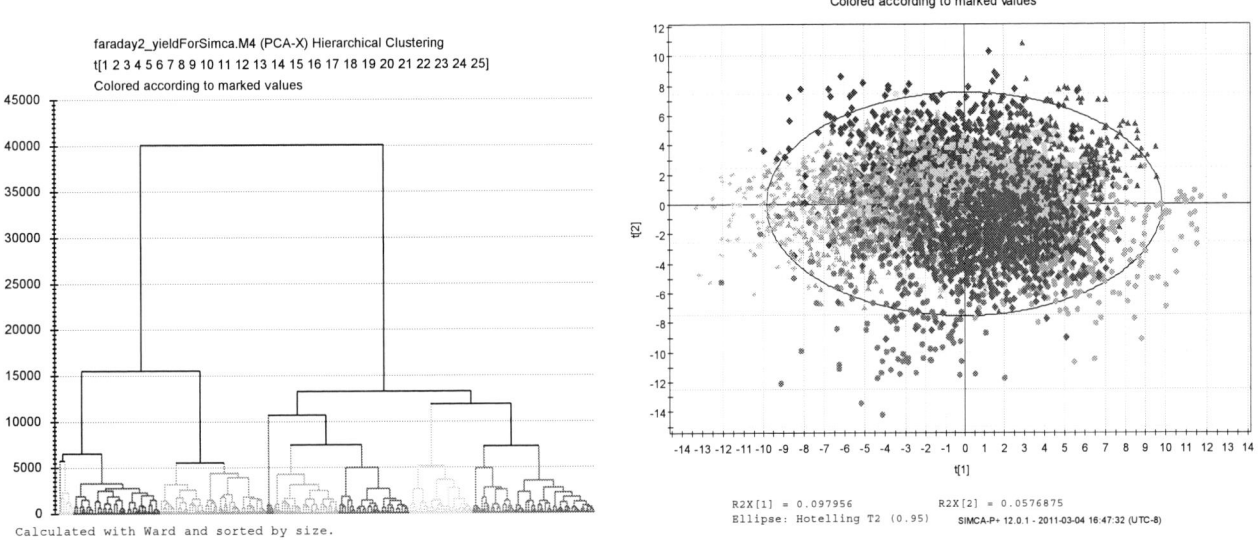

Figure 3. Hierarchical Cluster Analysis (HCA) is a bottom up approach where each observation starts as a separate group.

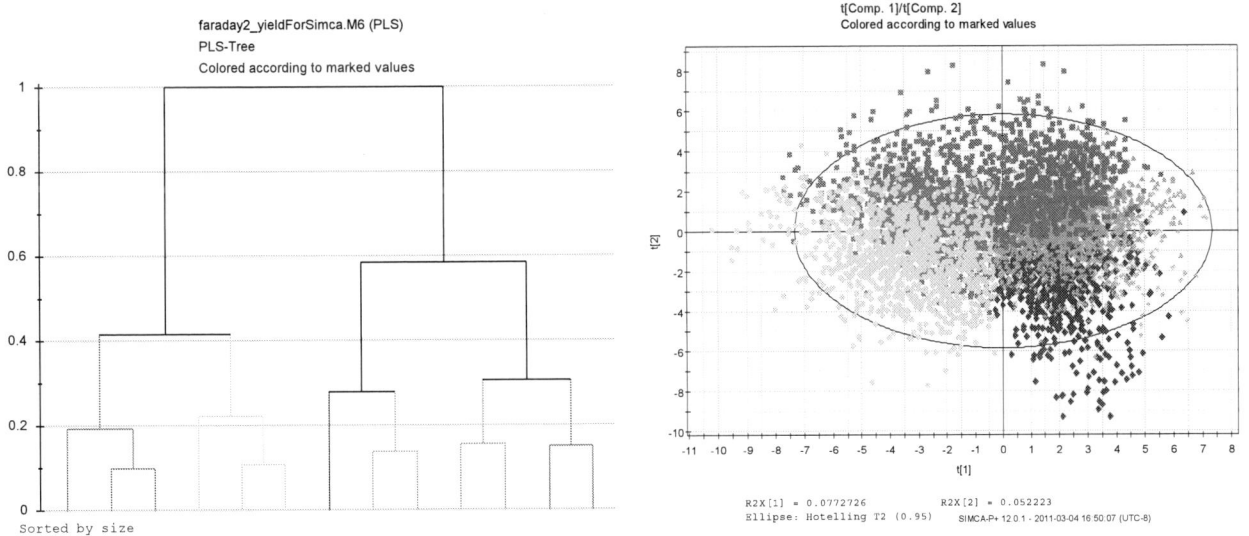

Figure 4. PLS-Trees is a top down approach where the entire population starts as a single group.

IV. MODEL VALIDATION

The most important, but often overlooked, part of model building is model validation. In many papers and studies, model fit (R2) is focused on, but model fit has no indicators if the model is overfit or if it will perform well with future data predictions. To ensure that the model is not overfit – i.e. modeling noise – the appropriate actions need to be taken depending on the model techniques used. When using traditional methods, like multiple linear regression (MLR), too many variables can cause the model to be overfit. To avoid this problem, non-influential variables may need to be excluded, and important information can be lost if those variables significantly deviate from the setpoint in the future. Unlike traditional methods, projection methods like PCA and PLS create more stable models when more variables are included, so there is no risk of missing a variable deviation. Because this is the opposite of many other well known methods, it is difficult to ignore instinctual impulses to reduce variables as much as possible, but it is important to do so for statistical as well as practical reasons. When highly correlated variables are included in a PCA or PLS model, the effect of noise can be minimized because the model can anticipate the actual values of the variables based on other variables correlated to it. From a practical sense, the more variables included in the model, the higher the chances of catching a deviation that could affect quality. Additionally, the impact of a malfunctioning sensor is minimized when other correlated sensors are present and can help the model to predict what the missing values would have been. Projection methods do however have the risk of being overfit, but it happens if too many components are taken. Each additional component calculates a new set of scores, and new components should only be taken when they are statistically significant. Cross validation can be used to determine the best number of components that can explain the systematic variation without also including the noise. The data are split into groups, we have found that seven groups is a good default, and one group is removed. Then the model is built with one component, and the missing group is predicted against the model and the error is calculated. This continues with each of the 7 groups and the errors are cumulated. This calculation results in the root mean squared error for one component. Then the process is repeated with two components, and three, and four, etc. The number of components is determined by which one produced the lowest cumulative error. Each of the multivariate methods, PCA, PLS, and PLS-Trees use cross validation to determine when to stop because the additional components (or groups) are no longer significant.

To make sure that the model will perform well in the future, the model should be tested. If you have enough data, you can use half of it to build a model, and the rest as a test set. Alternatively, you can use new data to test the model after the model is finished. Sometimes this is a better test because the process may have shifted a little, so you can make sure the model still performs well in the new operating space. Fig. 5 shows the model predicted values plotted against the actual measured values. The closer the observations are to the line, the more accurate the prediction is.

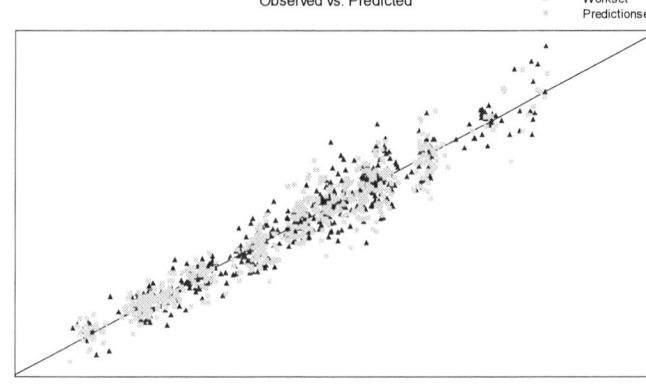

Figure 5.   Observed vs. predicted plot colored by predictions

The observations are colored based on if they were used to build the model, workset, or if they are being used to validate the model, prediction set. Here R2 is 0.93, meaning the predictions are 93% accurate, even considering the test set which consists of half of the data.

V. BATCH PROCESSES

Batch processes, like wafer processing – where there is a start and a stop for a given product - require special handling so time-dependent process evolution information is not lost [7]. There are types of data relevant to the wafer that can affect the final quality. The first is the transient behavior during the wafer processing. This trajectory is captured by collecting process trace data at small intervals, usually once per second in semiconductor industries. The second are initial conditions that don't change during the wafer, but that also can affect the final quality. Consumable age, type of resist, and the vendor name of the slurry provider, are all examples of initial conditions. The process data is a three dimensional dataset; there are variables, time points, and batches (wafer ID). The initial conditions are a two dimensional dataset; there are just variables and batches. Each of these types of data needs to be treated in a different way. To handle these different types of data, we build two models – an observation level model, and a batch level model. The observation level model calculates scores for each observation so you can see and monitor how the process trajectory is proceeding as shown in Fig. 6. After the wafer is finished processing, the batch level models how the initial

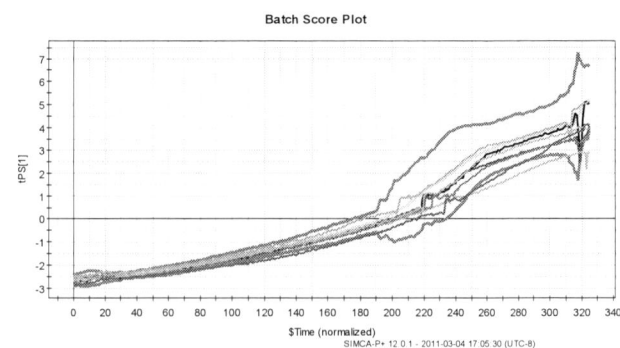

Figure 6.   Score trajectories for seven wafers after they have completed processing.

conditions in combination with the process data compare with other wafers. It is in this batch level model where virtual metrology and chamber matching applications can be created. This is also where PLS-Trees are the most useful because you can see if there are groups of wafers whose trajectories are different than the standard.

VI. DATA MINING EXAMPLE

A large semiconductor fab wanted to build a virtual metrology model for a rapid thermal deposition tool to predict thickness. The prediction accuracy was important because it was going to be used for two objectives. The first was to feed it into a real time run to run controller to adjust parameters to keep the thickness as near to target as possible. The second was to use the model to detect if there were any problems with the equipment process parameters that would affect the thickness. The fab didn't want to run a design of experiments on the equipment because it would be too expensive and getting the tool time would be too difficult. Instead, historical data, both process and metrology, were pulled for the chosen chamber recipe and a model was built. The model worked well for a while, but after a few months the prediction accuracy started to decline. An investigation of the data using PLS-trees determined that there were strong clusters, and the contribution plots showed they were from various temperature profiles. After some inquiries with process and equipment engineering, it was determined that the temperature profiles were behaving differently with different technologies even when the chamber recipes were identical. In the original PLS model, the predictions were basically predicting which technology was running from the process data, and then calculating a thickness based on that. Over time, when the chamber drifted a little, one technology with historically low thickness measurements started behaving like another with historically higher measurements, but the thickness prediction was still in the low range. This is because the differences between the technologies were larger than the differences within the technologies causing the shifts in thickness. You can see in Fig. 7, where the observed versus predicted plot is colored by technology, the shape of each cluster isn't diagonal like it would be ideally, but more circular indicating that the model isn't making accurate predictions within each cluster.

The fab decided to create chamber recipes by technology so that their systems could more easily distinguish between them and they built local models that performed much better for a long period of time. Fig. 8 shows the observed versus predicted plot for a single technology has the ideal diagonal shape. When it was tested with the new data, it performed much better than the previous model. With local models using individual technologies, the model is able to focus on those variables that actually affect the thickness and to make the prediction based on those. Additionally, the knowledge that the temperature profiles are significantly different based on technologies in this equipment helped to improve other control systems previously in place, including the multivariate fault detection models.

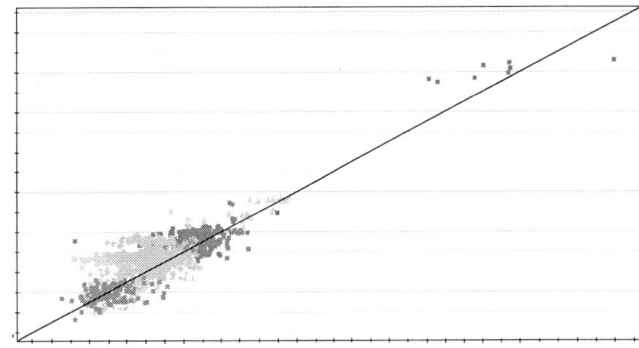

Figure 7.   Observed versus predicted thickness colored by technology.

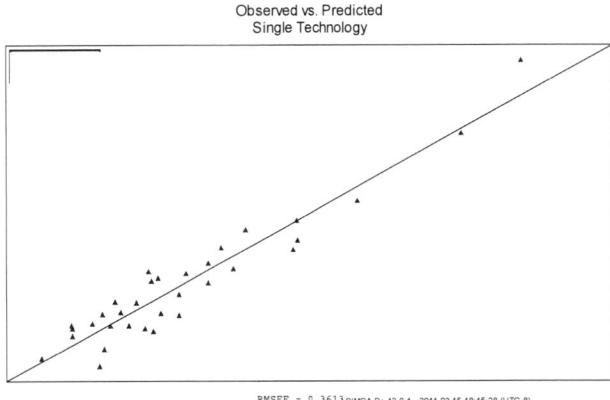

Figure 8.   Observed versus predicted for a single technology

VII. CONCLUSION

With such a large investment in gathering and storing process and quality data, powerful data mining methods should be leveraged to make the most use of it. Multivariate methods are a very efficient way to remove outliers, identify clusters, and then to build local models to either investigate further, or to use for real time process analysis.

REFERENCES

[1] Van Zant, Peter. Microchip fabrication, a practical guide to semiconductor processing. 4th ed. New York, NY: McGraw-Hill, 2000.

[2] S. Wold, M. Sjöström, and L. Eriksson. PLS-Regression: a basic tool of chemometrics. Chemom. Intell. Lab. Syst., 58 (2001) 109-130.

[3] S. Wold, K Esbensen, and P. Geladi. Principal Component Analysis. Chemometrics and Intelligent Systems, 2, 37 (1987).

[4] Kenneth R. Beebe, Randy J. Pell, Mary Beth Seasholtz. Chemometrics, a Practical Guide. New York, NY: John Wiley & Sons, 1998.

[5] L.Eriksson, E. Johansson, N. Kettaneh-Wold, J. Trygg, C. Wikstrom, and S. Wold (2006). Multi- and Megavariate Data Analysis. Umetrics Academy: Umetrics AB. 198.

[6] N. Kettaneh, S. Wold, Trends in chemometrics and data analysis, and software development implications, Umetrics User Forum, Boston, NNS, 2010.

[7] P. Nomikos, J.F. MacGregor. Multivariate SPC charts for monitoring batch processes. Technometrics, 1995, 37(1), 41-59.

The Effect of Bevel Film Removal on Wafer Warpage and Film Stress

Keechan Kim, Kwanwook Kwon, YS Kim
2300 Coronus & Flex team
Lam Research Cooperation
Fremont CA, US

Russ Dudley, David Marx
Technology Development
Tamar Technology
Newbury Park CA, US

Abstract—**The main objective of this study was to identify if there is any change in wafer warpage after bevel clean process, which might lead to higher film stress at device area in further processes. Correlation between bevel film Etch Rate (ER) and wafer warpage was also investigated. 25 kA thermal oxides (Tox) were etched on back side with various thicknesses remaining to generate a different amount of wafer warpage and thus different film stress on the front side of wafers. It was found that the thinner backside thickness leads to higher warpage with dome shape and film stress on front side. Test results indicate that bevel clean process did not increase the wafer warpage, on the other hand reduced the warpage about 1~4% compared to pre-bevel clean process in case of Tox. 3 kA silicon nitride (SiN) also showed the higher warpage of bowl shape with no films on back side compared to the un-etched one. No significant change in warpage was also observed in SiN. The correlation relationship of bevel ER and wafer warpage was significantly low, which provides a steady ER performance regardless of the warpage of monitor wafers.**

Keywords-bevel; warpage; stress; wafer edge; defect

I. INTRODUCTION

Recently, advanced chip makers are developing strategies to improve the yield of devices all the way to the wafer's edge in order to pursue the maximum yield of device. The reduction in edge exclusion from 3mm to 2mm currently and 1.5mm in the near future is worsening edge-related productivity issues, making edge-related defects a priority for IC manufacturers [1]. Therefore, the wafer edge is emerging as a key focus area in process optimization nowadays. Most chipmakers are implementing bevel clean processes into the process flow, using a dry, wet or polishing process. Plasma-based bevel clean is most popular approach, since it is easy to fit into the process flow for a wide range of application, such as post STI, gate, bit line, contact etch and dual-damascene etch.

The various films' mechanical stress could build up through a typical semiconductor process flow, and it can reach high enough to warp the substrate and crack the films in multilevel interconnect devices [2]. Therefore the film stress control in process integration is very critical to pursue the maximum yield of device. Although bevel clean process can remove multi-stack films at wafer edge, some of chipmakers have had concerns if bevel film removal can affect or change the wafer warpage. This could induce focusing issues at the subsequent photo processes and hence reduce the yield at device chips located at bevel area. Sometimes, it could cause a chucking issue during litho process.

The main objective of this study is to examine whether bevel clean process would have reverse impact on the wafer warpage and film stress. In addition, bevel ER was studied to investigate the correlation relationship with wafer warpage. Most chipmakers want to maintain a steady ER performance regardless of the warpage of incoming wafers for their Statistical Process Control (SPC).

II. EXPERIMENTAL PROCEDURE

25kA thick Tox at both front and back side of 300mm wafers were removed partially only on backside in a typical dielectric etch chamber to have 20, 15, 10, 5 and 0 (no remaining Tox) kA thick backside film, respectively. The partial etching on backside was intended to generate different amount of wafer warpage and stress on front side of the film by remaining film thickness control. 3kA thick SiN was also etched on back side to get 3 and 0 (no remaining SiN) kA thick back side film for the same purpose. Wafer warpage measurement was performed on these wafers using WaferScan™ from Tamar Technology. WaferScan™ measures wafer shape (bow and warp) by scanning the wafer with a non-contact chromatic confocal stylus. The wafers were supported at their edges, and all of the shape data was corrected for the effects of gravity. The film thickness was measured by F5x from KLA-Tencor. After pre- measurement of warpage, films at bevel were partially removed in Coronus 2300™ (plasma-based bevel clean chamber shown in Fig. 1) with fluorine containing chemistry to investigate the correlation between bevel ER and wafer warpage. These partially etched wafers again went through bevel clean process with extended etch time to completely remove films at wafer edge area and warpage measurement was done again to examine the change of warpage compared to pre-bevel clean process.

III. TEST RESULTS AND DISCUSSIONS

Fig. 2 exhibits the wafer warpage on Tox wafers depending on backside thickness before bevel clean process. The shape of wafer warpage is convex, which is caused by the compressive strength of Tox on Si substrate. On the other hand, SiN wafers showed concave shape, when backside film was completely removed as shown in Fig. 3. All of the warpage data was gathered following the semi-standard of using 3 point chuck

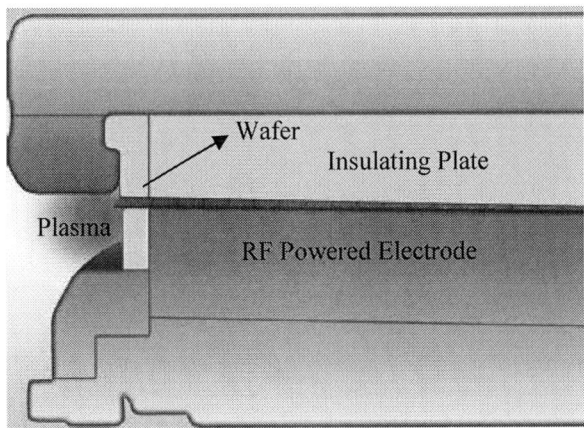

Figure 1. Schematic of Coronus 2300™ bevel clean chamber

with gravity compensation. Warpage amount and the corresponding film stress were plotted against the remaining backside thickness in Fig. 4 and Fig. 5. It clearly shows that wafer warpage and the film stress increases, as the backside thickness gets thinner. No remaining backside film (0 kA on back side) shows as high as ~430 μm bow (270 MPa stress) for Tox and ~-230 μm bow (143 MPa stress) for SiN, while back side thickness with no etch (25 kA for Tox and 3 kA for SiN on back side) shows ~0 μm wafer warpage (0 MPa stress).

The film stress was calculated using Eq. 1, the Stoney equation [3]:

$$\sigma = \frac{E\,D^2}{6\,(1-V)\,t\,R} \qquad (1)$$

Where, E = Young's modulus of the substrate

V = Poisson's ratio of the substrate

D = thickness of the substrate

t = thickness of the film

R = radius of curvature derived from bow x

with r = radius of the wafer:

$$R = \frac{r^2 + x^2}{2x} \qquad (2)$$

Both Tox and SiN films at wafer edge were removed during bevel clean process with the standard process and hardware configuration for FEOL applications. Films at bevel of both wafers were completely removed up to 1.1mm on the front side and 1.9mm on the back side as confirmed in the bevel inspection metrology scan data in Fig. 6 (Tox scan data). Re-measurement of warpage revealed that complete bevel film removal did not increase the wafer warpage and film stress as in Fig. 7. On the other hand, bevel clean reduced the wafer warpage by about 1~4% compared to pre-bevel clean in case of Tox wafers, possibly due to the release of stress by removing the film on the wafer edge area. SiN wafers also didn't show

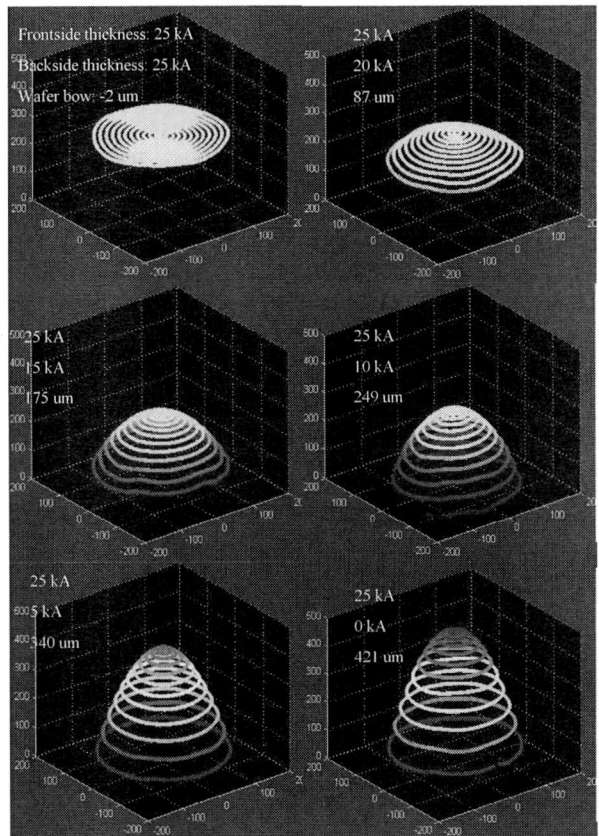

Figure 2. Warpage scan data of Tox depending on back side film thickness

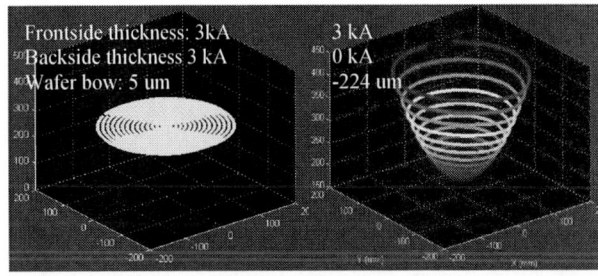

Figure 3. Warpage scan data of SiN depending on back side film thickness

significant change in warpage after bevel clean process. This test data assures that bevel film removal doesn't cause any film stress related issues and thus is a safe process with respect to film stress control in the process integration.

As indicated in the introduction, most chipmakers are monitoring the status of bevel clean chamber by tracking the SPC data of ER. If ER varies depending on the warpage of the SPC monitor wafer, then SPC ER data is not going to represent the tool status correctly. Therefore, the second part of this test was to investigate the correlation between ER and wafer warpage. Tox wafers, which were pre-measured in warpage and film thickness, were partially etched in bevel clean chamber to obtain the ER data. Two main chemistries for conventional FEOL applications (high fluorine flow and low fluorine flow) were evaluated. The normalized ER at 0.6mm

978-1-61284-408-4/11 $26.00 © 2011 IEEE

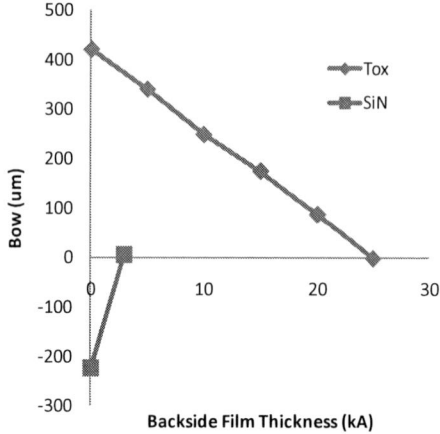

Figure 4. Film warpage as a function of back side film thickness

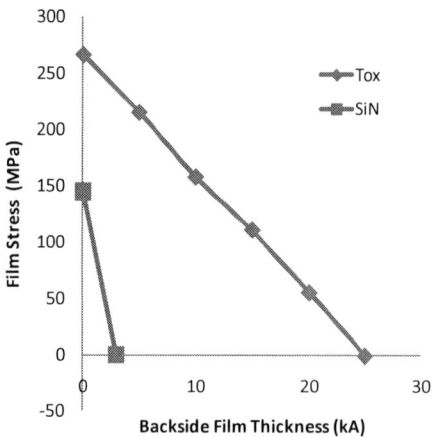

Figure 5. Film stress (absolute values) as a function of back side film thickness

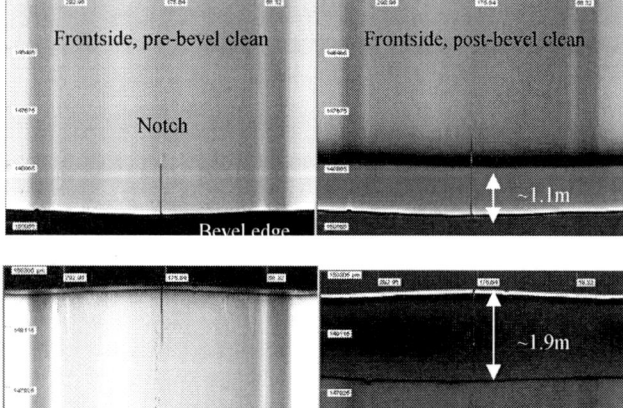

Figure 6. Bevel inspection scans on bevel of front and back side (pre and post Bevel clean process)

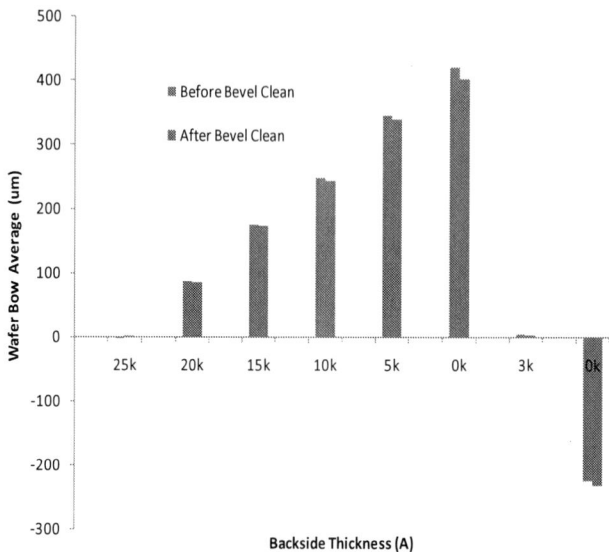

Figure 7. Comparison of wafer warpage between pre and post Bevel clean removal

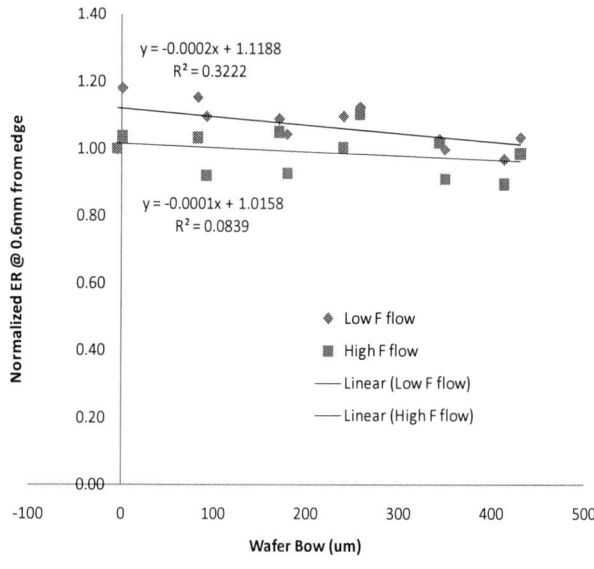

Figure 8. Normalized etch rate @ 0.6mm from edge as a function of warpage

from edge (149.4mm) was plotted against the warpage in Fig 8. The correlation factor (R^2 value) in the regression plots are less than 0.35, indicating there is no or little correlation between ER and warpage. This means that ER stability can be sustained regardless of the warpage of incoming SPC ER monitor wafers.

IV. CONCLUSION

Film stress of thermal Tox and SiN could be maneuvered by etching back side film with various remaining thickness. Tox showed convex shape warpage and SiN showed bowl concave warpage, as the backside film gets thinner. Bevel film removal didn't cause any change in warpage and turns out to be a safe process in terms of film stress control in the integration. Wafer warpage didn't exhibit correlation behavior with bevel ER, which provides ER stability during SPC ER monitor.

ACKNOWLEDGMENT

I would like to thank Elizabeth Pavel and Kyu Lee for their review and approval on this paper submission.

REFERENCES

[1] V. Vahedi, M. Srinivasan, and A. Bailey, "Raising the Bar on Wafer Edge Yield–An Etch Perspective," Solid State Technol., vol. 51, issue 11, November 2008

[2] M. Madou, Fundamentals of Microfabrication, 2nd ed. CRC Press, 2002, p. 268

[3] K. Seshan, G. McGuire, S. Rossnagel, R. Bunshah, Handbook of Thin Film Deposition, Process and Technology, 2nd ed. William Andrew, 2002, p.266

Eliminating a Polysilicon Hole Defect Created During Oxide Removal

Ikhoon Shin, Jason Doub, Keith Mortesen, Raymond Lappan

ON Semiconductor, Pocatello ID USA

Ikhoon.shin@onsemi.com

Abstract— **Gate Oxide failure analysis during technology qualification led to discovery of the polysilicon hole defects in large (>200K μ m^2) PMOS capacitors. In-line KLA inspections confirmed that polysilicon holes were formed during the salicide block process module. It is hypothesized that a three way interaction between the P+ source/drain implanted boron, heat added during salicide block mask deposition, and NH$_4$+ in the BOE causes the polysilicon hole. By replacing the BOE (Buffered Oxide Etchant) with a 100:1 HF solution, the creation of polysilicon holes was eliminated as confirmed by KLA and VBD testing.**

Keywords-: polysilicon void defects, GOI, Boron implant, Salicide block module, BOE

I. INTRODUCTION

Gate oxide failures are one of the major contributors to the reliability and yield loss of semiconductor devices. While most gate oxide issues are created within the gate oxide module, this paper will demonstrate that post gate processing can contribute to gate oxide defects. Specifically, defects formed during the salicide block module on polysilicon; a module that serves as a hard mask for selective salicide formation.

An in depth investigation revealed that failing gate oxide test results were due to polysilicon holes that formed in the salicide block module, during the pre-metal-deposition cleans. Subsequent silicide formation in the polysilicon hole caused electrical shorts. While not directly related to inherent oxide quality, the failures appeared as extrinsic gate oxide defectivity.

II. ISSUES AND IDENTIFICATION

As part of a high voltage 0.35um technology qualification, gate oxide integrity tests including TDDB and VBD were performed. TDDB and VBD tests showed that PMOS GOI capacitors exhibited high failure rates on PCSQ1 (large square active area), PCPE1 (poly edge) and PCBB1 (Birds Beak) structures, while all complimentary NMOS structures were relatively defect free. Data analysis indicated that the failures were Type-A extrinsic defects in nature. Excluding these failures otherwise indicated that the intrinsic lifetime for the gate oxide met the 10 year requirements. Table 1 outlines the test structure types, its dimensions, and it failure rate.

Table 1: Oxide qualification module

Structure Name and Size	PCSQ1	PCPE1	PCBB1
Active Area (um^2)	243,000	243,000	243,000
Edge length (um)	PE: 0 FE: 2,070	PE: 194,400 FE: 720	PE: 0 FE: 195,120
# of Defect (out of 116 caps)	15	28	41

FA (Failure analysis) by photo emission, SEM review, and SEM cross sections showed a round hole in the polysilicon top plate. These polysilicon holes were completely void of polysilicon and typically showed titanium silicide on the sidewalls and hole bottom. Figure 1 and 2 shows a typical appearance of these defects.

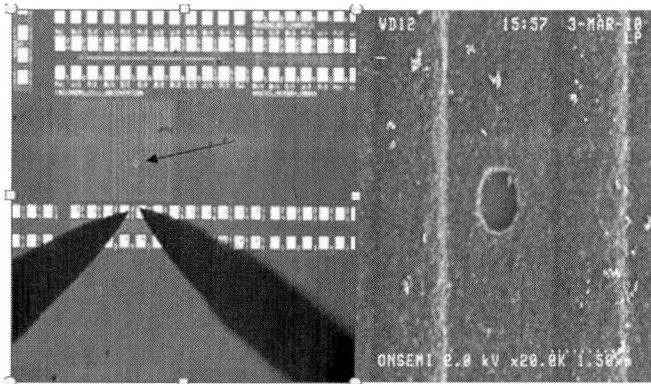

Photon emission   SEM: Top view

Figure 1: Failure analysis work by ONSEMI

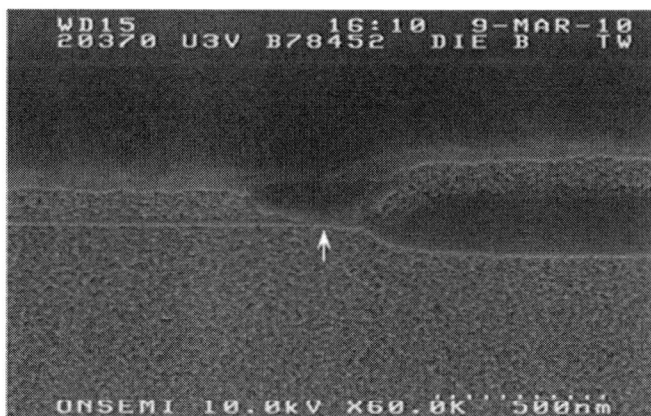

Figure 2: Polysilicon hole with Ti-silicide formation at the bottom of the hole.

978-1-61284-408-4/11 $26.00 © 2011 IEEE

Based upon the FA results, in-line product scans were intensified in the manufacturing line using a KLA2132 Brightfield wafer scan tool. The focus of these scans was in the salicide block module, as listed in Table 2.

Table 2: Salicide block processing module

Process Steps	Comments
SiO2 depsotion	LPCVD TEOS
Si3N4 deposition	LPCVD
Patterning / etch / ash / clean	
Pre-metal clean	BOE wet clean
Ti deposition	Endura deposition
TiSi formation	RTP
Ti Strip	SC1 strip

Stepwise scan recipes were developed to determine where in the salicide block processing module that the defect type was first formed. For these inline scans, a relative visual Defect Density (D_0) was determine through review of the defects found with KLA scans. D_0 is defined as number defects per square centimeter of area scanned. Through the in-line scans, the polysilicon defect was determined to be first detected after pre-metal clean. This clean is used to remove any native oxide prior to Ti deposition and can be considered as primary process step in the creation of polysilicon hole defects. With this discovery, additional experiments were performed to identify contributing factors to the defect formation and to understand the mechanism of polysilicon hole formation.

III. CONTRIBUTING FACTORS EXPERIMENTS

A short-loop flow was created based on salicide block module utilizing patterned implanted wafers. With the short-loop flow wafers, many experiments were designed to find the root cause for the defect of interest. Below are summaries of the important experiments that lead to determining a root cause for the polysilicon holes defects. All experiment results are based on visual defect counts.

A. Implant damage

One theory to the defect formation was that the P+ implant is physically damaging the poly-silicon and allowing the pre-metal clean to remove the damaged poly-silicon. An experiment designed to test this theory used B, BF2, and Ar as implant species with same dose and energy condition as the standard P+ source/drain implant used by the technology. The inert Argon gas was chosen to only study the physical damage effect of an implant on the polysilicon. Table 2 summarizes of the visual D_0 for the implant damage theory experiment.

Table 2: Effect of implant on polysilicon holes

	POE (BF2)	B	Ar only	B+Ar
Overall KLA visual defect counts	88	271	2.2	561
Polysilicon visual defect counts	70	211	0	505

The DOE results showed that any Boron base implant produced polysilicon holes, while Ar did not produced any holes.

B. Plasma charge damage

The second theory investigated was whether the P+ source/drain implant conditions were the initiator of the problem. Specifically, it was hypothesized that implant recipe arc current was too high and thus caused some type of ESD damage to the polysilicon which was later enlarged via the pre-metal clean. For this experiment, the arc current was varied to produce poor electron neutralization conditions with results are shown in Figure 3.

Scans showed that a polysilicon hole like defect was created, but differed in characteristic from the defect under study. The defect not only formed in the polysilicon but also in the Si substrate in the case of high arc current. Also the created defect had the appearance of a typical electrical discharge, and thus was not circular in nature. From these observations, it was thus concluded that the polysilicon hole defects were not due to wafer charging during ion implantation.

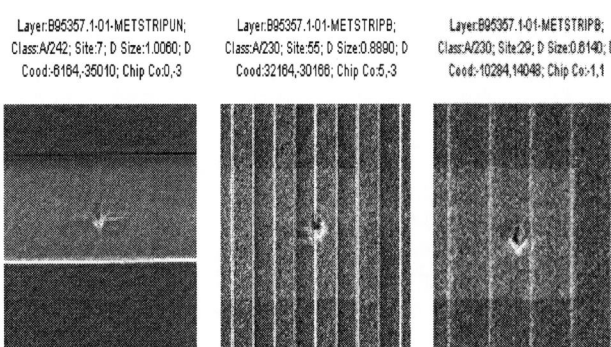

Figure 3: Implant charging damage defects at high arc current.

C. Pre-metal clean chemical

As KLA scans indicated that the polysilicon hole defects first appeared following the pre-metal clean, an experiment was centered around the clean conditions. The process of record pre-metal clean used a BOE based clean in order to remove native oxide prior to titanium deposition. An alternative of a 100:1 HF solution was proposed. A split was performed between these two chemicals and the resultant D_0 is shown in table 3. A significant reduction of polysilicon hole with HF was observed.

Table 3 Pre-metal Cleans

	HF	BOE
Overall KLA visual defect counts	6.5	105
Polysilicon visual defect counts	0	105

D. LPCVD vs PECVD vs HDP films

Summer F.C from SMIC claimed that silicon rich oxide with RI (refractive index) of 1.56 when used as hard mask for selective salicidation, also prevents polysilicon hole generation.[1] To determine if this was the case, a short-loop experiment was designed to test SiO_2 deposition method. In current process, both a SiO_2 (TEOS generated) and Si_3N_4 are used to form the salicide blocking mask. For this test, both LPCVD, PECVD and HDP oxide films and LPCVD, and PECVD nitride films were studied. The PECVD films employed standard operation condition as other intra-metal dielectric films used in the FAB, while the HDP wafer received a SiO_2 film having a Refractive index of 1.56. Table 4 shows defect density for this DOE. From the results it can be concluded that the HDP silicon rich film is capable of controlling the generation of the polysilicon hole defects but not to the same level as seen by changing the pre-metal cleans chemistry.

Table 4: Thermal budget and HDP experiment result.

	POE	HDP (Si Rich - RI of 1.56)	HDP + RTA	PECVD TEOS	PECVD TEOS + RTA	PECVD SiN	SiN removed via Hot Phos
Overall KLA visual defect counts	88	4.47	2.5	184	23.4	26	139
Polysilicon visual defect counts	70	3.6	2.1	184	23	26	125

E. Different salicide block module

The fact that there are salicide block module integration differences between 0.35μm processes in our FAB was noted. In-line and electrical testing that one integration scheme did not have the polysilicon hole defects while the other did. The major difference was that salicide block module included a Si_3N_4 layer on the process that had the polysilicon hole defects. A short-loop experiment was again used to determine if the factor was Si_3N_4 or other previous processing differences between the two flows. Splits were done around the salicide block module steps. Table 5 shows no defect creation when the Si_3N_4 layer is omitted from the flow.

Table 5: Salicide block module integration difference.

	POE	Thicker TEOS	Extened TEOS dep	No SiN layer	No SiN with BOE	STD with HF
Overall KLA visual defect counts	146	73	143	14	24	6.7
Polysilicon visual defect counts	146	43	143	0	0	0

IV. DISSCUSION AND RESULTS

Several hypothesis tests were developed based upon known failure mechanisms in an effort to resolve and understand the polysilicon hole defect issue. These hypotheses ranged from metal enhanced pitting of Si in HF solutions[2] to grain boundary enhanced HF etching. However, no known failure mechanisms completely explained the observed polysilicon hole defects. As a results, additional hypothesis test were performed internally. Many tests resulted in more questions, but several provided insight into the polysilicon hole formation mechanism.

From the observation that only the PMOS capacitor structures were impacted by this defect and later confirmation with ion implantation splits, it was determined that boron is a key component in this defect formation. From splits performed at the pre-metal clean, it was shown that BOE solution was a major factor. Additionally, by changing the salicide block integration and film stochiometry, the defects were modulated. This last result can be explained by a polysilicon grain boundary stuffing as hypothesized by Summer F.C, changes to film stress, or changes to the wafer thermal budget.

With these results, the authors hypothesize that the polysilicon hole defects arise from three way interaction between boron implant in polysilicon, subsequent thermal processing, and the BOE chemistry. During the salicide block depositions, the boron agglomerated in the implanted polysilicon due to the thermal energy in the LPVD deposition steps. This aggregated boron-rich-silicon is then etched by the BOE etch chemistry yielding a polysilicon hole defect. It remains unknown what property of BOE (vs HF) allows for this phenomenon: the difference in pH or the NH_4F in BOE solution. Further characterization is required to confirm this hypothesis and to understand the mechanism.

While the mechanism remains unknown, the solution of replacing the BOE based pre-metal clean with a 100:1 HF based pre-clean was implemented. Following implementation of this new pre-metal clean, no polysilicon hole defects have been detected with in-line inspections. Additionally, Vramp testing confirmed the D_0 results; HF based pre-metal clean shows improved Vramp performance (Figure 4).

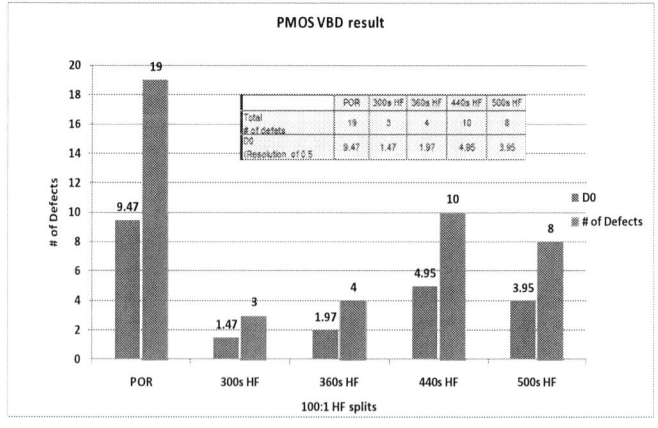

Figure4: VBD result on 100:1 HF

V. CONCLUSION

Reliability testing on a new process introduction identified a post-gate module defect issue impacting gate oxide integrity results. Stepwise product scan first identified the appearance of a polysilicon hole defects post the pre-metal clean. With the source identified as the pre-metal clean, short flow demonstrated that additional factors were necessary to the creation of these defects. These secondary factors include a boron doped polysilicon, and an oxide/nitride based silicide block module. It is theorized that aggregated boron in the polysilicon is etched in a BOE based solution; the final solution to the polysilicon hole defect was the replacement of a BOE clean with a HF based clean.

ACKNOWLEDGMENT

We would like to acknowledge contribution from Lieyi Sheng, Eddie Glines, Brett Williams, and Paul Porath for their support and contribution.

REFERENCES

[1] Summer F.C. Tseng, Wei-Ting kary Chien, Bing-Chu Cai, "Improvement of polysilicon hole induced gate oxide faiure by silicon rich oxidation," Microelectronics Reliability 43 (2003), pp. 713–724

[2] Kuiqing Peng, Juejun Hu, Yunjie Yan, "Fabrication of Single-Crystalline Silicon Nanowires by Screatching a Silicon Surface with Catalytic Metal Particles" Advanced Funcational Materials (2006), PP 387-394

Establishing Continuous Flow Manufacturing in a Wafertest-Environment Via Value Stream Design

Sophia Keil,
Germar Schneider,
Dietrich Eberts
Dept. of Automation,
Wafertest and Line Control
Infineon Technologies
Dresden GmbH
Dresden, Germany

Kristina Wilhelm,
Ingo Gestring

Faculty of Business
Administration
University of Applied
Sciences Dresden
Dresden, Germany

Rainer Lasch

Faculty of Business
and Economics
Dresden University of
Technology,
Dresden, Germany

Arthur Deutschländer

Faculty of Mechanical
Engineering,
University of Applied
Sciences Stralsund,
Stralsund, Germany

Abstract—**"Learning to see" is the message of the powerful Value Stream Mapping (VSM) technique, which was developed within the lean production paradigm to help practitioners redesign production systems, eliminate waste and create continuous flow manufacturing in a high-mix, low-volume manufacturing environment. It was originally developed mainly for disconnected flow lines of the automotive industry. In this contribution, both the adaption of this method to the characteristics of semiconductor manufacturing and its application in a Wafertest environment at Infineon Technologies, an important frontend site for microcontroller manufacturing, are shown. The article provides guidelines on how to implement continuous flow manufacturing into typical semiconductor production areas. The advantages of this production principle, as well as the special expenditures during the implementation, are demonstrated.**

Value Stream Design; continuous flow manufacturing; Wafertest; high-mix, low-volume, job shop; semiconductor industry

I. INTRODUCTION

"Technology is no longer king!" This was stated at the Advanced Semiconductor Manufacturing Conference (ASMC) already in 2007 [1] and suggested the increasing means of manufacturing logistics. Especially for mature 200 mm fabs which are on the "More than Moore" path in logic business, logistic innovation becomes more and more important. On the one hand, complexity of products is increasing. Therefore there is a development towards longer cycle times since product workflows become longer. On the other hand, in the produce-to-order business, the cycle time is part of the service apparent to the customer and, therefore, a basic competitive factor.

In general manufacturing theory, flow production is one of the best solutions to achieve short cycle times and low inventories [2]. Nevertheless, most of today's semiconductor fabrication facilities use job shop manufacturing, which offers high flexibility regarding changing product volumes and workflows with the price of long cycle times [3]. Both manufacturing organizations, flow production and job shop, offer assets and drawbacks. For future developments in the semiconductor industry it is important to combine advantages of both concepts.

The scope of the contribution is the development of a semiconductor specific "Virtual Time Based Flow Principle" (chapter II). To implement this, Value Stream Design (VSD) is a suitable means. Since VSD was originally developed mainly for disconnected flow lines of the automotive industry, it obviously has disadvantages in simply applying it within job shop environments, such as semiconductor production. Therefore, chapter III contains its adaption for complex production environments. The new approach for implementing Continuous Flow Manufacturing is evaluated within Wafertest area for selected automotive products in chapter IV. The contribution ends with a conclusion in chapter V.

II. VIRTUAL TIME BASED FLOW PRINCIPLE

Thinking about flow production, people often may have the topological aspect in mind. The installation of machines follows the product workflow. But in a flow production we have to divide into two aspects. The topological aspect alone does not lead to a continuous flow of materials. To realize a continuous flow, time matching and capacity balancing of consecutive process steps is needed. So a continuous flow is only possible when consecutive process steps have the same output in a defined time period. That means the topological aspect is not mandatory for using of flow production. The topological aspect leads to disadvantages like low flexibility regarding changing product flows, due to the need to rearrange facility layout. Especially for mature fabs it is, due to financial restrictions, nearly impossible. The advantages of flow production, such as low cycle times and inventories, come largely from the time aspect.

The main idea is to use only the time aspect of flow production and to logically overlay the job shop with the "Virtual Time Based Flow Principle". That means the topological characteristic of the job shop is not changed. The introduction of this new manufacturing organization is accomplished by using the approach of VSD, which is enhanced with respect to complex production environments and described in the following chapter.

978-1-61284-408-4/11 $26.00 © 2011 IEEE

III. ADAPTION OF VALUE STREAM DESIGN FOR COMPLEX PRODUCTION ENVIRONMENTS

The Value Stream Mapping (VSM) and Design technique was developed within the lean production paradigm by Taiichi Ohno [4]. A first full explanation was published by Rother and Shook, which is also the foundation of our contribution [5]. They define a value stream as all the actions currently required to bring a product through the main flows essential to every product: (1) the production flow from raw material into the arms of the customer, and (2) the design flow from concept to launch [5]. Within this contribution the focus lies on production flow. In contrast to traditional industrial engineering approaches, Value Stream Design covers the door-to-door production flow inside of a fab under consideration of the customer need. The goal is to establish a continuous, customer oriented and waste free value stream.

Within the classical approach of Value Stream Mapping and Design typically a small industrial engineering team develops a future state based on the assumption that the whole organization will follow and execute. Our experience shows that this procedure is not suitable for complex environments. Moreover, establishing flow production can be regarded as a paradigm shift within industry and necessitates organizational development [6]. The challenge is to change a mature fab with "well" established organizational structures, processes and tools. Changing skills, expertise, behavior and especially beliefs of all organizational members is a much more sophisticated process than changing organizational structure. Therefore, the scope of this new approach is an iterative convergence to the ideal of continuous flow manufacturing via organizational learning. But not only organizational aspects are relevant; also procedure and content need a facelift for adopting this approach. Figure 1 shows the new technique of Value Stream Mapping and Design.

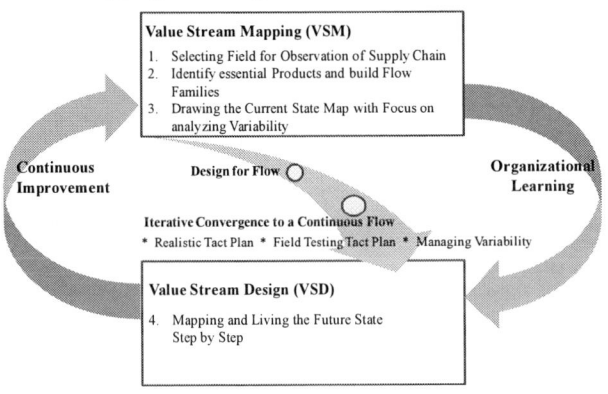

Figure 1. New Approach of VSM and VSD

1) Selecting Field of Observation of the Supply Chain

The main assumption of the classical value stream approach is the following: organizations can design a continuous flow for the whole value stream at one time. Manufacturing of an IC only within frontend requires up to 1000 process steps. Now imagine drawing a clear picture of this complexity, on only one large sheet of paper, as suggested in the classical approach, and then immediately establish the continuous flow for all those steps in parallel. Would it be possible? Referring to the ideas of the classical approach, the starting point should be the backend, being closest to the customer at the end of the supply chain. Afterwards, the upstream walk towards the starting point of feeding the wafer in the frontend takes place.

2) Identify Essential Products and Build Flow Families

In a high-mix, low-volume manufacturing environment, a new value stream for possibly hundreds of products would have to be designed. To reduce complexity, tools like ABC analysis or portfolio techniques [7] are used to identify important products. In contrast to classical VSD, we do not build product families, but rather flow families. A flow family is a united chain of consecutive single process steps which are similar within different product process of record's (POR), including the following similarities:

- sections of complete POR,

- same or replacing tool types with similar process times for single process steps, and

- length and sequence.

In contrast to flow families, product families within classical value stream design contain the whole product flow and belonging products are of one "class". A flow family can contain totally different products with similar sections of POR. Flow Families can be found for example within the area of cooper or aluminum metallization modules of POR`s.

3) Drawing the Current State Map with Focus on Analysing Variability

The basic concept of VSM is focused on metrics based on averages, for example average process time. Rother and Shook [5] suggest that the industrial engineer draws one snapshot of the current state during a line walk and then designs a future value stream. Neither variability in process time nor flow variability as barriers for continuous material flow are regarded. But variability is "the quality of nonuniformity of a class of entities" [2] and leads to non-compliance of production plans, long cycle times and is a detractor for capacity [7]. Without measuring and evaluation variability the introduction of a continuous material flow is nearly impossible in complex environments. But it is not sufficient to draw a value stream by only focussing on variability. Further essential influence parameters must be taken under investigation when drawing the new current state map which is displayed in figure 2:

- Decisions in the form of quality evaluations are done by engineers, which work isolated from the production floor in the office. The decoupled working style of engineers leads to stochastic process times for decisions. To demonstrate this, a separate level within the map is needed.
- Re-entrants have to be displayed because it is obligatory for the calculation of available capacity for future state mapping.
- Optional production steps also need an extra level, as those have a high impact on tact plans.

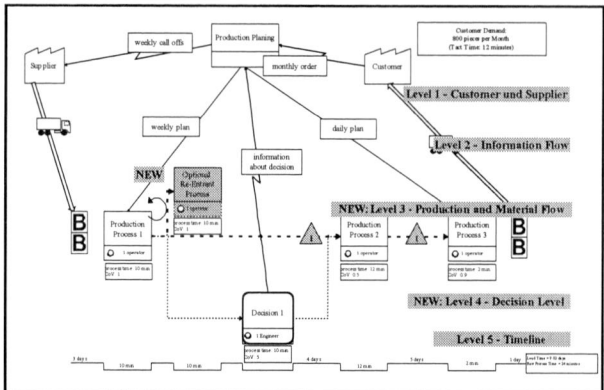

Figure 2. Extended Levels of Value Stream Map

3.1) Measuring and Evaluation of Variability

Variability can be quantified by the Coefficient of Variability (CoV) [2], [7]:

$$CoV = \frac{\sigma}{\mu} = \frac{standard\ deviation}{mean} \qquad (2)$$

The normalization allows comparison of variables with big and small averages. CoV's for e.g. customer demand, suplier deliveries, process-, material flow- and decision times should be computed for the evaluation of the flow family. After computing all necessary CoV's, variability is evaluated by variability classes which are shown in table 1.

TABLE I.   CLASSES OF VARIABILITY [12]

Coefficient of Variability	
CoV < 0.75	low variability
0.75 • CoV < 1.33	moderate variability
CoV • 1.33	high variability

The evaluation of variability of each single process step helps to prioritize activities of the next tasks. Figure 3 shows, for example, that manufacturing step 5 is the process which needs first attention, as opposed to step 4, which is the bottle neck when variability is not regarded.

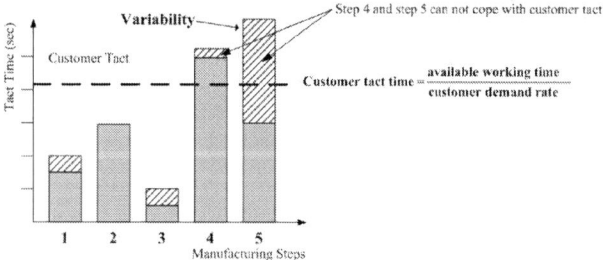

Figure 3. Tact Chart including Variability of Process Time

3.2) Classification of Processes

After measuring and evaluation of variability, processes are classified and characterized by the following criteria:

- frequency: A process of high frequency and high probability of breakdowns needs more attention than a process with a low frequency, e.g. optional process steps, because potential impact will be higher.
- duration: In processes with longer durations taking corrective actions in case of interruptions is more possible than in processes with shorter durations. Those processes definitely need preventive activities.
- responsibility: For defining suitable reaction schemes it is necessary to analyse responsibilities for single process steps.

3.3) Analyzing Interruptions

Besides external variability which is ocurred by customers (e.g. fluctuating demands) and suppliers (e.g. deviations in quality or quantity of needed materials), it is important to analyze company internal variabilities due to structure, organization, and operating policy of the manufacturing system. Focus lies especially on interruptions: frequency, duration, and possible causes and effects.

4) Mapping and LIVING the Future State Step by Step

Main steps within mapping and living the future state are design for flow and an approach which enables an iterative convergence to a continuous flow.

4.1) Design for Flow

The first step after VSM is to examine the regarded process steps of flow families in the manner of business reengineering. This means analyzing the arrangement of process steps of the whole workflow with respect to options regarding elimination, integration, parallelization, swapping, splitting and maybe enlargement due to quality issues. Besides the arrangement of process steps, the applied technology is of interest. It is suggested to question it with respect to robustness and mastering complexity.

4.2) Iterative Convergence to a Continuous Flow

Taking the initial current state map as the first available data, a realistic production schedule can be derived. The schedule itself has to mirror all current environment conditions. The approach is to plan for each process the necessary variability buffer individually. Within the automotive industry all single process steps of an assembly line are synchronized; this means every process step has nearly the same process time/output and works within the same tact time. In the semiconductor industry process times are defined, highly automated and can only be changed through process and capacity modifaction. Within frontend manufacturing, for example, process times for measurement steps can take seconds; meanwhile, the duration of furnace processes can be eight hours. Therefore, the solution is to summarize several different processes within a so called "Virtual Flow Unit". This unit is regarding its duration equal to the customer tact time and contains as many processes (with an adequate variability buffer) until the sum of process times reaches customer tact time. Then, an interative convergence to a continuous flow via several learning loops is needed. The above described content is schematically depicted in figure 4. As previously mentioned, the implemetation of continuous flow manufacturing should be done bottom up with the help of all organizational members. Therefore, living the future state step by step via three cluster of learning loops (see figure 4) is suggested. Every learning loop includes a field test and at the end a pilot production phase in which the production plan is executed over a longer time period. The field testing tact plan will then continuously be modified out of newly generated information about both process speed calculation (PSP), but also as a consequence of changes within the surrounding processes. This is to allow the variability buffer times to be decreased, as each substep can be evaluated and therefore individually reduced or even eliminated. In contrast to the traditional approch of simulating productional changes, we strongly believe that a real enduring change of manufacturing organization can only be achieved by practicing and involving the people. "Learning to See" becomes possible only through consideration of these boundary conditions.

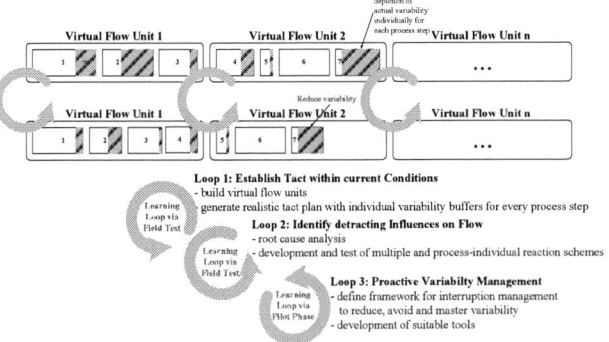

Figure 4. Approach of Iterative Convergence to a Continuous Flow

IV. APPLICATION OF CONTINUOUS FLOW MANUFACTURING FOR SELECTED AUTOMOTIVE PRODUCTS VIA VALUE STREAM DESIGN IN WAFERTEST-ENVIRONMENT

A. Proceeding of Application

1) Selecting Field for Observation of Supply Chain

Test steps within Wafertest are the last operations after frontend manufacturing before products are delivered to back-end plants or directly to the customer, which are flow lines in the case of microcontrollers. Product value is high and each waiting time is quite expensive. Therefore, a constant and just in time delivery of products with extremely short cycle times is absolutely mandatory and vital for commercial success. Figure 5 shows the manufacturing environment of the wafer test with the following impacting areas:

- frontend manufacturing (FE) as internal supplier,

- business units which are internal customers and finally distribute products to end customer

- internal departments like Wafertest Support, Yield Management and other manufacturing areas.

The mentioned areas are both supporting the continuous flow and hindering it because the tasks of many organizational members have to be synchronized. Departments of Test Program Engineering, for example, are located around the world and therefore working around the clock. They have to log on remotely on tools out of the regular production environment to implement important test process improvements. The alignment of those remote session requirements with the needs of production is hard to execute, as it has inherent trade-offs concerning tool utilization and manufacturing targets. That means it can not be excluded that a single productive lot must be interrupted for such "engineering sessions".

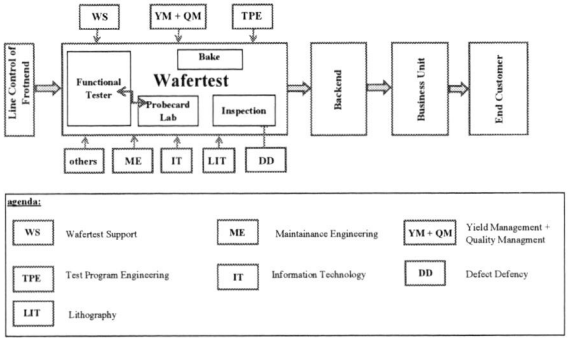

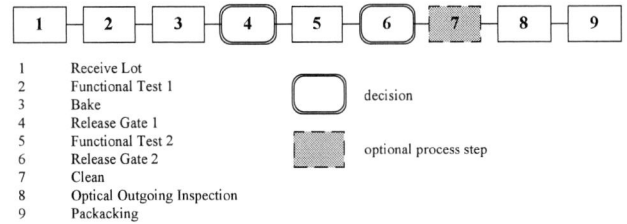

Figure 5. Manufacturing Environment of Wafertest

2) Identifying Essential Products and Build Flow Families

Now we proceed with application: essential products are identified and flow families are built. The flow of the identified family is shown in figure 6.

Figure 6. Arrangement of Process Steps

Within step 1 "receive lot" it is necessary to execute single wafer tracking. This means that a mapping step is needed to pick the right data about slot map from the host. Step 2 includes the first functional test, where the dies on the wafer are tested with respect to electrical functionality. In order to gather information about the aging of the microelectronic devices, a high temperature step (process 3) is applied between the two test insertions. Step 4 is the first release gate, where quality of dies is reviewed by the yield management group. This lot release is executed immediately if all product specifications are fulfilled. If not, additional product engineering analysis must be executed before final shipment release. Step 5 includes a second functional test, followed by another (final) release gate (step 6). Cleaning step 7 is implemented in the POR only for special products, where it migth be necessary after the functional test if the pads tend toward erosion.

978-1-61284-408-4/11 $26.00 © 2011 IEEE   115

Step 8 contains an optical outgoing inspection which detects optical defects on the wafer surface or embedded. Finally, the lot is packed in a moisture proofed manner.

3) Analysis of the Current State Map with Focus on Variability

The tact chart is a result of the current state mapping procedure and depicted in figure 7.

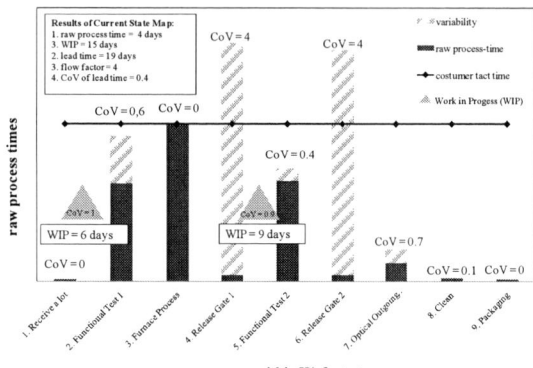

Figure 7. Tact Chart and Results of Current State

Here all nine process steps are compared to the customer voice (tact) of 1440 minutes. Without consideration of variability, industrial engineering would concentrate activities on step 3, the bake process. Variability evaluation reveals that both release gates cannot cope with customer tact when interruptions occur. Besides the fact that the raw processing time is four days, average lead time is 19 days, which results in a flow factor of four. As might not be initially visible, the bottle neck is not the bake process which is used in batch mode, but the functional tests with an average WIP in days of six and nine.

4) Mapping and LIVING the Future State Step by Step

4.1) Design for Flow

Creation of new POR's is often done on the basis of POR's of former product generations. In the described flow family (see figure 6) the DRAM-production period is reflected. The main scope was to bring as much material as possible through the facility with the highest capacity utilization. Based on this previous POR, the high temperature step (process 3) takes place before the lot release (process 4) is done which evaluates and gives feedback regarding first functional test (process 2). From the DRAM-time perspective, on a weekend the lot would get stopped because of the work-time-model of yield and quality engineers which are responsible for release gates. As a result, the bake sits in idle status. With the change of the Infineon business model from low-mix, high-volume towards high-mix, low-volume logic production where customer demands for single products for a week can be one lot or even one wafer, this can be hazardous. In case of a defect, most of the response time for feeding the fab with replacement material is lost. In the worst case this can lead to line stop the flow lines of the automotive industry.

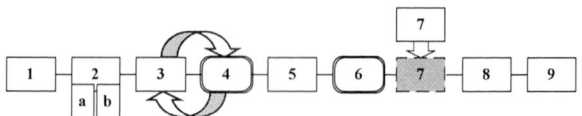

Figure 8. Design for Flow

To design a continuous flow for steps 2, 3 and 4, the following actions are recommended (see figure 8, above):

1. Swapping bake (process 3) and the release gate (process 4) which belongs to functional test 1 and solving the working-module issue via intelligent tact plans.

2. Splitting of functional tests in general: Because of their long process times, feedback is too late. A second aspect is that there are a lot of test insertions within one functional test which can be divided into stable (2 a) and unstable (2 b) sub processes. Through this clustering, pro-active managing of variability is easier.

3. Install a tool which enables early detection of interruptions and provides production controlling with fast feedbacks. Hereby, the surrounding processes, e. g. for the probecard logistics must be reviewed for appropriateness regarding flow production.

4. Establish cleaning process 7 as obligatory. A major portion of the WIP covered that step and suffered from delays up to 24 hours because of hold procedure and the fact that the cleaning tool was in another facility. Here organizational rule was – out of efficiency considerations - to extend the cycle of physical lot movement to that neighboring facility area, which was too rarely for a continuous flow. At this point rules must be designed for optimized lot dispatch.

4.2) Iterative Convergence to a Continuous Flow

Starting point was the legacy POR as it was prior to VSD.

Loop 1: Establish Tact within Current Conditions

For each step of the process, a first variability time estimation out of past data was added to the initial planned process time. This is to level the acceptable variability in each single unit and to provide reliable arrival times at the following unit. Via this, realistic cycle times are fixed within the initial schedule. The schedule itself is separated into five virtual units along the process plan, which all have one day of duration. The duration is determined by the pace maker step 3, bake (see figure 9). As also seen in figure 9, the release gates necessitates huge variability buffer times. During preparation of the second loop the POR was redesigned, as depicted in figure 8.

Loop 2: Identify Detracting Influences on Flow

As soon as the initial realistic tact plan is derived, the next learning loop with a field test can start. Now it becomes possible to monitor and evaluate the underlying variability and disturbances within the given plan in reality, as problems occur indeterminately. The monitoring tracks each lot in detail on its current operation in relation to the time stamp of the operation within the tact plan, in order to generate valid statistical data about timely deviations. Based on these findings, the tact plan will be modified here as well as in every other learning loop.

For this application, lot arrival rates within the test facility often bring first variability into the system. There are times of batch delivery out of the frontend-process as well as idle times, whereas the static planning anticipates average values. Another step worth mentioning is the reliability of the entire functional test system with its parts: probe cards, measurement programs, test heads and probers. If only a single part is unavailable, lot processing (measurement) will stop. Further, lot release and optical outgoing inspection are dependant on the individual lot quality.

With the new design of the POR (swapping release gate 1 and bake process), an additional waiting time was introduced with full awareness. The reason for this additional waiting time was to maintain the bake as pacemaker. Perspectively through elimination of variability in Virtual Flow Unit 1 and 2, the remaining step of Virtual Flow Unit 2 will be put into Flow Unit 1. Herby one out of the five units could be abolished, reducing overall lead time (learning loop 3).

On the other hand, it became obvious that the cleaning process (step 7) needed longer variability buffer times, due to unplanned lot transportation to another facility. In the initial tact plan step 6 was given a very long varibility buffer time. To bring this step together with step 5 – the second functional test - into only one Virtual Unit, these buffer times must be significantly reduced. Therefore, the above mentioned toolset for immediate interference detection now must be installed.

The two main elements of this toolset for "Interference Management" are immediate detection of all unplanned events at first and both immediate and adequate reaction at second, to resolve occurred interruptions without loss of time. For the first element, real-time analyis of all machine data must be provided, analyzed concerning flow requirements, e.g. disruption of lot measurement and trigger reactions within the organization. For this purpose, our tact plan controlling system was established, visualizing shift personnel the current state.

Figure 9. Iterative Convergence to a Continuous Flow

For the second element, each skill level within the organization has daytime specific orders on how to react, starting from standard tool operators and escalating further up to shift experts and even process engineers, see Figure 10. The daytime specific is crucial, as due to efficiency considerations, the most skilled colleagues are not available twenty-four hours a day.

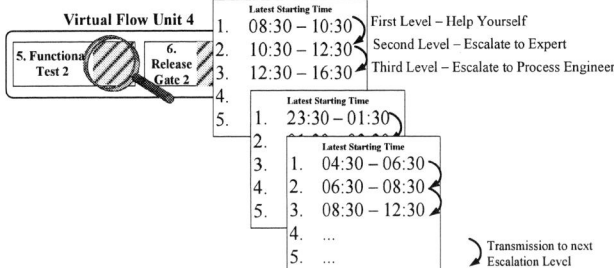

Figure 10. Time based Reaction Scheme

Finally it must be mentioned that also failing of assuring tact time - if all taken measures do not succed – is a possible option. In this case, a final acknowledgment toward the overall tact plan is mandatory to minimize the overall impact towards the customer.

Loop 3: Proactive Variability Management

The first effective term proactive measure here is to level the wafer delivery from the frontend to the wafertest. It shall be fixed to a continuous amount by using master storages, wherefrom exactly that amount is requested for the functional test, which is based on the customer tact. But here comes the next obstacle for "quick-fix" solutions: The realization of this leveled and continuous flow out of the storage must be aligned with surrounding business processes, going far outside of the encapsulated environement within te production site.

Further steps can then be evaluated from the generated statistics during the running field tests. Changes in product properties, for instance due to product maturation, may result in decreases of lot release times. Redefinition of test procedures can be used to group higher and lower variabilitites and so on for each single step within the POR.

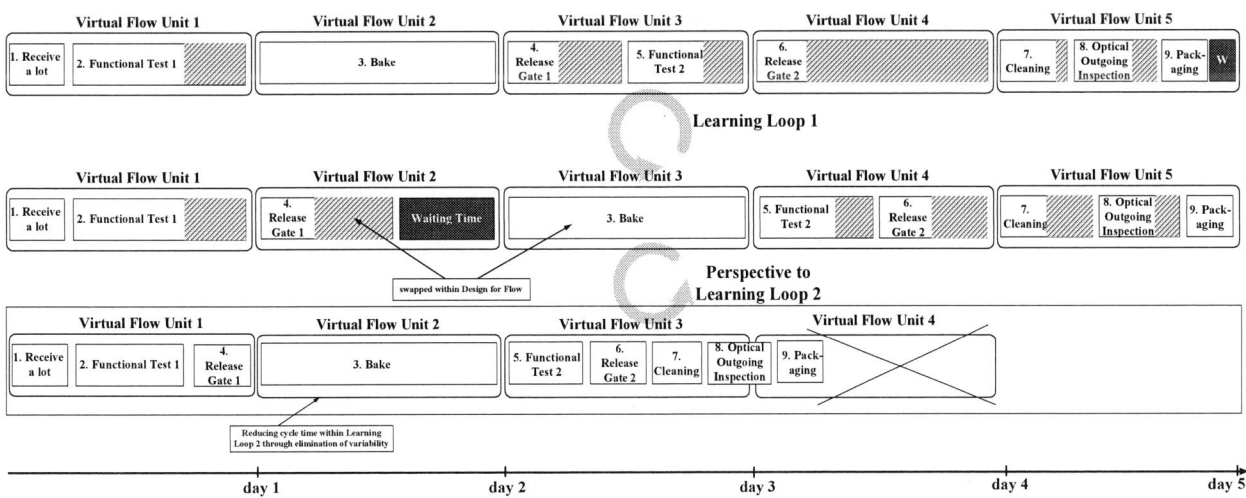

978-1-61284-408-4/11 $26.00 © 2011 IEEE

B. Results

For the selected flow family, the iterative approach of the Virtual Time Based Flow Principle started with a field test running 30 pilot lots. It was followed by a ten week pilot production phase exceeding the number of hundred lots. A third of these lots were running within the provided tact plan, without variability. The remaining lots were rescheduled due to interruptions.

The reduction of cycle time compared to the twelve months average of the legacy control mode was about 70%, without requiring additional capacity compared to the status quo. The observed capacity utilization was only 8% lower than in status quo mode. Besides the tact plan, there were no additional means as lot babysitting, or product prioritization. The main reasons for these results are the very specific (lot-wise) tact plans and the necessary close monitoring as well as proactive variability management to shorten reaction times. The observed reduction in capacity utilization must not be seen as a fixed loss, but as opportunity for further optimization through the application of more sophisticated tact plans in following learning loops. An option here could be additional loading with commodity products, which are not controlled with the Virtual Time based Flow Principle.

Variability is the main detractor when introducing Continuous Flow Manufacturing. The process specific variability time buffer allows standardized and limited time frames for reaction. Daytime specific predefined reaction schemes with respect to competence level enable very fast problem solving. Through the combination of these approaches, lots can be put back on the initial tact plan, reaching the next virtual unit on time. The underlying causes of variability now have to be addressed to eliminate the initial variability time buffers step by step. Therefore a classification of variability causes has been done, with the result of five main detractors:

- stability of entire test process: availability of e.g. probe card, test program, equipment status,

- usage of toolset: alignment between all departments,

- execution of the tact plan: awareness of lot start times and time based reaction schemes within organization,

- business processes (e. g. planning), networking: information flows across department borders and

- check for compliance with customer specifications.

Figure 11 shows the cycle time for 33 lots. We defined net cycle time as total cycle time, without the found interruptions, and cycle time with interruptions, leading to variability. The net cycle time remains within a relatively flat tube, whereas the variability cycle time is a line of peaks. Almost half of all disturbances which cause the variability tube are within the process itself, but the other half are the sum of organizational behavior: tester allocation, plan execution, availability, know-how and awareness within shift personnel. Therefore, the management should not only focus on the test technology, but with similar attention on the organizational behavior to optimize business processes, as they are independent from process stability issues.

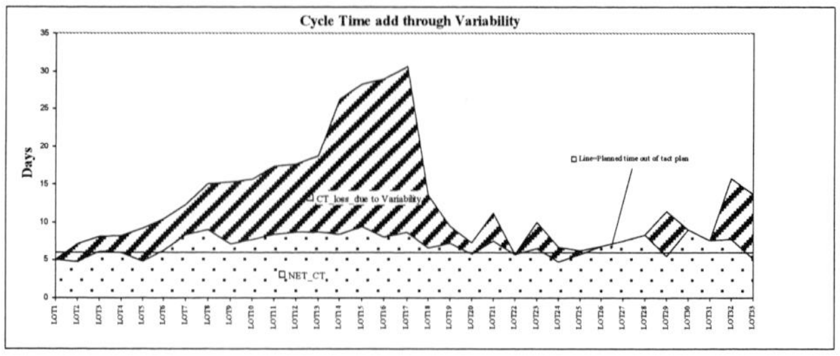

Figure 11. Cycle Time Loss due to Variability

V. CONCLUSION

"Technology is no longer king" could be extended to "Technology can remain king alongside the queen, logistics".

Establishing Continuous Flow Manufacturing in complex semiconductor environments requires as a precondition the mastering of the technological process. But the mastering of the surrounding business processes has risen to equal relevance. Therefore the technique of Value Stream Design is used and extended to the requirements of complex production environments. The focus here lies on design for flow necessities and variability management of all surrounding business processes. On the one hand, all members of the organization are taken into responsibility; on the other hand, via an iterative proceeding, they can become accustom to the organizational change step by step with its new structures, processes, and tools. Furthermore, they are empowered to redesign their own business processes within this paradigm change towards continuous flow manufacturing.

ACKNOWLEDGMENT

This work was financially supported by the Federal Ministry of Education and Research of the Federal Republic of Germany (Project No 13N10769). We would like to acknowledge also Jens Härtel and Heiko Stoll from Infineon Technologies Dresden GmbH for supporting this paper.

REFERENCES

[1] Peters, "Technology Is No Longer King," Semiconductor International, http://www.semiconductor.net/article/CA6454129.html, 2007.

[2] Wallace J.Hopp and Mark L.Spearman, "Factory Physics", McGraw Hill Companies Inc., US, 2001, p.252.

[3] D. Meyersdorf and A. Taghizadeh, "Fab Layout Design Methodology: Case of the 300mm Fabs," Semiconductor International, Issue 7, 1998.

[4] Jeffrey K.Liker, "The Toyota Way – 14 Management Principles from the World Greatest Manufacturer," CWL Publishing Enterprises Inc., US, 2004, p.10.

[5] Mike Rother and John Shook, "Learning to See: Value Stream Mapping to Add Value and Eliminate MUDA", Lean Enterprise Institute, US, 1999, pp.1-93.

[6] Keil, S. et al., "Flow Production in Semiconductor Industry – a Paradigm Shift in IC-Manufacturing", XVIII International Symposium, Research-Education-Technology, 2008.

[7] James P.Ignizio, "Optimizing Factory Performance – Cost Effective Ways to Achieve Significant and Sustainable Improvement", McGraw Hill Companies Inc., US, 2009.

FMEA for Lean Manufacturing

Michael E Lombardi
Technology Manufacturing Group Training
Intel Corporation
Chandler, Arizona, USA
Michael.e.lombardi@intel.com

Keywords-component; FMEA; Lean; Training

Abstract— With the rapid increase in the number of new product introductions, the volume & complexity of manufacturing changes has the potential to overwhelm the development, quality, coordination, and delivery of equipment training, which, in turn, could result in delays in these new product introductions. This abstract describes how an application of a Lean Manufacturing principle averted this risk and resulted in a 56% increase in productivity enabling Intel to meet a 46% increase in the number of training collateral with a 36% reduction in training resources.

I. INTRODUCTION

Prior to 2006, equipment development groups used a 5 year horizon to determine new equipment needs. Training departments, however, were notified whenever new equipment had been identified and approved by the engineering development groups, essentially providing training with a visibility of no more than 4-6 quarters out.

This reactive model was effective but not efficient and sometimes placed manufacturing schedules at risk due to training readiness.

Due to budget cuts, there was a decrease of 7% in training resources which nonetheless had to meet a 10% increase in the number of training collaterals delivered. The improvement method was "do more with less". This method was not sustainable.

CHALLENGE AND SOLUTION:

An equipment training lifecycle roadmap over a 5 year horizon was developed based on the equipment development group's roadmap. The roadmap identified a training need for 72 different manufacturing equipment types from 36 worldwide suppliers for 14 factories in 5 countries. Each equipment type has different training collateral requirements to support a manufacturing and maintenance

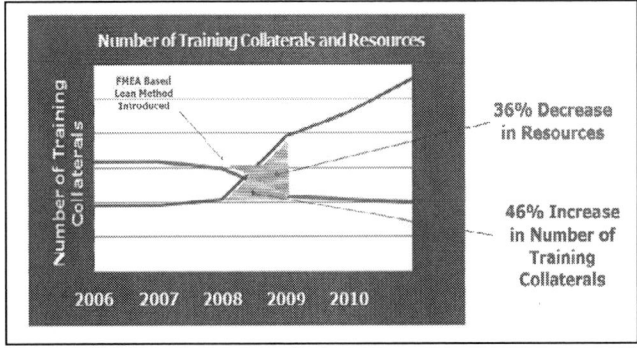

Figure 1 Number of Training Collaterals and Resources

audience making it extremely difficult to plan and resource the training groups. This represented a 46% increase in the number of training collaterals. In addition, there was a 36% reduction in training resources. The number and complexity of the changes, along with a reduction in resources had the potential to overwhelm the system managing the development, delivery, quality, and co-ordination of equipment training. See Figure 1.

The challenge was to develop a proactive and objective method to identify, quantify, and mitigate training risk so that training did not delay new product releases. A proven scientific method typically used in engineering, Failure Modes and Effects Analysis (FMEA), was applied. FMEA identifies the severity, likelihood, and difficulty to detect potential failures.

The business issues of resourcing equipment training did not fit a classical FMEA model, so the FMEA was modified to fit the business problems and to proactively identify and quantify training risk.

Four categories of problems were defined: Velocity (Schedule), Technology (Quality), Affordability, and Sustainability. Nine key variables were identified and mapped into one of the four categories. The severity and impact of each variable was weighted as high, medium, and low risk. Each variable was measured on a scale of low, medium, and high. The use of general terms, rather than finite numeric terms, allowed flexibility to account for the different requirements of each equipment type. It also allowed simple color coding (red, yellow, green) for tracking and reporting.

Velocity referred to having the right training available for the right people at the right time. The variables were Training Development Capability and Capacity, Training Delivery Capacity, Field Support. All the variables were weighted as high risk.

Technology referred to the quality of the training, it's ability to meet the targets. The variables were New Specialized Skills and Knowledge, Local Language Delivery, and Quality Indicators. Specialized Skills was weighted as high risk, Local Language and Quality Indicators were rated as medium risk.

Affordability referred to the costs associated with training, including travel costs. The key variables were Lowest Cost and Field Repairability. Lowest possible cost was weighted as low risk and Field Repairability was rated as medium risk.

Sustainability referred to ability to sustain the training over the course of the equipment lifecycle. The key variable was Self-Sustainability. It was weighted rated Low or Medium risk

depending on level of support available from equipment vendors.

The roadmap and FMEA were originally developed as a Microsoft Excel based spreadsheet. Macros were developed that identified overall risk to the critical path manufacturing schedule. The macros calculated the weight of the variable by the measured scale of each variable. For example, a variable with risk weighted as high and scale measured as high was interpreted as high risk to the manufacturing schedule. A variable with risk weighted as medium and scale measured as high would be interpreted as a medium risk to the manufacturing schedule.

The roadmap is focused on a rolling 6 quarter period with 5 year horizon. The roadmap is sustained by monthly updates to the FMEA to track progress and quarterly reviews with the equipment development groups to ensure alignment with future training needs. There was no direct monetary cost to develop the practice. 10% of one person's time was needed to develop the initial roadmap and FMEA. 5 – 8 hours per quarter is required to update the roadmap and revise the risk rating. The roadmap and FMEA were recently converted to a web based database that increased accessibility and improved reporting capability.

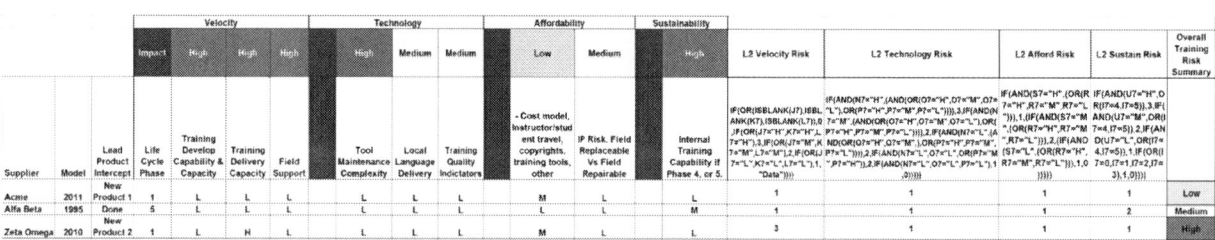

Figure 2 Modified FMEA and Macro

RESULTS: Resources were forecast and focused on reduction of high risk training items. The FMEA resulted in a significant reduction in high risk training items the first quarter it was used. Training development and delivery methods and capacity could be planned well in advance. Training management is able to use training and performance improvement methods more frequently that may have previously been ruled out due to time constraints such as: Human Performance Analysis, Web Based Training, Multimedia, Social Media, and Blended Learning.

Application of this Lean Manufacturing principle enabled Intel to meet a 46% increase in the number of training collateral with a 36% reduction in training resources. This represented a 56% increase in productivity. The Lean principle was sustainable with year over year 15% productivity improvements and a 10% reduction in resource requirements.

Application of a FMEA to an equipment training roadmap was an effective method to identify and quantify manufacturing equipment training risk.

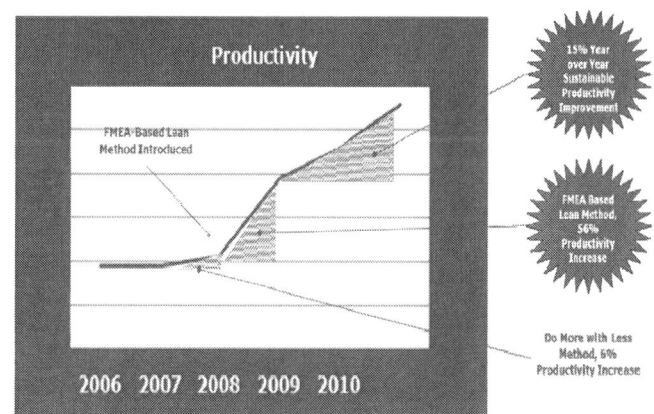

Figure 3 Productivity Increase

978-1-61284-408-4/11 $26.00 © 2011 IEEE

Laser Spike Annealing for Nickel Silicide Formation

Jeffrey Hebb, Yun Wang, Shrinivas Shetty, Jim
McWhirter, David Owen, Michael Shen, Van Le,
Jeffrey Mileham, David Gaines, Serguei Anikitchev,
Shaoyin Chen, Paul Bischoff
Ultratech Inc.
San Jose, California, USA
e-mail: jhebb@ultratech.com

Joe Lee
LeeDAC Consulting
Castro Valley, CA, USA

Abstract— Recent work has shown that laser annealing may have advantages over conventional RTP for nickel silicidation formation, such as lower leakage and better device performance [1, 2]. However, there are a number of requirements that must be met by any millisecond annealing tool to successfully bring this process to a high volume manufacturing environment. Ultratech's new low-temperature LSA system is designed to meet these requirements for middle-of-line processes such as nickel silicidation. Specifically, this system meets the requirements of within-die temperature uniformity, layout-independent processing, and closed-loop temperature control. Supporting data is presented for each of these requirements. Characterization data on blanket NiPt wafers is also presented which shows that a large step size can be used during the rastoring of the laser beam over across the wafer, enabling a production-worthy throughput of 70wph to be achieved by LSA.

Key Words: millisecond annealing, laser annealing, nickel silicide, pattern effects, temperature control

I. INTRODUCTION

As logic devices scale to the 28nm technology node and beyond, introduction of millisecond annealing (MSA) into the nickel silicide process is a promising approach for increasing device performance and improving yield [1, 2]. The conventional nickel silicidation process consists of two low-temperature rapid thermal processing (RTP) steps, with a selective etch inserted between the two anneals. The timescale of the RTP steps are on the order of 10 seconds. There has been recent work focusing on replacing the second RTP anneal with MSA. MSA has times scales on the order of 100's of microseconds, and temperatures several hundred degrees Celsius higher than RTP [1, 2, 3] for nickel silicidation. Two benefits have been reported which result from using MSA for this process step: 1) The short time scale of MSA minimizes the diffusion of nickel through the shallow extension, reducing junction leakage and improving yield [1, 2] and 2) The higher temperatures available from MSA can lower the source/drain series resistance and improve I_{on}/I_{off} performance [2]

In order to introduce any MSA tool into a high-volume manufacturing environment, the following requirements must be met:

A. Within-die uniformity on device wafers

The presence of device patterns on the wafer cause the heating energy to be absorbed unevenly across the die, causing potentially large within-die (WID) temperature variations referred to as pattern effects [4, 5, 6]. The hot/cold spots from pattern effects can degrade device performance and lead to parametric or catastrophic yield loss. Pattern effects can also cause high stress gradients and stress-induced leakage [7]. These kinds of problems can be extremely difficult and costly to solve in a manufacturing environment, and so the MSA system must be designed to minimize pattern effects.

B. Layout-independent process results

The same physical mechanism that causes pattern effects also causes the heating energy to be absorbed differently on average between device products, because different products will have different layouts and pattern densities. This phenomenon can cause process targeting problems when transferring a MSA process from one product to another. Pattern effects can also give inconsistencies in yield between products due to differences in WID temperature variations. The potential risk is especially severe for logic foundries, where there are always many different products going through the line.

C. Closed loop control of peak temperature

The first two requirements above address the issues of process non-uniformities at the die level. However, a production-worthy MSA tool must also be capable of minimizing process variations at the within-wafer (WIW) and wafer-to-wafer (WTW) levels. Since peak temperature is the most sensitive process parameter in MSA, the most optimal way to achieve acceptable process control is to directly measure the peak wafer temperature in real time, and use that measured temperature in a closed loop feedback system with the heat source as the controlled parameter. Such a system will address WIW uniformity, WTW and lot-to-lot repeatability, and day-to-day stability. A MSA system using an open loop processing approach puts production wafers at risk due to susceptibility to such variables as heat source drift, variations in device wafer optical properties, and long term degradation of

978-1-61284-408-4/11 $26.00 © 2011 IEEE

optical elements. Also, it is highly desirable that the system measures and controls the wafer temperature at multiple points over the wafer.

D. High productivity

No tool in an advanced logic manufacturing line can have a prohibitively high cost of ownership. Productivity, or throughput, is the largest technical factor that drives cost of ownership.

In this paper, we will describe the theory of operation of Ultratech's new low-temperature LSA (LT-LSA) system which is designed for middle-of-line processes such as nickel silicidation. Then, we will demonstrate how this system meets requirements A through C above. Finally, we will report on some basic process characterization of blanket Ni/Pt wafers using the LT-LSA system, and discuss the implications of the characterization results on tool productivity.

II.   LOW TEMPERATURE LSA

Figure 1 is a schematic illustration of the LT-LSA system. In a standard LSA system, typically used for high temperature ultra-shallow junction (USJ) activation, a single CO_2 laser beam ($\lambda=10.6\mu m$) is formed into a long, narrow "line beam" at the wafer plane. The combination of long wavelength, p-polarization, and Brewster angle minimizes WID reflectance variations, almost eliminating pattern effects [4]. The wafer sits on a heated chuck, typically set at $400^{\circ}C$ in order to thermally generate the free carriers required for the CO_2 laser to be absorbed by the wafer. The wafer is scanned rapidly under the beam, heating the wafer surface from the chuck temperature to the peak annealing temperature, which is typically $1100^{\circ}C$ to $1300^{\circ}C$ for USJ applications. Typical dwell times are in the range of 200-$1000\mu sec$, which is controlled by the stage speed (dwell time is defined as the amount of time that a point on the wafer spends scanning under width of the CO_2 beam, where the beam width is defined as the FWHM of the beam). The peak temperature is evaluated in real time by measuring the emitted radiation from the "hot spot", converting those measurements into a wafer temperature, and feeding this wafer temperature back to the laser which closes the loop at a 10kHz update rate to keep a constant wafer temperature [8, 9]. Through a combination of uniform absorption of the heating laser and real-time temperature control, the instantaneous peak temperature of the wafer surface is kept uniform and repeatable on production wafers during processing.

There are two main challenges that must be addressed in order to adapt the standard high-temperature LSA system for processing at lower temperatures required for nickel silicidation. First, the chuck temperature must be less than $250^{\circ}C$ so that silicide transformation does not take place from the chuck heating. At this low temperature, the wafer is semi-transparent to the CO_2 wavelength because there are not enough free carriers to absorb the long-wavelength photons.

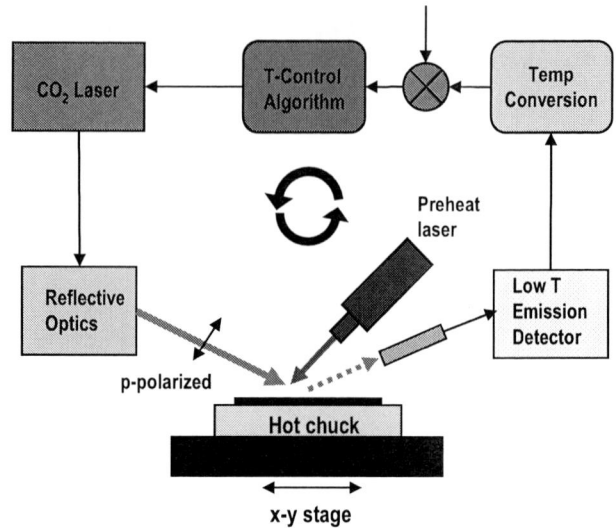

Figure 1: Schematic of low temperature LSA system

To resolve this issue, a second laser beam is introduced on the wafer, referred to as the preheat beam. The preheat beam can be absorbed by the wafer at low chuck temperatures [10]. This preheat beam overlaps the CO_2 beam at the wafer plane, and is much wider than the CO_2 beam in the scanning direction. The preheat beam provides localized heating which enables the CO_2 beam to be absorbed by the wafer, independent of chuck temperature. Investigations thus far have shown that a chuck temperature range of 150-$200^{\circ}C$ is appropriate for the application of LSA as a replacement for the second RTP nickel silicide step. In this temperature range, the thermal budget from the chuck does not cause any unintended phase transformation of the nickel silicide. For LSA as a replacement for the first RTP step, 50-$100^{\circ}C$ is an appropriate chuck temperature range. The preheat beam is used at a low power, tuned so that it heats the wafer just enough to allow the CO_2 laser to be absorbed. In this configuration, the CO_2 laser still does the majority of the wafer heating, and the pattern effect advantage of the standard design is still maintained. Data will be presented below to support this claim.

Secondly, the temperature measurement and control system must be optimized for the lower temperatures required for the nickel silicide process. Regarding the temperature range, it has been shown that the phase transformation temperatures for nickel silicide using LSA are several hundred degrees Celsius higher than they are for RTP [2, 3]. Hence, the operating range of the LSA temperature control system was chosen to be 500-$1000^{\circ}C$ for this application. A new low-temperature emission detector was selected based on analysis of the spectral distribution of the emitted radiation as well as other design criteria for optical pyrometry. The new detector provides a higher signal-to-noise ratio and has sufficient sensitivity for measurement and control in the temperature range of interest. Special digital filtering was also introduced in the control loop to suppress any reaction of the CO_2 laser power to perturbations in the emission due to micro-emissivity variations caused by the silicided areas of the device wafers.

III. PROCESS UNIFORMITY

A. Within-die Uniformity and Layout Dependence

As mentioned above, pattern effects in advanced logic processing can be a serious manufacturing problem, especially in an environment where there are many different products being manufactured. The advantages of long-wavelength LSA over short-wavelength millisecond tools for USJ applications have been well established in the literature [4, 5, 6], and validated in mass production of advanced logic devices [11]. Typically, the WID temperature range for LSA for USJ processes is on the order of 5-20°C. For short-wavelength MSA tools such as Flash Anneal (FA) or diode laser annealing (DL) the WID temperature range can be anywhere from 100-250°C, and is highly dependent on device layout. Solutions such as absorber layers [12] or optical dummification through design rules [13] are prohibitively expensive and/or too complex for manufacturing environments with many products.

Pattern effects on device wafers at the second step of nickel silicide formation have not been as extensively studied as those at the junction annealing step. At this point in the process, the wafer will have gone through the first RTP step and selective etch, leaving nickel-rich silicide in any areas where crystalline silicon or polysilicon is exposed. To quantify the effect of the nickel silicide on pattern effects, the within-die optical reflectance was mapped for three advanced logic device wafers pulled from the manufacturing process after selective etch, just before the second nickel silicide annealing step. All three wafers had different layouts. The WID optical reflectance was measured using an optical test stand that has a spatial resolution of approximately 100μm [10]. The WID reflectance maps were generated using two optical sources: 1) A diode laser (λ=850nm) at near-normal incidence, to simulate a full scale laser annealing tool with a similar heat source, and 2) A p-polarized CO_2 laser (λ=10.6μm) at Brewster's angle with a preheat beam, to simulate the heat source in a full scale LSA system. In both cases, the spot size of the illumination beam was approximately 100μm so that sufficient spatial resolution could be attained.

The reflectance measurements generate a 2D reflectance map having 100μm spatial resolution for an individual die and can be depicted in the form of a histogram, as shown in Fig 2. Figure 2a shows the reflectance histograms for a single die from each of the three wafers for the near-normal, short-wavelength diode source. The width of the distribution within a single die represents the WID reflectance uniformity, which will drive pattern effects (temperature variations) during laser annealing. Figure 2b shows the equivalent reflectance histograms for the long-wavelength LSA configuration. For the near-normal diode laser, the reflectance distribution within a particular die is relatively large, ranging from 20 to 50%, indicating that the temperature will be highly non-uniform during laser annealing. For the LSA configuration, the WID distribution is relatively narrow for all wafers (up to one order of magnitude more narrow), indicating that the temperature will be much more uniform during laser annealing. To quantify

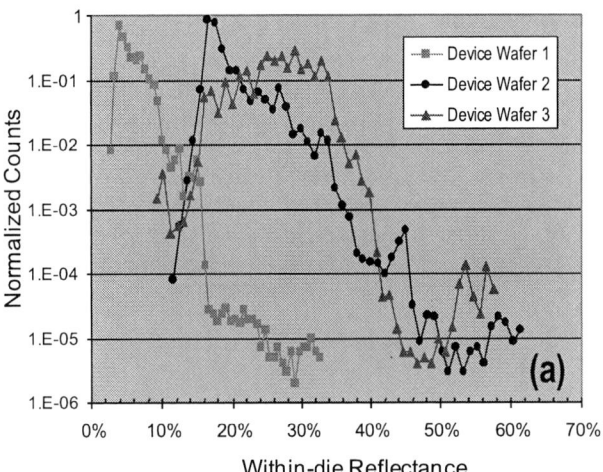

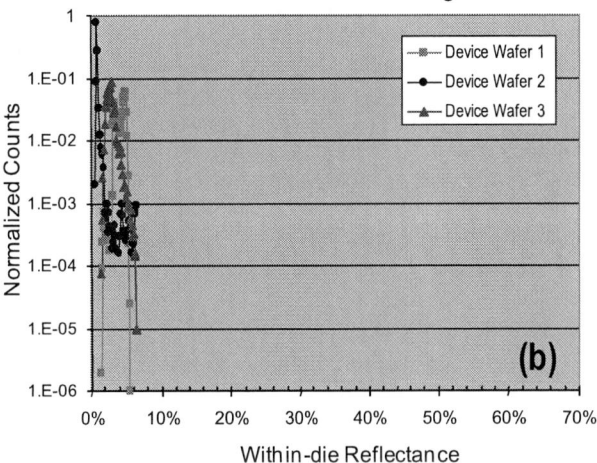

Figure 2: Histograms of within-die reflectance maps, measured for three different nickel silicide device wafers. a) Using diode laser at near-normal incidence for the illumination source. b) Using preheat + CO_2 laser at Brewster's angle for the illumination source.

the relationship between reflectance variations and temperature variations, Wang et al. [10] did detailed thermal simulations to convert similar WID reflectance maps to WID temperature maps for a typical second step laser annealing process at 850°C. The simulations predicted that the WID temperature range for a die with similar distributions as device wafer #2 for the laser diode case was over 100°C, and the equivalent temperature range for LSA was approximately 15°C. The trends of the reflectance histograms and temperature variations for LSA compared to short-wavelength alternatives show very similar behavior as for USJ processes [4, 5, 6], demonstrating that unique heating approach of LSA effectively suppresses pattern effects even in the presence of nickel silicide structures.

Hence, we conclude that the presence of the nickel silicide device structures do not appear to have a substantial impact on the effective optical properties of the wafer for the p-polarized CO_2 laser at Brewster's angle.

In the context of layout dependence, the three histograms in Fig 2a show large variations between them, both in average reflectance of each wafer and the distribution within a particular wafer. For a system that runs in open loop, i.e., constant power, differences in average reflectance will lead to differences in average processing temperature. In the case of Figure 2a, the difference in mean temperatures between Device #1 and Device #3 would be over 100°C. Furthermore, all three wafers would have different WID uniformities, which may lead to yield variations between the products. Fig 2b shows the layout dependence for LSA. The mean reflectances for all three wafers are within +/- 1.5% (leading to temperature differences of approximately 10°C) and the widths of the distributions for all three wafers are within a few percent of each other. This magnitude of layout dependence for LSA should lead to consistent yield from product-to-product, and process retargeting would generally not be necessary when transferring the LSA process to new device products.

As devices scale to smaller dimensions, the thermal process window for nickel silicidation will probably become more stringent due to the shorter distances that nickel atoms will have to diffuse to cause junction leakage. This could limit the extendibility of any thermal process which has large WID non-uniformities and/or issues transferring processes between products. For LSA, the small pattern effects and consistency between different layouts severely reduces the risk of pattern or layout dependent yield loss, and gives the tool extendibility to future device nodes.

B. Closed-loop temperature control

Fig. 3 shows a histogram of the measured temperature for a nickel silicide device wafer processed in closed loop on LT-LSA system, with a target temperature of 850°C. The average temperature for the entire wafer is 849.8°C, showing virtually no steady state control error. The temperature variation across the wafer is 4.2°C (1σ), demonstrating that the system has sufficient signal-to-noise ratio to maintain very stable control. Fig 4 shows a 2D temperature map of the same wafer, which can be generated for every wafer processed in high volume production. There is no correlation between the within-wafer temperature distributions and the device patterns, demonstrating that the system is effective at suppressing any system noise due to micro-emissivity fluctuations.

This temperature measurement and control system is very useful for process control in a manufacturing environment, serving the following functions:

- Its primary job is to provide WIW uniformity, WTW repeatability, and day-to-day stability, all of which should improve yield. The WIW wafer uniformity and long term stability of the temperature measurement and control system on the standard LSA system has been reported [8], and similar results are expected from the low temperature system.

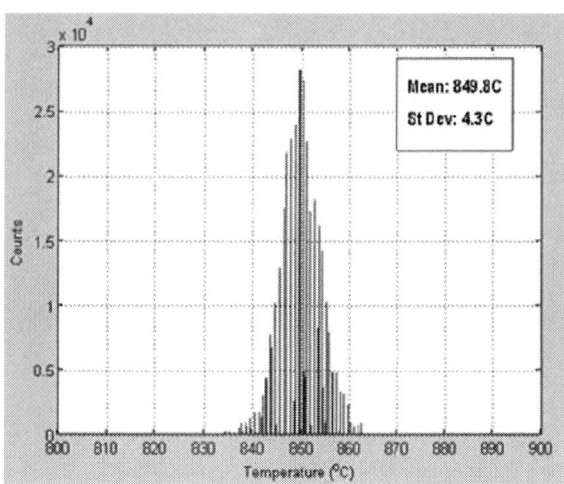

Figure 3: Histogram of measured temperature of a nickel silicide device wafer (after selective etch) processed with the closed loop temperature control system.

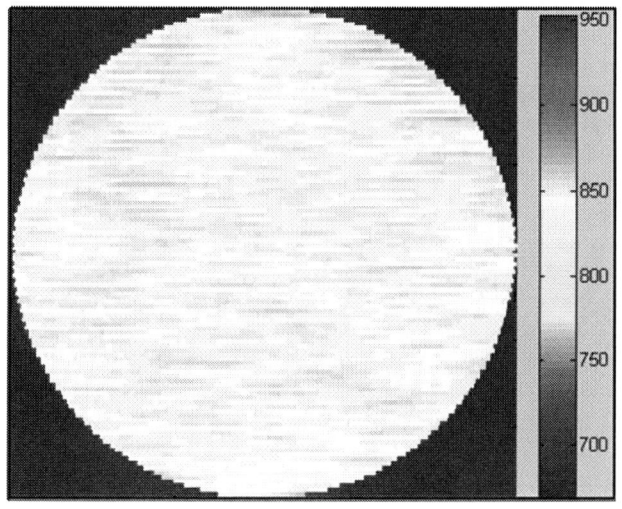

Fig 4: 2D temperature map (°C) of the same wafer represented in Fig 3. The x and y scales are each 300mm.

- A full wafer temperature map with thousands of points is generated for each device wafer, which can be used proactively or at a later date to compare against any anomalous electrical data.

- The system can also be useful for tool health monitoring and interdiction, where the measured temperature can be compared to warning and error limits on a per scan basis or per wafer basis, and process halted if the peak temperature goes out of control. The remainder of the wafer can be processed later after the issue is resolved so no wafer scrap is suffered. The worst case scenario is the loss of the die processed by the single interdicted scan.

IV. PROCESS CHARACTERIZATION

The key LSA process parameters for nickel silicidation are peak temperature, dwell time, and "step size", as all three of these parameters have a significant effect on the wafer thermal budget. The step size is defined as the distance which the stage moves perpendicular to the scan direction (the long direction) between subsequent scans "N" and "N+1", and it determines how much overlap there is between the CO_2 beam path in adjacent scans. The step size is stated as a percentage of the total length of the beam in the long direction. The step size has an effect on the thermal budget of any given point on the wafer. For example, for a process with 50% step size, each point on the wafer would pass under the CO_2 beam twice. For a process with 100% step size, each point on the wafer would pass under the CO_2 beam only once.

Silicon wafers with a blanket NiPt thin film were used to characterize the effect of these three LSA parameters on the phase transformation of nickel silicide. Most of this work was recently reported on in detail by Mileham et al. [14]. The data pertaining to the effect of step size will be examined more carefully below, as it has important repercussions for productivity. Figure 5 shows the Rs vs. Peak temperature for 50 and 100% step size. The dwell time for these wafers was 800µs. Near 850°C, the Rs on both curves rise sharply as transformation to the silicon rich phase ($NiSi_2$) and/or agglomeration take place. Interestingly, the curve with 100% overlap has a transition temperature which is significantly higher than the 50% overlap, which in principal would allow one to target a higher process temperature and potentially get better device performance. Based on this data, and the data reported by Mileham et al. [14], there is no apparent process disadvantage to using 100% step size as the best known method (BKM) for the nickel silicide process.

To further compare the performance of 50 vs. 100% step size, the WIW uniformity of sheet resistance (Rs) was measured on the same blanket NiPt wafers using a standard four point probe. The wafers were measured before and after the anneal, and the LSA-induced non-uniformity was calculated by subtracting the pre-measurement from the post-measurement. The results from two of the wafers are shown in Fig. 6. The wafer in Fig 6a was processed at 50% step size at 800°C, and 6b was processed at 825°C at 100% step size, both below the agglomeration temperature. LSA-induced non-uniformities for the 50% and 100% step wafers are almost identical at 0.74% and 0.78%, respectively. This data further supports the conclusion that 100% step size is an appropriate BKM for nickel silicidation.

V. PRODUCTIVITY

For LSA, the throughput is dominated by the actual processing time, i.e., the amount of time the wafer is scanned by the laser. This time is primarily determined by two parameters: 1) the number of scans it takes to fully anneal the wafer, and 2) the stage speed (which is set for a specified dwell time). For a given beam length (which is fixed), the number of

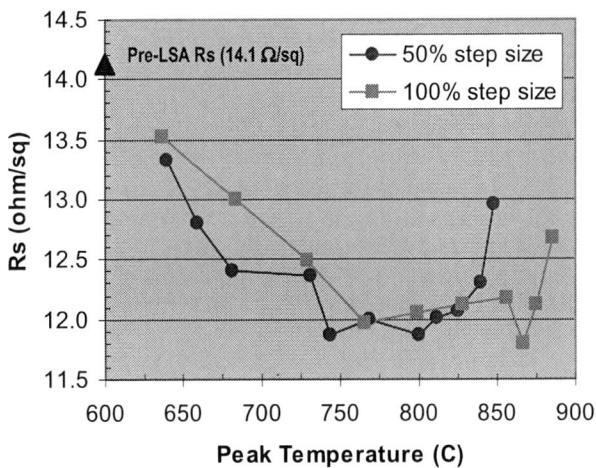

Fig 5: Nickel silicide transformation curves for a blanket NiPt film for 50% and 100% step size. Both curves were generated using a dwell time of 800 usec.

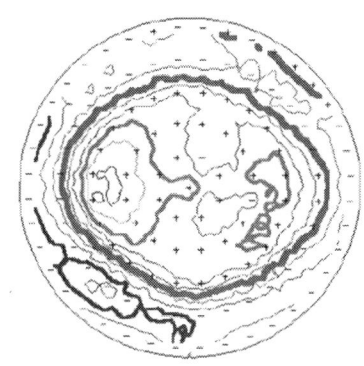

(a) 50% step size
Total NU = 1.74% (1s)
LSA NU = 0.74% (1s)

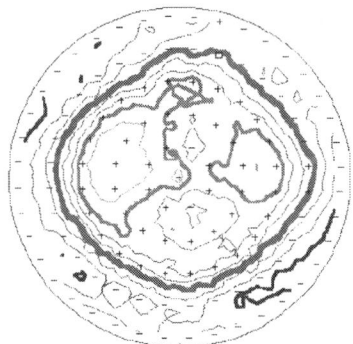

(b) 100% step size
Total NU = 1.78% (1s)
LSA NU = 0.78% (1s)

Figure 6: LSA-induced non-uniformity on blanket NiPt monitor wafers, comparing 50% and 100% step size. The wafers were processed at a)800°C and b)825°C.

scans is determined by the step size. Therefore the maximum throughput is obtained by maximizing step size (while maintaining process uniformity). For junction activation, the typical BKM is 50% step size, which was historically chosen to get the best WIW uniformity for critical front-end processes. For the case of nickel silicidation, the data presented above shows that a 100% step size is adequate, which will reduce the number of scans per wafer by a factor of two, and nearly double the throughput. For the most recent production model of LSA, a maximum throughput of 70wph can be obtained using a step size of 100%. This level of productivity is considered to be production-worthy when compared to other single-wafer thermal processing equipment.

VI. SUMMARY

Key design changes to the standard high-temperature LSA system enable processing at the lower temperatures required for nickel silicide applications. This new LT-LSA system meets several of the most critical requirements for successful advanced logic manufacturing.

- Pattern and layout dependence is minimized: Within-die reflectance measurements from multiple nickel silicide device wafers show that the low temperature LSA system is very effective is suppressing WID pattern effects, performing 5-10X better than a near-normal short wavelength laser diode source. The LT-LSA system also shows excellent layout-to-layout repeatability compared to the near-normal diode laser.
- Closed-loop temperature control: The new low-temperature measurement and control system demonstrates WIW temperature variation of 4.2°C (1σ) on device wafers, independent of device patterns. The combination of the above capabilities comprise excellent process control for nickel silicide processes, which has the potential to increase yield and device performance, and make LSA extendible to future device nodes where process windows are likely to shrink.
- High productivity: Process characterization on blanket NiPt wafers showed that phase transformation and Rs uniformity are not adversely effected by increasing the scanning step size to 100% of the beam length. This large step size enables one to achieve a production-worthy throughput of 70wph on the most recent LSA production system.

In summary, LT-LSA satisfies all major criteria to successfully implement the nickel silicidation step in high-volume production

REFERENCES

[1] C. Ortolland, E. Rosseel, N. Horiguchi, C. Kerner, S. Mertens, J. Kittl, E. Verleysen, H. Bender, W. Vandervost, A. Lauwers, P.P. Absil, S. Biesemans, S. Muthukrishnan, S. Srinivasan3, A.J. Mayur, R. Schreutelkamp and T. Hoffmann, "Silicide Yield Improvement with NiPtSi Formation by Laser Anneal for Advanced Low Power Platform CMOS Technology," IEDM Tech. Digest., 2009

[2] Y.W. Chen, N.T. Ho, J. Lai, T.C. Tsai, C.C. Huang, J.Y. Wu, B. Ng, A.J. Mayur, A. Tang, S. Muthukrishnan, J. Zelenko, and H. Yang, "Advances on 32nm NiPt Salicide Process," Proc. IEEE Int. Conf. on Adv. Thermal Processing Semi., 2009.

[3] J. Hebb, Y. Wang, S. Chen, M. Shen, S. Zhou, X. Wang, and D. Owen, "Expanded Application Space for Laser Spike Annealing of CMOS Devices," Proc. IEEE Int. Conf. on Adv. Thermal Processing Semi., 2009.

[4] L. M.Feng, Y. Wang, D.M. Markle, "Minimizing Pattern Dependency in Millisecond Annealing," Proc. Inter. Workshop Junction Tech., pp. 25-30, 2006.

[5] T. Miyashita, T. Kubo, Y. S. Kim, M. Nishikawa, Y. Tamura, J. Mitani, M. Okuno, T. Tanaka, H. Suzuki, T. Sakata, T. Kodama, T. Itakura, N. Idani, T. Mori, Y. Sambonsugi, A. Shimizu, H. Kurata, and T. Futatsugi, "A Study on Millisecond Annealing (MSA) Induced Layout Dependence for Flash Lamp Annealing (FLA) and Laser Spike Annealing (LSA) in Multiple MSA Scheme with 45 nm High-Performance Technology," IEDM Tech. Digest, 2009.

[6] R. Beneyton, A. Colin, F. Cacho, M. Bidaud, H. Bono, B. Dumont, P. Morin, K. Barla,"Origin of Local Temperature Variation during Spike Anneal and Millisecond Anneal," Proc. IEEE Int. Conf. on Adv. Thermal Processing Semi., pp. 183-193. 2008.

[7] D.M. Owen, C. Otten, H. Bu, Y. Wang, S. Shetty and J. Hebb, "Correlation of Device Performance to Die-Level Stress Variations," in Proc. Inter. Workshop Junction Tech., pp. 15-18, 2009.

[8] S. Chen, J. Hebb, A. Jain, S. Shetty, and Y. Wang, "Wafer Temperature Measurement and Control During Laser Spike Annealing," Proc. Int. Conf. on Adv. Thermal Processing Semi, pp. 239-244, 2007.

[9] J. McWhirter, D. Gaines, and P. Zambon, "Emission Feedback Control System for Sub-millisecond Laser Spike Anneal," Proc. IEEE Int. Conf. on Adv. Thermal Processing Semi, pp. 157-162, 2008.

[10] Y. Wang, S. Chen, M. Shen, X. Wang, S. Zhou, J. Hebb, D. Owen, "Dual Beam Laser Spike Annealing Technology," Proc. Inter. Workshop Junction Tech, 2010.

[11] R. Van Roijen, O. Gluchenkov, J. Willis*, and M. Hurley, "Application and control of Laser anneal at the 65 and 45 nm node," Proc. Adv. Semi. Man. Conf., pp.217-219, 2009.

[12] P. Morin, F. Cacho, R. Beneyton, B. Dumont, M. Bidaud, E. Josse, C. Gallon, R. Ranica, A. Villaret, R. Bianchini, T. Devoivre, E. Serret, R. Binger, K. Barla, M. Haond, A. Colin, H. Bono, C. Chaton, "Managing Annealing Pattern Effects in 45nm Low Power CMOS Technology," Proc. ESSDERC, pp. 288-291, 2009.

[13] K. J. Kuhn, "Reducing Variation in Advanced Logic Technologies: Approaches to Process and Design for Manufacturability of Nanoscale CMOS," IEDM Tech. Digest, pp., 471-474, 2007.

[14] J. Mileham, V. Le, S. Shetty, J. Hebb, Y. Wang, D. Owen, R. Binder, Rainer Giedigkeit, S. Waidmann, I. Richter, K Dittmar, H. Prinz, M. Weisheit, "Impact of Dual Beam Laser Spike Annealing Parameters on Nickel Silicide Formation Characteristics," Proc. IEEE Int. Conf. on Adv. Thermal Processing Semi., pp. 130-135, 2010.

Mechanical Properties of Si-C-O-H Low-k Dielectrics Prepared by Plasma Enhanced Chemical Vapor Deposition

Peter Woytowitz, Sassan Roham
Computational Modeling
Novellus Systems, Inc.
San Jose, CA 95134 USA

Dong Niu, Haiying Fu
PECVD Business Unit,
Novellus Systems, Inc.
Tualatin, OR 97062 USA

Abstract - Mechanical properties of low dielectric constant (low-k) materials are one of the key areas that need to be better understood in order to improve copper/low-k dual damascene integration. In this paper, mechanical properties of carbon doped oxide (CDO) films deposited using TMCTS and CO_2 as precursors are reported. Differences in Young's modulus, residual stress and fracture toughness for films prepared by two processes A and B have been observed and are correlated to carbon contents using Fourier transform infrared spectroscopy (FTIR). For similar carbon content, process B films show higher modulus and toughness and lower stress than process A films. Fracture toughness calculated from critical thickness measurement indicates bimodal behaviors for process B films. Analysis of detailed chemical bonding structures is needed to further understand the mechanical properties of low k dielectrics. While this work was performed using a particular CDO formulation, the methodology and considerations are applicable to a range of nano materials. Therefore, these methods and results will prove to be useful to a wide range of industries interested in integrating nano-materials into complex structures.

I. INTRODUCTION

As the aggressive scaling of microelectronic devices in the integrated circuit (IC) industry continues, tremendous challenges are being faced in back-end-of-the-line (BEOL) technologies, [1 – 3]. With the feature device dimension continually shrinking, signal propagation delay, inductive crosstalk noise between metal lines, and power dissipation become much more pronounced. In order to minimize these effects, enormous efforts have been made in implementing low dielectric constant (low k) materials as interlayer dielectrics (ILD) in replace of SiO_2. The ideal low k materials properties include excellent thermal stability, chemical stability, high mechanical strength, and excellent electrical properties. Potential candidates are carbon doped SiO_2 (CDO or Si-O-C-H), hydrogen or methyl-silsesquioxane (HSQ or MSQ), organic polymers, and amorphous carbon (*a*-C:F), [4 – 14]. Among them,

carbon doped SiO_2 prepared by plasma enhanced chemical vapor deposition (PECVD), also known as organo-silicate glass (OSG), has attracted much attention due to its compatibility with conventional processing techniques, high thermal and chemical stabilities, and good mechanical strength. Still, issues such as delamination and film cracking during chemical mechanical polishing (CMP) and packaging remain [15].

As the main focus of this paper, film cracking is closely related to mechanical properties of films. In order to understand and control film cracking behaviors, it is crucial to understand (1) how film cracking is determined by mechanical properties such as hardness, modulus, and fracture toughness and (2) how mechanical properties are related to chemical bonding structure and can be improved to prevent cracking from occurring during subsequent processing.

In this paper, we report the mechanical properties of Si-O-C-H films prepared by plasma enhanced chemical vapor deposition (PECVD). Fracture toughness is exacted from film modulus, residual stress and critical thickness. By comparison of films deposited at various conditions, mechanical properties are correlated to chemical structures and compositions. We compare measured fracture properties to expected trends and compare these to those of amorphous silicon dioxide (SiO2).

II. EXPERIMENT

CDO films (Si-O-C-H) with thicknesses from 500 nm up to 10 μm were deposited on p-type 200-mm Si(100) substrate using tetramethylcyclotetrasiloxane (TMCTS) and CO_2 as precursors in a Novellus Concept Two SEQUAL™ reactor . TMCTS, a liquid compound, was introduced into the chamber through an Integrated Liquid Delivery System (ILDS). TMCTS and CO_2 flows were in the range of 0.5-6 ccm and 1000-10000 sccm, respectively. The process chamber pressure varied from 2 to 10 torr. The substrate temperature was set at 400°C. High frequency (HF) or both HF and low frequency (LF) RF plasmas were used. Film thickness and refractive index were measured using a

978-1-61284-408-4/11 $26.00 © 2011 IEEE

Therma-Wave Opti-probe 5340C or KLA-Tencor ASET-F5x. The dielectric constant (k) was measured using a SSM mercury probe 5130. The k values are in the range of 2.7-3.4 in this paper. The Fourier Transform Infrared (FTIR) spectra, taken on a Brad-Rad QS-312 with the scan range from 400 to 4000 cm^{-1}, were used to determine Si-O, Si-CH$_3$, and Si-H bonding concentrations. The stress was measured using a Tencor Flexus-5400. Cracks were monitored using a Jenatech optical microscope.

III.   RESULTS & DISCUSSION

In the following section, we divide the TMCTS-based CDO films into two groups: process A films and process B films, since the two groups of films demonstrate very distinct mechanical properties, as will be shown in the following section. As the main focus of this paper is mechanical properties of Si-O-C-H films, dielectric properties and their relationships with process parameters will be discussed elsewhere.

In order to improve the mechanical strength, it is essential to correlate mechanical properties with chemical structure and composition of low-k films. For CDO materials, it is possible to modulate the amount of methyl groups and therefore affect the relative number of Si-O bonds relative to the number of Si-CH$_3$ bonds. This is quantified experimentally here through the bond area ratio. A representation of the bond differences between amorphous oxide (SiO2) and CDO materials is shown in Figure 1. The theoretical bond strength in the SiO$_2$ structure can be inferred from theromodynamic data (enthalpy of formulation) which can also be approximated using the average bond energy, [16]. For one example from [16] the bond energy for an Si-O bond is 452 kJ per mole of bonds, while the average bond energy for an Si-C bond is only 360 kJ per mole of bonds. Therefore the addition of the Si-CH$_3$ groups is expected to reduce the energy required to rupture these bonds (say during fracture), and therefore, would indicate lower mechanical strength properties (especially fracture toughness) for the CDO materials compared to the SiO$_2$. Experimental measurements of fracture toughness bear this out. Here we present quantitative data related to these effects.

Figure 2 shows the films Young's Modulus (E$_f$) as a function of the area ratio of [Si-CH$_3$] to [Si-O]. The areas of Si-CH$_3$ and Si-O bonding are calculated from peaks near 1270 and 1040 cm^{-1} in FTIR spectra (spectra not shown). Open triangles are process A films; closed squares are process B films. These symbols are applied to the same groups of films throughout the paper. The two groups of films both show monotonic decrease of modulus with increasing Si-CH$_3$ concentration, indicating that the addition of CH$_3$ group into the films decreases the film modulus. On the other hand, for the same Si-CH$_3$ concentration, process B films have a higher modulus than process A films for most cases. This indicates that Si-CH$_3$ concentration alone cannot account for the mechanical property difference among different processes. Detailed chemical bonding structure must be considered. The films in discussion here are quite dense, and effects of pore size and distributions are not considered. It is interesting to note that if the modulus is plotted against dielectric constant k, the differences between process A and B films are much less pronounced (data not shown), implying that the governing factors for k and modulus are quite similar. Figure 2 also shows elastic modulus for another typical CDO film (of unknown bond area ratio) which shows similar average values for modulus. [17].

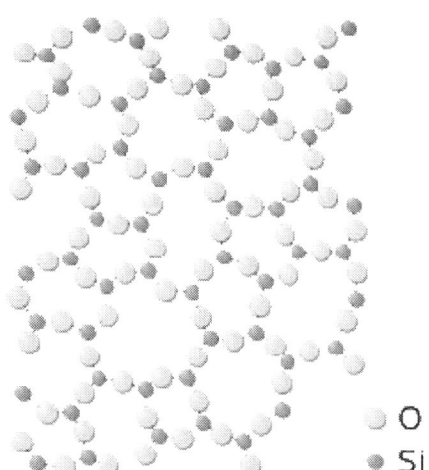

Amorphous SiO2 Structure, Density = 2.1 gm/cm^3

[18] Typical CDO Structure, Density = 1.2 – 1.4 gm/cm^3 [19]

FIGURE 1:   BOND DIFFERENCES REPRESENTATION BETWEEN AMORPHOUSE OXIDE (SIO2) AND CDO.

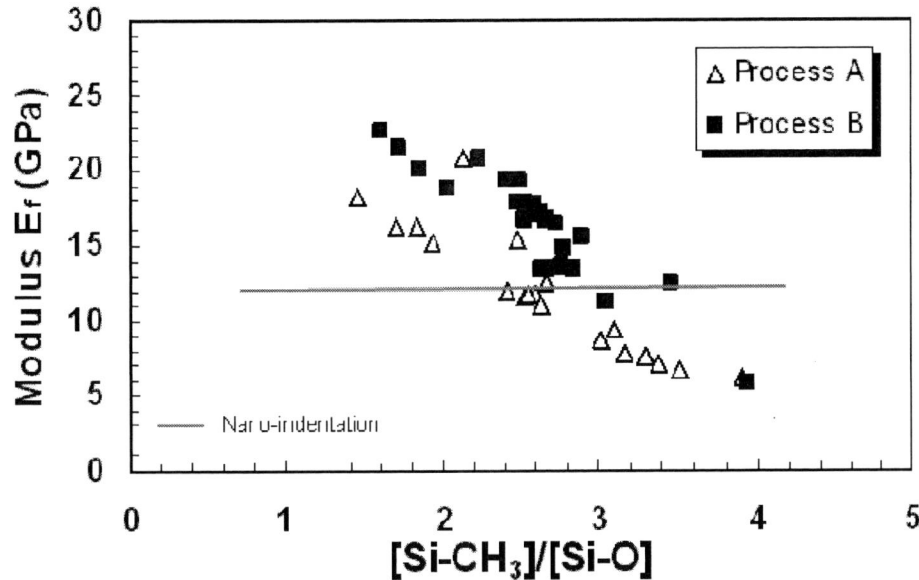

FIGURE 2: MODULUS E_f AS A FUNCTION OF AREA RAIO [SI-CH3] TO [SI-O]. THE AREA RATIO IS CALCULATED FROM FTIR SPECTRA..

Figure 3(a) shows the critical thicknesses (CTs), also known as cracking limit, for the two groups of films. This is the film thickness at which tensile residual film stress driven cracking is observed. (The use of this data to determine critical energy release rate,will be discussed later.) After deposition, the film was left in the cleanroom for three days before it was put under a microscope for crack inspection. In this way, with a magnification of 500, a 10 μm crack corresponding to a crack propagation velocity of 3.86×10^{-13} m/s can be readily identified. The critical thicknesses were determined by averaging the upper limit of the film thickness where definite cracks were observed and the lower limit where no cracks were observed. The error bar is one half of the distance between the upper and lower limits. In general, CTs increase with increasing modulus. However, it is apparent that there are distinct differences between the two groups of films. The two lines are for eye guidance only. For the same modulus, process B films show much higher CTs than process A films, especially in high modulus regime. The slope of the line for process B is also steeper. It is worth noting that for a process A film with the modulus of ~25.7 MPa, the CT is still below 3.3 μm. These phenomena can be partially explained by the film residual stress, as shown in Figure 3(b). The stress for process B films stays fairly constant at low modulus; at high modulus, the stress shows bimodal behaviors, with one group of films staying constant and the other group decreasing with the modulus. The films with lower stress show higher critical thickness in the high modulus regime. Stresses of process A films generally are higher than those of process B films, which may explain why the CTs of the process A films are lower than those of process B films.

Fracture toughness is an important mechanical property, which is defined as the work required to create a unit area of fracture surface from the original material, [20]. Based on Griffith's theory, the toughness Γ can be computed by

$$\Gamma = 2\gamma \tag{.1}$$

where γ is fracture surface energy. Since fracture surface energy is difficult to measure directly, one of the methods for toughness measurement is to measure critical fracture energy release rate G_c. For plane strain, the fracture energy release rate for channel cracking is defined as

$$G = Z \frac{(1-v^2) \cdot \sigma_f^2 \cdot h_f}{E_f} \tag{2}$$

where h_f is film thickness (see Figure 4), v is the Poisson's ratio of the low film, Z is the driving force number which is a function of the crack pattern and the modulus difference between the film and the substrate, [21]. It has been shown that Z of 1.62 for channel cracking is reasonable for low k films of interest whose moduli are much smaller than that of Si, [22]. Note Z is only a weak function of the modulus of low k film, therefore, the assumption that it is a constant will introduce only some minor error in the calculation, and general trends derived will still hold. The value of v is set at 0.25. Cracking occurs when fracture energy release rate is equal or greater than fracture toughness. Critical thickness (CT) or h_c is defined as the film thickness at which the fracture energy release rate is equal to the toughness:

$$\Gamma = G_c = Z \frac{(1-v^2) \cdot \sigma_f^2 \cdot h_c}{E_f} \tag{3}$$

Therefore, CT measurements provide a convenient way to calculate film toughness.

The calculated toughness using equation (3) is plotted against the area ratio of [Si-CH₃]/[Si-O], as shown in Figure 5. For process A films, Γ slightly decreases with increasing [Si-CH₃]/[Si-O] area ratio. For process B films, again, the films show bimodal behaviors. While toughness of process B film with higher stresses stays fairly constant, toughness for films with lower stresses increases almost linearly with increasing carbon content. These differences indicate that fracture toughness depends on not only carbon content but also on detailed chemical structures. A detailed analysis of the chemical structures is needed.

The ultimate goal of mechanical strength measurements of low-k dielectric films is to predict whether the films can stand subsequent processes including thermal treatments, chemical erosion, and external forces such as normal stress and shear stress without cracking.

This requires that the low-k films undergo similar conditions to subsequent processes in order to make a reliable prediction. In this sense, critical thickness measurement is a simulation of the external normal and shear stress since equations 2 and 3 hold only for mode 1 cracking, assuming that elastic linear cracking at crack tip dominates and deformation zone due to plasticity can be neglected.

Critical thickness measurement is very useful in low k dielectric development. Figure 5 shows that critical thickness is proportional to the ratio of modulus to residual stress square for both process A and process B films. Due to the high stresses, process A films with k values of interest generally show low critical thickness. However, for a similar value of modulus/stress², the cracking limit can vary in a large range (which is also illustrated by the difference in toughness, shown in Figure 6). Figure 6 also shows that channel cracking G_C results compare favorably with G_c values obtained in Ref [18] using nanoindentation (bond area ratio for Ref [18] results is unknown).

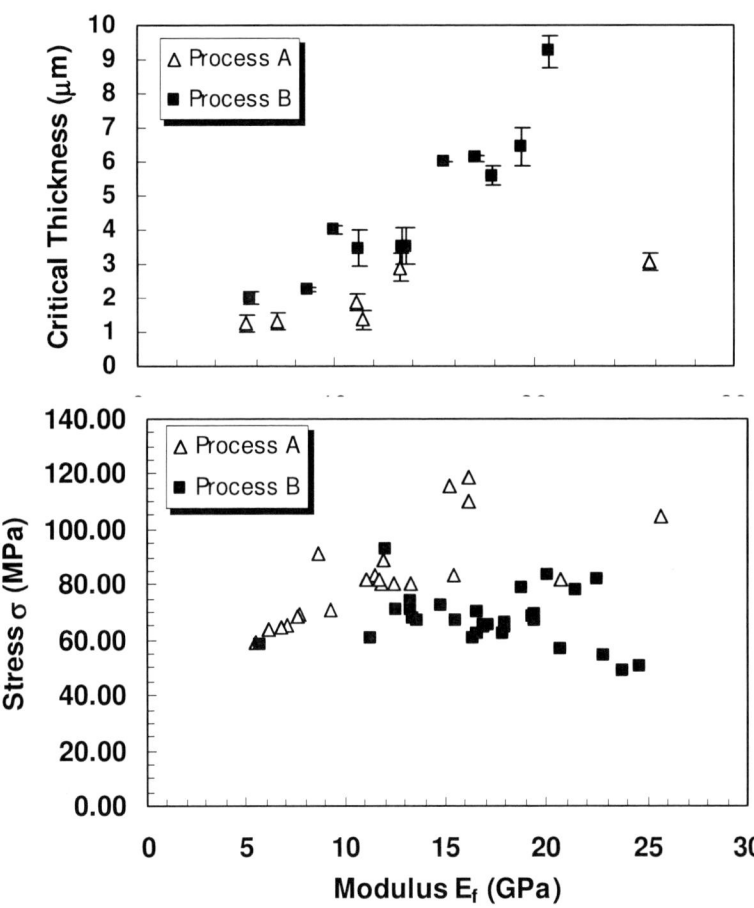

FIGURE 3: (A) CRITICAL THICKNESS OF THREE GROUPS OF FILMS VS. MODULUS. (B) FILM RESIDUAL STRESS AS A FUNCTION OF MODULUS.

978-1-61284-408-4/11 $26.00 © 2011 IEEE   130

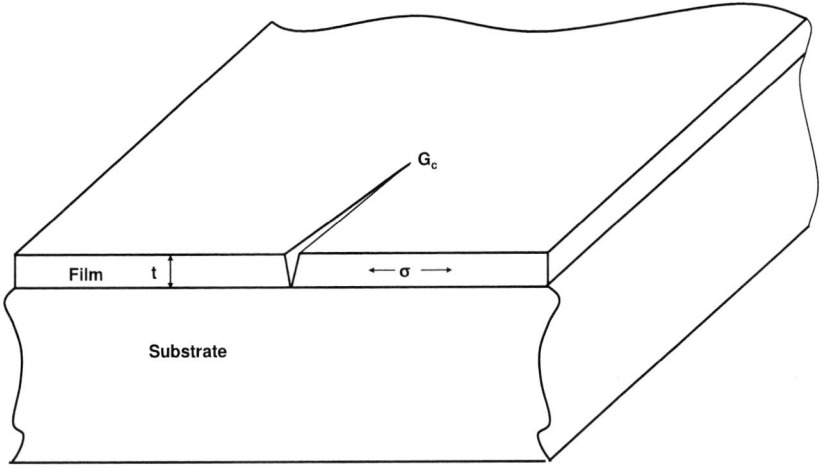

FIGURE 4:   CHANNEL CRACK OF A THIN FILM ON A SUBSTRATE.

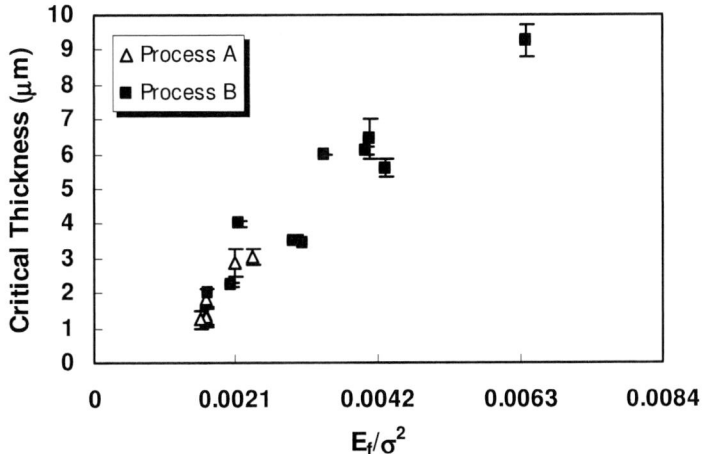

FIGURE 5:   CRITICAL THICKNESS OF TWO GROUPS OF FILMS VS. MODULUS/STRESS.

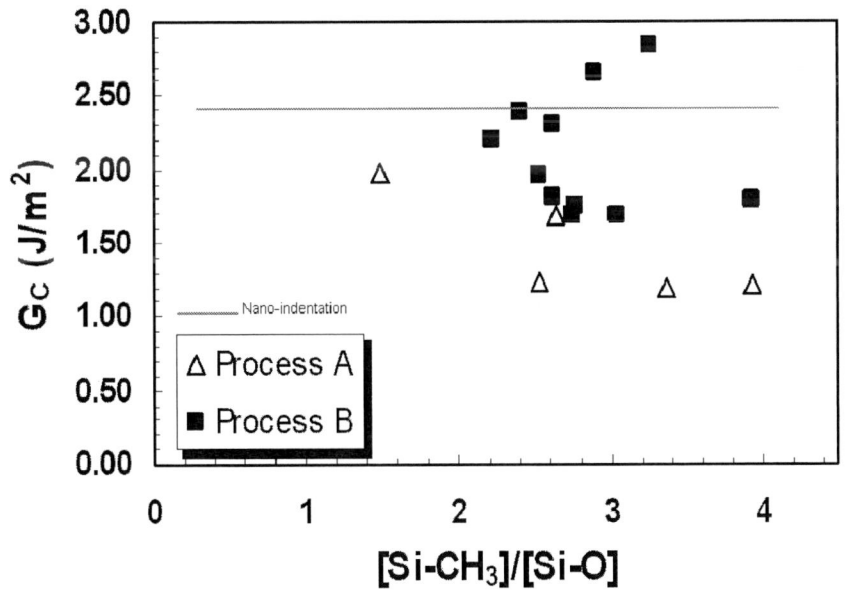

FIGURE 6:   FRACTURE TOUGHNESS AS A FUNCTION OF THE AREA RATIO OF SI-CH3 TO SI-O BONDING.

The low toughness of process A films and bimodal behaviors of process B films indicate detailed chemical bonding is likely as important as the carbon content in the low k film. Considering the residual stress of process B films is also bimodal, this may imply the same mechanism contributes to both the stress and fracture toughness. On the other hand, Equation 3 shows that film modulus, residual stress, and fracture toughness are three crucial properties of low k films which all play important roles in low k film cracking. It is also noted that use of the mechanical testing (channel cracking) has identified possible bond structural differences and therefore provides for an additional method of indentifying such microstructural structural differences.

IV. CONCLUSION

Mechanical properties of Si-O-C-H films deposited using TMCTS and CO2 are presented. Distinct differences in modulus, residual stress and toughness between process A and B films are correlated to carbon contents using FTIR. For similar carbon content, process B films generally show higher modulus and toughness and lower stress than process A films. Critical thickness is proportional to the ratio of film modulus over film residual stress square for both process A and B films. Toughness calculated from critical thickness measurement indicates bimodal behaviors for process B films. Analysis of detailed chemical bonding structures is needed to further understand the mechanical properties of low k dielectrics.

REFERENCES

[1] S. I. Association, The International Technology Roadmap for Semiconductors, 2001 edition (Austin, TX, 2001).

[2] M. Morgen, E. T. Ryan, J. H. Zhao, C. Hu, T. H. Cho, and P. S. Ho, Annual Review of Materials Science **30**, 645-680 (2000).

[3] K. Maex, M. R. Baklanov, D. Shamiryan, F. Iacopi, S. H. Brongersma, and Z. S. Yanovitskaya, Journal of Applied Physics **93**, 8793-8841 (2003).

[4] A. Grill and V. Patel, Journal of Applied Physics **85**, 3314-3318 (1999).

[5] L. C. M. Han, J. S. Pan, S. M. Chen, N. Balasubramanian, J. N. Shi, L. S. Wong, and P. D. Foo, Journal of the Electrochemical Society **148**, F148-F153 (2001).

[6] Y. S. Mor, T. C. Chang, P. T. Liu, T. M. Tsai, C. W. Chen, S. T. Yan, C. J. Chu, W. F. Wu, F. M. Pan, W. Lur, and S. M. Sze, Journal of Vacuum Science & Technology B **20**, 1334-1338 (2002).

[7] J. Lubguban, T. Rajagopalan, N. Mehta, B. Lahlouh, S. L. Simon, and S. Gangopadhyay, Journal of Applied Physics **92**, 1033-1038 (2002).

[8] Q. G. Wu and K. K. Gleason, Journal of Vacuum Science & Technology A **21**, 388-393 (2003).

[9] D. Shamiryan, K. Weidner, W. D. Gray, M. R. Baklanov, S. Vanhaelemeersch, and K. Maex, Microelectronic Engineering **64**, 361-366 (2002).

[10] A. Grill, Journal of Applied Physics **93**, 1785-1790 (2003).

[11] T. Furusawa, D. Ryuzaki, R. Yoneyama, Y. Homma, and K. Hinode, Journal of the Electrochemical Society **148**, F175-F179 (2001).

[12] S. Z. Yu, T. K. S. Wong, K. Pita, X. Hu, and V. Ligatchev, Journal of Applied Physics **92**, 3338-3344 (2002).

[13] V. Ligatchev, T. K. S. Wong, B. Liu, and Rusli, Journal of Applied Physics 92, 4605-4611 (2002).

[14] J. M. Shieh, K. C. Tsai, S. C. Suen, and B. T. Dai, Journal of Vacuum Science & Technology B **20**, 1388-1393 (2002).

[15] A. E. Braun, Semiconductor International **26**, 52-56 (2003).

[16] Zumdahl & Zumdahl, Chemistry, Houghton MIfflin Co. (2003), ISBN 0-618-22158-1.

[17] Roham, S., Woytowitz, P., Niu, D., Fu, H., 47th Annual Meeting, Society of Engineering Sciences, Oct. 4 – 6, Ames, IA.

[18] Wikipedia http://en.wikipedia.org/wiki/Silicon_dioxide

[19] Nanoindentation on low-k Dielectrics, 3rd European Symposium on Nanomechanical Testing, Sywert Brongersma, IMEC 2002.

[20] Z. Suo, Journal of Vacuum Science & Technology a-Vacuum Surfaces and Films **11**, 1367-1372 (1993).

[21] J. W. Hutchinson and Z. Zuo, Advances in Applied Mechanics **29**, 63-191 (1992).

New Methods for Improved SRAM Detection through Scattered Light Collection

Reuven Barel, Keren Shachar, Yakir Bechler, Nir Horesh

Applied Materials Israel, PDC, Rehovot, Israel
reuven_barel@amat.com

Hsien-Tsung Chiang, To-Yu Chen
Taiwan Semiconductors Manufacturing Company (TSMC)
Hsinchu, Taiwan ROC

Abstract - **Accelerated design rule (DR) shrinkage is introducing new challenges to the world of process control, yield monitoring and wafer inspection (WI). One of the main challenges of WI is the detection of small defects below the optical resolution limit of the inspection systems. This paper present a new optical scheme for inspection of advanced SRAM patterns and demonstrates the significant value of the transition from pattern contrast detection to dark background detection by applying tailored light illumination and collection techniques. This work demonstrates the sensitivity improvement of this approach on two advanced 2xnm DR devices. This scheme, compared with current methods, demonstrates enhanced defect signal to noise ratio (SNR), enhanced defect capture rate, and higher throughput.**

Keywords: SRAM, Wafer inspection, DUV, Scattered light, Pattern suppression, Filtering

I. INTRODUCTION

Accelerated design rule shrinkage over the past years has continuously introduced new challenges to the world of yield monitoring and optical wafer inspection. One of the main challenges is the detection of ever smaller defects of interest (DOI) below the optical resolution limit of the WI systems. Evident to any large scale manufacturer today, these DOI, typically in the size range of the technology node, are correlated to yield, and their detection is therefore crucial to the production yield.

Traditional bright field (BF) systems detection sensitivity depends on the optics resolution [1] and pattern contrast, which in turn is correlated with λ/NA (λ is the illumination wavelength and NA is the numerical aperture of the objective lens). With advanced DR, when the pitch size is much smaller than the optical spot size on the wafer, BF resolution reaches its limit and certain DOI cannot be detected by traditional BF inspection.

Scattered light detection, which involves collection of scattered light outside of the illumination cone, and detection on a dark background, overcomes the resolution barriers, since the detection is determined by the Signal to Noise Ratio (SNR) and not by pattern contrast, which is impacted by the optics resolution.

Unresolved dark scenarios are typical to repetitive dense arrays, where distinct diffraction lobes at specific locations are formed and blocked by spatial filtering [2]. Typically, when such lobes can be blocked by a spatial mask as shown in Figure 1, dark background detection scenarios are enabled. This is usually the case of advanced DR, line-space structures, such as DRAM and Flash, where the lobes are either outside of the collection aperture or are filtered out. In such cases, this detection method offers higher sensitivity than traditional BF [3] as shown in Figure 2.

However, on logic patterns, this method is typically not applied due to the pitch size and the pattern complexity formed by multidirectional random features. In general, logic devices consist of two main pattern types, which require different inspection strategies. The first pattern type is the logic circuitry, and is typically characterized by random patterns. The second type of pattern is the SRAM, which is typically characterized as a repetitive pattern. Scattered light detection on SRAM patterns, based on scattered light and pattern suppression was typically not performed due to the plurality of the diffraction lobes and of the complexity of the required filters. On the other hand, BF imaging has limited resolution for detection of small defects in the size range of the DR. This situation often leads to reduced sensitivity on critical structures.

In this work, we present a new scheme involving tailored illumination and collection to enable effective pattern suppression. To demonstrate the new scheme, we used a laser-based DUV system having both BF and scattered light collection channels (UVision™, Applied Materials). BF light is collected through the illumination aperture and scattered light (grayfield - GF) is collected outside of the illumination aperture.

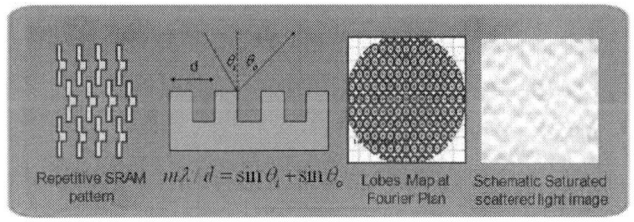

Repetitive SRAM pattern $\quad m\lambda / d = \sin\theta_i + \sin\theta_o \quad$ Lobes Map at Fourier Plan $\quad$ Schematic Saturated scattered light image

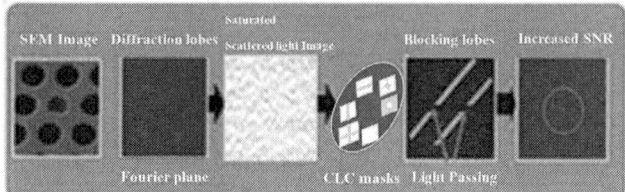

Figure 1. Pattern suppression with spatial filtering. Without filtering, the diffraction lobes saturate the image and typically carry no defect information. With filtering, the pattern is effectively filtered, leading to a dark background and higher defect SNR.

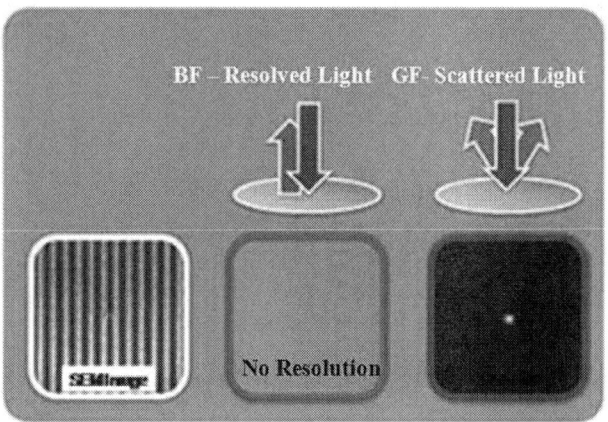

Figure 2. BF signal vs. scattered light signal on a 2xnm DR device.

II. EXPERIMENT

Two 28nm DR wafers at two process steps, Via layer after resist development (Via ADI) and SiGe deposition at the gate stage (SiGe DP) were analyzed in this work. In the Via ADI layer, the SRAM areas were inspected in scattered light mode, with pixel size (PS) of 100nm and 140nm using tailored masks. In the SiGe DP layer, the SRAM areas were inspected using scattered light mode with PS of 100nm, BF mode with PS of 100nm and 70nm, and in the new SRAM mode using PS of 140nm and 190nm. Table 1 summarizes the configurations used for the two layers. Random defects from all inspection maps were reviewed on a SEM tool.

Configuration	Via ADI	SiGe DP
Scattered light (100nm PS)	√	√
BF (100 nm PS)	-	√
BF (70nm PS)	-	√
New SRAM mode (140nm PS)	√	√
New SRAM mode (190nm PS)	-	√

Table 1. Inspected configurations used for the Via ADI and SiGe DP layers.

III. RESULTS AND DISCUSSION

Via ADI – In order to compare the SNR between the BF and the scattered light configurations, two particle defects in the size of ~50nm which were detected by both configurations were chosen. The SNR was calculated based on the inspection images, as shown in Figure 3.

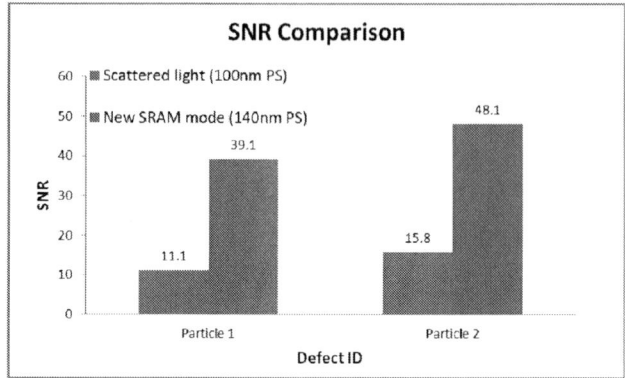

Figure 3. SNR comparison for two particle defects, for the scattered light @100nm PS (left columns, Blue) and new SRAM mode @140nm PS (right columns, Red).

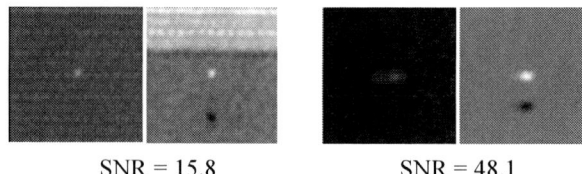

SNR = 15.8   SNR = 48.1

Figure 4. Scan and difference images of particle #2 of Figure 3. Left image of each pair correspond to scattered light @100nm PS. Right image of each pair correspond to the new SRAM mode @140nm PS.

From the comparison in Figure 3 it can be clearly seen that the SNR is significantly higher in the new SRAM mode compared to the scattered light mode. The SNR difference is understood by comparing the scan images in Figure 4. While both images have a similar defect signal, the pattern, seen in the scattered light mode images, is effectively filtered in the new SRAM mode, hence improving the SNR.

SiGe DP – SRAM pattern areas were inspected using the configurations shown in Table 1. Capture rates of three defect types were compared between the inspection configurations: 1) Missing pattern (Figure 6, left), 2) Bridge (Figure 6, right), 3) Particles.

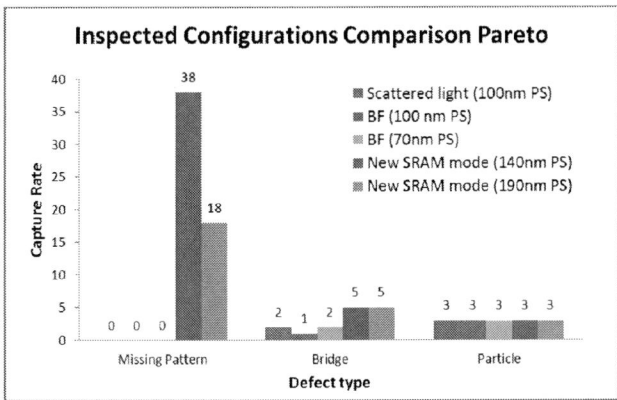

Figure 5. Capture rate comparison of Missing Pattern, Bridge and Particle for all the inspected configurations described in table #1.

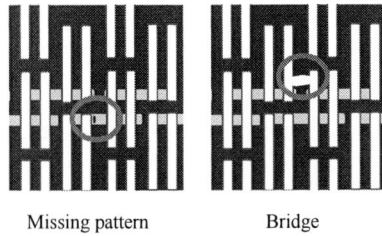

Missing pattern   Bridge

Figure 6. Missing pattern and bridge defects illustrations.

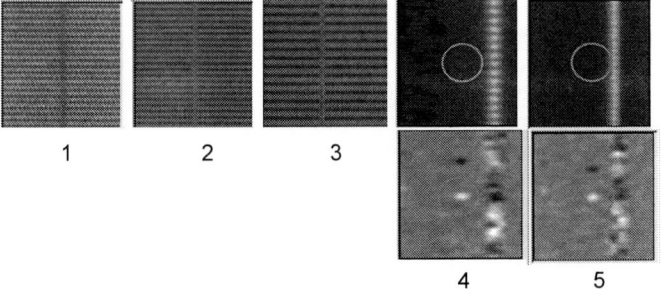

Figure 7: Scan images of one of the missing pattern defects using: 1) Scattered light (100nm PS) 2) BF (100 nm PS), 3) BF (70nm PS), 4) New SRAM mode (140nm PS) Scan and diff image. Defect location is marked by the red circle. 5) New SRAM mode (190nm PS) Scan and diff image. Defect location is marked by the red circle

From the comparison Pareto shown in Figure 5, it can be clearly seen that the new SRAM modes are the only modes to capture the missing pattern defect type, and that these modes also have a better CR of the bridge defect type compared to the other modes.

The difference in the CR between the resolved configurations to the new SRAM modes can be explained by analyzing the missing pattern defect scan images shown in Figure 7. The defect signal is low, hence reducing the background as in the new SRAM modes enhances the SNR.

The new SRAM modes have additional benefits in addition to the increased CR and SNR. The ability to use larger PS (140nm and 190nm) vs. 70nm and 140nm, as demonstrated in the two examples above, enables higher WI productivity and therefore lower cost of ownership.

IV. CONCLUSIONS

A new inspection scheme was proposed and demonstrated for enhanced sensitivity on SRAM patterns for 28nm DR, Via ADI and SiGe DP layers. The scheme is based on tailored illumination and collection. The new inspection scheme has demonstrated higher sensitivity over scattered light and BF inspection. The ability to filter SRAM pattern increases the defects' SNR. The SNR increase enhances the CR of these defect types and enables capturing defects with lower signal. Using the new SRAM modes enables not only higher sensitivity, but also allows using large PS, thereby enabling higher throughput and productivity.

V. REFERENCES

1. M. Born and E. Wolf, Principles of Optics, Ch. 8, Cambridge University Press, Cambridge UK, 7th ed. 2001.
2. J. W. Goodman, Introduction to Fourier optics, Ch. 4 & Ch. 8, McGraw-Hill, Boston Massachusetts, 2nd ed. 1996.
3. D. Meshulach, I. Dolev, Y. Yamazaki, K. Tsuchiya, M. Kaneko, K. Yoshino, T. Fujii, Advanced lithography: wafer defect scattering analysis at DUV, SPIE Proceedings Vol. 7638, Metrology, Inspection, and Process Control for Microlithography XXIV.

Reduction of CMP-Induced Wafer Defects through *In-Situ* Removal of Process Debris

S. J. Benner, G. Perez, D. W. Peters, K. Hue[a], and P. O'Hagan[a]

Confluense

7277 William Ave., Suite 300

Allentown, PA 18106 USA

[a]Particle Sizing Systems

Port Richey, FL 34668 USA

Abstract - Rotary chemical-mechanical polishing (CMP) tools are widely used for integrated circuit (IC) manufacture. Logic device manufacturing has required CMP for front end (FEOL) and back end (BEOL) processes, such as shallow trench isolation (STI), high-K metal gate (HKMG), dielectric (PMD, ILD, Contact), and interconnect metallization (Cu). Recently, manufacturers of memory ICs adopted Cu metallization which increased their CMP requirements. The normal mode of operation for a rotary CMP tool is one of dilution since fresh slurry is over-fed to a polishing pad that is saturated with spent slurry and process debris in an attempt to achieve pseudo-equilibrium for material removal rates. As a consequence, slurry consumption is significantly higher (by 2X or 3X) than that needed if the spent slurry and process debris were removed *in-situ*, resulting in a replenishment mode of operation. Changing the CMP tool's mode of operation from dilution to replenishment has been demonstrated to significantly reduce slurry consumption while achieving higher material removal rates (MRR), in agreement with predictions. Rotary CMP tools recycle spent slurry and process debris under the wafer during the polish process. The maximum concentration of process debris occurs at end of polish when one is trying to achieve a high quality surface finish, which is precisely when debris will do the most damage. Much has been reported about the effect of long slurry residence times which result in kinetic decay of the process. However, virtually no data has been published on the impact of process debris residence time on wafer defects, due to the fact that until recently the ability to vary the debris residence time in a controlled manner was not available. Exhausting spent slurry and process by-products *in-situ* allows removing debris that can cause wafer defects. Several attempts have been made to modify CMP tools and associated components (i.e., slurry delivery arm, pad, wafer retaining ring, etc.) to reduce slurry consumption; however, only one CMP tool modification provides control of both slurry and debris residence times, namely Pad Surface Manager (PSM). PSM, in its simplest mode of operation, can be described as a vacuum cleaner for the pad. Previously published data demonstrated 50% cost savings from reductions in slurry consumption, water consumption, and extended conditioning disk and pad life. This paper will present data that show the impact of PSM on CMP-induced wafer defects.

CMP-induced scratches are typically caused by particles, such as agglomerated abrasive, pad debris, or materials of construction (i.e., plastic) larger than 1μm. A flare-up of microscratching can cause both yield loss and reliability issues, bringing a production line to a halt and resulting in premature failure of IC devices in the field. The forces exerted during CMP are more than sufficient to cause particle agglomeration in alumina- or silica-containing slurries. *In-situ* pad conditioning produces very large particles (e.g., 2 to 200μm) of polyurethane that has been identified as a source of chatter marks, scratches, and microscratches. Accusizer 780 particle size distribution data was acquired for commercially available slurry before and after oxide polishing with samples collected at 10 second intervals using PSM. The mean particle size shifted from 0.6μm to 1.6μm after 10 seconds of polishing and particles larger than 50μm were detected in the used slurry. The number of particles per ml larger than 1μm in the control sample of slurry was 0.13×10^4. After 10 seconds of polishing with in-situ pad conditioning, the number of particles per ml larger than 1μm increased to 5.7×10^4, more than a 40X increase. After 80 seconds of polishing, the number of particles per ml greater than 1μm increased to $>33 \times 10^4$, a more than 250X increase over the control. As the number of particles per ml larger than 1μm exceeds 2.0×10^4, the density of microscratches increases exponentially. A typical slurry residence time has been reported to be around 40 seconds. Based on the above results, debris residence time apparently may be considerably longer. Very large particles (i.e., >100μm) from filtered effluent were examined and identified as silica encrusted polyurethane pad particles using energy dispersive spectroscopy (EDS). A 50% reduction in light-point defects on polished thermal oxide wafers was achieved with PSM operated at a moderate vacuum. No attempt was made to differentiate between particles and scratches as the source of light-point defects.

I. INTRODUCTION

The efficiency of CMP processes on rotary tools has been studied by numerous tool and material suppliers, as well as in academia [1-3]. It has been generally agreed that the process is rather inefficient in that a significant fraction of delivered slurry does not participate in the polish process and is sent directly to waste, pads are over dressed leading to a shorter operational life, and tools recycle polish by-products and pad debris under the wafer leading to poor surface finish (i.e., scratches, microscratches, chatter marks, etc.) and defectivity (i.e., particles, residues, etc.). The typical rotary polishing tool uses a slurry dispense system to deliver fresh slurry to a pad surface that is already saturated with water or spent slurry and polish by-products. A conditioning arm with a diamond abrasive disk is used to mix fresh slurry with materials already resident on the pad, resulting in over dressing the pad and decreasing its lifetime. The net effect is that a typical process of record (POR) will flood the pad with slurry in an attempt to reduce the impact of dilution by spent slurry, reaction products, and debris. To underscore the inefficiency of this approach, it has been reported that only about 5% of the slurry delivered to the pad actually participates in the polishing process [1].

After achieving a quasi-steady state material removal rate (MRR) with a higher than needed slurry flow rate, it has been determined that about 70% of the slurry contacting the wafer is spent [3]. The mean residence time for slurry has been reported to be a significant portion of the total polishing time [2]. The low slurry turnover rate has been shown to result in a reduced MRR and higher within wafer non-uniformity (WIWNU) [2]. The long slurry residence time in conjunction with an increasing concentration of by-products leads to an inherently unstable process and continuous kinetic decay. The maximum concentration of CMP by-products occurs near the end of the polish process, when one is trying to achieve a high quality surface finish. The high concentration of agglomerated slurry particles and pad debris results in microscratches and chatter marks, some of which lower yield while others decrease device reliability [4]. A buildup of Cu reaction products, leading to pad staining, has been shown to have a deleterious effect on MRR, WIWNU, dishing, erosion, and selectivity [5]. Removing process waste (i.e., spent slurry, reaction products, pad debris, etc.) *in situ* has the potential to lower wafer defectivity and improve process stability.

Confluense has developed a Pad Surface Manager (PSM) which consists of a replacement conditioning arm, electronics cabinet, vacuum and fluid controls, and a computer which allows *in situ* vacuuming of material from the polishing pad and delivery of water or process fluids to the pad, independent of that delivered by the CMP tool [6]. Vacuum and fluid delivery capabilities are plumbed into the PSM arm to allow exhausting material or delivering liquids through the arm's end effector. The PSM arm is bolted onto the conditioning arm mount on the rotary CMP tool and has an external tube which is used to direct exhausted waste from the pad to a separator which spins fluid and solids from the vacuum. Various sensors (e.g., pH, conductivity, ion specific electrode, light or acoustic spectroscope, etc.) can be inserted in the effluent stream for real time measurements of the waste's characteristics [6]. The PSM arm can deliver water or process fluids to the pad for rinsing or *in situ* process modifications. Control of the conditioning arm motion is retained by the CMP tool. The down force for the abrasive disk is controlled through the PSM computer and can be adjusted to take into account the level of applied vacuum, which exerts additional down force on the abrasive disk. PSM can be used to vacuum waste from the pad with zero or negative abrasive down force to significantly extend pad life while maintaining MRRs.

Implementation of PSM has allowed investigating and controlling CMP processes in real time. Analysis of the effluents can be used to detect endpoint or generate feed-forward control signals for tool parameter adjustment and slurry composition modifications to continuously tune the CMP process to meet required targets [6]. Figure 1 contains a picture of a PSM mounted on an IPEC 372 developmental

Fig. 1. A picture of a Mirra-type Pad Surface Manager (PSM) mounted on a modified IPEC 372 polisher.

tool. PSM data reported here were generated with TBW Industries abrasive disks; however, the efficacy of PSM is not strongly dependent on the physical characteristics of the abrasive disk. Some slight variations in PSM end effector design or use of a disk mounting adaptor allows accommodating abrasive disks from nearly all suppliers. End users typically specify their current supplier's abrasive disk. Detailed abrasive disk designs developed and tested are available for conditioning disk suppliers to help maximize the effectiveness of their abrasive products. Previous reports showed that PSM has a significant impact on lowering the cost of ownership (CoO) and cost of consumables (CoC) by improving throughput, reducing slurry and water consumption, and extending pad and abrasive disk life [7-10]. Previous evaluations of the change in slurry particle size during a complete polish cycle have indicated a significant increase in mean particle size and in particular, a significant increase in the number of particles larger than 1μm [1,8]. This paper will next examine how controlling the process debris residence time can affect wafer defectivity by extracting samples during the polish process to evaluate the impact of polish time on changes in the slurry particle size distribution. In additional defect experiments, oxide wafers were polished and inspected to determine the impact of debris removal on CMP-induced wafer light-point defects.

II. EXPERIMENT

Data presented here were generated on a modified IPEC 372 polisher using a Mirra POR (i.e., 7psi down force, 1.5psi back pressure, 65rpm for the carrier, and 83rpm for the platen). Oxide polishing data consisting of MRR and light-point defects were generated on 200mm silicon wafers with 500nm of thermal oxide. Polishing was performed with a K-grooved Dow IC1010 pad. Pad conditioning was performed with a 4-inch diameter TBW Grid-Abrade™ abrasive. Standard pad break-in was performed using slurry and a conditioning disk. The oxide slurry used was Cabot SS11. The slurry flow rate was set at 200ml/min. The respective POR *in-situ* conditioner parameters (i.e., sweep rate and abrasive rpm) were used for pad conditioning, controlled by the polishing tool. Abrasive down force and vacuum were controlled by the PSM. Slurry samples for patterned STI wafer polishing were collected at 10 second intervals and analyzed for their particle size distributions. The samples were collected using 9in. of Hg on the PSM. The samples were analyzed on a Particle Sizing Systems Accusizer[TM]780. Undiluted samples varying in volume from 50µl to 1ml were injected into the analyzer. Some samples were injected multiple times to determine the repeatability of the data. Counts for particle sizes from 0.5µm to 400µm were recorded for each sample run. Silver mesh with 800nm pores was used to filter effluent obtained during an oxide polish process. The captured particles were then examined in a scanning electron microscope (SEM) and analyzed with energy dispersive spectroscopy (EDS). Polishing tests evaluating the impact of PSM vacuum on oxide wafer defectivity were conducted using a Mirra POR with *in-situ* pad condition and PSM vacuum levels varying between 0 and 12in. of Hg on low defect oxide wafers. After polishing, the wafers were cleaned with DI water and dried in a Semitool SRD. The wafers were given a post-post-CMP clean in SC1/SC2 in an Akrion wet bench before measuring the number of light-point defects on a Tencor 6220. No attempt was made to differentiate between particles and scratches in the resultant optical defect data.

III.DISCUSSION

Figure 2 contains a plot of measured slurry particle sizes for a sample of Cabot SS11 slurry before use. The mean particle size was 0.6µm and the number of particles per ml that were larger than 1µm was 0.13×10^4, well below the projected threshold for microscratching [1]. Furthermore, the distribution does not show a tail much beyond 1.5µm. Figure 3 contains a particle size analysis of a sample of SS11 that was used for 10 seconds while polishing an STI wafer. The distribution now shows a tail that goes to about 50µm. Furthermore, the mean particle size has increased to 1.6mm and the number of particles per ml that were larger than 1µm

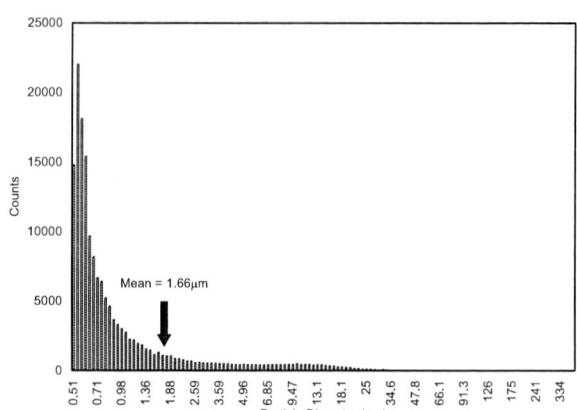

Fig.3. A plot of measured particle sizes for a sample of oxide slurry used for 10sec. The mean particle size was 1.6µm and the number of particles per ml larger than 1µm was 5.7×10^4, a more than 40X increase relative to the control.

increased to 5.7×10^4, an increase of more than 40X and well above the projected threshold for producing microscratches. The change in particle size distribution is assumed to be due to slurry agglomeration and pad debris, both of which can be influenced by the CMP tool process parameters.

Additional slurry samples were taken at 10 second intervals and Figure 4 contains a plot of the number of particles per ml that are larger than 1µm for seven samples. As can be seen, there is a general trend of an increase in the number as a function of polish time, as one might expect. At 80 seconds, there was a greater than 250X increase in the number of particles that produce microscratches, a factor of more than 15X over the projected threshold for microscratching. The rate of increase of large particles may be even greater had samples not been removed at shorter intervals (i.e., if the only sample removed was after 80 seconds of polish) [8]. Previous work to evaluate slurry particle size distribution changes after polishing had been compromised by high levels of dilution with rinse water since samples were not obtained directly from the pad but from the tool's waste drain [1]. Use of PSM allows

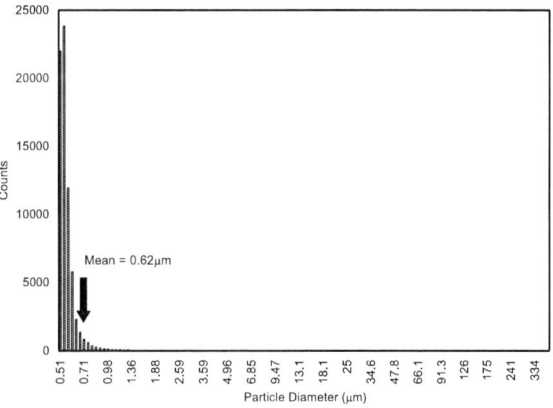

Fig.2. A plot of measured particle sizes for a control sample of oxide slurry. The mean particle size was 0.6µm and the number of particles per ml larger than 1µm was 0.13×10^4.

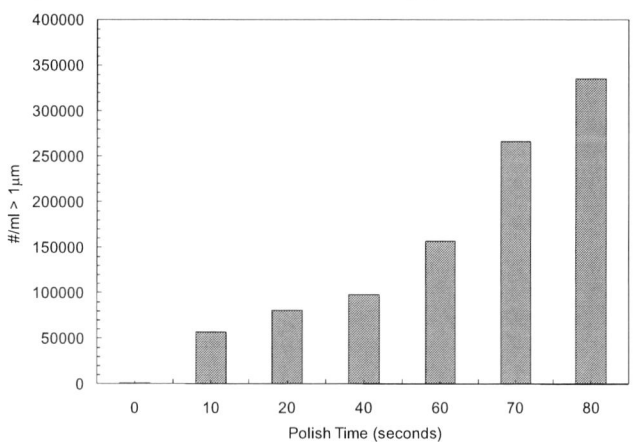

Fig.4. A plot of the number of particles per ml larger than 1µm as a function of polish time. The slurry sample taken at 80 seconds had 33.5×10^4 particles larger than 1µm per ml. That was an increase of more than 250X compared to the control sample shown at time 0.

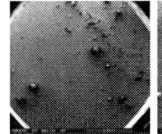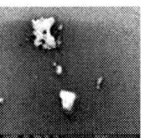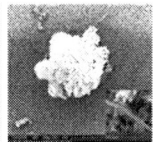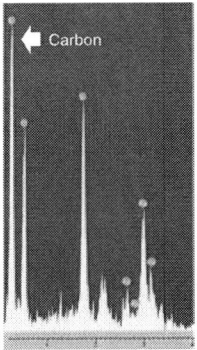

Fig.5. SEM images of several large particles and EDS analysis of a large polyurethane particle encrusted with silica. The particle was obtained from filtered effluent during an oxide wafer polish.

obtaining used slurry samples directly from the pad prior to dilution with rinse water and any resultant changes in emulsion characteristics which may enhance agglomeration. Additional experiments are underway to investigate particle size distributions for used slurry samples obtained at longer time intervals and under varying process conditions.

Examination of filtered effluent obtained using PSM during oxide polishing with *in-situ* pad conditioning indicated that polyurethane particles >200mm in size were produced. EDS analyses of these polyurethane particles showed that they were silica encrusted; greatly altering their ability to produce surface scratches and chatter marks [4].

Figure 6 contains a plot of light-point defects measured using a Tencor 6220 on polished oxide wafers for different levels of PSM vacuum. The data were normalized to that observed without vacuum. As can be seen, as the PSM vacuum level was increased, CMP-induced wafer defects decreased. A nearly 50% reduction in light-point defects was obtained. No attempt was made to identify the source of the light-point defects (i.e., were they a result of particles or scratches?). Additional experiments are underway to attempt to correlate the number of particles per ml larger than 1µm with scratch density on oxide wafers.

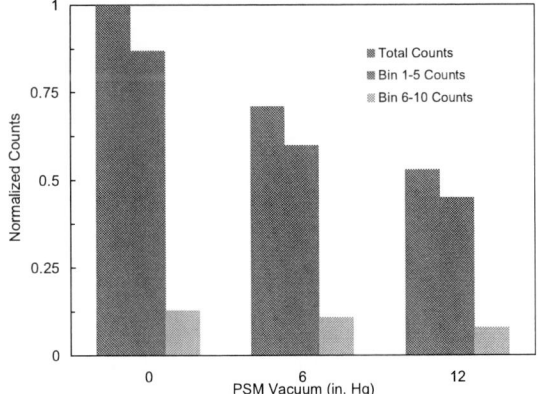

Fig.6. A plot of the dependence of light-point defect counts, measured with a Tencor 6220 on oxide wafers, as a function of PSM vacuum level. A reduction of nearly 50% was observed at a PSM vacuum of 12 in. of Hg.

IV. CONCLUSIONS

Use of PSM to convert the mode of operation of a rotary CMP tool from dilution to replenishment can favorably alter the tool's characteristics. Reducing the residence time of the slurry improved the material removal rate (MRR), as previously predicted, and reduced slurry consumption. Measurable changes in the slurry particle size distribution were observed over time during the polish process. The mean particle size increased and the number of particles per ml larger than 1µm also increased. After only 10 seconds of polishing, with *in-situ* pad conditioning, the number of particles per ml larger than 1µm was well above the projected threshold for microscratching. Examination and analysis of solids from filtered PSM effluent indicated silica encrusted polyurethane particles larger than 200µm were produced. Reducing the number of large particles on the pad by removing them *in-situ* during pad conditioning will also reduce the density of microscratches. Slurry agglomeration, pad debris, and material removed from the wafer can all contribute to wafer defectivity by causing scratches. In a typical polishing process without PSM, debris is at its maximum concentration at end of polish where it will have the greatest negative impact on the surface finish. Consequently, reducing the residence time of debris by removing it *in-situ* will decrease CMP-induced wafer defectivity. PSM allows a programmable way to control the residence time of the slurry and process debris resulting in improvement in both the CoO and CMP-induced wafer defects.

ACKNOWLEDGMENT

The authors wish to thank TBW Industries, Inc. for supplying abrasive disks.

REFERENCES

[1] A. Philipossian, et al, "Analytical and Functional Evaluation of Fresh, Spent, and Reprocessed Fumed Silica Slurries in ILD CMP", presented at First International Workshop on Nanoscale Semiconductor Devices (2004).

[2] A. Philipossian and E. Mitchell, "Mean Residence Time and Removal Rate Studies in ILD CMP", J. Electrochem. Soc., 151(6), pp. 403-407 (2004).

[3] G. Muldowney, "On the Relationship of CMP Wafer Nanotopography and Groove Scale Slurry Transport", MRS Sympo. Proc., 867, 2005.

[4] Y. Gotkis, "A Couple of Considerations on the Dynamics of Defectivity Generation in CMP Technology", presented at NCCAVS CMPUG meeting (April 2007).

[5] J. Han, S. Han, Y. Kang, and J. Park, "Effect of Polish By-Products on Copper Chemical Mechanical Polishing Behavior", J. Electrochem. Soc., **154** (6), pp. 525-529 (2007).

[6] S. J. Benner and D. W. Peters, "CMP Optimization and Control through Real-Time Analysis of Process Effluents", presented at NCCAVS CMPUG meeting (July 2009).

[7] S. J. Benner and D. W. Peters, "The Greening of CMP: Improvements in Chemical and Material Efficiencies and Life Cycle Management", Future Fab International, Issue 31 (November 2009).

[8] C. Burkhard, J. Zhao, P. Wu, M. Fox, S. V. Babu, and Y. Li, "Wafer Characterization and Spent Slurry Evaluation with a Novel Pad Conditioner", presented at CMP-MIC (2004).

[9] S. J. Benner and D. L. Dance, "CMP Productivity Improvement Using Pad Surface Management", presented at ISMI Symposium on Equipment-Related Productivity Improvement Activities (March 2006).

[10] S. J. Benner, G. Perez, D. W. Peters, and Y. Li, "Pad Surface Management as a Strategy to Reduce the Cost of Ownership of CMP", Proc. ASMC, pp. 232-235 (2010).

978-1-61284-408-4/11 $26.00 © 2011 IEEE

Sampling Process Information from Unstructured Data

J. Popp, D. Ortloff

Process Relations GmbH
Dortmund, Germany
jens.popp@process-relations.com

T. Schmidt, K. Hahn, M. Mielke, R. Brück

Naturwissenschaftlich-Technische Fakultät
Universität Siegen
Siegen, Germany
kai.hahn@uni-siegen.de

Abstract— **Process Data Management and information governance are important tasks in product engineering. Formal process descriptions, quick access to technology data, check for consistent process flows, and the easy reuse of technology data are the main drivers of these developments. This paper introduces new methods for systematically collecting and storing technology data preserving the multidimensional context in which it had been created. Furthermore new methods and tools are introduced reconstructing the context of historical data, therefore creating information from the "heap".**

Process development execution system, process data management

I. INTRODUCTION

Designing and monitoring silicon-based manufacturing processes is a complicated and often tedious task. Interactions between process steps, materials and manufacturing equipment often lead to unexpected situations challenging even the most experienced process engineers. Timely detection, correct analysis, and future avoidance of such situations to a large degree depend on efficient access to detailed information on the manufacturing technology. In most settings this information is hidden in a hodgepodge of excel sheets, images, diagrams, and other digital data that has been collected during previous research and development projects. Finding the required piece of information in such a big heap of mostly unstructured digital data is in itself a tedious and time consuming task.

Even if the specific data is found, it may be useless without the context it was generated in. For instance, without detailed knowledge about the applied manufacturing process, a SEM-image or a diagram depicting analysis results conveys only a very limited amount of information. Currently more or less complicated hierarchies in file systems or extensive use of meta data document management systems are used to add this kind of context information to the data. But these tools do not support to maintain the relations between the data points and metrology results. Additionally searching capabilities are limited by the one dimensional search criteria on non-formalized data provided by the file servers. Parameters kept as numbers in meta data or in Excel sheets are of limited value, e.g. because searching for values in an interval is not supported. Additionally the meta data is managed in a non-formalized manner, meaning that the application does not know about the physical quantities a number is representing.

E.g., searching for a process step operating within a certain pressure range where the pressure could be measured in *Torr, Pascal, Bar, …* is difficult at best.

This paper addresses those issues by introducing a new method for collecting and storing of technology data together with the multidimensional context in which it has been created. The information is stored in the *XperiDesk* Process Development Execution System (PDES) that has sophisticated visualization capabilities and allows efficient data-analysis of information with multidimensional context. Furthermore tools are introduced allowing the reconstruction of the original context of historical data [8] [9].

To keep the overhead as low as possible and reduce the burden of data collection and reporting, most of the collection and contextual linking procedures have been automated. The engineers are relieved from tedious collection and arrangement tasks and searches and reports become a matter of a few button clicks. Additionally, engineers can opt to continue using their established data collection procedures and storage means including file system storage and Excel worksheets, at least for a transition period. Keeping the old procedures and documentation approaches unchanged can limit the long term gains in efficiency of directly using the XperiDesk tools due to a less insightful and detail rich documentation fostered by the structured guidance inside the XperiDesk GUI.

To clarify the context of the work, the paper starts with a description of historical technology process management for MEMS based on thin film techniques and its requirements. In a second step *XperiDesk* is introduced. It offers comprehensive capabilities of designing, managing, checking, and storing technological information.

The conversion of technology data into information is described in the subsequent chapter. Two approaches implemented in *XperiDesk* are discussed in the last chapters. An Excel client and a file loading client were developed to easily access, convert and restructure the appropriate data.

II. PROCESS DEVELOPMENT EXECUTION SYSTEMS

The functionality of thin film devices is usually defined by their 3-dimensional structure. Since the vertical structure is the result of the manufacturing process, modern technologies like advanced microelectronics or MEMS need flexible tools and techniques to configure and modify manufacturing processes. This task is performed by manufacturing process management.

Process Management covers more than just the ordinary formats and technology handling tools within the frameworks that are available today. Foremost this includes a central management capability for process information and knowledge handling the whole range of process related information and the complex relationships between this information in a flexible manner. Based on this information management specific tools supporting the specific process management tasks can be implemented.

The presented approach to process management has been developed at the University of Siegen. It was made commercially available by a spin-off company. The capabilities of this approach go far beyond earlier approaches that were only able to cover a subset of the process management tasks.

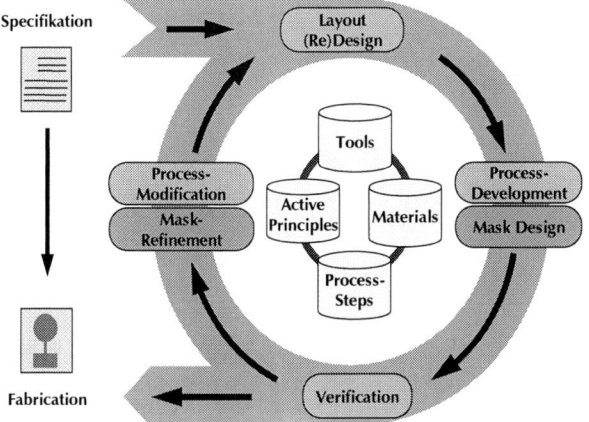

Figure 1: Circle model for process and mask design

As a basic functionality process management requires a comprehensive knowledge base containing all relevant information about fabrication processes and process steps. It needs to be capable of supporting a complex network of verification and synthesis tools. The following list can only give a few examples of the categories of data to be supported by the process management knowledge base:

- Documentation data: tables, images, free text descriptions.

- Process object data: process steps, processing parameters, materials, rules, simulation parameters.

- User group specific data: process step designer, process sequence designer, staff responsible for service processes.

- Various views: which machine is capable of ...? Which process step can produce ...? Which process sequence holds condition ...?

Data of all these types and various others are available for fabrication processes and process steps and must be coded into a complex knowledge network to cover all relevant relationships.

The interdependency between manufacturing technology (process data) and device properties is one of the major challenges of thin film product engineering [1]. The EU FP6 project PROMENADE (IST 507965) tackled this challenge with the development of an software environment for manufacturing process development [3]. The environment introduced central information management capabilities for process knowledge. Design tools using the knowledge base for specific process design tasks like consistency check and process simulation have been directly integrated into the environment. Additionally, the environment addressed multi-user and multi-site development problems by supporting a collaborative access approach.

The environment is now commercially available as the XperiDesk PDES. The XperiDesk product extends the former PROMENADE functionalities with automated data collection methods, information and data visualization functions and comprehensive search and other retrieval means. The following sections outline the main functionalities of the environment

A. Information Management

The information management is the base of a PDES. All process design related data (such as process steps, process sequences, materials etc.), parameters, and simulation models are stored in a central database system and can be accessed by multiple users via various specialized graphical user interfaces. Process recipe data, like process steps and materials are structured using sophisticated inheritance and property propagation mechanisms. Those mechanisms allow properties and constraints to be shared between related design elements in combination with efficient structuring of the process recipes.

One of those mechanisms is the ability to define abstract design elements, e.g. elements that represent a group of elements that share certain characteristic properties. For instance a group "etching" can be used as an abstract element for all etching techniques. When used for process design, these elements allow an incremental design approach, e.g. defining

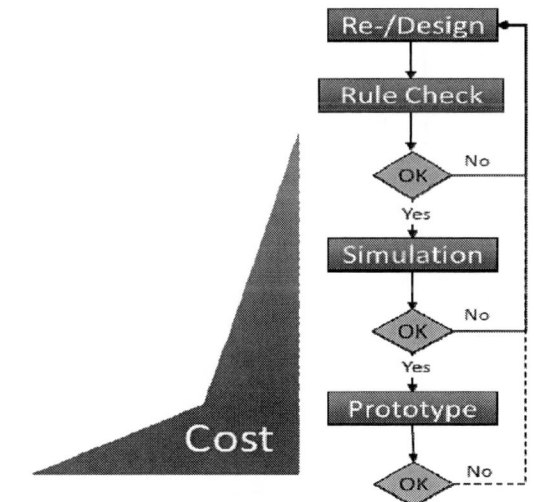

Figure 2: Three level verification

only that a wet etch process with certain properties but leaving the concrete process recipe for later design stages. Initial idea verification, including initial geometry and process design on an abstract level as well as a geometry based three dimensional shape emulation, can therefore be done with minimum effort.

B. Process Verification

The system follows a three level verification approach with each level becoming more complex and cost intensive as depicted in Figure 2. The first level is a rule based consistency check for assessing the feasibility and manufacturability of the process. The consistency checker uses previously collected rules / constraints regarding the compatibility requirements of process steps. Common mistakes like using invalid combinations of process steps that are likely to damage deposited layers or are leading to contamination of manufacturing equipment can be avoided.

The second level is a virtual manufacturing run using an interface to commercial TCAD and other calculation tools. This interface allows efficient simulation and emulation of the process based on models stored in the database [7]. This level gives first feedback on the device geometries and can be used by every process engineer without detailed knowledge of the simulators.

The third verification level supports the execution of lab-experiments and prototype runs. This includes Design of Experiments (DoE) support, automated experiment documentation and graphical information exploration. The overall goal is to store the full context of an experiment together with all data dimensions in all formats. Together with

the design data and the potentially available simulation results a calibration loop optimizing the simulation and the manufacturing results can be achieved [5].

C. Data Exploration and Information Exchange

Comprehensive data export and import modules provide the basis for data, information and knowledge exchange between various legal entities and development stakeholders, like fabless design houses or cooperation partners. The modules support technology transfer projects by providing a reference base preventing costly re-developments induced by differing equipment and calibrations. Today experimental development data is typically stored on file servers. This type of data repository only provides a one dimensional search criterion following the folder structure. With the search capabilities offered by XperiDesk it is possible to answer requests like, list all wafer which have seen a certain material or certain process step (combination). The graphical information representation, as presented in Figure 3, allows visual exploration of the dependencies and relations between the different data objects. The data objects are visualized as nodes in an information network. The edges represent the relations between the data objects. Nodes can be selected as new exploration center for visual navigation through the information network [6]. Together with the graphical representation of node types this allows the selection of subsets of data for export.

III. CONVERTING TECHNOLOGY DATA INTO INFORMATION

The insufficient internal information and knowledge management resulting in recurring engineering "déjà vu's" is

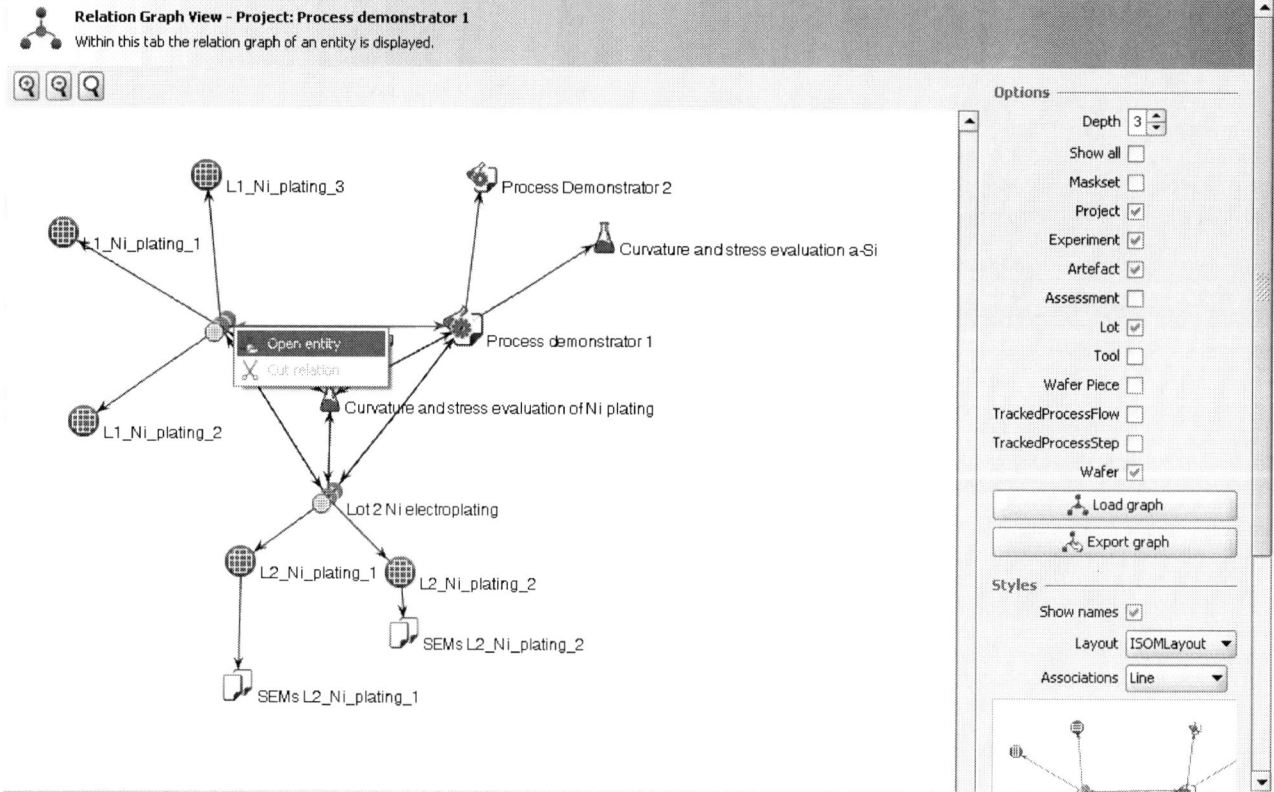

Figure 3: Formalized data with relation in the XperiDesk Graph View

one of the main issues experienced in today's development organizations. Experiences gained by previous developments, scientific papers, and old lab-books provide the major contribution to the realization of new product ideas. Having no or only insufficient structure in this data causes a lot of trouble and double work. Experts in semiconductor process development estimate that at least 10-15% of failed and double experiments could be avoided, if previous results would be accessible in an easier way. This observation ties in with issues arising from engineer fluctuation between different projects. Moving the project expert into a different project might jeopardize the previous project while engineers moving into a running project are flooded with lots of unstructured information.

Additionally cluttered result data storage and important data on local disk drives, often referenced as "the heap", cause tedious and error prone manual data collection and sometimes even data loss. Furthermore often only the pure data points or result data sets are stored with limited or no context information. Having only limited context poses problems when trying to reproduce previously seen effects or result in drawing the wrong conclusions from cause-effect analysis. These circumstances produce "déjà vu's" in the form of "once we had a result ..." that can be very annoying and cost intensive.

Documenting and reporting the development progress can be tedious at best. Cluttered results storage puts major manual effort onto the development engineers requiring them to manually collect data from diverse machinery. The assembly of the collected result data into reports and the evaluation can take a major part of engineering time. Reporting on the development status is often times more a manual assembly of the reports than an automated process. The input data is often not up to date so that the Work In Progress (WIP) status is not necessarily precise. The impacts of these effects are even aggravated by quality assurance, information governance, and compliance demands such as ISO 900X, CMMI, SOX etc. Because those apply more and more in development as well as in production, there is a strong demand to fulfill the imposed documentation requirements and to structure "the heap".

A. Addressing the problem

Interviews with process engineers at several large and small MEMS companies revealed that simple spreadsheet tools like Microsoft Excel are commonly used in process development for data management and analysis. Spreadsheets proved to be very versatile and can be used to very quickly convert raw numbers into diagrams for visualization.

However, those tools are not designed to replace a database. Even relatively simple searches that take into account more than one criteria are quite challenging. If there is need for context information like how the data relates to previous results, the limits of those tools are reached. Of course, the tables could be extended to include that kind of information, but then again it would become complicated very fast. Additionally, copy and paste being the preferred method for connecting data from different spreadsheets is a huge source for errors and inconsistencies leading to non-conclusive results.

Besides these disadvantages, spreadsheets still have one big advantage that makes them so successful: Most engineers are familiar with these tools and many software products are able to import data from spreadsheet files. Integrating the benefits of using spreadsheets with the power of database functionalities introduces many advantages to a R&D organization:

- The data from existing spreadsheet files can be accessed from comprehensive search algorithms

- Existing spreadsheet templates can still be used for data entry and visualization

- Established procedures of data acquisition can be kept enabling a smooth change to new procedures

- Other tools that export or import spreadsheet files can be integrated into the environment without much overhead

The data stored in spreadsheets is still only a small part of all data that is collected in course of a development project. The spreadsheets are accompanied by a big heap of digital data in the form of images, analysis result files, diagrams and other formats. Many hours are spent: first to archive this data and later on to extract information from it. Traditional approaches like storage on file servers or even document management systems do not account for the complexity of research data. In many cases, the files alone are useless. Without knowing the manufacturing process a SEM image does not carry significant information. Result diagrams are also not of much value without knowing the conditions under which they were created. For proper storage of this data it is therefore essential to capture as much of the context as possible.

The following two paragraphs introduce two standalone clients for the XperiDesk system addressing the issues motivated above. The so called Excel-Client has the capability to import table based data into XperiDesk, formalize the data on the fly and relate it to the pre-existing data and therefore building information. The second standalone client is the File Loading Client. It is capable of importing all types' files into the system, add meta data to them and relate them to the pre-existing items or create new items other than files from the information in the file paths. Managed files will be indexed as well, so that fast searching in text files is enabled.

IV. EXCEL CLIENT APPROACH

The purpose of the Excel-Client is in making the often significant amount of historical data stored in spreadsheets accessible to the XperiDesk PDES. With the help of a graphical user interface the content of a work sheet can be analyzed and modified while loading the data into XperiDesk. If the work sheet has been used as a template this operation can be automated to import huge amounts of legacy data.

During the analysis parameters and units are assigned to all numerical values in the worksheet, thus enabling sophisticated queries. For instance searches for everything with a resistance of less than $3k\Omega$ are possible.

978-1-61284-408-4/11 $26.00 © 2011 IEEE

Another benefit comes from the automated generation of relationships. For instance, the ID of the wafer an experiment was done for can be extracted from anywhere in the spreadsheet or even from the file or path name. This information is used to automatically link the experimental result to the specific wafer involved in the experiment. The new information together with the relationships to existing data is generated on the fly during the import.

It is not just a raw data import as it provides a searchable structure for the data converting it into information that can be used for effective decision-making. The user gains a structured repository of the historical data. Data from different sources are now related and can be searched using the relationships. If, for instance, two work groups did measurements on the same device, the data can now be merged and referenced.

For automated loading of data a batch mode has been implemented. Once a job is defined using the graphical user interface, the client can be started on the command line or using a scheduling service. This way the Excel files can be checked for changes at given intervals and those changes are automatically imported into the PDES. The described method can also be used to import external data from project partners. This is extremely valuable for users who outsource experiments or parts of experiments to 3rd parties, but who want to analyze the resulting data internally. New questions can be asked to the system in seconds instead of the hours it took before to collect the data scattered across several spreadsheets.

V. FILE LOADING CLIENT APPROACH

The File Loading Client follows similar principles like the Excel Client. The main difference is that it doesn't analyse the contents of the files but the file system in which the files are stored. For this purpose it creates file containers, called artefacts, containing the raw files and more sophisticated entities like wafers, lots or experiments based on the context

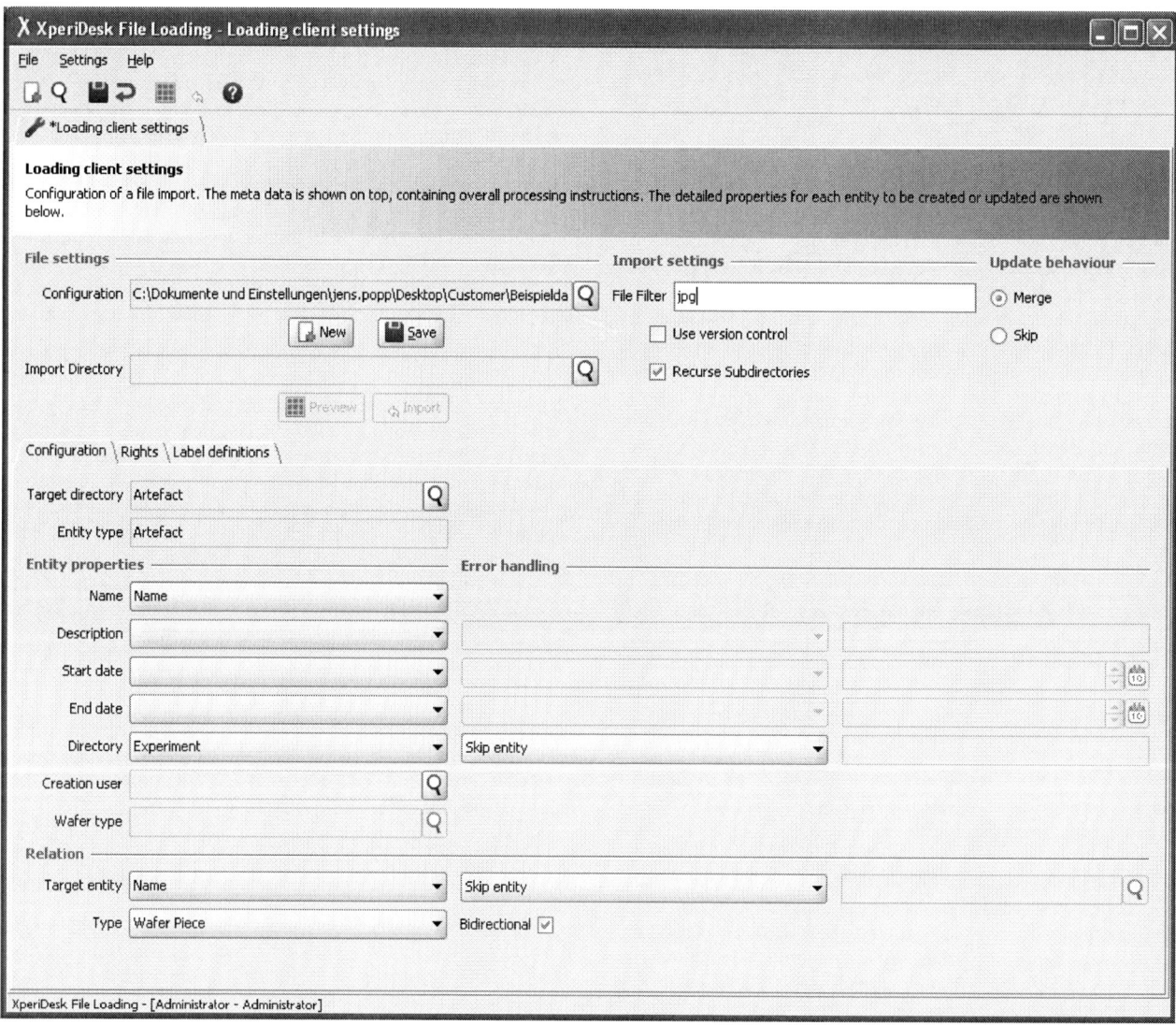

Figure 4: The File Loading Client

information given by the file system, e.g. file-names and directories. All these entities are linked, thus generating a searchable context. To do so the client uses customizable patterns. These patterns can be used on any file and path name to extract meaningful data used to create new entries in the database, to update existing ones and to attach files to newly created or existing wafers, experiments and other entities.

Like the Excel Client the File Loading Client is a standalone tool that can be deployed to multiple machines. Thus it is possible to centralize file data from different sources. Additionally, new views of the data become available to everyone working with the system. Users are now able to look at the result files of other departments (if security permits it) making the information accessible to other research projects throughout the company.

The File Loading Client can also be run in batch mode. A graphical user interface is used to define the loading configurations and then updates can be run regularly using a scheduler. Changes in files or the file system are detected and recorded using an internal versioning system. By establishing a certain hierarchical structure in the file system and by using standardized names for the files, the File Loading Client is able to load and link these files. No additional overhead is necessary. Users can continue to use the tools and file servers as they are used to.

VI. CONCLUSION

The manufacturing of nanoelectronic devices or MEMS based on modern thin film technologies requires a state of the art process management. Process design and configuration tools can accelerate the development of new devices and build a core task within the thin film product engineering. The set-up of new processes often generates huge amounts of process data to be stored within process management systems. With respect to information reuse the storage of technology data is only meaningful if the context is considered. Based on the process development execution system XperiDesk new means for collecting and storing correlated technology data were presented These new means allow generating process information from previously unstructured "heap" of file and spreadsheet data.

ACKNOWLEDGMENT

These developments had been carried out and supported by an international multi-site research project (CORONA - funded by the European Commission CP-FP 213969-2).

REFERENCES

[1] E.K. Antonsson, Structured Design Methods for MEMS, Workshop Report, Caltech, Pasaden, CA. National Science Foundation, 1996.

[2] H. van Heeren, Appearance of a moore's law in mems? trends affecting the mnt supply chain, MEMS, MOEMS, and Micromachining II 6186(1), 2006.

[3] D. Ortloff and J. Popp and T. Schmidt and R Brück, R., Process development support environment: a tool suite to engineer manufacturing sequences, *Int. J. Computer Mater. Sci. and Surf. Eng.* 2(3/4), 312–334, 2009.

[4] K. Hahn and A. Wagener and J. Popp and R. Brück. Process Management and Design for MEMS and Microelectronics Technologies, In *Proceedings of SPIE: Microelectronics: Design, Technology, and Packaging,* volume 5274, 2003.

[5] D. Ortloff and B. Veenstra and F. Cooijmans, A Systematic Approach towards Reproducibility and Tracking of MEMS Process Development, in Proc. of 10th Int.Conf. on Commericalisation of Micro and Nano Systems (COMS 2005), 2005

[6] S. Langenhuisen and D. Ortloff. An Approach to Generate Knowledge to Support Silicon Based MEMS Development. In Proceedings of ICME: 5th CIRP Inter-national Seminar on Intelligent Computing in Manufacturing Engineering, Ischia, 2006.

[7] T. Schmidt and D. Ortloff and J. Popp and K. Hahn and R. Brück and A. Hössinger. Verification of thin film processes in a virtual fabrication environment, In proceedings of SPIE: Micromachining and Microfabrication Process Technology XII, 6462, 2007.

[8] P.J. Bruening and K.K. Waterman, Data Tagging for New Information Governance Models, IEEE Security & Privacy Volume 8 Issue 5.

[9] Yi Sun and Xueliang Xia and Yalang Mao, Modeling based on structural characteristics integrated with product engineering quality information, Workshop on Advanced Computaional Intelligence (IWACI), 2010.

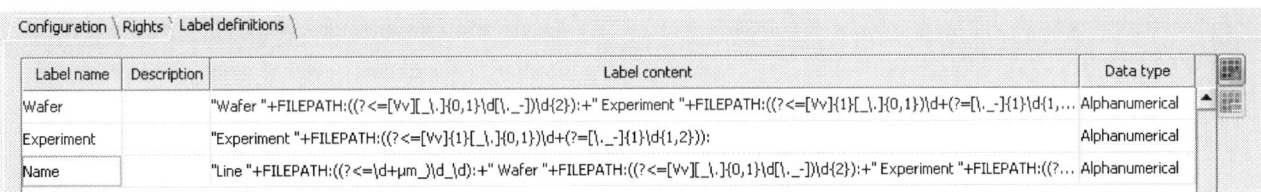

Figure 5: Extraction of data from filepath using pattern matching

Substrate Cleaning using Ultrasonics/Megasonics

Mohammad Kazemi, Helmuth Treichel and Rito Ligutom

Xyratex International Inc
46831 Lakeview Blvd
Fremont, California 94538
Mohammad_Kazemi@us.xyratex.com

Abstract—**The adhesion and detachment forces acting on spherical particles attached to a substrate is investigated. It is found that alumina particles larger than 50 nm and glass particles larger than 10 nm can be successfully dislodged by acoustic waves in an ultrasonic tank. Inspection of detachment forces reveals that the dominant detachment force is acoustic cavitation force for particles larger than 35 nm ($f = 950$ kHz).**

Keywords- Ultrasonics/Megasonics, Particle Removal, Van der Waals Force, Acoustic Cavitation, Acoustic streaming

1. Introduction

The feature-sizes in the semiconductor industry is shrinking, pushing down the size of the contaminating particles that need to be removed. Similarly, disk drive OEMs face increasingly stringent cleaning requirements that are mainly driven by the areal density growth and flying height reduction. Failure to meet those requirements results in severe problems with drive yield and reliability. To this end, advanced cleaning techniques need to be developed to meet future cleaning challenges.

Ultrasonic/Megasonic cleaning (USC/MSC) appears to be promising in meeting the future demands for cleaning substrates and disks [1]. Despite wide usages of USC/MSC, the underlying physics of particle dislodgement is not well understood [2]. Hence, the present paper focuses on exploring the physics of particle dislodgement subject to acoustic waves generated by piezoelectric transducers.

Van der Waals (VDW) force, ionic double layer force and hydrophobic force are the major forces of adhesion between a particle and a substrate (see Fig. 1). Those adhesion forces depend on particle diameter, hardness of particle and substrate, chemical composition of particle and substrate and the environment in which the contaminated substrate is located. Acoustic pressure (AP), acoustic streaming (AS) and acoustic cavitation (AC) are the detachment forces in USC/MSC system. Those detachment forces depend on particle diameter, transducer frequency, and power intensity of transducer and the chemical composition of the solution.

In this study, the adhesion and detachment forces of spherical particles with a diameter smaller than 400 nm are investigated. Two different cases are studied herein: a) glass particle on glass substrate; and, b) alumina particle on alumina substrate. Those cases correspond to practical substrate cleaning of glass and aluminum substrates in the disk drive industry. The detachment of such particles subject to the acoustic waves generated by a piezoelectric

transducer is investigated. For this purpose, two different transducer frequencies are studied, commonly used in USC/MSC tanks: 430 kHz and 950 kHz.

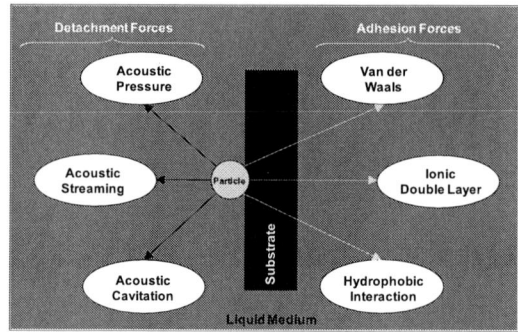

Fig. 1. Schematic of adhesion and detachment forces acting on a particle attached to a substrate in a liquid medium.

2. Particle Dislodgement Methods

A particle attached to a substrate can be dislodged if the detachment forces overcome the adhesion forces. With reference to Fig. 2, there are three methods through which a particle can be dislodged [3]:

- Rolling dislodgment
- Sliding dislodgment
- Lift-off dislodgment

Rolling dislodgment occurs when the detachment force creates a moment that exceeds the moment generated by the adhesion force. Under this circumstance, the particle rolls off and is detached from the substrate. The minimum detachment force required for rolling dislodgement can be calculated as follows:

$$F_{Det} \times (d/2 - b) \geq F_{Adh} \times a \qquad (1)$$

In this equation, F_{Det} and F_{Adh} are the resultant detachment and adhesion forces, respectively; d represents the particle diameter; a stands for the radius of contact and b is the particle penetration depth.

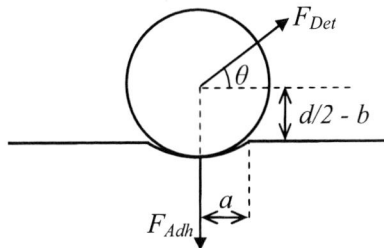

Fig. 2. Diagram of equivalent adhesion and detachment forces acting on a spherical particle that is attached to a substrate.

Sliding dislodgment occurs when the detachment force exceeds the coulomb friction force that is caused by the adhesion force. The minimum detachment force required for sliding dislodgement can be calculated as follows:

$$F_{Det} \geq \mu \, F_{Adh} \qquad (2)$$

In this equation, μ stands for the coulomb friction coefficient. The lift-off dislodgement occurs when the detachment force exceeds the adhesion force causing the particle to be directly pulled away from the substrate surface. The minimum detachment force required for lift-off dislodgement can be calculated as follows:

$$F_{Det} \geq F_{Adh} \qquad (3)$$

Among these dislodgment methods, rolling dislodgement is the most likely mechanism of dislodgement as it requires the lowest level of detachment force.

3. Adhesion Forces

There are three major types of the adhesion forces that a particle may experience in a liquid medium:

- Van der Waals (VDW) force
- Ionic double layer force
- Hydrophobic force

Van der Waals force is a short-range force that decays rapidly as the separation distance between the particle and substrate increases [4]:

$$F_{VDW} = \frac{A_{132} d}{12 z_0^2} \qquad (4)$$

In this equation, A_{132} stands for the Hamaker constant between the particle (1) and substrate (2) in the medium (3) and z_0 represents the separation distance between the particle and substrate and is typically set to 0.4 nm for smooth surfaces. Table 1 summarizes the values of Hamaker between an alumina particle on an alumina substrate and a glass particle on a glass substrate in both air and water media. Inspection of data shown in table 1 reveals that: a) Hamaker constant in water is less than that in air; and, b) Hamaker constant for alumina-on-alumina is larger than that for glass-on-glass.

		in Air	in Water
A_{132}	Alumina-on-Alumina	1.400E-19	2.703E-20
	Glass-on-Glass	3.400E-20	6.437E-22

Table 1. Hamaker constants for alumina-on-alumina and glass-on-glass in both air and water media.

The VDW force between a particle and a substrate causes the deformation at the interface. The deformation increases the area of contact and consequently the VDW force which can be evaluated using the following equation [4]:

$$F_{VDW} = \frac{A_{132} d}{12 z_0^2} \left[1 + \frac{2 a^2}{d z_0} \right] \qquad (5)$$

The radius of contact can be calculated using the JKR theory [5]:

$$a = \left[\frac{3 \pi W_A d^2}{8 E_{Comp}} \right]^{1/3} \qquad (6)$$

Where W_A represents the work of adhesion for an undeformed interface and E_{Comp} stands for the composite Young's modulus of elasticity at the interface. Work of adhesion and composite Young's modulus of elasticity can be calculated as follows [3,5,6]:

$$W_A = \frac{A_{132}}{12 \pi z_0^2} \qquad (7)$$

$$E_{Comp} = \frac{4}{3} \left[\frac{1 - \gamma_P^2}{E_P} + \frac{1 - \gamma_S^2}{E_S} \right]^{-1} \qquad (8)$$

In this equation, E_P, E_S, γ_P and γ_S represent the Young's modulus of elasticity and Poisson ratio of the particle and substrate, respectively.

Figure 3 shows plots of VDW force for the cases of glass-on-glass and alumina-on-alumina, both in water medium. As can be seen, the VDW force for alumina-on-alumina is larger than that for glass-on-glass which purely originates from the difference in Hamaker constants (see Table 1). It should be mentioned that the increment in the VDW force due to the deformation at the interface is less than 8% for alumina-on-alumina and less than 1% for glass-on-glass (for d ≤ 400 nm).

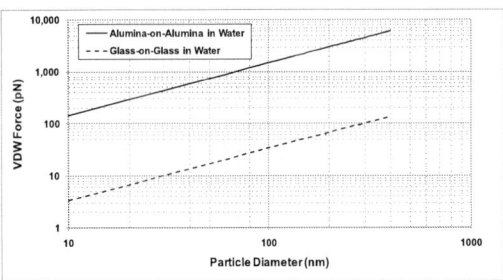

Fig. 3. Plot of VDW force versus particle diameter for glass-on-glass and alumina-on-alumina in water medium.

Ionic double layer (IDL) is a structure that appears near the outer surface of an object when it is placed in a liquid and consists of two parallel layers of ions (see Fig. 4). The first layer is the surface charge (either positive or negative) and the other layer is in the fluid, and electrically screens the first layer. The zeta potential (also known as electrokinetic potential) is used for characterizing of IDL charge. Zeta potential is the potential measured at the slipping plane as shown in Fig. 4. The zeta potential indicates the degree of repulsion/attraction between similarly/oppositely charged objects that are adjacent to each other.

The Zeta potential is a function of the solution pH and the material composition of the particle/substrate (see Fig. 5). Interaction force between a particle and substrate can be attractive or repulsive depending on the zeta potential of particle and substrate. If the particle and the substrate have the same chemical composition, their zeta potentials would be the same regardless of the solution pH. For the glass-on-glass and alumina-on-alumina, the IDL force is repulsive therefore it does not pose a challenge for particle dislodgement.

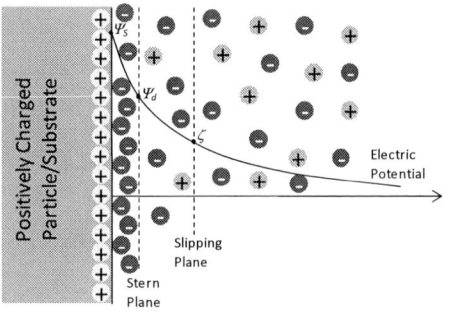

Fig. 4. Schematic of the electric potential variation in a liquid medium in the proximity of a solid object.

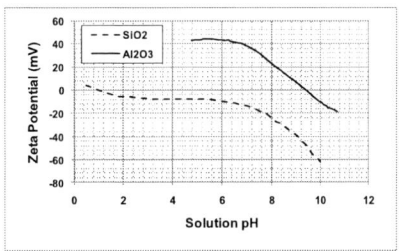

Fig. 5. Plot of zeta potential of alumina and silica as function of solution pH [7].

A particle attached to a substrate in a liquid medium may experience an additional adhesion force if the interface between the particle and substrate is hydrophobic. The hydrophobicity of the interface depends on the average contact angle of the interface as calculated below [8-9]:

$$\cos(\theta_{avg}) = (\cos(\theta_P) + \cos(\theta_S))/2 \qquad (9)$$

In this equation, θ_P and θ_S represent the contact angle of the particle and substrate, respectively. If the average contact angle (θ_{avg}) is larger than 90 degrees, then the interface is hydrophobic and the hydrophobic adhesion force can be calculated using the expression below [9]:

$$F_H = \frac{K_{132}d}{2 z_0^2} \qquad (10)$$

Where K_{132} is the hydrophobic force constant between particle (1) and substrate (2) in the medium (3) and can be calculated as [9]:

$$\log(K_{132}) = -7.0 \times \cos(\theta_{avg}) - 18.0 \qquad (11)$$

For glass and alumina, the contact angles are less than 45 degrees hence the interface between particle and substrate is hydrophilic for glass-on-glass and alumina-on-alumina.

4. Detachment Forces

A USC/MSC system uses a batch of piezoelectric transducers to generate acoustic waves in a tank that is filled by a liquid solution (DI water and a diluted amount of some chemicals). The frequency of the transducers is typically between 150 kHz and 600 kHz for the USC systems and is larger than 600 kHz for the MSC systems. As can be seen in Fig 6, the piezoelectric transducers generate acoustic waves that propagate in the solution, causing the formation of acoustic streaming. Such flow will create a boundary layer when it passes over a substrate surface. The thickness of the acoustic boundary layer can be evaluated as follows [10-11]:

$$\delta = \left(\frac{2 v}{2 \pi f} \right)^{1/2} \qquad (12)$$

In this equation, v is the kinematic viscosity of the solution and f represents the transducer frequency.

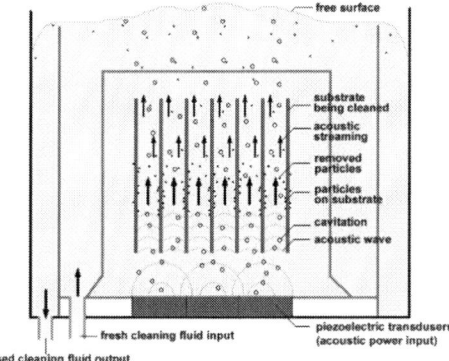

Fig. 6. Schematic of USC/MSC tank and the physical phenomena induced by the piezoelectric transducers [4].

Inspection of equation 12 reveals that the thickness of the acoustic boundary layer can be reduced by either lowering the kinematic viscosity of the solution, for instance by increasing temperature, or increasing the transducer frequency. It should be noticed that lower thickness of acoustic boundary layer allows for exposures of particles to larger flow velocity and consequently increases the chances of particle dislodgment. Figure 7 shows the variation of acoustic boundary layer thickness with the transducer frequency. As can be seen, the thickness of the acoustic boundary layer drops from 640 nm to 430 nm (at 50 C) when the transducer frequency is increased from 430 kHz to 950 kHz.

978-1-61284-408-4/11 $26.00 © 2011 IEEE   148

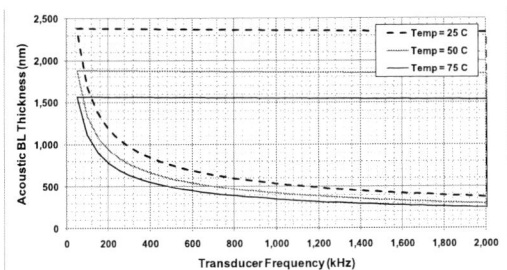

Fig. 7. Plot of acoustic boundary layer thickness versus transducer frequency for various solution temperatures.

The present study focuses on particles smaller than 400 nm that are exposed to transducers with frequencies of 430 kHz and 950 kHz (δ > 400 nm). Therefore, the particle will be within the acoustic boundary layer and exposed to boundary layer flow as shown in Fig. 8. The velocity profile within the boundary layer can be expressed as follows [10-11]:

$$u = u_e \cos(2\pi f t)\left[1 - \exp\left(\frac{-y}{\delta}\right)\cos\left(\frac{-y}{\delta}\right)\right] \quad (13)$$

In this equation t stands time and u_e represents the velocity magnitude outside of the acoustic boundary layer and can be estimated as follows [12]:

$$u_e = \left(\frac{2I}{\rho C}\right)^{1/2} \quad (14)$$

Where I stands for the power intensity of transducer; ρ represents the solution density and C is the speed of the sound in the solution (~ 1475 m/sec).

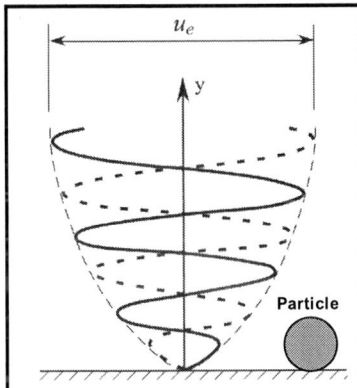

Fig. 8. Schematic of the velocity profile within an acoustic boundary layer that a particle experiences. The sketch shows the velocity profile at two instants separated by one-half period [11].

The aerodynamic drag force (F_D) and moment (M) acting on the particle can be calculated as follows [3-4]:

$$F_D = 1.7 \times \left(3\pi\mu dV\right) \quad (15)$$

$$M = 0.370 F_D d \quad (16)$$

In these equations, μ represents the dynamic viscosity of solution and V is the flow velocity at the particle center ($y = d/2$), calculated from eq. 13. The total acoustic streaming force can then be calculated as follows:

$$F_{AS} = F_D + M/\left(d/2\right) = 1.74 F_D \quad (17)$$

Figure 9 shows a plot of acoustic streaming force versus particle diameter. As can be seen, the acoustic streaming force increases when the transducer frequency increases, which is due to the reduction in the acoustic boundary layer thickness.

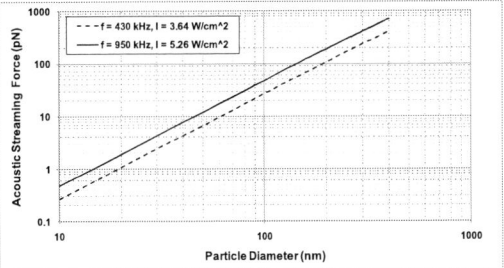

Fig. 9. Plot of acoustic streaming force versus particle diameter for two different transducers: a) a transducer with $f = 430$ kHz, $I = 3.64$ W/cm²; and, b) a transducer with $f = 950$ kHz, $I = 5.26$ W/cm².

Oscillation of a transducer causes a pressure variation wave that propagates in the solution with the speed of sound. When the pressure variation wave passes over a particle, the particle experiences a force due to the pressure gradient across the particle (see Fig. 10). This force is known as acoustic pressure force and can be calculated by integrating the pressure over the particle surface [13]. Since the wavelength ($\lambda = C / f$) is of the order of 1 mm and is much larger than the diameter of a submicron particle, the pressure gradient can be treated as constant across the particle surface [13]:

$$F_{AP} = \frac{dP}{dx} \times \left(\frac{\pi}{6}d^3\right) \quad (18)$$

The pressure variation wave is a harmonic wave that can be stated as:

$$P = P_0 \sin\left(\frac{2\pi x}{\lambda}\right) = P_0 \sin\left(\frac{2\pi f x}{C}\right) \quad (19)$$

In this equation, P_0 is the amplitude of pressure variation wave and can be described as [12]:

$$P_0 = \sqrt{2\rho I C} \quad (20)$$

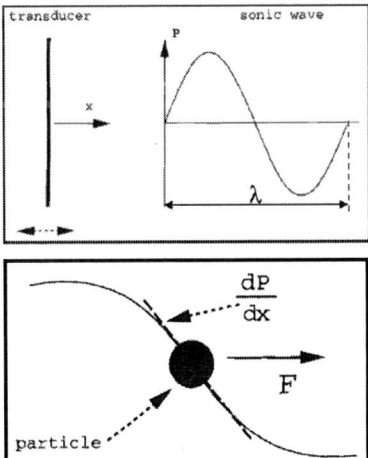

Fig. 10. Schematic of a pressure variation wave induced by an oscillating transducer (top) and the pressure gradient across a particle (bottom) [13].

By substituting eq. 19 and 20 into eq. 18, one can obtain an expression for the maximum acoustic pressure force that a particle can experience:

$$F_{AP} = 2\pi f \left(\frac{\pi}{6} d^3\right) \sqrt{\frac{2\rho I}{C}} \qquad (21)$$

Figure 11 illustrates the variation of acoustic pressure versus the particle diameter. Inspection of this figure reveals that acoustic pressure force: a) increases by increasing the particle diameter and transducer frequency; and, b) is more than one order of magnitude smaller than the acoustic streaming force.

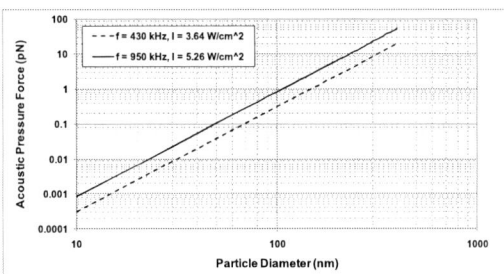

Fig. 11. Plot of acoustic pressure force versus particle diameter for two different transducers: a) a transducer with $f = 430$ kHz, $I = 3.64$ W/cm²; and, b) a transducer with $f = 950$ kHz, $I = 5.26$ W/cm².

The pressure variation wave causes high and low pressure in the solution as shown in Fig. 10. The low pressure half-cycle can cause expansion of pre-existing bubbles and, potentially, formation of new bubbles. The high pressure half cycle can cause compression or, potentially, implosion of bubbles. To this end, two classes of acoustic cavitations are defined: a) steady cavitation that is a process in which bubbles oscillates in size and shape; and, b) transient cavitation in which bubbles implodes producing a shock wave. In a USC/MSC, the steady cavitation is the desired type of cavitation and will be the focus of the present study.

For a bubble of diameter d_B, the natural frequency is given by Minnaert formula [14]:

$$f = \frac{1}{\pi d_B} \left[\frac{3k}{\rho} \left(P_0 + \frac{4\sigma}{d_B} \right) - \frac{4\sigma}{\rho d_B} \right]^{1/2} \qquad (22)$$

In this equation, P_0 (=101.325 kPa) is the atmospheric pressure, k is the polytropic gas constant ($1 \leq k \leq 1.4$) and σ is the surface tension of the solution. For a given transducer frequency, the active bubbles are the ones that resonate when exposed to ultrasound waves. In other ones, active bubbles are the ones that have a natural frequency equal to the transducer frequency. Figure 12 shows a plot of active bubble diameter. As can be seen, the active bubble size drops from 16 μm to 8 μm when the transducer frequency increases from 430 kHz to 950 kHz. It should be noticed that those bubbles are more than one order of magnitude larger than the particle sizes studied herein ($d \leq 400$ nm).

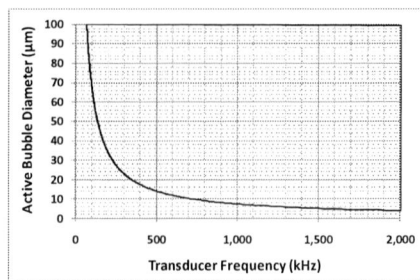

Fig. 12. Plot of active bubble diameter versus transducer frequency.

The detachment force applied on a particle due to radial oscillation of steady bubble (acoustic cavitation force) can be estimated as follows [2]:

$$F_{AC} \approx \frac{\pi}{6} \rho \omega^2 d^3 d_B \left(\frac{R_B}{r} \right)^5 \qquad (23)$$

In this equation, ω (=2πf) is the angular frequency; R_B (= d_B / 2) is the radius of active bubble corresponding to the transducer frequency f and r is the distance between the bubble center and the particle surface, as illustrated in Fig. 13. Inspection of eq. 23 reveals that the acoustic cavitation force declines with the fifth power of distance between bubble center and particle (r). Hence, the maximum acoustic cavitation force occurs when the oscillating bubble comes into contact with the particle (i.e. $R_B/r = 1$).

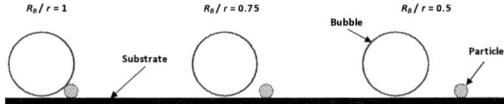

Fig. 13. Schematic of position of a bubble relative to a particle for three different R_B / r values.

Figure 14 shows the plot of acoustic cavitation versus particle diameter for $R_B/r = 1$. As can be seen, acoustic cavitation force increases by increasing the transducer frequency. It can be also observed that the acoustic cavitation force is about two orders of magnitude larger than the acoustic pressure force.

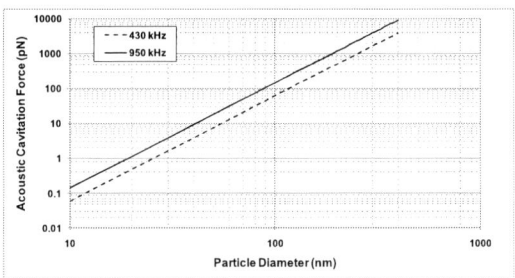

Fig. 14. Plot of acoustic cavitation force versus particle diameter for two different transducers frequencies: 430 kHz and 950 kHz.

5. Particle Dislodgement Inspection

In this section, the detachment of glass/alumina particles from glass/alumina substrates is investigated. As mentioned in section 2, rolling dislodgment is the most likely method of dislodgment for a particle. Hence, a scaling factor is introduced to scale the adhesion force (in this case, the VDW force) as follows:

$$F_{VDW}^* = F_{VDW} \times \left(\frac{a}{d/2 - b} \right) \qquad (24)$$

Where $a / (d/2 - b)$ is the scaling factor that originates from eq. 1 and is smaller than 0.05 for both glass and alumina particles (10 nm $\leq$ d $\leq$ 400 nm). Rolling dislodgement occurs if the detachment force exceeds the scaled VDW forced expressed in eq. 24.

Figure 15 illustrates plots of detachment forces and the scaled VDW forces versus particle diameter. Inspection of Fig. 15 reveals that: a) for the case of alumina-on-alumina, particles larger than 50 nm can be dislodged; b) for the case of glass-on-glass, particles larger than 10 nm can be dislodged; and, c) the dominant detachment force is acoustic cavitation force for particles larger than 35 nm (for f = 950 kHz) and 43 nm (for f = 430 kHz).

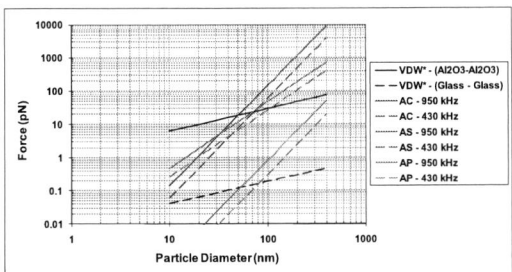

Fig. 15. Plot of detachment forces and scaled VDW forces for the cases of alumina-on-alumina and glass-on-glass.

6. Conclusions

The adhesion and detachment of spherical particles with diameter less than 400 nm was investigated for the cases of glass-on-glass and alumina-on-alumina. It was found that the VDW force is the dominant force of adhesion that needs to be overcome by detachment forces induced by transducers in a USC/MSC tank. The acoustic pressure force appeared to be much smaller than the acoustic streaming and acoustic cavitation forces. It was also found that the dominant detachment force is acoustic cavitation force for particles larger than 35 nm (for f = 950 kHz) and 43 nm (for f = 430 kHz).

Inspection of the adhesion and detachment forces revealed that the detachment threshold limit is 50 nm for alumina particles whereas glass particles as small as 10 nm can be successfully dislodged. It should be mentioned that the analysis conducted herein is based on the assumption that the acoustic waves are not blocked by any structure to cause shadowing effect. If shadowing is an obstacle for substrate cleaning then either: a) the structures supporting the substrate need to be redesigned; or, b) a mechanism needs to be provided to allow for rotation of the substrate so that the substrate surface is uniformly exposed to the acoustic waves.

References

[1] Vereecke, G., Parton, E., Holsteyns, F., Xu, K., Vos, R., Mertens, P.W., Schmidt, M. O., Bauer, T., Siltronic, W., "Investigating the role of gas cavitation in megasonic nanoparticle Removal," Micromagazine.com Article.

[2] Kim, W., Kim,T. H., Choi, J. and Kim, H. Y. "Mechanism of particle removal by megasonic waves," Appl. Phys. Lett., vol. 94, 081908, 2009.

[3] C. Toscano, G. Ahmadi, "Particle removal Mechanisms in Cryogenic surface cleaning," Journal of Adhesion, vol. 79, pp. 175-201, 2003.

[4] Busnaina, A., "Nanomanufacturing Handbook," CRC Press, First Edition, 2006.

[5] Johnson, K., Kendall, K. and Roberts, A. D., "Surface Energy and Contact of Elastic Solids," Proc. R. Soc., London, 324, pp. 301-313, 1971.

[6] Zhang, F., Busnaina, A. A., Fury, M. A., Wang, S. Q, "The Removal of Deformed Submicron Particles from Silicon Wafers by Spin Rinse and Megasonics," Journal Of Electronic Materials, (2000).

[7] Brunelle, J. P., "Preparation of Catalysts by Metalic Complex Adsorption on Mineral Oxides," Pure & Appi. Chern., Vol. 50, pp. 1211—1229.

[8] Birdi, K.S., "Handbook of Surface and Colloid Chemistry," CRC Press, Third Edition, 2009.

[9] Yoon, R.H., Flinn, D. H. and Rabinovich, Y. I., "Hydrophobic Interactions between Dissimilar Surfaces," Journal of Colloid and interface Science, vol. 185, 363–370 (1997).

[10] Schlichting, H. and Gersten, K., "Boundary Layer Theory," Springer Publication, 8th Edition, 2000.

[11] Culick, F. E. and Kuentzmann, P., "Unsteady Motions in Combustion Chambers for Propulsion Systems," NATO Research and Technology Organization, Originator's Reference # RTO-AG-AVT-039 AC/323(AVT-039)TP/103.

[12] Keswani, M. , Raghavan, S., Deymier, P. and Verhaverbeke, S. "Megasonic cleaning of wafers in electrolyte solutions: Possible role of electro-acoustic and cavitation effects," Microelec. Eng., vol. 86, pp. 132-139, 2009.

[13] Olim, M. "A theoretical evaluation of megasonic cleaning for submicron particles," J. Electrochem. Soc., vol. 144, no. 10, pp. 3657-3659, 1997.

[14] Edmonds, P. D., "Methods of Experimental Physics," Volume 19, Ultrasonics, Academic Press, 1981.

Use of Neural Network to Model the FTIR Spectra of PECVD Silicon Nitride Films for Cardiovascular Pressure Sensor Applications

Thongchai Thongvigitmanee[a,b], Wisut Titiroongruang[a]
[a]Department of Electrical Engineering

King Mongkut's Institute of Technology Ladkrabang
Bangkok, Thailand
Thongchai.thongvigitmanee@nectec.or.th

Arckom Srihapat[b], Amporn Poyai[b]
[b]Thai Microelectronics Center
National Electronics and Computer Technology Center
Chachoengsao, Thailand

Abstract—**In this paper, the empirical process models based on neural network are applied to discover the relationship between inputs and outputs of the plasma enhanced chemical deposition (PECVD) silicon nitride process. The design of experiments are based on a 2^{6-2} fractional factorial experiment with four center replicate on six factors which are 1) the SiH_4 flow rate, 2) the NH_3 flow rate, 3) the N_2 flow rate, 4) the chamber pressure, 5) the radio frequency (RF) power/distance between the wafer base and shower gas, and 6) the deposition temperature. Once these experiments are performed, different neural networks are applied to identify these six inputs to the chemical bonding information from the FTIR measurements. The best performances of neural networks for each response are selected based on the smallest prediction error. Then the three-dimensional surface plots are generated to qualitatively interpret factor effects. The corrosion testing in saline solution based on a potentiostatic measurement of an aluminum film with the protective PECVD silicon nitride is demonstrated. This measurement could be used as a tool to identify the best dense PECVD silicon nitride film in order to improve the protective performance of the existing recipe. The silicon nitride film is used as the final passivation layer of the cardiovascular pressure sensor.**

Keywords-Modeling; neural networks; plasma-enhanced chemical vapor deposition (PECVD); silicon nitride film; FTIR

I. INTRODUCTION

The PECVD silicon nitride is used as a final passivation layer for a medical sensor, especially a cardiovascular blood pressure catheter measurement [1]. The ability of this film which has a 200nm thickness to protect water permeability is crucial for device performance, such as a moisture-induced drifting problem and a wet isolation of metal conductor. The optimization of PECVD silicon nitride deposition having a dense film will be desirable in order to minimize these problems. In order to find the better deposition recipe than the existing one, the accurate and efficient modeling of this PECVD process is necessary for recipe optimization in the later phase. In addition, a metrology for quantifying the water permeating into the film was also needed.

The characteristics of PECVD silicon nitride film depends on the complex physical and chemical processes inside the plasma chamber. Neural network modeling was chosen and applied to reveal the relationship between inputs and outputs as an empirical approach [2]. In this paper, the authors continue to build the neural networks to discover the relationship between inputs and outputs of chemical information from the FTIR measurement. With prior research work on the neural network modeling for refractive property [3], these process modeling will be used to search for the optimized recipe in order to have a denser silicon nitride film in later phrase.

In this paper, the potentiostatic measurement of an aluminum metallization with the protective PECVD silicon nitride film in saline solution is studied. This method will be use to classify the water permeating property of the protective film.

II. EXPERIMENTAL DETAILS

A. Design of Experiment

The PECVD process was characterized by the 2^{6-2} fractional factorial experiment with four center replicate. The resulting twenty experiments were performed. The six input factors and their respective ranges are listed in Table 1. The SiN films were deposited by using an Applied Materials Precision 5000 Mark II CVD single-wafer parallel-plate system. Each SiN experiment used a forty second deposition time on top of a 130nm thick thermal oxide film on a six inch silicon substrate. The Bruker IFS66v FTIR was used to measure the chemical bonding information.

B. Neural Network Modeling

To model the process response, the data from the DOE were used to train feed forward neural networks using the error back-propagation (BP) algorithm. Recently, it has been shown that the ability of neural networks to discover input/output relationships from limited data is useful in semiconductor manufacturing, where numerous highly nonlinear fabrication processes exist and experimental data for process modeling are expensive to obtain. Fig. 1 shows the general structure of a feed forward neural network. These networks consist of several layers of individual processing elements. These processors (called "neurons") are interconnected in such a way that knowledge is stored in the weight of the connections between them. The activation level of a neuron is determined by a sigmoidal activation function. The activation function endows the network with the ability to generalize with an added degree of freedom not available in statistical regression techniques. To model the PECVD process, neural networks were used to quantitatively relate the input parameters to the refractive index. Using the BP algorithm, the network begins with a random set of weights. Sets of input vectors are then presented to the network, and the calculated outputs are compared with the measured outputs. The squared differences between these outputs determine the system error. This process is continued until the desirable system error is met.

C. Corrosion Testing

AUTOLAB model PGSTAT12 is used for the potentiostatic measurement under saline solution. The corrosion testing area was 1 cm^2. The dimension of the test device equals to 1 inch^2. There are two test structures. One structure has a 1 μm aluminum alloy (98.5%Al,1%Si, 0.5%Cu) thickness on a silicon substrate. The other structure has a 200nm SiN thickness over an aluminum metallization silicon substrate.

TABLE I.   INPUTS VARIABLES AND RANGES

Parameter	Range	Unit
SiH$_4$ flow	50-190	sccm
NH$_3$ flow	50-90	sccm
N$_2$ flow	2500-4500	sccm
Pressure	4-6	torr
RF power/Gap	0.6-1	W/mm
Temp	300-400	^{0}C

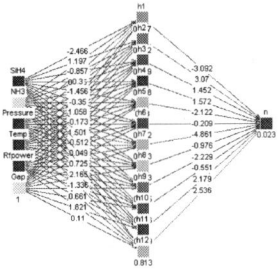

Figure 1. Example of neural network with six input neurons, twelve hidden neurons, and one output neuron

III.   RESULTS AND DISCUSSION

A. Neural Network Modeling of FTIR Spectra

Examples of FTIR spectra from three experiments are shown in Fig. 2. By examining data from twenty samples, we could classify the five-peak locations with their corresponding wave number within their ranges in Table 2. Note that some data from twenty samples do not have Peak 4, such as the red plot of Experiment #2 in Fig. 2. Since Peak 4 is corresponding to Si-H bonding which has a wave number between 2000 to 2260 as shown in Table 4; therefore, we will assign the value of 2000* for modeling purpose to represent the data which do not have Peak 4. During the experiment, there are four experiments yielding unacceptable high fit error of thickness and refractive index due to the instability of silicon nitride film. The remaining of sixteen data is divided into two groups. The training data set for neural network consists of fourteen data, and the validating data set consists of two data.

To realize the BP neural network training algorithm for developing neural process models, the Java Neural Network Simulator (JavaNNS) was used. The JavaNNS is developed at the Wilhelm-Schickard-Institute for Computer Science (WSI) in Tubingen, Germany. Each peak was modeled using six inputs and one output network. The numbers of hidden neurons were varied between six and twenty in order to find the network structure with the smallest validating mean-square error (MSE). The BP parameter consists of learning rate which varies between 0.1 and 0.2, and the error tolerance equals to 0.05. The neural network training is stopped when the training MSE and validating MSE are comparable the same figure. The neural network structures which have a smallest prediction error for each peak response are shown in Table 3.

Modeling results for each peak response are graphically displayed in Figs. 3-7 when setting other inputs at their midpoint. Each plot shows the results of varying SiH$_4$ flow and NH$_3$ flow on a given response. For modeling of Peak 1, decreasing NH$_3$ flow provides shifting to a higher wave number, while altering SiH4 flow has a slight effect on shifting of wave numbers. For modeling of Peak 2, decreasing NH$_3$ flow provides shifting to a higher wave number, while increasing SiH$_4$ flow during high NH$_3$ flow provides shifting to a lower wave number. In contrast, increasing SiH$_4$ flow when NH$_3$ flow is low has no effect on shifting of wave number. For modeling of Peak 3, decreasing NH$_3$ flow provides shifting to a higher wave number, while altering SiH4 flow slightly affects shifting of wave numbers. For modeling of Peak 4, increasing NH$_3$ flow provides shifting to a higher wave number, while altering SiH4 flow when NH$_3$ flow is high slightly affects the shifting of wave numbers. In contrast, while NH$_3$ flow is low in the range of 50 to 60 sccm, there is a high probability that Peak 4 disappears. Please note that the wave number near 2000 corresponds to no Peak 4. Since the training error of the neural network for Peak 5 does not converge, there is no interpretation for Fig. 7.

From the literature review [4-5], the absorption peaks of silicon nitride is shown in Table 4. It can be shown that Peak 1 is a broad spectrum with its range covers the Si-H and Si-N bonding. Peak 2 corresponds to Si-N bonding. Peak 3 corresponds to N-H bonding. Peak 4 corresponds to Si-H bonding. Peak 5 corresponds to N-H bonding.

B. Corrosion Testing

The potentiostatic measurement for the aluminum alloy on silicon substrate is shown in Fig. 8. From this Tafel plot, the corrosion potential (E_{corr}) is determined to be -0.9081 volts, and the corrosion current (I_{corr}) is determined to be 6.01×10^{-7} mA/cm^2. The Tafel plot for the SiN passivation which has 200nm thickness over the aluminum film on silicon substrate is shown in Fig. 9. The E_{corr} equals to -0.6988 volts, and I_{corr} equals to 11.483×10^{-7} mA/cm^2. Usually, for a good protective film, the E_{corr} will shift to a more positive potential and I_{corr} will shift to a lower current. Unfortunate, the I_{corr} from the protective file is higher. Thus, more measurements are underway to investigate this discrepancy.

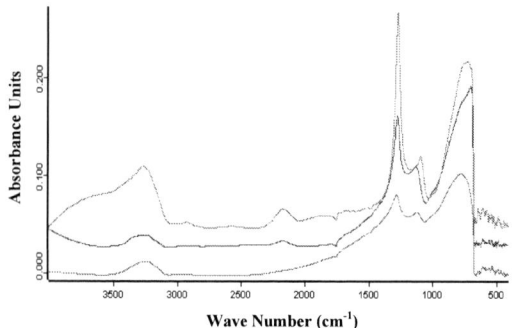

Figure 2. FTIR spectrum from three different recipes (blue: Experiment #1, red: Experiment #2, pink: Experiment #7)

TABLE II.   PEAK ASSIGNMENT VS. WAVE NUMBER

Peak number	Wave number (cm^{-1})	
	Minimum	*Maximum*
Peak 1	677	783
Peak 2	1033	1133
Peak 3	1270	1284
Peak 4	2000*	2166
Peak 5	3255	3256

* represents no Peak 4 in the modeling

TABLE III.   NEURAL NETWORK WITH SMALLEST PREDICTION ERROR FOR EACH PEAK RESPONSE

Peak response	Num. of hidden neural	Training MSE	Validating MSE
Peak 1	12	0.0677	0.0740
Peak 2	6	0.0579	0.0060
Peak 3	9	0.0283	0.0268
Peak 4	9	0.0234	0.0275
Peak 5	16	0.1748	0.1892

TABLE IV.   ABSORPTION PEAKS OF SILICON NITRIDE FROM FTIR

Wave Number (cm^{-1})	Peak assignment
450-462	Si-N wagging
650-670	Si-H wagging
750-1050	Si-N stretching
800-900	TO Si-N stretching
1100	LO Si-N stretching
1150-1200	N-H bending
1550	N-H$_2$ bending
2000-2260	Si-H stretching
3300-3400	N-H stretching
3445	N-H2 stretching

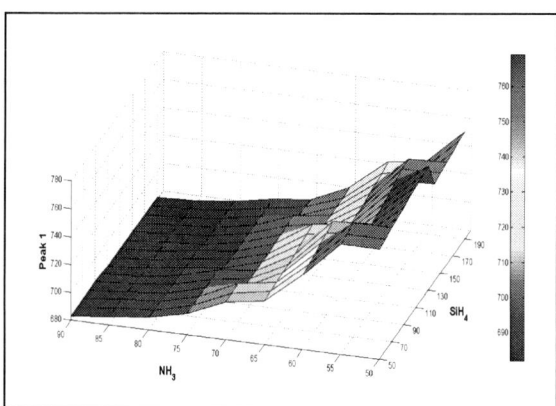

Figure 3. Peak 1 vs. SiH$_4$ flow and NH$_3$ flow

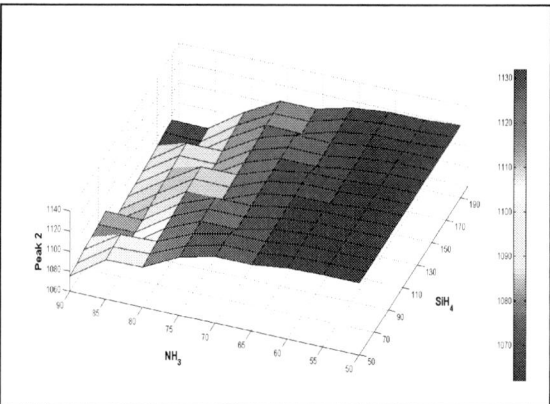

Figure 4. Peak 2 vs. SiH$_4$ flow and NH$_3$ flow

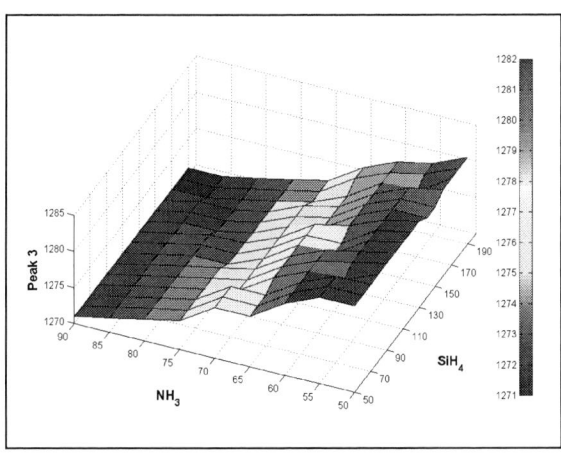

Figure 5. Peak 3 vs. SiH₄ flow and NH₃ flow

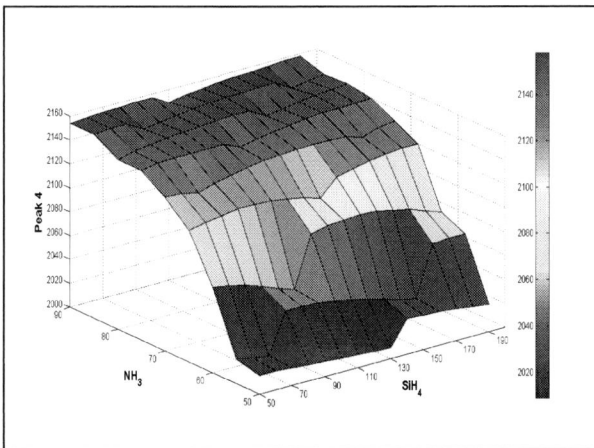

Figure 6. Peak 4 vs. SiH₄ flow and NH₃ flow

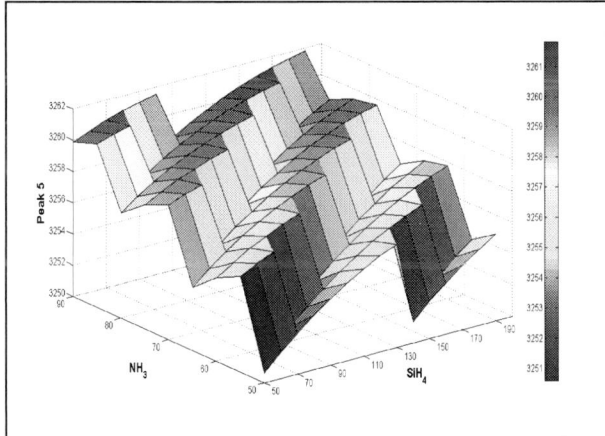

Figure 7. Peak 5 vs. SiH₄ flow and NH₃ flow

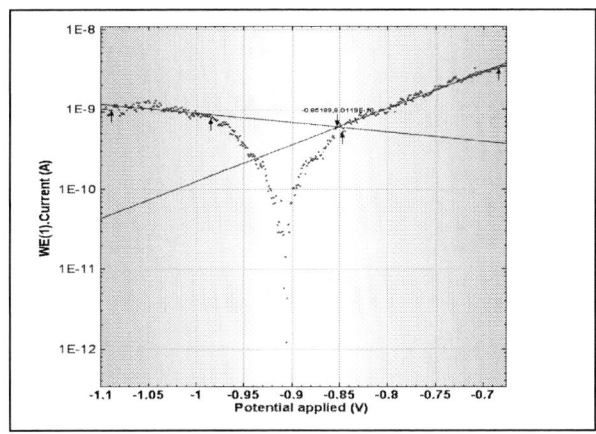

Figure 8. Potentiostatic measurement of aluminum film on silicon substrate

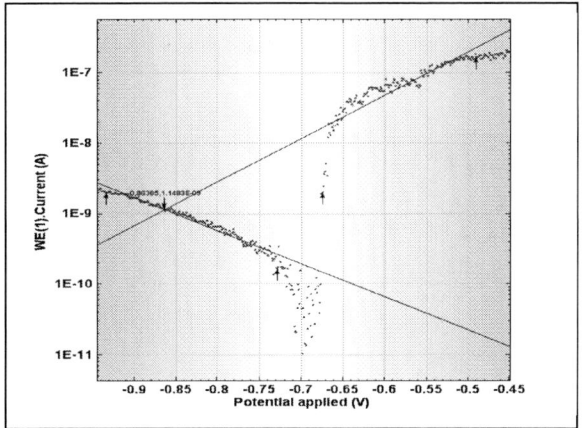

Figure 9. Potentiostatic measurement of silicon nitride passivation on aluminum film on silicon substrate

IV. CONCLUSION

The neural networks have successfully modeled the chemical bonding from FTIR measurement in order to understand its variation with various silicon nitride deposition recipes. The potentiostatic measurement needs to further investigate with more samples in order to verify that the measurement result agrees with theoretical result.

ACKNOWLEDGMENT

The authors thank the Thai Microelectronics Center, National Electronics and Computer Technology Center, National Science and Technology Development Agency for support of this research. We are grateful to Dr. Supavadee Kiatisevi at the Department of Chemistry, Faculty of Science, Mahidol University for FTIR measurement; and Dr. Sutha Sutthiruangwong at the Department of Chemistry, Faculty of Science, King Mongkut's Institute of Technology Ladkrabang for potentiostatic measurement.

REFERENCES

[1] E. Kalvesten, L. Smith, L. Tenerz, and G. Stemme, " The first surface micromachined pressure sensor for cardiovascular pressure measurements," Proceedings the 11th Annual International Workshop on MEMS, pp. 574-579, 1998.

[2] C. Himmel and G. May, "Advantages of plasma etch modeling using neural networks over statistical techniques," IEEE Transaction on Semiconductor Manufacturing, vol. 6, no. 2, pp. 103-111, May, 1993.

[3] T. Thongvigitmanee, A. Srihapat, C. Khompatraporn, A. Jiraprayuklert, W. Titiroongruang, "Use of Neural Network to Model the Refractive Property of PECVD Silicon Nitride Films Used to Prevent Water Permeability of Piezoresistive Pressure Sensor," ECS Transactions, Vol.27, No.1, 767-772, 2010.

[4] T. Stapinski and B. Swatowska, "a-Si:C:H and a-Si:N:H thin film films obtained by PECVD for applications in silicon solar cells", journal of electronic materials, vol. 37, No.6, 2008

[5] V. Tolstoy, I. Chernyshova, and V. Skryshevsky, Handbook of Infrared Spectroscopy of Ultrathin Films. New Jersey: Wiley, 2003.

978-1-61284-408-4/11 $26.00 © 2011 IEEE

Virtual Metrology Models for Predicting Average PECVD Oxide Film Thickness

Ariane Ferreira, Agnès Roussy, Christelle Kernaflen
Department of Manufacturing Science and Logistics
Ecole Nationale Supérieure des Mines de Saint-Etienne
Gardanne, France
aferreira@emse.fr

Dietmar Gleispach, Günter Hayderer
austriamicrosystems AG
Unterpremstätten, Austria
dietmar.gleispach@austriamicrosystems.com

Hervé Gris, Jérôme Besnard
PDF Solutions
Montpellier, France
herve.gris@pdf.com

Abstract— **The semiconductor industry is continuously facing four main challenges in film characterization techniques: accuracy, speed, throughput and flexibility. Virtual Metrology (VM), defined as the prediction of metrology variables using process and wafer state information, is able to successfully address these four challenges. VM is understood as definition and application of predictive and corrective mathematical models to specify metrology outputs (physical measurements). These statistical models are based on metrology data and equipment parameters. In this paper, two VM models based on industrial data are presented. The objective of this study is to develop a model predicting the CVD oxide thickness (average) for an IMD (Inter Metal Dielectric) deposition process using FDC data (Fault Detection and Classification) and metrology data.**

Keywords— *Advanced Process Control, CVD Oxide thickness, Partial Least Squares Regression, Tree Ensembles, Semiconductor Manufacturing, Virtual Metrology*

I. INTRODUCTION

The semiconductor manufacturing industry has a large-volume multistage manufacturing system. To ensure high stability and high production yield, reliable and accurate process monitoring is required [1]. Advanced Process Control (APC) is currently deployed for factory-wide control of wafer processing in semiconductor manufacturing. The APC tools are considered to be the main drivers to guarantee a continuous process improvement [2].

However, most APC tools strongly depend on the physical measurement provided by metrology tools [3]. Critical wafer parameters are measured, such as, for example, the thickness and/or the uniformity of thin films. If a wafer is misprocessed in an early stage but detected at the wafer acceptance test, unnecessary resource consumption is unavoidable. Measuring every wafer's quality after each process step could avoid late wafer scraps but it is too expensive and time consuming. Therefore, metrology, as it is employed for product quality monitoring today, can only cover a small fraction of sampled wafers. Virtual metrology (VM) in contrast enables prediction of every wafer's metrology measurement based on production equipment data and previous metrology results [4]-[7]. This is achieved by defining and applying predictive models for metrology outputs (physical measurements) as a function of metrology and equipment data of current and previous steps of fabrication [8]-[10].

Of course it is necessary to collect data from equipment sensors to characterize physical and chemical reactions in the process chamber. Sensor data will constitute the basis for the statistical models that will be developed. A typical Fault Detection and Classification (FDC) system collects on-line sensor data from the processing equipment for every wafer or batch. They are called process variables or FDC data. Reliable and accurate FDC data are essential in VM model [11]. The objective of a VM module is to develop a robust prediction that can provide estimation of metrology and which is able to handle process drifts whether they are induced by preventive maintenance actions or not.

This paper deals with the prediction of PECVD (Plasma Enhanced Chemical Vapor Deposition) oxide thickness for Inter Metal Dielectric (IMD) layers using FDC and metrology data. Two types of mathematical models are studied to build VM modules for PECVD processes. Partial Least Squares Regression (PLS) [12]-[13] and a non-linear approach based on Tree ensembles [14]-[16] are considered. The technical challenge and innovation are to build a single robust model, either with PLS or Tree ensembles, which is valid for several products, different layers and two different chucks. The alternative would be to make a model per layer, chuck and product, but we strongly believe that the maintenance of many single models, in our case 12 different models, is not compatible with the constraints of the industry.

Section II deals with fabrication process. In section III we present the mathematical background to build VM models. Results and model comparison are described in sections IV. The section V concludes this paper with a summary and a discussion of future work.

978-1-61284-408-4/11 $26.00 © 2011 IEEE

II. FABRICATION PROCESS

The film layer under investigation for thickness modeling is part of the IMD used in the Back-End of Line (BEOL) of a 0.35μm technology process. This oxide layer is used three times during the production of a four metal layer device. PECVD USG (Undoped Silicon Glass) films are commonly used to fill the gaps between metal lines due to their conformal step coverage characteristics. However, as the device geometry is shrinking, the gap fill capability of USG films is no longer sufficient. State of the art technique is the combination of HDP (High Density Plasma) and USG films to provide a high-productivity and low-cost solution. HDP is used to fill the gap just enough to cover the top of the metal line and then the USG is used as a cap layer on top of the HDP oxide film [17].

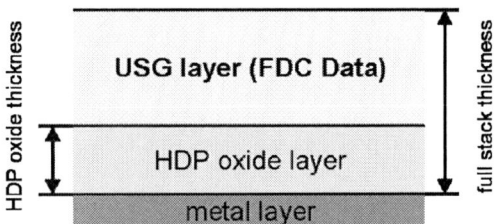

Figure 1: Layer structure for inter metal dielectric

Figure 1 shows the layer structure for one inter metal layer just before the Chemical Mechanical Polishing (CMP) step. The process steps are identical for all three stages. We use identical equipment production recipes, identical metrology setup and identical cleaning procedures for all three stages in the process flow. After metal deposition and structuring (lithography and etch) the HDP oxide is deposited. The HDP oxide film thickness is then measured by ellipsometry, using a 9-site template recipe. FDC data for VM modeling are collected during the USG deposition right after the HDP process. The full oxide stack is measured by the same ellipsometer tool also using a 9-site template recipe.

A schematic drawing of the process flow can be seen in figure 2. To guarantee the collection of a proper set of data within a few weeks, ten wafers per lot are measured before and after the USG process. The objective of the VM model is to use the predictive results as an input parameter for the following CMP process step. The CMP tool uses this input data for calculating the polishing time of each wafer and therefore the integrated layer thickness measurement could be skipped.

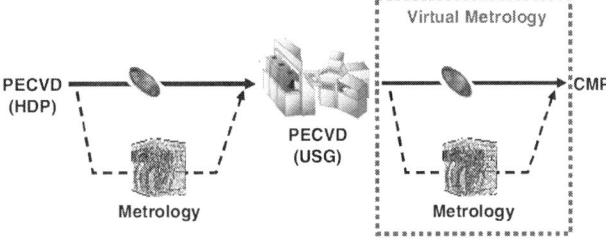

Figure 2: Process flow

Wafers are processed in a twin-chamber of a PECVD tool. The same deposition recipe is used for the deposition of different inter-metal layers and several products. During wafer processing, the relevant process parameters that characterize the PECVD process, such as gas flows, pressure, temperature plasma parameters, etc., are gathered. These temporal data are then consolidated with statistical methods. The temporal data (sensors) are collected at a sampling rate of 0.5 second. If too many samples are missing during the data collection, the data are discarded and the wafer is not used in the VM modeling. The temporal data are then transformed into the so-called FDC Indicators. A FDC Indicator is the summarization of temporal data into a single point, based on a given algorithm (mean, range, maximum, minimum, slope, etc). A data set consists of data from production equipment (input data X for VM modeling) and metrology equipment (output data Y for VM modeling). To assure the quality and effectiveness of VM models it is necessary to do preliminary quality studies of process and metrology equipments like variance analysis or repeatability and reliability studies (R&R studies). In addition, context information like layer, product and chuck is essential as categorical input for VM modeling.

Input data X consists of 24 indicators and three contextual variables. The output variable Y represents the average of the PECVD oxide thickness of each wafer.

III. MATHEMATICAL MODELS

A. Notation

The following notation conventions are used in this paper: scalars are designated using lowercase italics. Vectors are generally interpreted as column vectors and are designated using bold italic lowercase (i.e. x). Matrices are shown in bold italic uppercase (i.e. X), where x_{ij}, with ($i=1,\ldots,I$) and ($j=1,\ldots,J$), is the ij^{th} element of $X(I\times J)$. Let X of ^{p} be an input data set and Y of ^{m} be arranged in the following way:

$$X = \begin{pmatrix} x_1^T \\ x_2^T \\ \vdots \\ x_I^T \end{pmatrix} \quad Y = \begin{pmatrix} y_1^T \\ y_2^T \\ \vdots \\ y_I^T \end{pmatrix}$$

where $x_i^T = \left(x_{i,1},\ldots,x_{i,p} \right)$ and $y_i^T = \left(y_{i,1},\ldots,y_{i,m} \right)$. The characters I, J, N, p, q, m and n are reserved for indicating the dimension of vectors and matrices of data.

B. VM Modeling

There are some important points when designing the mathematical models and a methodology that should be considered. In this section we propose two successive stages to deploy mathematical models in order to build a VM Module for an individual process:

1) Data partitioning: Training set and Test set

Let X ($I \times p$) and Y ($I \times m$) be the available data set (cleaned and normalized) respectively from production and metrology process. The data set partitioning consist in the extraction of two units: a unit of 70% of the data set for the training-validation (training and cross validation) and a unit of 30% of the data set for the test. Let X_N ($N \times p$) and Y_N ($N \times m$) be the training-validation data set, and let X_n ($n \times p$) and Y_n ($n \times m$) be the test data set with $N+n=I$. It is possible to split the available data set in a temporal way (chronological selection) without loss of representativeness. In this case study we have chosen this type of data partitioning before the application of the three mathematical models.

Alternatively, the Kennard-Stone method [15] can be used to perform the data set partitioning. The inputs variables domain X of ^{p}, $x_i^T = \left(x_{i,1}, \ldots, x_{i,p} \right)$ is considered for the Kennard-Stone method. It is a sequential method to select a training set uniformly which covers the entire space of input variables X. The selection criteria use the Euclidean distance.

2) Mathematical Modeling

A linear regression model of a given process can be written as:

$$Y = XB + E \qquad (1)$$

where X is the matrix of input data, Y is the matrix of output data, B is the matrix of regression coefficients and E is the matrix of errors whose elements are independently distributed with mean zero and variance ^{2} [18]-[19]. Linear or non linear regression methods can be applied to the matrices X and Y to compute the coefficient matrix B. The regression model is built in two levels: the Training-validation level with the training-validation data set and the test level with the test data set. The training of models that are linear with respect to their parameters (such as linear regressions, polynomials models) can be performed easily with the traditional least-squares method, whereas the training of models that are nonlinear with respect to their parameters (such as neural networks) requires more complex methods. More details about the training of mathematical models can be found in [16].

Global approaches to model selection in the training-validation level are Cross-validation [20] and Leave-One-Out, methods for estimating generalization error based on resampling [21]. It is obvious to perform the model selection on the basis of the Validation Root Mean Square Error on the Training-validation data set (VRMSE). The VRMSE is given by equation (2):

$$VRMSE = 100 \left(\frac{1}{N} \sum_{i=1}^{N} \left(\frac{y_i - \hat{y}_i}{y_i} \right)^2 \right)^{1/2} \qquad (2)$$

where y_i is the measured output value, $\hat{y}_i$ is the estimated output value from the model, and N is the size of the training data set. The VRMSE is often used for comparing various models. In n-fold Cross-Validation the data are divided into n subsets of (approximately) equal size. The net is trained n times, where one of the training subsets is left out. Only the omitted subset is used to compute the error criterion of interest.

If n is equal to the sample size it is called leave-one-out cross-validation. The prediction performance of the selected model is estimated using the test data set. The performance indicator is the Test Root Mean Square Error of Prediction (TRMSE) computed on the test data set:

$$TRMSE = 100 \left(\frac{1}{n} \sum_{i=1}^{n} \left(\frac{y_i - \hat{y}_i}{y_i} \right)^2 \right)^{1/2} \qquad (3)$$

where y_i is the measured output value, $\hat{y}_i$ is the estimated output value from the model, and n is the size of test data set.

C. PLS Models

Consider a set of historical process data consisting of an ($I \times p$) matrix of process variable measurements (FDC data) X and a corresponding ($I \times m$) matrix of metrology data Y. Projection to Latent Structures or Partial Least Squares (PLS) can be applied to the matrices X and Y to estimate the coefficient matrix B in (1).

$$\hat{Y}^{PLS} = X\hat{B}^{PLS} + E \qquad (4)$$

where $\hat{Y}^{PLS}$ is the PLS estimate of the process output Y. PLS modeling consists of simultaneous projections of both the X and Y spaces on low dimensional hyper planes of the latent components. This is achieved by simultaneously reducing the dimensions of X and Y, by seeking q ($< p$) latent variables which mainly explains covariance between X and Y. Therefore this method is useful to obtain a group of latent variables which explain the variability of both, Y and X. The latent variable models for linear spaces are given by Equations (5) and (6) [12]:

$$X = TP^T + E \qquad (5)$$

$$Y = TQ^T + F \qquad (6)$$

where E and F are error terms, T is an ($I \times A$) matrix of latent variable scores, and P ($p \times A$) and Q ($m \times A$) are loading matrices that show how the latent variables are related to the original X and Y variables. The sample covariance matrix is $X^T Y Y^T X$. The first PLS latent variable $t_1 = Xw_1$ is the linear combination of the X-variables that maximizes the covariance between t_1 and the Y space. The first PLS weight vector w_1 is the first eigenvector of the sample covariance matrix $X^T Y Y^T X$. After the scores for the first component have been computed, the columns of X are regressed on t_1 to give a regression vector:

$$p_1 = \frac{X^T t_1}{t_1^T t_1} \qquad (7)$$

In NIPALS (Nonlinear estimation by iterative Partial Least Squares) algorithm [13] the second latent variable t_2, orthogonal to t_1, is calculated from the new matrix of covariance $X_2^T Y_2 Y_2^T X_2$, where X_2 and Y_2 are calculated by the equations (8) and (9):

$$X_2 = X - t_1 p_1^T \qquad (8)$$

978-1-61284-408-4/11 $26.00 © 2011 IEEE

$$Y_2 = Y - t_1 q_1^T \qquad (9)$$

q_1 is obtained by regression of the columns of Y in t_1, i.e.:

$$q_1 = \frac{t_1^T Y}{t_1^T t_1} \qquad (10)$$

The second latent variable is computed by the equation $t_2 = X w_2$, where w_2 is the first eigenvector of the sample covariance matrix $X_2^T Y_2 Y_2^T X_2$, and so on. The new latent vectors or scores (t_1, t_2,...) and the weight vectors (w_1, w_2, ...) are orthogonal. The final models for X and Y are given by Equations (5) and (6).

Latent variable models assume that both the process and metrology data spaces are observed with error and that both are effectively of very low dimension (i.e. non-full rank). The dimension A of the latent variable space is often quite small compared with the dimension of the process data space, and it is determined by cross-validation or some other procedure. Effectively, these models reduce the dimension of the problem through a projection of the high-dimensional X and Y spaces onto the low-dimensional latent variable space T, which contains most of the important information [12].

D. Tree Ensemble Models

It has been shown by Breiman *et al.* [22], in the classification case, that under reasonable assumptions, an ensemble procedure allows getting accurate models. Indeed, if the base model has a low-bias and high variance under some random perturbation of the learning conditions, then aggregating a large family of such models give birth to a low-bias, low variance aggregated model, that is more accurate than the individuals models [15].

To allow such results to hold, it is critical that the individual models are as independently built as possible, while maintaining low bias. Tree base learners, either based on algorithms such as CART [22] or C4.5 [23], are known to have a low bias when fully learned (no pruning) [24]. In order to be able to build families of trees that have a low correlation to one another, from a finite dataset, several methods have been proposed: Bootstrapping the learning set (also known as bagging methods), Random splits, Injecting random noise in the response or building random artificial features as (linear) combinations of the existing ones. All these ideas aim at learning trees that are as uncorrelated as possible.

Following [22], we use here a combination of bagging from the base learning set, and random splits as our main ensemble method. Base learners are regression trees, following a modified CART algorithm for tree learning. Given X, a set of ($I \times p$) FDC data, and a corresponding Y ($I \times m$) metrology, and 2 parameters q (random selection among features at the individual split level) and *nTrees (number of trees grown and aggregated)*, the algorithm is as follows:

1. Iterating over the m responses:

2. Looping 1 -> *nTrees*:

a) Build a bootstrap sample ($I \times p$) X^b and corresponding Y^b ($I \times 1$) response

b) Build a fully-grown tree , following modified CART algorithm

    i. randomly selecting $q<m$ candidates for a given split inside a node

    ii. Select the best split among the q candidates as the one that reduces most the residual variances over the 2 children nodes

    iii. Recursively until stopping criterion is reached, i.e. node is pure (internal variance equal 0)

3. Average the predictions, i.e.

$$\hat{y} = \frac{1}{nTrees} \sum_{i=1}^{nTrees} {}_i (X) \qquad (11)$$

Bagging allows the calculation of the so-called out-of-bootstrap prediction, which is very similar to cross-validation or leave-one-out, since predictions on the learning set are derived by averaging, for each individual, the set of trees in which this individual is not in the bootstrap sample. Hence bagged ensembles have an internal estimation of their generalization error. A well-known property of trees is their inability to model linear effects, which is why, when a strong linear effect from one or several parameters is discovered in the data, we build the tree ensemble on the residual from the main linear effect.

Finally, tree ensembles have internal estimations of the importance of each of the feature that are calculated by averaging their out-of-bootstrap contribution in the prediction for each tree. More precisely, one can estimate the increase in out-of-bootstrap error that scrambling one parameter would produce, over several repetition of the scrambling procedure. This allows dropping low importance features from the model by comparing features importance to probes (random features). In the end, the model will be:

$$\hat{y} = A X^{lin} + \frac{1}{nTrees} \sum_{i=1}^{nTrees} {}_i (X^{nlin}) \qquad (12)$$

where *lin* and *nlin* are disjoint subsets of the initial set of indicators. R^2 is defined by equation (18):

$$R^2 = 1 - \frac{\frac{1}{N} \sum_{i=1}^{N} (y_i - \hat{y}_i)^2}{S_y^2} \qquad (13)$$

where y_i is the measured output value, $\hat{y}_i$ is the estimate output value by model, N is the size of training data set and S_y^2 is the variance of y. This metric is estimated from the out-of-bootstrap predictions.

IV. RESULTS AND MODELS COMPARISON

In this section we present the results of two different models (PLS and tree ensembles) for prediction of the PECVD oxide film thickness.

978-1-61284-408-4/11 $26.00 © 2011 IEEE

A. PLS models with 3 qualitative variables

A PLS model is built without input variable selection. The model is calibrated with 306 wafers of the training validation data set. The 168 wafers of the test data set are used to validate the model. The input variables of the PLS model are the 24 quantitative FDC indicators and the three qualitative variables which are chuck, layer and product. The PLS model with four principal components is selected by cross-validation. Q^2(cum) is increasing for the first four principal components and decreasing for the fifth principal component. Actually, using the first four principal components, 46.5% of the X variability (quantitative and qualitative variables) can explain 89.7% of the output Y (average thickness) variability (see table I). Therefore the best statistical model is achieved by using only four principal components.

Table I: Result table for the principal components of PLS

Principal Component	R^2X(cum)	R^2Y(cum)	Q^2(cum)
1	16.2	75.6	74.2
2	29.5	85.5	83.7
3	39.0	88.3	85.7
4	46.5	89.7	86.1

Analyses have been done on each parameter to quantify its importance. In table II the five most important variables can be found.

Table II: Variable importance of PLS model

Variable	VIP (4PC)
X24	2.16
X10	1.89
X3	1.77
X17	1.61
X25	1.60

The trained model is applied to the test data set. In figure 3 the measured and the predicted average oxide film thickness for PLS model are shown. The *VRMSE* and *TRMSE* are around 0.53% and 0.58% of the average thickness, respectively. Figure 3 shows the predicted average oxide film thickness using the results of PLS model versus the measured average thickness.

B. Tree ensembles model

Modeling is done using the learning set of 306 wafers. After one iteration of the algorithm, one indicator (X10) is selected for the prelinear part; the remaining 26 are left for tree ensemble modeling, including the three qualitative parameters chuck, product and layer. The second iteration of the algorithm provides a model that selected five parameters: X10 in the linear part of the model, and four in the tree ensemble model (X9, X8, X24 and X25). X24 and X25 are two qualitative parameters (see table III). R^2 is estimated to be 0.84, being defined as equation (13).

Table III: Ranking of indicators for the tree ensemble model

Rank	Variable	model
1	X10	Linear
2	X24	Tree ensemble
3	X25	Tree ensemble
4	X9	Tree ensemble
5	X8	Tree ensemble

This metric is estimated from the out-of-bootstrap predictions. Other model quality metric include VRMSE, estimated also from the out-of-bootstrap predictions, measured at 0.69% of the average thickness for this model. The model is then used to predict average oxide film thicknesses for the test set. Figure 4 shows the result of the tree ensembles model. The TRMSE is comparable to the VRMSE and is equal to 0.59% (see table IV). R^2 is calculated to be 0.84 also, for the test set.

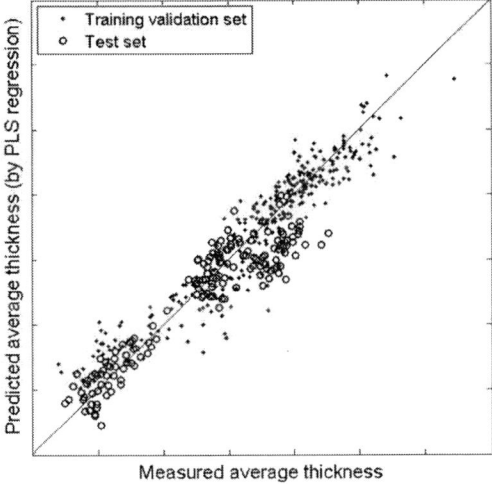

Figure 3: Predicted average thickness (by PLS model) versus measured average thickness representation

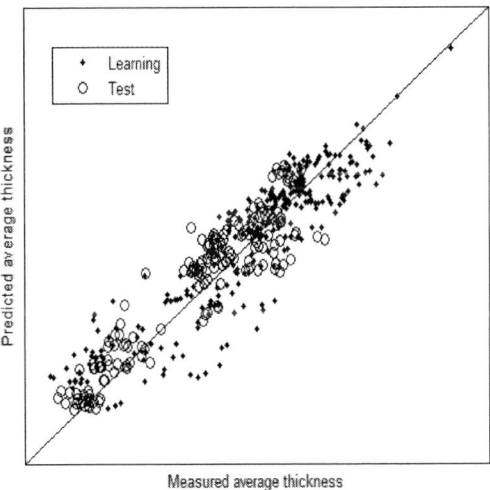

Figure 4: Predicted average thickness (by tree ensemble model) versus measured average thickness representation

TABLE IV: Tree ensemble model summary

Quant. variables	Qual. variables	Selected	R²	VRMSE (%)	TRMSE (%)
24	3	5	0.84	0.69	0.59

C. Model comparison

We have used two different algorithms on the same dataset, with identical learning and test sets. In table V the metrics for the PLS models and the tree ensemble model are presented.

TABLE V: Model comparison

	Quant. variables	Qual. variables	VRMSE (%)	TRMSE (%)
PLS	24	3	0.53	0.58
TE	24	3	0.69	0.59

All models have good predictive power. PLS and tree ensemble show equivalent performance on the test set, but PLS shows slightly better results on the learning set. PLS uses four principal components, which are based on all the variables. The tree ensemble model uses five variables only. Three out of the top five most important variables of PLS are used in the tree ensemble model.

V. CONCLUSION AND PERSPECTIVE

This paper describes three mathematical models that have been used to develop virtual metrology for predicting the average oxide film thickness deposited during a PECVD process. The predictive results are in excellent agreement with the measured data. In addition, we have shown as a novelty in virtual metrology that it is possible to create a single model for different layer, different products and for one chamber with two different chucks.

In order to run such a model online for predicting the CMP polishing time we will have to expand the modeling to the HDP tool and also predict the full stack film thickness uniformity along with the average value. We will also need to add a predictive quality index to the current VM models in order to guarantee the accuracy and the robustness of the model. This quality index could be compared to the GOF (goodness of fit) available in a physical measurement.

ACKNOWLEDGMENTS

This work was supported by ENIAC project IMPROVE (Implementing Manufacturing science solutions to increase equipment pROductiVity and fab pErformance). Funding by the EU, the FFG and the MINEFI is gratefully acknowledged.

REFERENCES

[1] A.C. Diebold, "Overview of metrology requirements based on the 1994 National Technology Roadmap for semiconductor". *Advanced Semiconductor Manufacturing Conference and Workshop 1995*, ASMC 95 Proceedings. IEEE/SEMI 1995, November 1995, pp. 50–60.

[2] J.R. Moyne. "Making the move to fab-wide APC", Solid State Technology, vol. 47, no. 9, September 2004, pp. 47-52.

[3] S. J. Qin, G. Cherry, R. Good, J. Wang and C. A. Harrison. "Semiconductor manufacturing process control and monitoring: A fab-wide framework". Journal Process Control, vol. 16, no. 3, 2006, pp. 179–191.

[4] Y.-J. Chang, Y. Kang, C.-L. Hsu, C.-T. Chang and T. Y. Chan. "Virtual Metrology Technique for Semiconductor Manufacturing", *Proceedings of the Conference on Neural Networks,* Sheraton Vancouver Wall Center Hotel, 2006, pp. 5289–5293.

[5] P. H. Chen, S. WU, J. Lin, F. Ko, H. Lo, J. Wang, C. H. Yu, and M. S. Liang. "Virtual Metrology: A solution for wafer to wafer advanced process control", *Proceedings of the IEEE International Symposium on Semiconductor Manufacturing*, September 2005, pp.155–157.

[6] F.-T. Cheng. "Researching Strategy and Development Proposal of e-Manufacturing". *Automation Division of National Science Council*, Taiwan, R.O.C, October 2004.

[7] F.-T. Cheng, H.-C. Huang, and W.-M. Wu "Dual-Phase Virtual Metrology Scheme", IEEE Transactions on Semiconductor Manufacturing, vol. 20, no. 4, November 2007, pp. 566–571.

[8] M.-H. Hung, T.-H. Lin, P.H. Chen and R.-C. Lin. "A novel virtual metrology scheme for predicting CVD thickness in semiconductor manufacturing", IEEE/ASME Transactions on Mechatronics, vol. 12, no. 3, June 2007, pp. 364–375.

[9] A.A. Khan, J.R. Moyne and D.M. Tilbury. "An Approach for factory-wide control utilizing virtual metrology", IEEE Transactions on semiconductor Manufacturing, vol. 20, no. 4, November 2007, pp. 364–375.

[10] T.-H. Lin, M.-H. Hung, R.-C. Lin and F.-T. Cheng. "A virtual metrology scheme for predicting CVD thickness in semiconductor manufacturing". *Proceedings of the 2006 IEEE International Conference on Robotics and Automation*, May 2006, pp. 1054–1059.

[11] Y.-C. Su, T.-H. Lin, F.-T. Cheng, and W.-M. Wu. "Accuracy and Real-Time Considerations for Implementing Various Virtual Metrology Algorithms". IEEE Transactions on Semiconductor Manufacturing, n. 21, vol. 3, August 2008, pp. 426–434.

[12] T. Kourti, "Application of latent variable methods to process control and multivariate statistical process control in industry," Int. J. Adapt. Contr. Signal Process, vol. 19, no. 4, 2005, pp. 213–246.

[13] M. Tenenhaus, *La Régression PLS Théorie et Pratique*, Editions Technip, Paris, 1998.

[14] L. Györfi, M. Kohler, A. Krzyzak, and H. Walk, *A Distribution Free Theory of Nonparametric Regression*, Springer-Verlag, New York 2002.

[15] R.W.Kennard and L.Stone. "Computer Aided Design of Experiments", Technometrics, no. 11, 1969, p.137-148.

[16] G. Dreyfus. *Neural Networks Methodology and Applications.* Hardcover, 2002.

[17] S. P. Muranka, M. Eizenberg, A.K. Sinha. *Interlayers Dielectrics for Semiconductor Technologies*, Elsevier edition, 2003.

[18] S. Chen, S. A. Billings, and W. Luo. "Orthogonal least squares methods and their application to non-linear system identification", International Journal of Control, vol. 50, no. 5, 1989, pp. 1873 – 1896.

[19] A. C. Rencher, *Methods of multivariate Analysis*, Hardcover 2002.

[20] M. Stone, "Cross-validatory choice and assessment of statistical predictions", Journal of the Royal Statistical Society, B 36, 1974, pp 111-147.

[21] B. Efron and R.J. Tibshirani. "Improvements on cross-validation: The bootstrap method", Journal of the American Statistical Association, vol. 92, 1997,pp.548-560.

[22] L. Breiman, "Arcing classifiers", The Annals of Statistics, vol. 26, no.3, 1998, pp. 801-849.

[23] J. R. Quinlan, *C4.5: Programs for Machine Learning*, Morgan Kaufmann, 1993.

[24] J. H. Friedman and P. Hall. "On Bagging and Nonlinear Estimation" . Journal of Statistical Planning and Inference, vol. 137, no.3, March 2007, 669 – 683.

Wet Etch step modelling to help Shallow Trench Isolation module control

A.Roussy[1], M.Gedion[1]
[1]EMSE-CMP Georges Charpak
880 Avenue de Mimet,
13541, Gardanne FRANCE
roussy@emse.fr

N.Crousier[2], J.Pinaton[2], K. Labory[2]
[2] *STMicroelectronics,*
Zone Industrielle de Rousset,
13106 Rousset cedex, FRANCE.

Abstract— **We propose a method to model the wet etch process within the Shallow Trench Isolation (STI) module in the CMOS technology. To model a process is the first step in the design of a run to run system, in order to reduce for example the lot to lot variability (a lot equals 25 wafers). The developed predictive model is based on a Design Of Experiments (DOE).**

Keywords: Shallow Trench Isolation, Wet Etch process, Design of Experiments, Run to Run

I.   INTRODUCTION

The semiconductor manufacturing industry has large-volume multistage manufacturing systems. To insure the high stability and the production yield a good control is required [1]. The Advanced Process Control (APC) is currently deployed in factory-wide control of Front End Of Line (FEOL) processing in semiconductor manufacturing. The APC tools are the main way to ensure a continuous process improvement [2]. The process disturbance usually includes shift, drift, and other noise, such as random noise. A key for the application of Advanced Process Control (APC) is the availability of metrology tools to measure the key wafer, equipment, and process parameters in a suitable time frame. In addition to the measurement data, some analysis methods are required. A Run to Run (R2R) controller can help to reduce the process variability and then, minimize the cost. Frequently, R2R controllers are model-based. During qualification of a new process, experimental and statistical techniques are used for obtaining initial models and an initial optimal recipe. The development of mathematical models that relate the process inputs to the corresponding outputs (responses) is based on statistical methods such as Design Of Experiments (DOE) and Response Surface Methods (RSM) which are widely used in industry for process characterization and optimization. The R2R controller tunes or adjusts these initial models and recipes, trying to keep the process at the optimized level (which becomes the target value in case targets were not previously available).

In ultra-large scale integration (ULSI) and for some technology nodes below 0.25μm, it is crucial to be able to reduce the size of the active isolation region. To help, the Shallow Trench Isolation (STI) process has been developed

[3]. STI filling-in is one of the critical processes during CMOS device fabrication due to its tremendous impact on transistor performance [4-5]. The typical STI process is described as follows. First, a sacrificial oxide is realized. Then, a silicon nitride thin film is deposited. The shallow trenches are defined with Photolithography and Etch processes, before being filled by a High Density Plasma Chemical Vapor Deposited (HDP-CVD) oxide that will be polished by the Chemical Mechanical Polishing (CMP) Process. At the end of the CMP process, the tops of oxide and nitride levels are aligned. Finally, the step between the active area and the top of isolation is defined as shown figure 1. This critical step is called Step Height. The oxide height in trench, which is at the same level than nitride layer after CMP process, is adjusted by an HF Wet etching process. Then the nitride layer is completely removed without oxide consumption. At the end of the process, the topology is controlled by metrology tool.

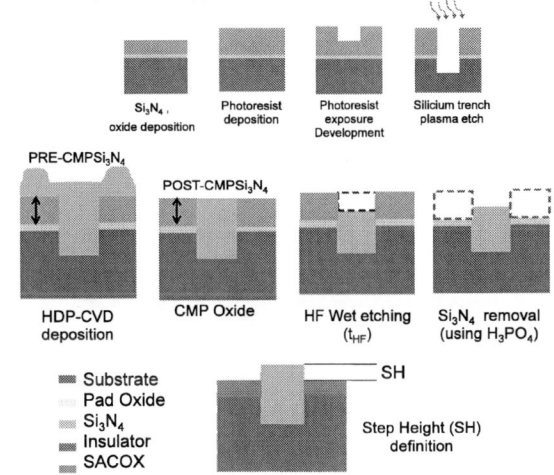

Figure 1. STI Fabrication Steps

The Step Height is a very important parameter for the transistor behaviour. On the one hand it must be very stable because of his topology impact on different next steps in flow. On the other hand it has a big influence on the final electrical A Step Height variation will cause an unstable electrical circuit and further a loss of quality and yield. So, to get the expected electrical parameters and the right active area, the

978-1-61284-408-4/11 $26.00 © 2011 IEEE   163

SH value should be maintained as close as possible to a predefine target value. A way to control the variability of this parameter around a target value is to implement a regulation by R2R loop.

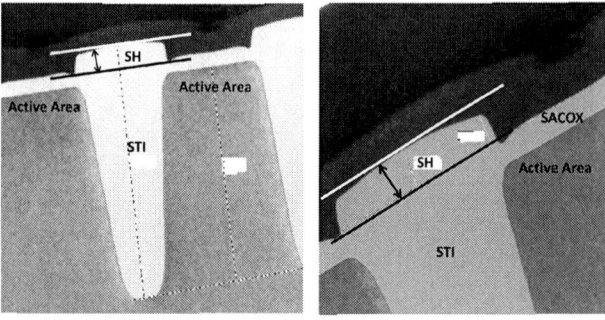

Figure 2. Step Height (SH) SEM cross section

To keep the SH value as close as possible from the target value, the post CMP nitride thickness, controlled by POST-CMPSi$_3$N$_4$ measurements, can allow to calculate how to compensate the oxide thickness with the HF wet etching step process. Modifying the wet process parameters for example the etching time t$_{HF}$, the oxide level and consequently the step height are adjusted. A feedforward R2R loop between the post CMP measurement and the wet-etch process will allow an automatic adjustment from the Wet Etch parameter as showed figure 3.

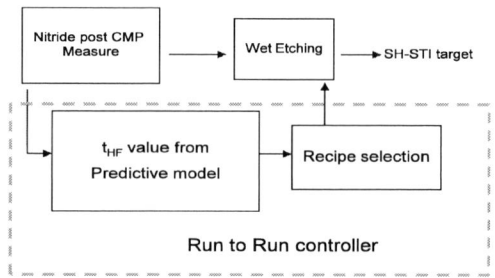

Figure 3. CMP Wet-etch feedforward regulation loop.

The objective of this study is to be able to compensate the oxide level related to post CMP nitride thickness using a feedforward regulation loop. Depending of the POST-CMP measurements, the regulation loop will, within a predictive model, calculate which etching time t$_{HF}$ should be used to compensate the oxide step. From this predictive etch time, the right Wet Etch recipe will be chosen, to reach the goal and the right SH thickness.

First analyses have been done, based on 15 lots data (this study is based on a new technology in development so there is not a lot of historic data available) to classify the type of variability and then to know which type of regulation loop will be implemented. In the case of this study a lot to lot variability has been light up based on POST-CMPSi$_3$N$_4$ measurements.

So, the goal is to create a lot to lot control strategy based on the available measurement. The final objective is to be able to shrink the dispersion, then to get closer to the SH target and to be able to compensate a drift from previous steps, specially, the CMP process via the Wet Etch process. To reach this, it is necessary to well understand this last process step mechanism to be able to choose the right and adjustable parameter. In the second section the wet etch process will be presented. It will be followed by a section on statistic modelling, one about the regulation loop industrialization and then the conclusions.

II. WET ETCHING PROCESS

Wet chemical etching is used extensively in semiconductor processing. Prior to thermal oxidation or epitaxial growth, semiconductor wafers are chemically cleaned to remove contamination that results from handling and storing. Wet chemical etching is especially suitable for blanket etches (over the whole wafer surface) of poly-silicon, oxide, nitride, metals, and III–V compounds. For example, silicon oxide etch is achieved using a diluted fluorhydric solution. In semiconductor production lines, highly uniform etch rates and selectivity with other material are important. Etch rates must be uniform across a wafer, from wafer to wafer, from run to run, and for any variations in feature sizes and pattern densities [6].

In our case wafers are immersed in a reactive solution (etchant) as illustrated figure 4. The layer to be etched is removed by chemical reaction or by dissolution. From the wet etching equipment, two etching baths are important. The HF etching bath where a part of the oxide is removed [7]. This des-oxidation step is very important to define the step height because it helps to determinate the trench oxide thickness. The second important etching bath is the H$_3$PO$_4$ etching solution. The wafers are immersed into a H$_3$PO$_4$ bath to remove the Si$_3$N$_4$ layer with a very good selectivity with the oxide layer.

Once the wet etch process step is described, it is interesting to focus on the process parameters that effect a possible Step Height thickness dispersion through statistical analysis.

III. STATISTICAL MODELLING

The Wet Etch statistical modelling has been done following different steps: first the input critical parameters are selected, then, some running experiments are done (Design Of Experiments) to establish a physical model. In order to validate this model, a comparison between experimental data and simulated data has been done. Statistic analyses help to identify process problems. They will make it possible to light up and to explain the SH variations. First, using SAS software, a lot to lot dispersion has been light up (STI Mean dispersion contributions: 74.2% for lot-to-lot compare to 25.8% for the wafer to wafer contributions). So, this is a lot-to-lot regulation loop which will be implemented (as already mentioned section I).

978-1-61284-408-4/11 $26.00 © 2011 IEEE

A. Multivariate analysis

Multivariate methods [8] are used as a solution to compensate the limitations of univariate methods where variables are followed by individual control charts [9]. The final goal using multivariate methods in this paper is to be able to analyze the correlations between the variables and to reduce the number of parameters to be considered for the DOE. Multivariate analysis have been done on production data to help to determinate the parameters with the most influence on the SH variation [8]. The best way is to use some statistic tools such as the PLS Partial Least Squares and the VIP Variable importance in the projection.

The Wet etch steps are characterized by bath concentration, temperature and time. The results are based on 15 lots. Usually to get results 30 lots are necessary. In the case of this study the work has been done on a new technology so with much less available lots (less historical data). Already a study done on 15 lots can help to get a good idea of which parameters are more critical than the others. At the beginning all the parameters are taken into account then of course to get the easiest model the less important parameters won't be take into account.

B. Partial least square regression

PLS regression [10] is an extension of the multiple linear regression models. In its simplest form, a linear model specifies the (linear) relationship between a dependent (response) variable Y and a set of predictor variables

$$Y = b_0 + b_1 X_1 + b_2 X_2 + \cdots + b_p X_p \ (1)$$

where b_0 is the regression coefficient for the intercept and the b_i values are the regression variables' weight on PC (for variables 1–p) computed from the data. The loadings $b_1, b_2, \ldots, b_i$ belong to [0, 1]. If this value is closer to 0 then the variable has not much influence on the output, in this case the SH Mean thickness (MeanThickness). After discussion with the process expert, a PLS model has been made. The parameters that are involved in the SH thickness variations are the PRE-CMPSi$_3$N$_4$ thickness, the POST-CMPSi$_3$N$_4$ thickness, the initial pad-oxide thickness and the SACOX thickness illustrated figure 1.

The model based on these parameters gives a R²Y(cum) of 0.8, so 80% of the SH variation is explained using these parameters. The parameter Q² is about 0.72 which means that the model can predict 72% of the SH variations. Once this first model is done, the next step is to find out which are the parameters that are the most involved and so the most critical in this model to make it simpler.

The figure 4 shows the different degree of contribution of the different parameters. Higher than 1 is the VIP, more it is involved in the response variation. The critical parameters that are able to better explain the SH (Step Height) variations are

the POST-CMP-Si$_3$N$_4$ thickness and the PRE-CMP-Si$_3$N$_4$ thickness. Previous studies have been done using a constant HF wet etching time, not recipe dependant [11]. But, what should be not forgotten is that, in the case of a variation of the etching time, the etching layer thickness will be different (recipe dependant). So a third parameter will be taken into account in this study: the etching time t_{HF} time necessary to remove a part of the oxide. A simple regression (POST-CMPSi$_3$N$_4$=b$_1$.PRE-CMPSi$_3$N$_4$+b$_0$) with a R²(Y)=97.8% showed that The POST-CMPSi$_3$N$_4$ and the PRE-CMPSi3N4 thicknesses are strongly correlated. The POST-CMP Si$_3$N$_4$ has been chosen instead of the PRE-CMPSi$_3$N$_4$ because it is directly connected to the HF wet etch step input parameter and its VIP is higher than the one from the PRE-CMPSi$_3$N$_4$ as illustrated figure 4.

Thus, to regulate, two parameters will be used:
- the POST-CMPSi$_3$N$_4$ thickness
- the wet etching time t_{HF}

Once these critical parameters lighted up the study can be focused on the statistical method used to find the predictive model which will feed the regulation loop and then, the way to implement the regulation loop.

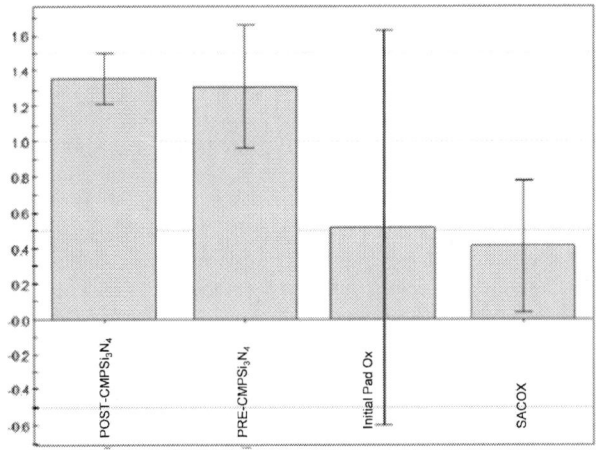

Figure 4. VIP representation

C. Design of Experiments (DOE) and Response Surface Methodology

DOE is based on the fact that the quality of results is not directly linked to the number of experiments. It is used in process improvement because it enables gathering a maximum of information with a minimum of experiments. Thus, the cost factor is taken into account in DOE approach. DOE is a set of designed experiments selected and performed to model the relation between inputs and outputs [12]. A specialized application helps in selecting optimal experiments and in conducting analysis. DOE methodology is based on: problem description, choice of input factors, choice of design, realization of experiments, data Statistical analysis and modeling, model validation. The DOE model will have to

978-1-61284-408-4/11 $26.00 © 2011 IEEE

reach the following goals: getting the SH value as close as possible from the target. To summarize, the Wet-Etch model embedded in the process control strategy shall control the step height SH value using the input parameters POST-CMP Si_3N_4 measurement and the wet etch time t_{HF} (the two inputs: target and limit values).

RSM is a collection of mathematical and statistical techniques that are useful for modeling and analysis in applications where a response is influenced by several variables and the objective is to optimize this response. Linear RSM are often used for many processes in semiconductor manufacturing R2R controllers [13]. A polynomial model can be expected to provide a good predictive capability within the experimental design space. Even within the design space, though, the error in the prediction increases as one moves away from the center point. A second order model will estimate first order effects and all second order interactions. As for the central composite design, three dimensions space is defined by three inputs: $Nc = 2^k$ points (with k the input numbers) are placed at the cube corners, $Ns = 2k$ points with coordinates $(\pm\alpha, 0, \ldots, 0)$, $(0, \pm\alpha, \ldots, 0)$, $(0, 0, \ldots, \pm\alpha$ and No center points with coordinates $(0, 0, \ldots, 0)$. α and No depend on design choice. Thus, the Wet Etch process design will be a response case, a face centered central composite design with 2 inputs is well adapted (with $\alpha = 1$, points outside of the domain will not be experiment and $No = 6$ which is a sufficient repeatability test). The DOE application, after the selection of the design, edits the experiments. For centered faces central composite design 14 runs must be processed varying the two inputs $k = 2$ and N(number of experiments) = $Nc + No + Ns = 2^k + 6 + 2k = 4 + 6 + 4 = 14$. Data capture in the application permits an easy data processing: model construction and variance analysis.

As seen previously, a linear regression is a mathematical model linking a random variable Y (output) to k independent variables (inputs), this model being built to predict and optimize Y. The general form of the function can be represented by a polynomial function, which is a good approximation for a small study area. For k=2:

$$Y = a_1x_1 + a_2x_2 + a_{11}x_1*x_1 + a_{12}x_1*x_2 + a_{22}x_2*x_2 + \varepsilon \quad (1)$$

$a_1, a_2, \ldots a_{22}$ are the coefficients, ε is the random error, x_1, x_2 are the input variables

Analysis of variance (ANOVA) is used to validate the model to fit the results. Applied to Multiple Linear Regression (MLR), some statistical tests allow to validate the regression significance and to extract negligible coefficients. ANOVA is used to validate the model using Fisher test which compares explained model variance and the variance's residual. The result of this test gave a statistically significant MLR with 5% risk. For the predictive model, only terms with $P_{value} < 0.05$ are selected (Table I) because these are significant with again 5% risk. All the contributions, as shown in table II are significant ($P_{value} < 0.05$). As the contribution of $t_{HF}*POST\text{-}CMPSi_3N_4$ is far less significant compared to the others, the decision has been taken to simplify the model by not taking it into account. Then the predictive model can be written as:

$$SH = a + b*PostCMPSi_3N_4 + c*t_{HF} + d*t^2_{HF} \quad (2)$$

a and b are the constants, SH the step height STI-value, POST-CMPSi$_3$N$_4$ the POST-CMPSi$_3$N$_4$ thickness and t_{HF} the wet etch time.

TABLE I.   PVALUES CONTRIBUTION

Contributions	P value
Constant	5.58201e-32
POST-CMP-Si$_3$N$_4$	5.69896e-18
t_{HF}	1.08007e-16
$t_{HF}*t_{HF}$	3.22917e-12
$t_{HF}*POST\text{-}CMP\text{-} Si_3N_4$	0.0274748

Then, a second test is done to get the model coefficients' significance. This is the student test. The results can be found in Table I. Only the first four parameters are significant with again 5% risk. Then, a new ANOVA is done to validate the new model based only on the four variables with significant coefficient parameter. Adj-$R2$ is the fraction of the total variation in the data explained by the model. The Adj-$R2$ values imply excellent fit for the data by the models, see results Table II. This function can now be represented as an area as illustrated figure 8. In the present study, according to the mathematical function, the bigger the HF etching time is, the lower is the SH value. This representation helps to understand which ranges of etching time and POST-CMPSi$_3$N$_4$ have to be chosen to get the SH target value (normalized as value 1).

TABLE II.   DOE RESULTS

	R²	Adj-R²	Q²	Model Validity	Reproductibility
Step Height	0.9940	0.9927	0.9890	0.8441	0.9927

To be valid, the model should be tested on production lots to check if the model can predict the SH which will be measured.

D. Model validation:

Some Step Height values are predicted using a few production lots. The goal is to validate the model (equation 2). Some known POST-CMP-Si$_3$N$_4$ thickness and wet etching time t_{HF} are injected into the model. Then, the calculated SH is compared to the experimental SH value. We found a difference of about 5%, as illustrated figure 6. So, the predicted model build, which is a first model, is ok.

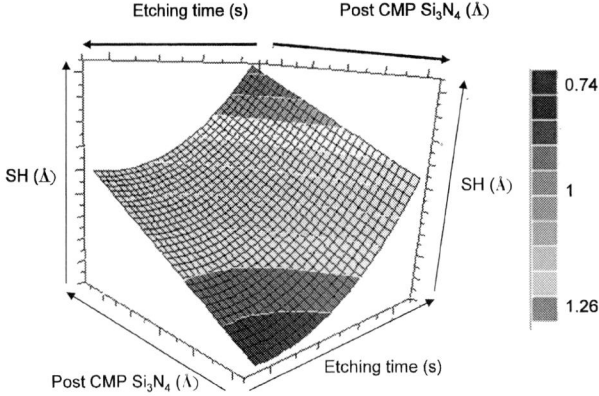

Figure 5. Model Surface response.

Once the model is validated, the next step is the choice of the recipe which is needed to feed the regulation loop.

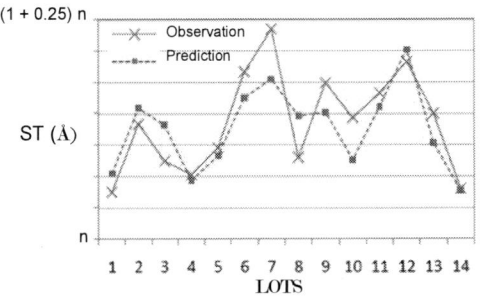

Figure 6. Observation versus prediction.

E. Recipe creation and selection:

To start, only one recipe was available with a fixed time HF process. To implement the regulation loop it is necessary to create new wet etch recipes with different calculated times. More recipes were created; the harder it will be to keep the regulation under control. The created recipes have a different etching times but as soon as the "right" etch time is chosen, to reach the SH target, it will be fixed all along the regulation. It is useful to write down the HF wet etch time equation to be use to get the expected SH target value (from the SH equation -2-).

$$t_{HF} = \frac{-c - \sqrt{c^2 - 4d(a + b.POSTCMPSi_3N_4 - SH_{target})}}{2d} \quad (3)$$

SH_{target} is the SH target value. The equation above shows the required HF wet etching time to get the SH_{target}. From the equation 2, for one value of step height, two t_{HF} are possible. Only one value should be kept. If we plot the equation 2, an increased and decreased ramps appear. The increased ramp is impossible because as more insulate layer is etched, SH will be smaller. So, the time T_{HF}^+ can't be used. From the equation 3 and the historical POST-CMPSi$_3$N$_4$ thickness data available

it is possible to determinate different wet etch recipes which might be created. Within the definition range of our model six wet etch recipes have been needed to have a stable and controlled step height With these recipes it will be possible to cover the full range of etching time necessary to reach the SH target within the control limit values (around the target). Thus, once a t_{HF} would be calculated the right recipe will be chosen. The software Process Works from Rudolph Technology will take care of the application of the right wet etch recipe to the regulation loop. In the next part, the SH regulation loop industrialisation will be presented.

IV. REGULATION LOOP INDUSTRIALIZATION

A regulation loop is managed by a Run to run controller. This controller will select the right recipe (the best calculated choice to reach the SH target knowing the POST-CMPSi$_3$N$_4$ value) and will send it to the process equipment. How does the controller work? ProcessWorks software from the company Rudolph Technology is one of the leader software in Run to Run control [14]. Strategies, created by the engineers and based on the SH predictive model presented previous section (equation 2) are used. The software calculates what has to be adjusted, analyse the previous data (from the process) and adjust the process model to be as close as possible to the target value. The R2R controller works as illustrated figure 7.

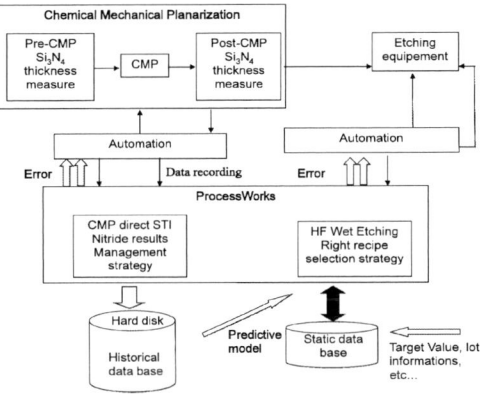

Figure 7. Run to Run controller: ProcessWorks Software.

For the Step Height (SH) regulation loop, ProcessWorks can use two different strategies:
- CMP
- HF Wet Etching

This paper is focused on the HF WET Etching strategy. From the CMP strategy, POST-CMPSI$_3$N$_4$ thicknesses data are collected and recorded. Then, these data are sent to the second strategy HF Wet etching. Thanks to the predictive SH thickness model (presented previously), the HF wet etching strategy can calculate the exact HF etching time and then will be able to select the right recipe with the closest etching time.

978-1-61284-408-4/11 $26.00 © 2011 IEEE

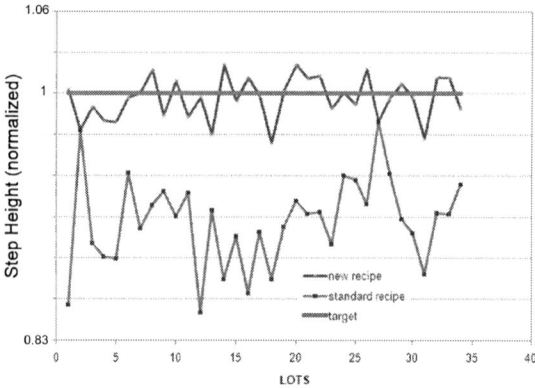

Figure 8. Step Height regulation example.

The HF wet etch strategy makes the communication between the process and the metrology equipment possible. It allows the recipe adjustment. Once the recipe selection is done, the strategy sends a message through the automation to the etch equipment. Thus the step height regulation loop is applied to a lot. If no POST-CMPSi$_3$N$_4$ measurement is available the strategy will take the target value from a historical data base and then, will calculate the HF etching time. Some first results are illustrated figure 8. Using the created recipes helps clearly to get closer from the SH target. Of course these are first results and some improvements have still to be done specially to take into account the errors we still get using some strategy with a missing measurement or some lot with a POST-CMPSi$_3$N$_4$ thickness outside the limits of specification.

V. CONCLUSION

The use of multivariate analyses such as partial least squares methods is very useful to treat the huge data of the equipment and process parameters. A good knowledge of the process physical mechanisms is required to select only the most critical parameters in order to get a simple model. A modeling technique through a DOE of the Wet Etch process for STI was proposed. First experimental results showed that the proposed method is effective and can reduce the lot to lot SH thickness variability observed. Relevant information must be provided to R2R controller in order to increase the reliability of regulation loops. It will take some more time to be able to really quantify the impact of this regulation on the production yield.

VI. ACKNOWLEDGMENT

This research was financially supported by: *Communauté du Pays d'Aix*, *Conseil Général des Bouches du Rhône* and *Conseil Régional Provence Alpes Côte d'Azur,* through the focused research program called "Rousset 2003-2010" in partnership with STMicroelectronics

VI. REFERENCES

[1] A. Diebold, Overview of metrology requirements based on the 1994 National Technology Roadmap for semiconductor. *Advanced Semiconductor Manufacturing Conference and Workshop 1995. ASMC 95 Proceedings. IEEE/SEMI*, pp. 50–60, 1995.

[2] J. Moyne, *Making the move to fab-wide APC. Solid State Technology*, n. 47, vol. 9, September 2004, pp. 47, 2004.

[3] S. Gaddam and M.W. Braun, Etch chamber condition-based process control model for shallow trench isolation trench depth control. *IEEE/SEMI Advanced Semiconductor manufacturing conference and workshop*, pp. 17-20, 2005.

[4] W. Y. Lien, W.G. Yeh, C. H. Li., I.H.Chang., H.C. Chu., W.R. Liaw., H.F. Lee., H.M. Chou., C.Y. Chen and M.H. Chi. A manufacturable shallow trench isolation process for sub-0.2μm dram technologies. *IEEE/SEMI Advanced Semiconductor manufacturing conference and workshop* , pp. 11-16, 2002..

[5] K. J. Kuhn, D-H. Mei., I. Post and J. Neirynck. Scaling challenges for 0.13μm generation shallow trench isolation. *International Symposium on Semiconductor Manafacturing*, pp. 187-190, 2001.

[6] Fuller, L., "Wet Etch For Microelectronics", http://people.rit.edu/lffeee/wet_etch.pdf, 2006

[7] S. M. Sze, Semiconductor devices: Physics and Technology, 2nd Edition Wiley, 2002.

[8] F. Bergeret, Y. Chandon, and C. Le Gall, "De la statistique dans l'industrie: Un exemple à FREESCALE," *J. Soc. Française Stat.*, vol. 145, no. 1, pp. 71–95, 2004.

[9] P. Besse. *Data Mining, II: Modélization Statistique and Apprentissage* [Online]. Available: http://www.lsp.ups-tlse.fr/Besse/

[10] M. Tenenhaus, *La Régression PLS: Théorie et Pratique*. Paris, France: *Edition Technip*, Aug. 1998.

[11] D. Belharet, Etude et validation de boucles d'asservissement permettant le contrôle en microélectronique : Application à l'étape d'isolation par tranchées peu profondes en technologie CMOS, Ph.D. dissertation, EMSE-CMP, France, 2009.

[12] L. Delachet, "Run-to-run process optimization on HDP-CVD tools," STMicroeletronics, Rousset, France, Mastere Rep., 2006.

[13] S.Adivikolanu and E. Zafiriou, "Robust run-to-run control manufacturing: An internal model control approach," *in Proc. Am. Control Conf*, pp. 3687–3691, 1998.

[14] *Rudolph Technologies* [Online]. Available: http://www.rudolphtech.com

Yp – Ypk: Product Test Yield and Yield Dispersion Indicators

Matthias T. Bostelmann
Characterization Department
ALTIS Semiconductor
Corbeil-Essonnes, France
Matthias.Bostelmann@Altissemiconductor.com

Abstract—**Product test results deliver multiple data about yield and detractors, comprising their centering and dispersions. They present a challenge for information delivery, in particular when the data ranges imply more than 3 orders of magnitude. To visualize them in a "at-a-glance" or "one-shot" dashboard, we present Yp and Ypk as a practical solution. This set of indicators Yp and Ypk is calculated relative to a best-can-do real-world reference. First examples illustrate the effectiveness of the method.**

Keywords—DM Data Management and Data Mining Tools; new visualization method ; Yp Ypk; dashboard

I. INTRODUCTION

In our semiconductor manufacturing context, ease at yield and detractor readout is essential in monitoring product test results. For any product, yield models [1] define yield targets but mostly no model for detractor paretos nor specifications for detractor impacts are available. This paper presents a newly implanted representation of relative yields and detractors called Yp and Ypk easy to understand as analogy to well known cp - cpk and pp - ppk. The purpose of Yp - Ypk is to define a general rule usable for any product / any process, compare yield centering and variability referenced to best lots, and track evolution of small detractors. Yp and Ypk calculations are referenced to test data already available used as a base line to account for dispersion of real life reference population. As actual yields can be obtained with different detractor breakouts, and specific test problems may overlay with process bound signatures, identification of abnormal behavior in graphical representation is needed. But what is «regular» behavior of yield and detractors (comprising e.g. perfect and redundancy fixed chips)?

II. PRODUCT TEST YIELD AND YIELD DISPERSION INDICATORS

A. Limitations of current indicators

In the context of our mature 0.13 µm C11RF2 CMOS RFIC process widely used for wireless SOC RF front-ends, wafer level product test contain many potential detractors, resulting from several 100 tests resumed in sorted Hardbin

key groups and Softbin subgroups [2] to sort KGD and KBD (known good and bad dies).

Some relevant parametric tests results could be used for cp / cpk monitoring such as Idd leaks, Vdd currents, band gap voltage, speed monitors, VCO frequency, etc. Formally, this does not apply for detractors attributes expressed in % fallout that are even not subject of a specification. Classical graphical yield and detractor representations (Fig. 1) as cumulative frequency plots, BEX plots [6], Boxplots, Barchart stacks of Hardbins / Softbins, multiple detractor trends or text definitely allow monitoring, but are not always suited for covering >3 orders of magnitudes nor «one-shot» visibility.

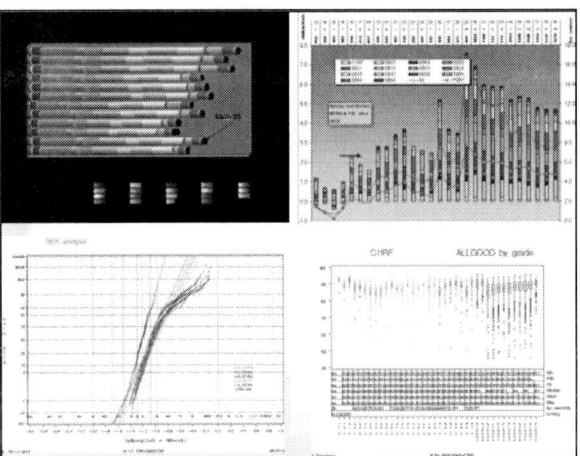

Figure 1. Standard charts yield break out: stacked bar charts, absolute or relative fallouts, BEX plots [6] or boxplots, offer limited visual simultaneous captions for a wide variety of yields and detractors ranging 3-4 orders of magnitude.

B. Definition of Yp - Ypk - Reference System

Reminding that generally no specifications exist for test yield detractors, we proceed in two steps to build a reference system per product:

First step: Assess initial reference for yield and yield dispersion targets as based on mean and standard deviations obtained on best lots manufactured and tested so far.

Second step: These best lots are than used to build a detractor specifications based upon detractor means and standard deviations of these same lots.

With these target references for means and dispersions, we than calculate Yp and Ypk for test data of any population we want to compare and visualize against the initial best lots breakout of yields and detractors:

TABLE I.   YP AND YPK FORMULA

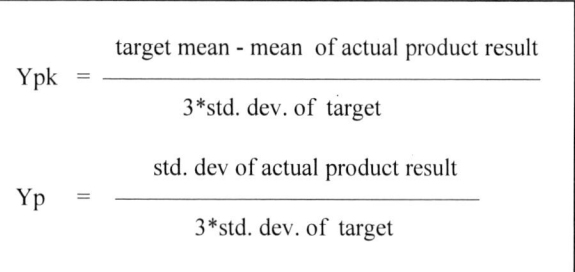

With Yp and Ypk, we represent yields and detractors and their respective dispersions versus this best lot population as target. By this referencing, the new indicators enhance the visibility of small variations of minor contributors amid major yield drivers. This benefits to visual early alert for less common fallouts. More than 3 orders of magnitude 0.01…to > 10% range appearing in test data are easily visualized in a single chart (Fig. 2). In contrast to cp - cpk, high values of Yp or Ypk indicate a problem, i.e. lower yield, or higher impact, and wider dispersion respectively. Yp=0 or Ypk=0 result from equivalence to reference population, while negative values indicate yields, detractors, and dispersions even better than reference

C. Reinitializing the reference system

At the first occurrence of generating a Yp - Ypk read out for a given product tested, the reference system is initialized upon availability of first test results. For any following iteration, this reference system per product is updated dynamically on weekly, monthly, or quarterly period etc. depending on test volume, product ramp or maturity phase, and read out frequency. Any product test changes as test coverage, test flow, or test specs should not be mixed up and the reference system should be re-initialized.

D. Maintaining the reference baseline and refinements

We build the reference base line for Yp and Ypk dynamically by classification of lead production lots by lot yield-classes of 1% steps: < 89, 89, 90, 91, 92, 93, 94 etc. We than build a spectrum per lot yield-class attributes Hardbins and Softbins and verify the distributions of all attributes by BEX plots that are lognormal plots [6]

- for yield in [%]: BEX = log10(- ln (yield))

- for loss in [%]: BEX= log10(- ln (100-loss))

respectively.

This would allow to identify and suppress, if needed, lots incompatible with a straight slope set in BEX plots, that give indication for a random defect density driven base reference product yield and absence of parametric or other systematic issues not representative of lots considered for the base line.

III.   PRACTICAL USE OF YP – YPK

The best lot yield class and its associated Hardbin / Softbin spectrum are evaluated to retain mean, median, standard deviations per yield or loss attributes resulting in the base line reassessment.

Once this reference is established, we can run Yp - Ypk and compare populations visually in graphs based on the proposed formula.

A. Example 1: Production monitoring

In our first illustration (Fig. 2), we will for use one and the same product, and compare three production months to base line target built on best lots class. For this purpose, the data are obtained with same test release and test flow, same test platform and probing, and are strictly comparable.

Hardbins, Softbins, Allgoods (i.e. sum of perfect and fixable chips) are compared per Yp and Ypk charts.

The first example's Yp chart shows 2 abnormal peaks:

- SBIN11 fixable rate corresponding to defect density event induced increase of redundancy used during cm01 period.

- SBIN08 4 maverick wafers process problem (fail usually flat= 0% fallout) in period labeled cm03.

Two other detractors , even though reducing over time

- SBIN43 and SBIN69 show rather stable impacts versus target

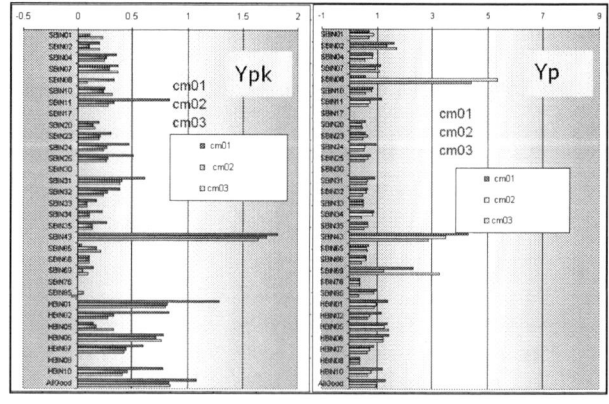

Figure 2.   Yp – Ypk application example: More than 3 orders of magnitude < 0.1 … to > 10% range appearing in test data.

B. Example 2: Product test monitoring

In this example (Fig. 3) we present a tester related SBIN95 fallout event that occurred during a partial test offload to an other test site. SBIN95 never gave such fails before at in-house testing. The impact starts well below < 0.5% level per wafer as precursor of degrading test quality. Corrective actions for the test floor were induced very shortly while experimenting with Yp and Ypk.

Following up the review of test setup and retest conditions, a new early alert and retest threshold set to 0.2% fallout were implemented. The fail was DUT dependent and recovered by retest after probe clean and new probe setup. Much higher retest rates or yield losses were immediately stopped before piling up.

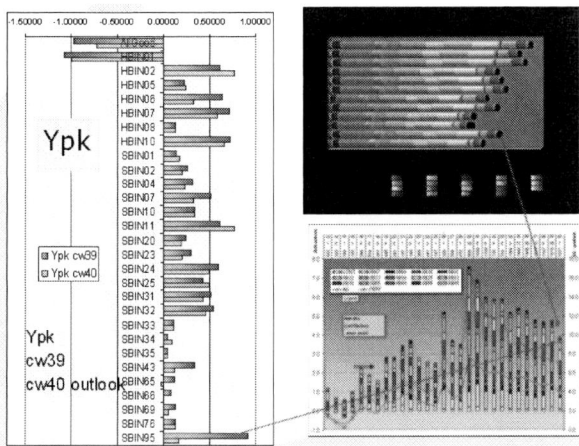

Figure 3. Ypk application example: Ypk visualization exceeding classical representations: SBIN95 fails cw39 enhanced visibility in Ypk only.

C. Example 3: Fab comparison

As third example of Yp - Ypk use we show the comparison of a same SOC RF IC produced and tested in different fabs (Fig. 4). For most evident reasons, common test platform, test program and flow, test probe design (DUTs, test board, probe cards) grant such a comparison across wafer fabs.

As the environments differ by fab related tool set constraints and defect density, distinct yield and detractor situations result, i.e. yield and detractor behavior are site dependent. Here, the Yp - Ypk graphs reveal typical signatures in

- perfect HBIN01 yield

- redundancy repaired yield (SBIN11, HBIN02)

- ratio of HBIN02 to HBIN01

- embedded ROM fails classified in SBIN23 and SBIN33

- Idd fails resumed in SBIN43

The common reference target used here is built on FAB A lots > 93% for this product as for both preceding examples. The inspection of example 3 Ypk and Yp highlight

Ypk:

- FAB C: suffers clear offset in SBIN33, i.e. ROM fails at Vddmin conditions

- FAB A is impacted by offset SBIN43, SBIN65 & SBIN69, SBIN01, SBIN07 and SBIN24

- FAB C is distinct by a factor 2x fixable percentage, and a typical "Perfect to Fixable"-ratio change with respect to FAB A, also visible in better SRAM SBIN24.

Yp:

- FAB A has higher dispersion in most detractors except SBIN33 ROM at Vddmin.

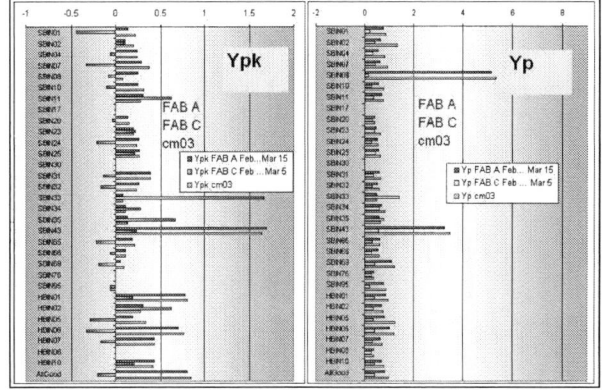

Figure 4. Fab comparison for same product, same test: reference base line = FAB A's lots > + 93%.

IV. DISCUSSION

The basic idea for Yp and Ypk is the recycling of best lots results into a reference target for yields and detractor Hardbins and Softbins. These relative product yield indicators allow easy visual one-shot inspection of a large range of data from subtle fallouts to high yields. Yp and Ypk are intended as visualization tools. They have already proven helpful in detecting test problems (Fig. 3), defining retest criteria, comparative analysis of multiple DUT tests, and fab comparisons (Fig. 4).

V. CONCLUSION

The visual resolution limit is better than 3 orders of magnitude and oriented to mid and long term considerations. Our practical experience has demonstrated the ability to create pertinent visual information pointing quickly even to minor detractor appearance. The increased high dynamic range eases the "one-shot" inspection of more or less dispersed yields and detractors compared to the product related baseline.

REFERENCES

[1] C. H. Strapper, "Small-area fault clusters and fault-tolerance in VLSI systems", IBM Journal on Research and Development, vol. 33, pp. 174–177, Mar. 1989.

[2] Standard Test Interface Language (STIL) 1450.4 Test Flow Extension Working Group Issues and Resolutions Documents, http://grouper.ieee.org/groups/1450/dot4/index.html

[3] Douglas C ; Montgomery, "Introduction to Statistical Quality Control", 3rd Editon John Wiley&Sons (1997) ; p190 ff ; p 430 ff

[4] Domart, F.; Saout, D.; Degand, F.; Tran, A.; Chauvet, A.; Grolier, J. L.; Richard, O. , "An efficient methodology for Electrical Monitoring in manufacturing environment" in Advanced Semiconductor Manufacturing Conference, 2008. ASMC 2008. IEEE/SEMI; pp117-122

[5] ISO document 3534-2/2006: Part 2. "Applied Statistics" (1993), http://www.iso.org/iso/iso_catalogue/catalogue_ics/catalogue_detail_ics.htm?csnumber=40147

[6] Melzner, H.; Darnhofer, T., "BEX - a novel wafer yield distribution model and analysis", in Advanced Semiconductor Manufacturing, 2004.ASMC '04. IEEE Conferenc

Neural Network Modeling of Fabrication Yield Using Manufacturing Data

Z. N. Mevawalla, G. S. May
School of Electrical and Computer Engineering
Georgia Institute of Technology
Atlanta, USA
gary.may@ece.gatech.edu

M. Honjo, M. W. Kiehlbauch
Dry Etch
Micron Technology, Inc.
Boise, USA
mkiehlbauch@micron.com

Abstract—**This paper describes the creation of artificial neural network models using production line data and illustrates their usefulness for process control in semiconductor manufacturing. Three artificial neural network models are created. The first models a high aspect ratio etch process. The other two are created to predict yield metrics from inline critical dimension (CD) measurements. One model predicts the number of faults on a die, and the other predicts the probability of die failure at probe. The high aspect ratio etch model has an average prediction error of 3.9%. The average prediction error for the number of faults on a die is 14.9%, and the average prediction error for probability of die failure at probe is 21.8%. A sensitivity analysis is performed on each model to illustrate how they can be used to judge the relative impact of each input.**

Neural networks; semiconductor manufacturing; advanced process control; production line

I. INTRODUCTION

The purpose of this paper is to describe the creation of artificial neural network models using production line data in semiconductor manufacturing. The nature of semiconductor manufacturing processes makes modeling from first principles very difficult. However, the large amounts of data available in fabrication facility databases (both on-line process data and in-line post process measurement data) allow for the creation of empirical process models [1]. Artificial neural networks are one way to encode such empirical knowledge [2], and they have shown themselves to be effective in modeling several semiconductor manufacturing processes [3]–[10].

An artificial neural network is a mathematical modeling tool that is inspired by neural networks found in biology. Artificial neural networks are highly parallel, nonlinear computational systems [11]. Like their biological analog, they consist of simple processing elements called neurons that are connected to each other to form a network. The strength of the connections between these neurons varies and is represented by a weight. Artificial neural networks can learn the relationship between a set of input data and output data using learning algorithms. In learning the mapping from the input of a process to its output, a neural network becomes a model for that process [12]. This learning property allows the user to create

sophisticated models without making assumptions about the data used or without deep knowledge of the process being modeled [13]. Furthermore, artificial neural network models typically outperform other more traditional statistical modeling methods given equal amounts of training data. Himmel and May demonstrated neural network models that vastly outperformed (38.3% reduction in prediction error) models created by response surface methodology, for an etch process [6].

Typically artificial neural network models of semiconductor manufacturing processes are trained with data from designed experiments [5] or a mixture of data from designed experiments and production line data [4]. The models created for this research were trained with production line data only.

Three artificial neural network models were created. The first was to model a high aspect ratio etch process. The other two were created to predict yield metrics from inline CD measurements. One predicted the number of faults on a die, and the other predicted the probability that a die would fail a test of functionality at probe. The high aspect ratio etch model had an average prediction error of 3.9%. The average model prediction-error for the number of faults on a die was 14.9%. And the average model prediction-error for probability of die failure at probe was 21.8%. A sensitivity analysis was performed on each model to illustrate how they can be used to judge the relative impact of each input. The ultimate goal is to use these models for process optimization and for process control.

II. THEORY

Artificial neural networks are mathematical modeling tools inspired by biological neurological systems. They are complex, parallel, nonlinear computational systems made up of simple processing elements called neurons [11]. A neuron passes the sum of its inputs, through an activation function, to its output. The hyperbolic tangent was chosen as the activation function in this research because it has been shown that when using a function like the hyperbolic tangent, neural networks can capture a very large range of input-output mappings [19].

The authors thank Micron Technology, Inc. for supporting this research.

978-1-61284-408-4/11 $26.00 © 2011 IEEE

Interconnection in these networks enables certain functionality [15]. The rules determining how neurons are positioned relative to each other in a network and how they are allowed to connect to other neurons in the network determine the network architecture. Network architectures have significant effect on the performance of the network and the type of function the network is best suited for. This research uses the multilayer feed-forward neural network, which is the most commonly and successfully used architecture [11].

The neurons in a multilayer feed-forward neural network are, as the name suggests, arranged in layers. One input layer, one output layer and one or more hidden layers between them are typical. Neurons in one layer can only connect to neurons in the subsequent layer. Thus data can travel in only one way through the network. Fig. 1 provides a schematic of an example multilayer feed-forward neural network. There are three neurons in the input layer, three in the hidden layer and three in the output layer. This network would be abbreviated as a 3 – 3 – 3 network. The number of neurons in the input layer is determined by the number of inputs to the process being modeled, and similarly, the number of output-layer neurons is determined by the number of outputs to the process. The number of hidden layers and the number of neurons in each hidden layer is variable. In this research, the number hidden layer neurons was determined by trial and error.

The lack of a strong theoretical framework for choosing neural network topology is considered one of the drawbacks of using neural networks [18]. On the other hand, it has been proven that given enough hidden layer neurons, a network can learn any nonlinear input output mapping [19]. Additionally, no knowledge or assumptions about the distribution of the data is necessary to make a neural network model. This makes them attractive for dealing with the complexities of semiconductor manufacturing processes. Ideally, a neural network model will have just enough neurons that it can properly capture the input-output mapping of the process being modeled [20]. Too many hidden layer neurons lead to "over-fitting," in which the model starts learning the noise and the outliers within the training data, leading to larger errors in predictions.

Training is the process whereby the network learns based on data provided to it. A network learns by adjusting the values of the weights of the connections between neurons. The methodology used to adjust the weights using training data is the learning algorithm. The algorithm used to train the networks in this research is the error back-propagation algorithm. This type of learning is categorized as supervised learning, because the network parameters are adjusted so the network emulates a specific input-output mapping. This is in contrast to adjusting network parameters to minimize a cost function (unsupervised learning).

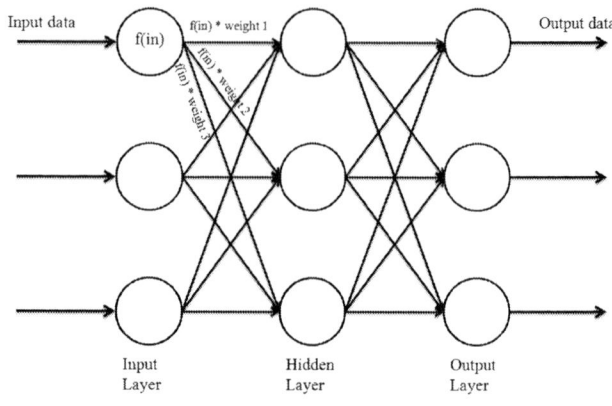

f(x) is any differentiable function
f(in) is the function f(x) performed on the input to that neuron

Figure 1 A feed-forward neural network

With error back-propagation, the network is presented with data in the form input-output pairs. The inputs are measured process variables (gas flows, RF powers, etch times, etc.), critical dimensions, and other variables that can be controlled and are known to affect the process being modeled. The outputs are critical dimensions measured after the process. Initially, the network weights are randomly set to small values [21]. Then the input is entered into the network and the resulting network output is compared to the measured outputs. An error is calculated, in this case a sum of squared error, and the weights are adjusted so as to decrease the error. The nature of the algorithm is to adjust the weights layer by layer moving from the output layer through the hidden layers to the input layer, hence the name error back-propagation. This process of passing measured input through the network and back-propagating the error continues until the error decreases to the level required or until it converges. The algorithm is parameterized by a learning rate and momentum that effect the rate at which error converges. A more thorough description of the algorithm is presented in [22].

The fidelity of artificial neural network models is tested on datasets other than the ones used to train it. These are labeled testing datasets. The literature concludes that neural networks tend to over-fit training data [13],[18],[21], so it is important to test networks on data they have not been trained on and ideally not too similar to training data [21]. This allows judgment on the "generalizability" of the network.

Creating and training neural networks is computationally intensive and takes more time than regression modeling [13][18][23]. However the most cited disadvantage of neural network modeling is their "black box" nature [18], or the inability to judge the relative importance of an input to the model. One workaround for this concern is sensitivity analysis, which is outlined in a later section. In spite of these disadvantages, neural networks remain attractive modeling tools because of their ability to model any input-output mapping, their tolerance to noisy data, the ease with which they can be updated [18], and because their predictive accuracy has been shown to be greater than that of statistical modeling methods such as regression modeling [12].

III. EXPERIMENT

A. Data Collection

Data collection is an important part of the modeling process. The quality and amount of training data used greatly affects the performance of neural network models. In past studies, the data used typically originated from designed experiments [5], [6], [9], and only occasionally from industrial fabrication processes [3],[4]. Ironically, the operations of semiconductor fabs generate very large amounts of useful data. Although the data is typically generated for purposes other than process modeling [1], some of it – like on-line sensor data (gas flows, RF power, in-line CD measurements, etc.) – is nevertheless relevant. The data generated are usually stored in large databases that can be queried [4].

The advantage of using production data for modeling is that it saves the cost and time spent to carry out designed experiments [1]. However, there are some disadvantages. Artificial neural network models are much better at interpolating than extrapolating [14]. With production data, there is much less variation in the process variables. The resulting models thus may not be as useful for optimization purposes, as one has to assume that the optimal recipe for a process is within that narrow range of inputs [14]. The other problem with using production data is the effort needed to get the data in a format ready for modeling. There can be inconsistencies in the way the data is collected, labeled, stored, etc. [14]. Training neural network models with designed experiments provides more variation in the training inputs, but incurs additional cost and will typically generate less training data. Less training data potentially limits the accuracy of the network models that can be created.

B. Models

The models were created using the Object Oriented Neural Network Simulator (ObOrNNS), a program developed by the Intelligent Semiconductor Manufacturing Group at Georgia Tech. It allows for rapid creation, training and testing of neural network models [5].

1) High Aspect Ratio Etch

The purpose of the high aspect ratio etch is to create an array of high aspect ratio holes in an oxide layer. These holes act like a mold for device structures created in subsequent fabrication steps. Therefore, the dimensions of these holes are important. Of particular importance are the distances between these holes in the x-direction and y-direction. They are the CDs we are interested in controlling and thus they form the output of the model (Fig. 2). The overall etch is made of three smaller etches each with its own set of process parameters like gas flows, RF powers, bias powers, and charge species tuning unit (CSTU) settings. These, together with the measurements of the critical dimensions in the masking layer, the wafer location of these measurements, and the chamber in which the wafer is processed, form the model inputs.

A 16-16-2 network was created to model the process. Data from approximately 1100 wafers was used to train the model and data from another 50 wafers was used to test it.

2) Yield Metrics

There were two models created to predict yield metrics. In-line CD measurements, the location of these in-line CD measurements, and an indicator of the process chamber form the inputs to these models (Fig. 3). The in-line CD measurements are the same as those in the output to the oxide etch model. This fact is important for future work involving the optimization for yield because the models can be connected as in Fig. 4.

a) Defect Counts: The first of the yield models predicts defect counts on the wafer. The model makes separate predictions for defects that are caused by over-etch and under-etch during the high aspect ratio etch step. The model was created with a 5-6-12 neural network. Data from approximately 650 wafers was used to train the model and data from another 50 wafers was used to test it.

b) Probability of Die Failure: The second of the yield models predicts the probability that a die will fail a test of functionality at probe as a result of non-ideality in the high aspect ratio etch step. A larger amount of preprocessing had to be performed for this model. This is because the probabilities of failure was not directly available from the database and had to be calculated and arranged properly into the modeling datasets. The model was created with a 5-5-6 neural network. Data from approximately 650 wafers was used to train the model and data from another 27 wafer was used to test it.

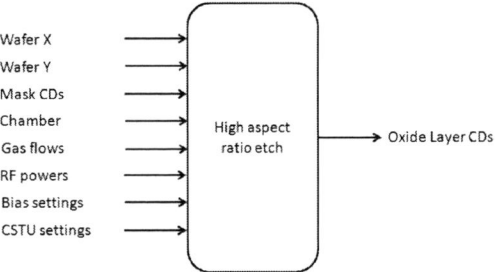

Figure 2 Overview of model for High Aspect Ratio Etch

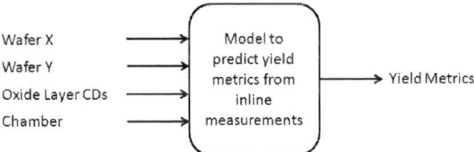

Figure 3 Overview of Yield Metrics models

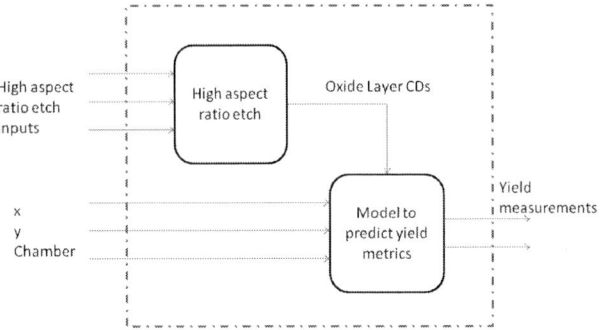

Figure 4 Overview of sequential neural network model

IV. MODELING RESULTS

A. High Aspect Ratio Etch

The model performed very well with an average prediction error of 3.9%. Results are presented graphically as error contour plots. This allows comparison of both the values of the output and its variation across the wafer. Fig. 5 shows the results for x-direction CD and Fig. 6 shows the results for y-direction CD. The CD values have been normalized using the range of the measured CD values. The results presented are the average over 47 wafers.

The model has accurately captured the distribution of the CDs as seen from the similarity between the plots of measured CDs (left) and the model-predicted CDs (center). The rightmost contour plots in Fig. 5 and Fig. 6 are of percentage error between measured CDs and model-predicted CDs. The error remained below 5% across most of the wafer for both predictions.

B. Yield Metrics

1) Defect Counts

The average prediction error for defect counts was 14.9%. Fig. 7 shows the results for defects as a result of under-etch. The model accurately captured the distribution of the defects as seen in the similarity between the plots of actual defects (left) and predicted defects (center). The prediction error remained mostly below 15% as seen in the plot on the right.

Fig. 8 shows the results for defects as a result of over-etch. The model accurately captured the distribution of the defects as seen in the similarity between the plots of actual defects (left) and predicted defects (center). The prediction error remained mostly below 25% as seen in the plot on the right.

2) Probability of Die Failure

The average prediction error for probability of die failure was 21.8%. Fig. 9 shows the results for probabilities of die failure as contour plots. The model captured the distribution of the defects as seen in the similarity between the plots of actual defects (left) and predicted defects (center). There was a deviation in the predictions for area of high probability of failure on the right part of the wafer. It is slightly higher up on the wafer in the predictions (center) than the actual distribution (left). The prediction error remained mostly below 30% as seen in the plot on the right.

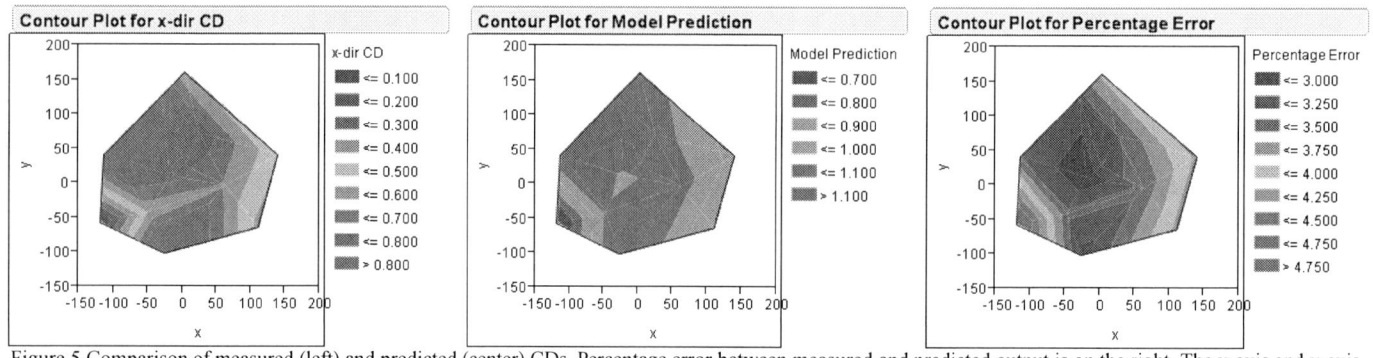

Figure 5 Comparison of measured (left) and predicted (center) CDs. Percentage error between measured and predicted output is on the right. The x-axis and y-axis represent wafer location in mm.

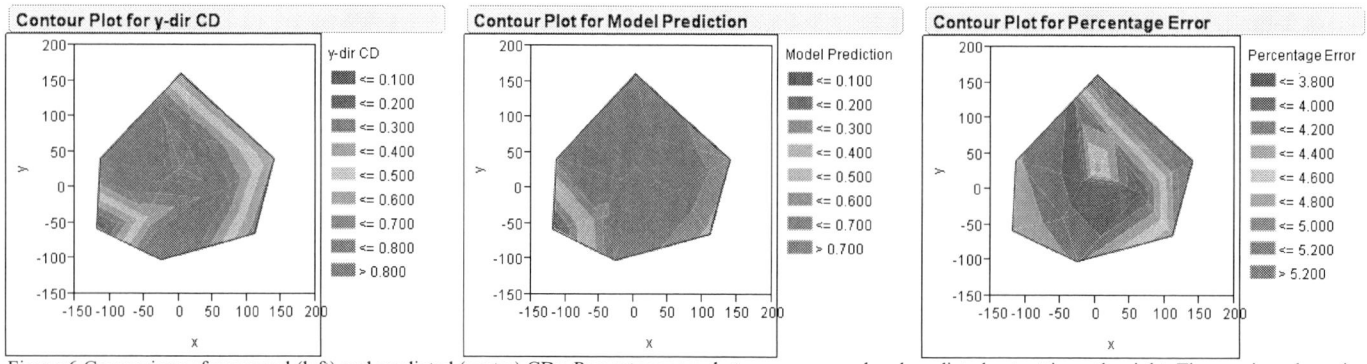

Figure 6 Comparison of measured (left) and predicted (center) CDs. Percentage error between measured and predicted output is on the right. The x-axis and y-axis represent wafer location in mm.

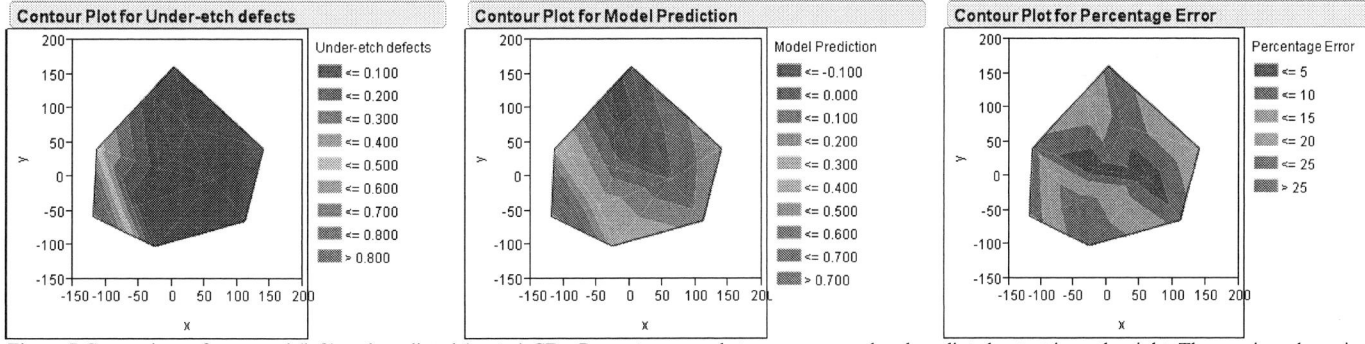

Figure 7 Comparison of measured (left) and predicted (center) CDs. Percentage error between measured and predicted output is on the right. The x-axis and y-axis represent wafer location in mm.

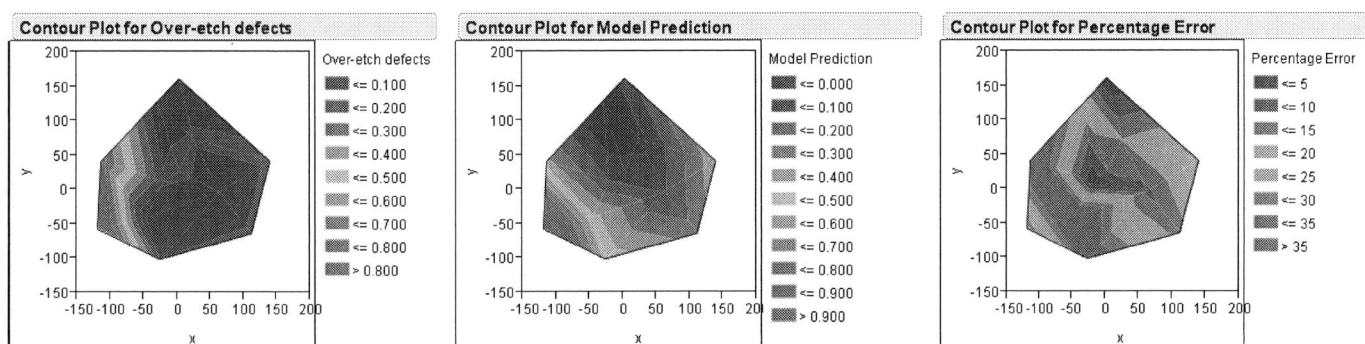

Figure 8 Comparison of measured (left) and predicted (center) CDs. Percentage error between measured and predicted output is on the right. The x-axis and y-axis represent wafer location in mm.

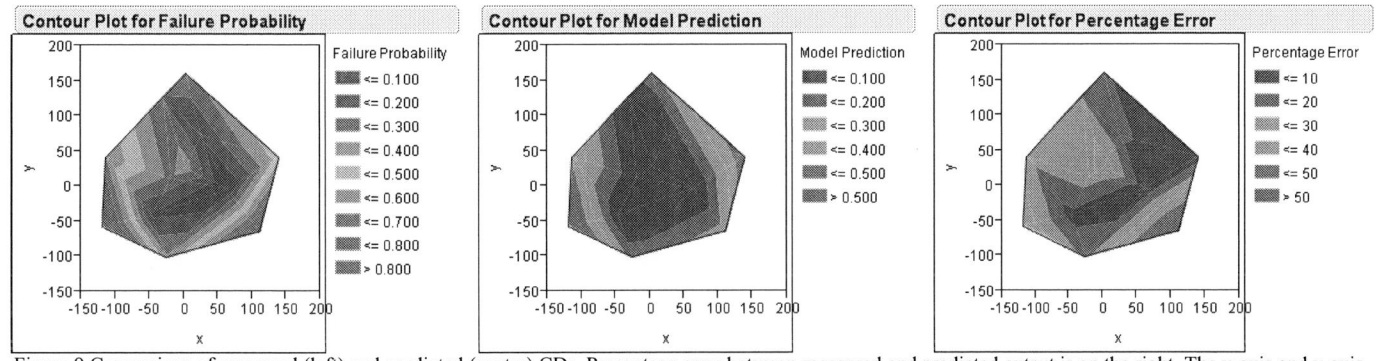

Figure 9 Comparison of measured (left) and predicted (center) CDs. Percentage error between measured and predicted output is on the right. The x-axis and y-axis represent wafer location in mm.

V. SENSITIVITY ANALYSIS

If a model has accurately captured the input-output relationships of a particular process, then performing a sensitivity analysis using the model can further the understanding of that process. A sensitivity analysis is a systematic way of studying how the output of a process responds to changes in each of the inputs individually.

Starting at a point of interest in the input space, a process recipe for example, a small change is made in one of the inputs keeping the rest the same. A small change would be approximately 10% of the range of the variable in the input space. The output of the model given the modified input is compared to the output of the model given the unmodified input (original process recipe).

The analysis can be repeated for all input variables in a process step, allowing their relative effects on the output to be studied. A representative example would be to look at the relative effects the inputs of the defect counts model have on under-etch related defects (see Fig. 10).

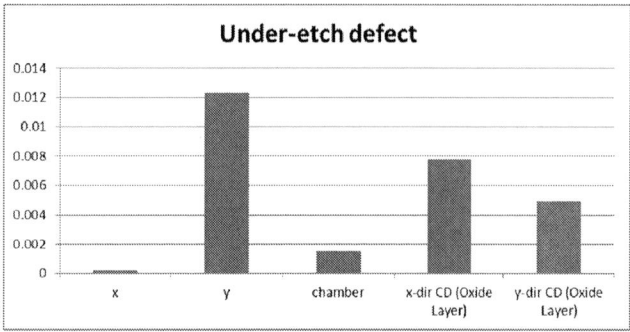

Figure 10 Sensitivity analysis. The bars represent the percentage change in under-etch defects in response to a uniform change in their respective inputs

VI. CONCLUSION

Artificial neural networks were able to model a fabrication process accurately using production line data only. They were also able to make predictions about yield using inline CD measurements. They captured the uniformities of the fails across wafer particularly well.

Future work will be to use genetic algorithms on the models to find optimal processing conditions [10] and to incorporate the models into a supervisory controller [8].

ACKNOWLEDGMENT

The authors thank T. S. Kim for his input.

REFERENCES

[1] W. Sukthomya, J. Tannock, "The training of neural networks to model manufacturing processes," Journal of Intelligent Manufacturing, vol. 16, no. 1, pp. 39-51, 2005

[2] H. White, "Learning in Artificial Neural Networks: A Statistical Perspective," Neural Computation, vol. 1, no. 4, pp. 425-464, Winter 1989

[3] C. T. Su, T. L. Chiang, "Optimizing the IC wire bonding process using a neural networks/genetic algorithms approach," Journal of Intelligent Manufacturing, vol. 14, no. 2, pp. 229-238, 2003

[4] E. A. Rietman, E. R. Lory, "Use of Neural Networks in Modeling Semiconductor Manufacturing Processes: An Example for Plasma Etch Modeling," IEEE Transactions on Semiconductor Manufacturing, vol. 6, no. 4, pp. 343-347, Nov. 1993

[5] T. S. Kim, and G. S. May, "Sequential modeling of via formation in photosensitive dielectric materials for MCM-D applications," IEEE Transactions on Semiconductor Manufacturing, vol.12, no. 3, pp. 345 – 352, Aug. 1999

[6] C. D. Himmel, G. S. May, "Advantages of plasma etch modeling using neural networks over statistical techniques," IEEE Transactions on Semiconductor Manufacturing, vol. 6, no. 2, pp. 103 – 111, May 1993

[7] G. S. May, "Manufacturing ICs the neural way," IEEE Spectrum, vol. 31, no. 9, pp. 47 – 51, Sep. 1994

[8] T.S. Kim, G. S. May, "Intelligent control of via formation process in MCM-L/D substrates using neural networks," International Symposium on Advanced Packaging Materials: Processes, Properties and Interfaces, 1999., pp. 106 – 112, Mar. 1999

[9] S .S. Han, M. Ceiler, S. A. Bidstrup, P. Kohl, and G. May, "Modeling the properties of PECVD silicon dioxide films using optimized back-propagation neural networks," IEEE Transactions on Components, Packaging, and Manufacturing Technology, vol. 17, no. 2, pp. 174 – 182, Jun. 1994

[10] S. Han, G. S. May, "Recipe synthesis for PECVD SiO2 films using neural networks and genetic algorithms," Proceedings 46th Electronic Components and Technology Conference,1996., pp. 855 – 860, May 1996

[11] S. Haykin, Neural Networks: A Comprehensive Foundation, New York: Macmillan College Publishing Company, Inc., 1994

[12] G. S. May, C. J. Spanos, Fundamentals of Semiconductor Manufacturing and Process Control,New Jersey: John Wiley & Sons, Inc., 2006

[13] J. V. Tu, "Advantages and Disadvantages of Using Artificial Neural Networks versus Logistic Regression for Predicting Medical Outcomes," Journal of Clinical Epidemiology, vol. 49, no. 11, pp. 1225 – 1231, Nov. 1996

[14] D. Svozil, V. Kvasnicka, J. Pospichal, "Introduction to multi-layer feed-forward neural networks," Chemometrics and Intelligent Laboratory Systems, vol. 39, no.1, pp. 43-62, Nov. 1997

[15] J. J. Hopfiled, "Neural networks and physical systems with emergent collective computational abilities," Proceedings of the National Academy of Sciences USA, vol.79, pp. 2554-2558, Apr. 1982

[16] F. D. Palma, G. D. Nicolao, G. Miraglia, E. Pasquinetti, F. Piccinini, "Unsupervised spatial pattern classification of electrical-wafer-sorting maps in semiconductor manufacturing," Pattern Recognition Letters, vol. 26, no. 12, pp. 1857-1865, Sep. 2005

[17] C. T. Su, T. Yang, C. M. Ke, "A neural-network approach for semiconductor wafer post-sawing inspection," IEEE Transactions on Semiconductor Manufacturing, vol. 15, no. 2, pp. 260-266, May 2002

[18] A. Vellido, P. J. G. Lisboa, J. Vaughan, "Neural Networks in business: a survey of applications (1992– 1998)," Expert Systems with Applications, vol. 17, no. 1, pp. 51-70, Jul. 1999

[19] K. Hornik, M. Stinchcombe, H. White, "Multilayer Feedforward Networks are Universal Approximators," Neural Networks, vol. 2, no. 5, pp. 359 – 366, 1989

[20] R. Reed, "Pruning algorithms - a survey", IEEE Transactions on Neural Networks, vol. 4, no. 5, pp. 740 – 747, Sep. 1993

[21] A. Krogh, "What are artificial neural networks?," Nature Biotechnology, vol. 26, no. 2, pp. 195 – 197, Feb. 2008

[22] D. E. Rumelhart, G. E. Hinton, R. J. Williams, "Learning representation by back-propagating errors," Nature, vol. 323, pp. 533-536, Oct. 1986

[23] K. Y. Tam, M. Y. Kiang, "Managerial Applications of Neural Networks: The Case of Bank Failure Predictions," Management Science, vol. 38, no. 7, pp. 926-947, Jul. 1992

[24] R. J. Schalkoff, Artificial Neural Networks,New York: McGraw-Hill, 1997.

[25] Z. N. Mevawalla, M. W. Kiehlbauch, G. S. May, "Neural Networks for Advanced Process Control," IEEE/SEMI® Advanced Semiconductor Manufacturing Conference, 2010, pp.137-142, 2010

[26] Z. N. Mevawalla, M. W. Kiehlbauch, G. S. May, "Neural Network Modeling for Advanced Process Control Using Production Data," to be published

Reducing Environmentally Induced Defects While Maintaining Productivity

R. van Roijen, S. Conti, R. Keyser, R. Arndt, R. Burda, J. Ayala, R. Henry, J. Levy, J. Maxson, E. Meyette, W. Steer, K. Tabakman and C. Yu

IBM Systems and Technology
Hopewell Junction, NY 12533, USA

Abstract— In Semiconductor manufacturing we expect the cause of defects to be process or tool related. However, at recent technology nodes we find that defects can be caused by issues related to the wafers environment, such as processing of other wafers in the same tool or in the same carrier, or by seemingly innocuous actions. One result is the rapid proliferation of queue time restrictions and batching rules. In this work we show defects which are caused by the environment and several ways to reduce the sensitivity to environmental factors. Process and tool changes are found to eliminate yield detractors. We also present a workaround that has helped to reduce the impact of queue time restrictions on cycle time.

Keywords-defect; queue time; contamination; foreign material; defect classification

I. INTRODUCTION

When we consider defects in semiconductor manufacturing, we typically expect the cause to be process related, e.g. a process step leaves a foreign material or fails to remove a material that causes defects in subsequent processing, either because of inherent process properties or faulty tool conditions. However, at recent technology nodes we increasingly find that defects, besides those coming from traditional sources, can originate when wafers are not being processed. This can occur by processing of other wafers in the same tool or in the same carrier, by actions as simple as opening the carrier door or by contamination that increases with time as a wafer waits to be processed[1,2]. One of the consequences of this trend is the rapid proliferation of queue time restrictions and other handling restrictions at the 45nm node and beyond. By imposing restrictions we can control the level of defects, but at the price of reduced productivity.

In this work we will show a number of instances of defects affected by the wafers environment and the solutions that have been applied. Defects are detected by inline inspection, electrical test and failure analysis. We also discuss some generally applicable concepts that can be used to reduce the impact of the increased sensitivity to environmental factors.

When defect density is found to increase with time, a queue time restriction is implemented. However, queue time restrictions can sometimes be an impediment to throughput. We will describe a work-around for queue time restrictions where the exposure to defects is reduced while maintaining throughput.

II. CONDENSATION DEFECTS

At a number of process steps we find so called condensation defects. We recognize defects as condensation defects by their appearance (typically round, droplet like) and by the fact that they are removed by a water rinse or that they disappear spontaneously over time. Depending on their composition and the subsequent process step, we distinguish three cases: condensation is observed at inspection, but can be ignored because it leaves no permanent defect; it can be removed by relatively simple steps (e.g. deionized water rinse) or it has significant impact at subsequent processing and should be completely eliminated.

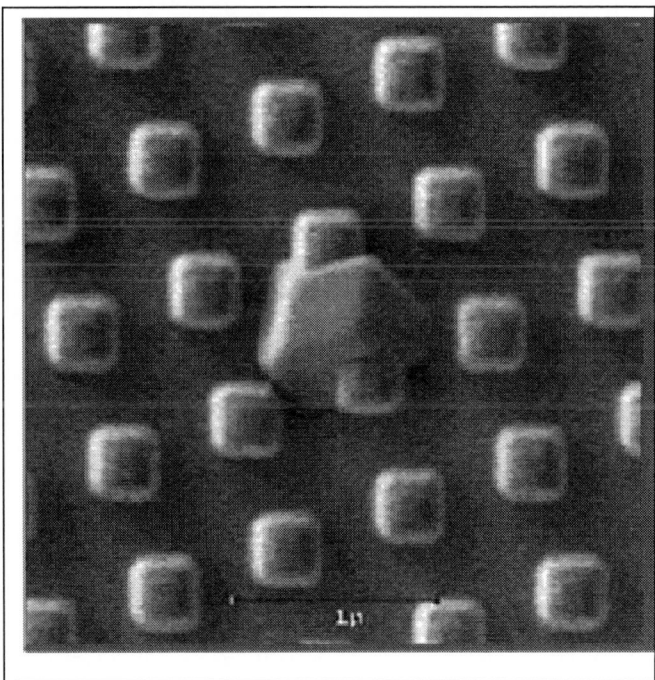

Figure 1. A condensation defect over an active area fill structure

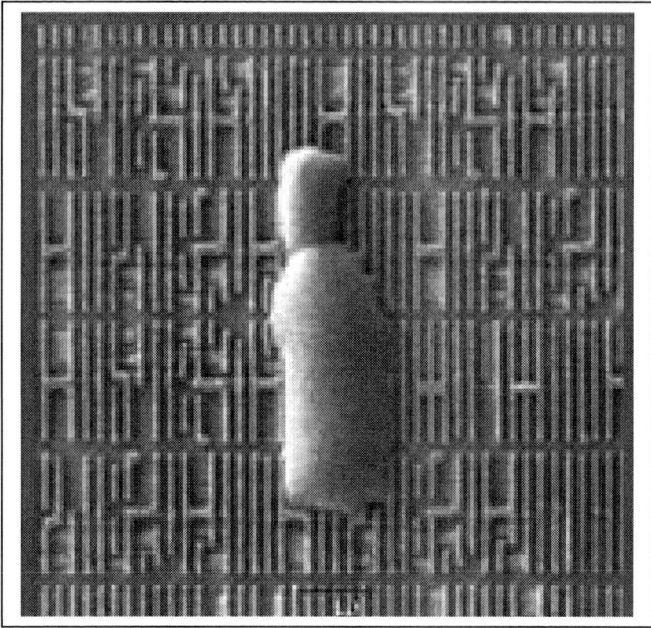

Figure 2.   A condensation defect over an area with polysilicon gates.

A defect classified as condensation, which is found after silicon reactive ion etching (RIE), is known to contain bromine. It often has a faceted shape such as shown in fig. 1. This defect is water soluble and is easily removed by a water rinse. After wet process steps (e.g. a wet etch) we sometimes find what are thought to be water droplets on the surface (fig. 2). The composition of these droplets is difficult to confirm since they typically disappear before analysis can be performed.

A process step that is particularly sensitive to surface contamination and thus to condensation defects is epitaxial growth of SiGe, used for strain engineering of pFET devices (embedded SiGe or eSiGe)[3,4]. SiGe epitaxy is very sensitive to surface conditions. In particular when removing oxide from the silicon surface we have to take care not to deposit any contaminants before the wafers enter the epitaxy tool[5].

There are several mechanisms that lead to so-called blocked growth (BG) of SiGe, but the most persistent was related to condensation. Defect inspection of wafers after the epitaxial growth step found a significant density of blocked growth defects, where no SiGe growth occurs without visible foreign material (FM). An example of this defect in scanning electron micrograph (SEM) review is shown in fig. 3.

To find the cause of blocked growth we inspected wafers before they enter the epitaxy tool. Before entering the epitaxy tool wafers are cleaned in dilute hydrofluoric acid (HF) to remove the native oxide on the surface, a process step we call the preclean. We found a significant density of the water-like condensation defects. By inspecting the location of the condensation defect after epitaxy, we established that the BG defects would occur at the location of the previously observed condensation defect (fig. 4).

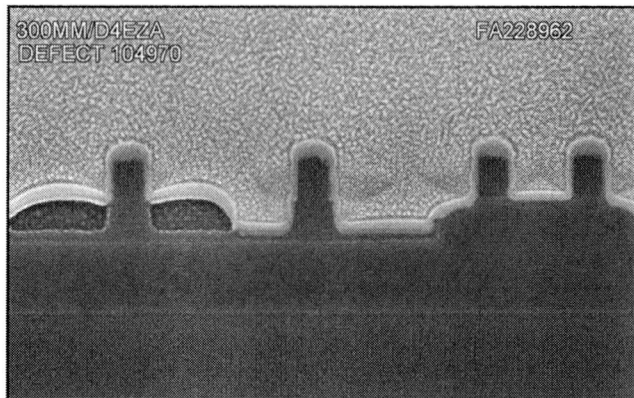

Figure 3.   A cross section of a BG defect. The two areas to each side of the central gate structure are missing SiGe on the thin silicon layer. This we call blocked growth.

Another observation related to the BG defects is the dependence on slot position. Defect density would typically increase with slot position, but fall to a low value on the last wafer in the lot. The front side of the wafer in last position faces away from the other wafers, indicating that the condensation defects are affected by the proximity of another wafer. This suggests that eliminating the nearby wafer surface during processing might reduce defect density.

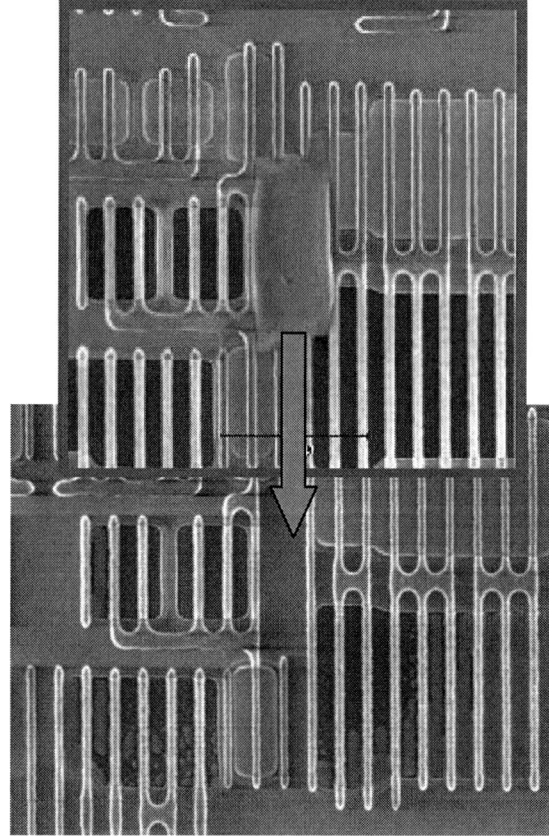

Figure 4.   Pictures of the same area on a wafer before and after epitaxy. The condensation of the picture before epitaxy (top) results in areas of blocked growth after epitaxy (bottom).

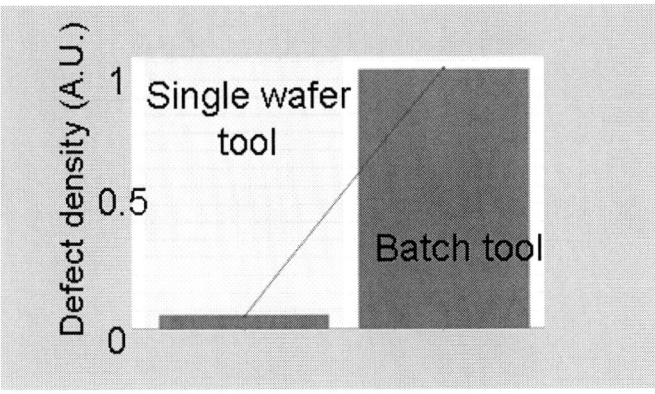

Figure 5. Defect density of blocked growth defects after using the single wafer clean tool vs. a batch clean tool for the preclean step.

The preclean step was performed in a conventional batch tool, loading 25 wafers in a tank for the Wet clean steps. By switching to a single wafer clean tool we avoid all wafer to wafer interaction during processing. We developed a dilute HF process removing a comparable amount of silicon oxide in the short processing time typical for single wafer tools by significantly increasing the concentration of HF. We achieved a dramatic reduction of the number of defects by switching to the single wafer clean tool (fig. 5).

III. OTHER ENVIRONMENTALLY INDUCED DEFECTS

A. Defects at Polysilicon RIE

At polysilicon RIE it was found that the wafer carrier can transmit contamination from a preceding process step to exposed wafers. The sequence of process steps here is a

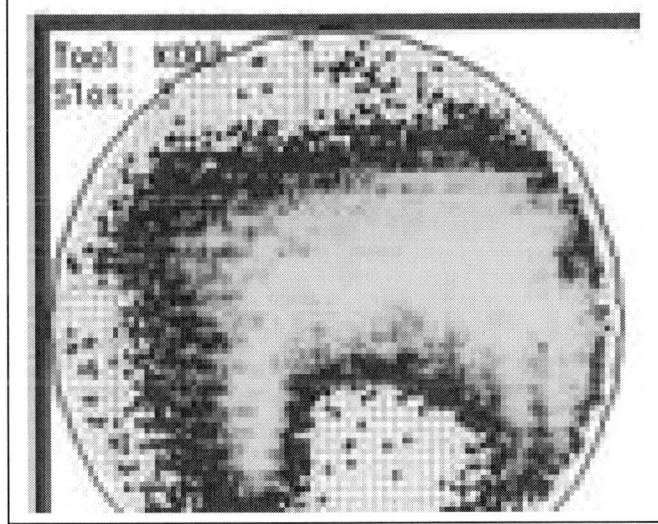

Figure 6. A high density of defects (a blue dot indicates a single defect, a green dot many defects) is observed on wafers which have completed processing through RIE and resist strip and which share a carrier with wafers processing through RIE later

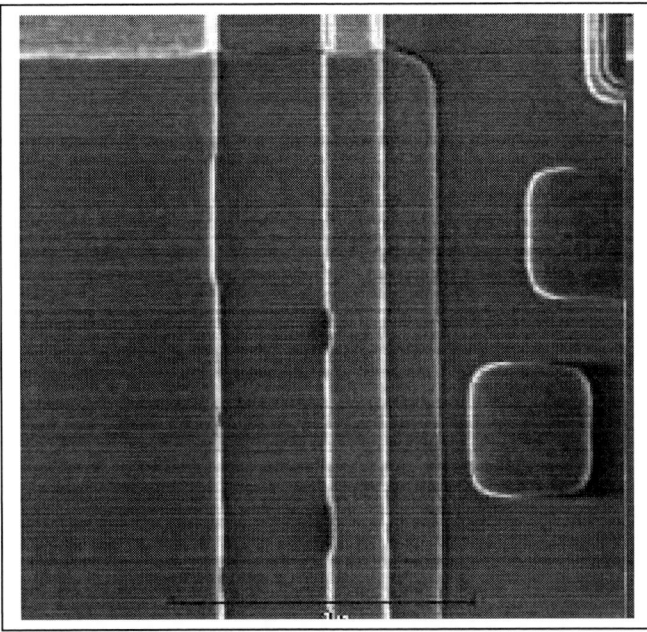

Figure 7. Erosion of the spacer after spacer RIE

lithography step, followed by RIE etching to define the gate width, followed by a Wet strip of the remaining photoresist.

Sometimes a wafer or wafers in a carrier are etched and inspected, while other wafers in the same carrier are not processed. On subsequent processing of the wafers that had remained, we found that the wafers that had already been etched and stripped showed a high density of defects (fig. 6). At electrical test, it was found the defect limited yield of these wafers was degraded by 1% on average, compared to the rest of the population. A second observation was that these defects did not appear if wafers were RIE etched, but not stripped before the other wafers were etched.

Apparently the carrier and wafers are contaminated by remnants from the etch process which damages the wafers if the contamination remains. The strip process removes the contaminants, which explains the second observation. In cases where we have to etch wafers, while other wafers in the same lot remain, the remedy is to move the wafers that are to be etched to a different carrier and only merge the lot after all wafers have completed the etch and strip process.

B. Defects at Oxide Spacer RIE

At spacer RIE it was found that the spacer sometimes erodes, which we call missing spacer defect (fig. 7). The defect density increases as wafers wait for the subsequent strip process. These defects form gradually over time, so a queue time limit was imposed to preclude damage. Though it is considered detrimental to throughput to impose a queue time restriction, it was decided in this case not to investigate possible process changes to reduce the defect level, since a sufficient number of tools is usually available to handle the wafers from the RIE tool.

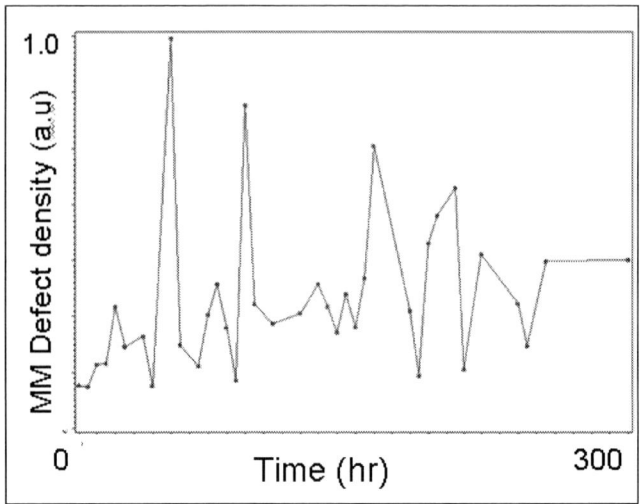

Figure 8. Defect density at polysilicon gate RIE as a function of the time wafers wait to get processed from lithography through RIE

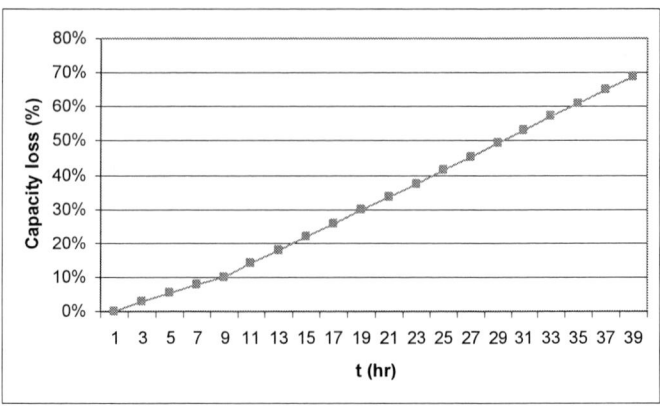

Figure 9. Capacity loss due to queue time restriction as a function of the time from lithography to epitaxy

IV. QUEUE TIME RESTRICTIONS

The spacer RIE step from the previous paragraph is but one example of the need to impose queue time restrictions. We are finding increased levels of FM and micromasking (MM) defects with time at a number of process steps, such as at gate formation and SiGe epitaxy mentioned earlier. These increases are typically not catastrophic but do result in yield degrade.

An example is shown in fig. 8. When we plot the density of MM defects vs. the time that it takes wafers to complete processing from gate lithography through polysilicon RIE, we find a gradual increase. A similar trend is found if we look at defect density after SiGe epitaxy vs. the time from SiGe lithography to epitaxy. In response to such increase, we might prefer to institute a queue time restriction between the process steps where the increase is found. However, placing many queue time restrictions, especially those that cover a number of process steps, affects productivity. The reason productivity is affected is because of the way scheduling software protects us from queue time failures. The scheduling software will evaluate the tools available for processing at any given time and dispatch lots accordingly. When a queue time window encompasses many process steps, the scheduler will have to take into account any delays that might occur along the way. Tools which are unavailable will lead it to reduce the number of lots entering the queue time window. If the window covers a large number of process steps and long process time, any tool down situation will cause it to reduce the number of lots entering. The probability of such events increases with the number of process steps and the length of the window. The impact of all these factors on capacity is shown in fig. 9 for the example of the sequence of steps from lithography, RIE and SiGe epitaxy.

To avoid severe capacity loss while still protecting our hardware we developed an alternative method to limit the time the wafers spend between sensitive process steps. Instead of imposing a queue time limit, we limit the number of wafers between operations. When the number of wafers between the lithography and epitaxy step reaches a certain threshold, we stop additional wafers from entering the window. By optimizing the number of wafers in the window we have managed to allow a very small impact on capacity and utilization while also keeping the time wafers spend in the window within acceptable limits. In fig. 10 we show the capacity loss as a function of the number of wafers in the window.

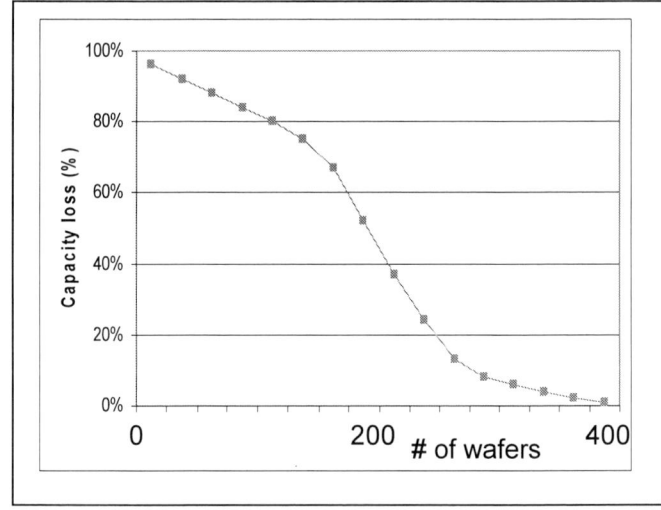

Figure 10. Capacity loss as a function of the number of wafers allowed in the sectors between lithography and epitaxy

V. SUMMARY

Defects are found to be caused by environmental factors, i.e. they are caused by factors not related to the actual processing of the wafers themselves. These environmental factors can be non-processing operations, the processing of other wafers in the same carrier or the gradual formation of defects over time, while wafers are waiting for the next process step. We have shown instances where a change in processing eliminates these defects, as well as cases where a queue time restriction is imposed to avoid defects.

Where queue time restrictions were found to be necessary, but harmful to productivity, we have implemented an alternative method to limit the time wafers wait for processing to start.

ACKNOWLEDGMENT

The authors are grateful to K. Giewont and S. Sankaran for their support. We also like to acknowledge the contributions from W. Brennan.

REFERENCES

[1] M. Okazaki, S. Niehoff, N. Alkurjy, C. Cai and C. Ngai, "VOC Induced Particle Generation during Wafer Transportation and Its Solution", proc. ASMC 2010, pp.196-199 (2010)

[2] V. Pandit and E. Kuo, "Reduction of Electrostatically Adhered Particles on Wafer Backside using Ionizers", proc. ASMC 2010, pp. 200-203 (2010).

[3] W.H. Lee et al, "High Performance 65 nm SOI Technology with Enhanced Transistor Strain and Advanced-Low-K BEOL", Electron Devices Meeting, 2005 IEDM Technical Digest. IEEE International Publication Date: 5-5 Dec. 2005, pp. 59-63

[4] R. van Roijen et al. "Control of etch and deposition for embedded SiGe," Advanced Semiconductor Manufacturing Conference, ASMC 2010, pp. 133-136.

[5] K. Umezawa et al, "Advanced Surface Cleanness Evaluation Technique Using Epitaxial Silicon Germanium (SiGe) Process beyond 32nm Node", International Symposium on Semiconductor Manufacturing, (2007) pp. 1-4.

Cost Effective and Robust Nickel Silicidation Process Qualification and Chamber Matching in Rapid Thermal Processing Tools

Weihua Tong, K. Suresh, Miowchin Tan, Peter
Benyon, Vish Srinivasan
Fab7 GLOBALFOUNDRIES
Singapore
tongwh@globalfoundries

Jinping Liu
US-IROQUIOS, GLOBALFOUNDRIES
NY, the United States
liuj@globalfoundries.com

Abstract— **A new technique in qualification of low temperature nickel silicide process is studied. Bare silicon wafers are first oxidized to form a thick film of thermal oxide, followed by nickel and titanium film stack deposition. The samples are then annealed at low temperature using a rapid thermal processing tool. It is shown that the nickel and titanium film stack forms an alloy above the thermal oxide layer. This alloy's sheet resistance depends on the nickel/titanium film thickness, nickel to titanium thickness ratio, annealing temperature and time. Using the high temperature sensitivity compared to that in conventional technique, this new technique offers an accurate, reliable, and cost effective approach for process qualification, rapid thermal annealing chamber matching and daily monitoring.**

Keywords-nickel silicide; low temperature; nickel and titanium film stack; process qualification; chamber matching

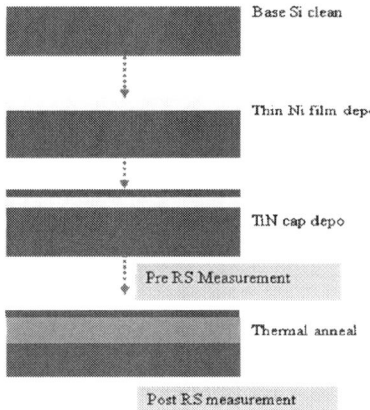

Figure 1 Schematic flow chart for the conventional approach

I. INTRODUCTION

Rapid thermal processing (RTP) technology has been widely used in the semiconductor industry for decades [1]. One of the major RTP applications is metal silicidation. As the industry moves towards smaller technology nodes, metal silicide has been driven from titanium silicide (TiSi2) to cobalt

silicide (CoSi2). The low resistance and low silicon consumption requirement for ultra shallow junction has pushed the metal silicide further to nickel silicide (NiSi) and its annealing temperature down to around 300C and below [2-7]. Extensive studies have been reported for the electrical and mechanical properties of nickel silicide, the mechanisms for its formation in narrower lines and/or with various impurities [4, 8]. However, there are not many publications on low temperature nickel silicidation process qualification or chamber matching in RTP tools.

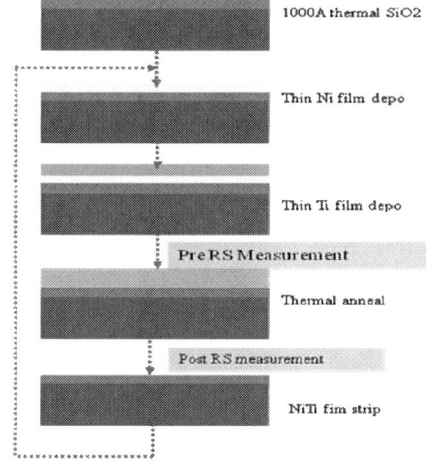

Figure 2 Schematic flow chart for the new approach

The metal silicide transformation point is the most widely used method for RTP based new process qualification and annealing chamber matching [5, 9, 10]. In this conventional method, some high quality bare silicon wafers are cleaned with HF to remove the native oxide. Then, a thin film of nickel is deposited followed by a thin cap layer of titanium nitride deposition. Pre sheet resistance (Rs) measurement is performed. The wafers are then annealed with temperatures ranging from 260C to 800C with same dwell time. Post annealing Rs is measured to obtain the conversion curve. The most sensitive temperature is determined as the conversion point, and it is used for silicidation process tuning or annealing chamber matching. The schematic process flow of this conventional approach is plotted in Figure 1.

As shown in Figure 2, a new method of using nickel and titanium thin film stack on thick thermal oxide technique is studied. Its properties after annealing, and application on low temperature nickel silicidation process qualification and chamber matching in RTP tools, are also investigated.

II. EXPERIMENTS

Firstly, bare silicon wafers are oxidized in a furnace to provide a thick silicon oxide (around 1000Å) to serve as an insulator for protection of substrate silicon from acting with other materials. Secondly, a thin film of nickel is deposited, followed by a thin film of titanium deposition. Different thickness film stack and thickness ratio for nickel film over titanium film are employed to study the effect on temperature sensitivity. Pre metal deposition Rs is then measured. Thirdly, all prepared wafers are annealed in nitrogen ambient with the same dwell time at temperatures between 300C to 400C. After the annealing process, post Rs is measured again. The delta Rs between the pre annealing and post annealing measurement is plotted in Figure 3.

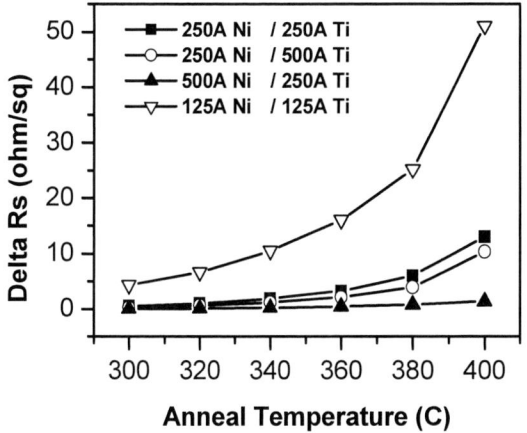

Figure 3 Change in Ni/Ti Stack Rs vs anneal temperature

For comparison, the traditional method of metal silicide transformation point is studied as well. In this study, an 80Å nickel thin film and a 50Å titanium nitride cap layer are deposited on the clean silicon wafers. The delta Rs between the pre annealing and post annealing measurement is plotted in Figure 4.

As shown in Figure 4, a sudden Rs increase is observed at the low temperature range of 260C to 280C. The Rs maintains relatively stable at temperature range of 310C to 650C. The steep Rs increase is observed again at temperature range of 650C to 750C. However, the biggest Rs mean and sigma change are observed in the temperature range between 290C and 300C. Based on the theory of transformation, the Rs increase in low temperature of 260C to 280C indicates the nickel starts to diffuse into the silicon, and the Rs increase after 650C indicates the high resistance phase of NiSi2 formation. The transformation temperature point is then determined to be between 290C and 300C. To further

determine the exact conversion temperature and verify the repeatability, another two different sets of wafers are used to draw the silicide transformation curve in a small temperature range from 285C to 305C. The results are shown in Figure 5.

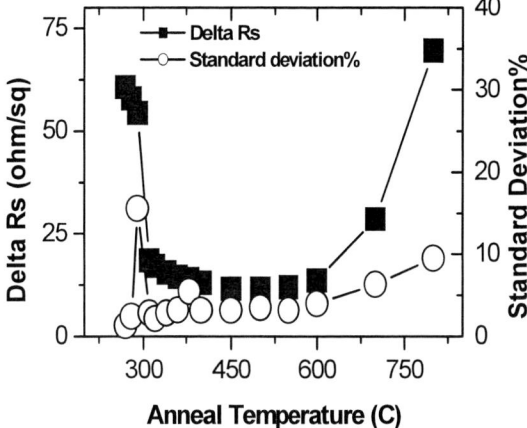

Figure 4 NiSi transformation curve of an 80A Ni and 50A TiN cap film for soak anneal time of 60 seconds.

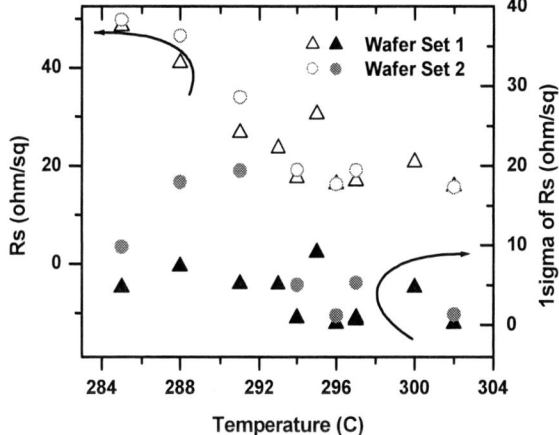

Figure 5 Smaller temperature range transformation curve with same film stack but different set of wafers

III. RESULTS AND DISCUSSION

As shown in Figure 3, the nickel/titanium film stack delta Rs increases with increasing annealing temperature for all the four different nickel/titanium film stacks. However, the highest thickness (total 750Å) thin film stack of 500Å nickel plus 250Å titanium and 250Å nickel plus 500Å titanium show the lowest delta Rs increase trend, followed by a less thick (total 500Å) film stack of 250Å nickel plus 250Å titanium. The steepest delta Rs increase trend is seen on the thinnest (total 250Å) film stack of 125Å nickel plus 125Å titanium. A nickel/titanium alloy is believed to has been formed during the annealing process.

978-1-61284-408-4/11 $26.00 © 2011 IEEE

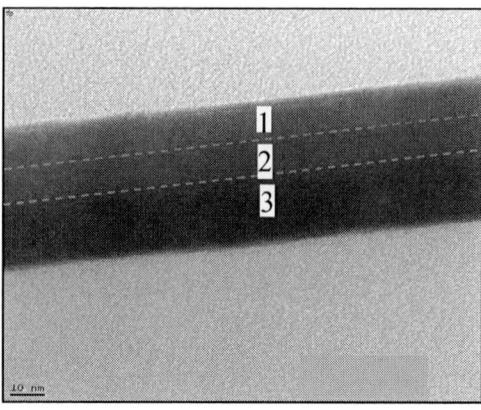

Figure 6 TEM on annealed Ni/Ti film stack shown a new film layer has been formed

To verify the assumption, transmission electron microscopy (TEM) and energy-dispersive X-ray spectroscopy (EDX) are performed on one of the annealed nickel/titanium film stack. As shown in Figure 6 of the TEM, the original two-layer nickel/titanium film stack becomes a three-layer film stack. A new film (labeled as "2" in the figure) has been formed between the original deposited nickel (labeled as "3") and titanium (labeled as "1") films after the thermal annealing process. As shown in Figure 7 of the EDX, the major component of the top film is titanium. It is originally deposited before the annealing process. In the middle film, the peaks of components of titanium and nickel are almost the same height, confirming a new film of nickel/titanium alloy has been formed during the annealing process. Although different peaks are observed in the bottom film, the major one is nickel. It is the un-acted nickel material deposited before the annealing process.

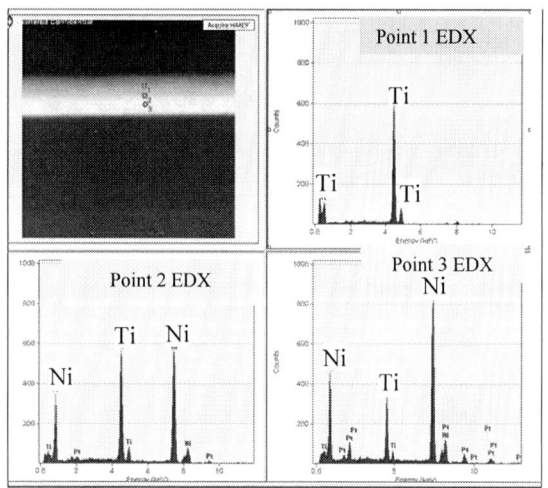

Figure 7 EDX shows the new film layer is a Ni/Ti alloy

Please note that the EDX does not show any silicon component in any of the three layers. It is an extremely important special feature of this study which will be discussed in more details in part IV.

Different nickel/titanium film stack thickness and thickness ratio yield different delta Rs against annealing temperature. A thinner nickel/titanium film stack shows a steeper delta Rs increase at the experiment temperature range than thicker nickel/titanium film stacks. It indicates that thinner nickel/titanium film requires less thermal budget to complete the form of the alloy. Higher nickel over titanium thickness ratio shows flatter delta Rs increase than that of lower nickel over titanium thickness ratio. It might indicate the major diffusing element is nickel during the nickel/titanium alloy process

In silicide annealing process qualification and chamber matching practice, the most important consideration is the Rs sensitivity over temperature change. It is defined as delta Rs (post annealing measurement minus pre annealing measurement) divided by the temperature change. The higher Rs sensitivity over temperature, the more accurate of silicidation process matching or annealing chamber matching. The Rs sensitivity of the this study of the 125Å nickel plus 125Å titanium film stack on 1000Å silicon oxide together with Rs sensitivity of the traditional method of 80Å nickel with 50Å titanium nitride on bare silicon wafer are plotted in Figure 8. It is very clear that the Rs sensitivity of the traditional approach is almost zero at annealing temperature range of 300C to 380C. On the other hand, this new work of nickel/titanium thin film stack on thick oxide shows a significant Rs sensitivity increase for the annealing temperature range of 320C to 380C. This work's higher Rs sensitivity over temperature change provides a real advantage in silicidation process qualification and annealing chamber matching.

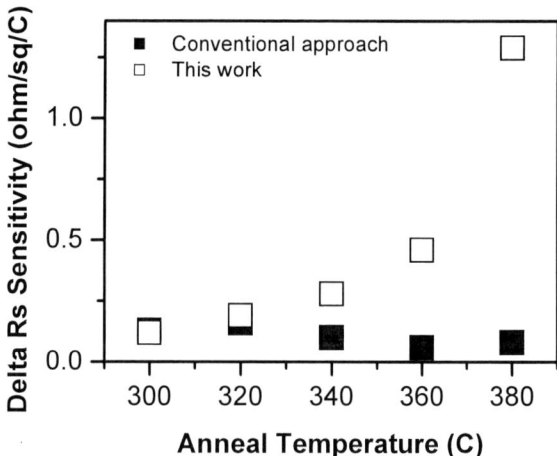

Figure 8 Ni/Ti alloy and NiSi Rs sensitivity vs. anneal temperature

IV. REUSING WAFERS AND REPEATABILITY

As mentioned in part III in discussion of the TEM and EDX of the new method of nickel/titanium thin film stack on thick silicon oxide, no silicon component is observed in the metal films even after thermal annealing process. That means no kind of silicide has been formed during the annealing process. The thick thermal oxide layer, which is formed before the nickel/titanium film stack deposition, has successfully prevented nickel to diffuse into the substrate of silicon or the silicon to diffuse into the nickel/titanium film. It is also one of the initial intentions to grow this thick thermal oxide.

The thin films of titanium, nickel/titanium alloy and nickel can easily be removed by wet etch without damage to the substrate silicon and silicon oxide. After wet etch, the wafers are left with only thermal oxide on them. Thus, they can be deposited with nickel/titanium film stack again and reused for new nickel silicidation process qualification or annealing chamber temperature matching again. This special feature makes it possible to recycle the wafers again and again without degrading the quality.

To demonstrate this new method's Rs repeatability of the recycle wafers, different batches of wafers with same thermal oxide thickness and nickel/titanium film stack are annealed with the same temperature and time in different nickel silicide annealing chambers. Post annealed wafers are recycled with same wet etch condition. Different recycles are also employed in this study. Figure 9 shows the delta Rs for part of the tested wafers. It is well repeatable as the delta Rs is well within 0.2ohm/sq range which is equivalent to 1.0C in temperature control in this study. Thus nickel and titanium thin film stack on thick oxide approach is very robust and cost effective. Therefore, it is the most promising method for low temperature nickel silicide process qualification or chamber matching in RTP tools.

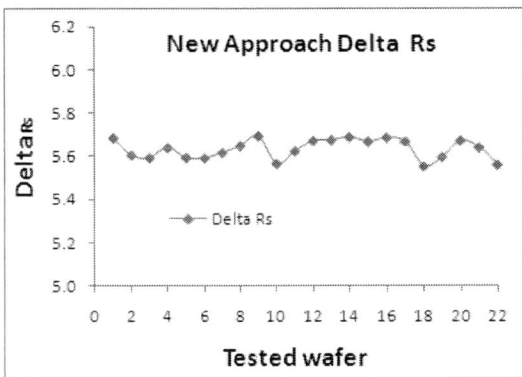

Figure 9 Repeatability of the new technique

On the other hand, as discussed in part II and part III, it is very difficult to achieve similar Rs results using different sets of wafers by conventional methods. The reason is that the silicide Rs at conversion temperature range is too sensitive for the wafer batch, which is related to wafer source, pre metal deposition clean, delay time from clean to metal deposition and metal sputter tool. To overcome the above drawbacks, the process qualification or chamber matching has to be completed using the same batch of wafers. It is very inefficient to use it as daily process monitoring due to poor repeatability of transform point of temperature. The requirement of high quality of new silicon wafers for each test also leads to high cost for its application in daily chamber monitoring.

V. CONCLUSION

Nickel/titanium thin film stack on thick silicon oxide film forms nickel/titanium alloy after thermal annealing process. The alloy's Rs depends on the nickel/titanium film stack thickness, nickel to titanium thickness ratio, annealing temperature and time. This nickel/titanium thin film stack on thick thermal oxide approach can obtain much higher temperature sensitivity than traditional metal silicidation approach in the 300C to 400C temperature range. The used wafers of this approach can be recycled for many times by stripping the metal layers via wet etch. It is robust and cost effective. Therefore, it is the most promising method for low temperature nickel silicide process qualification or chamber matching.

VI. ACKNOWLEGEMENT

The authors would like to thank Arunachalam, Valli, Brett Williams for their invaluable suggestions in drafting of this paper. We express our thanks to Yushuang Xiang and Henry Lim in preparing some of the experiment samples. Sean Dion, Laura Brown, Jasmin Tan would be also acknowledged for proof reading of the final paper.

REFERENCES

[1] J. Niess, S. Paul, S. Buschbaum, P. Schmid, W. Lerch Materials Science and Engineering B114-115 (2004) 141-150;

[2] S. Wolf, Silicon processing for the VLSI ERA, volume 4, Deep-submicron process technology, Lattice Press;

[3] C. Y. Chang, S. M. Sze, ULSI technology, McGRAW-HILL international editions;

[4] L. J. Chen, Silicide technology for integrated circuits, Department of material science and engineering, National Tsing Hua University;

[5] P. G. Vermont, X. Pages, E. H. A. Granneman, K. Vanormelingen, K. Verheyden, S. Mertens, Microelectronic Engineering 84 (2007) 2572-2574;

[6] F. Deng, R. A. Johnson, P. M. Asbeck, S. S. Lau, W. B. Dubbelday, T. Hsiao, J. Woo, J. Appl. Phys. 81 (12) 15 June 1997. 8047-8051;

[7] Y. L. Jiang, A. Agarwal, G. P. Ru, X. P. Qu, J. Poate, B. Z. Li, W. Holland, Nickel silicidation on n and p-type junctions at 300C, Applied physics letters, Vol. 85, No.3, 2004, 410-412;

[8] K. Ohuchi, K. Adachi, A. Hokazono, Y. Toyoshima, Source/Drain engineering for sub 100-nm technology node, 2002 IEEE;

[9] M. Tinani, A. Mueller, Y. Gao, E. A. Irene, Y. Z. Hu, S. P. Tay, In situ real-time studies of nickel silicide phase formation, J. Vac. Sci. Technol. B 19(2) Mar/Apr 2001, 376-382;

[10] PennWell Corporation, Solid state technology, October 2004 edition.

A New Device for Highly Accurate Gas Flow Control With Extremely Fast Response Times

Kevin Boyd*, Adam Monkowski[†], Jialing Chen[†], Tao Ding[†], Ray Malone[†] and Joseph Monkowski[†]

*IBM Corporation
Hopewell Junction, New York
[†]Pivotal Systems Corporation
4683 Chabot Drive Pleasanton, California 94588

Abstract—**This paper presents a new type of control scheme and device for controlling gas flow into semiconductor process chambers. The key component of the Gas Flow Controller (GFC) is a high-precision valve with an integrated position sensor, which is used to maintain a constant flow rate. A map lookup scheme is employed to adjust the valve position to accommodate the upstream pressure, including any changes or disturbances. The layout of the flow controller also allows for the incorporation of a pressure-volume-time-temperature-based flow measurement, which is a primary standard flow measurement, to confirm and maintain flow accuracy throughout the device's lifetime. The fast response time of the sensors and the high sampling rate in the control loop enables the control of the gas flow within the order of tens of milliseconds.**

I. INTRODUCTION

In plasma-etch and chemical vapor deposition processes, accurate metering of gas flow into the process chamber is critical because, beyond the process wafer, all materials that participate in the etch or deposition are introduced in gas form. In a majority of these processes, two or more of these gases react to produce the essential film or passivation layer and even slight deviations in gas flow—even on the order of 1%—can cause the process to fail. While numerous technologies have been developed to accomplish gas flow metering, the semiconductor market has focused largely on two: the thermal-based mass flow controller (MFC) and the more recently introduced pressure-based flow controller.

These two technologies have served the industry well for a number of years; however, as critical dimensions continue to decrease, process envelopes become tighter, and process steps become shorter, these technologies are reaching their limit. For the semiconductor industry, an ideal gas flow control device will respond instantly ($\ll$ 1 sec) with no overshoot, and will produce a flow that is accurate and does not drift over time. Further, flow accuracy will be maintained in the presence of significant (up to several hundred Torr) downstream pressure and will remain accurate in the presence of fluctuations in supply pressure.

Figure 1 shows the basic control loops present in thermal and pressure-based flow controllers. Thermal MFCs rely on a flow sensor, shown in Fig. 1-A, that was developed by Benson, et al., in the 1960s [1]–[3]. Since its development, the thermal flow sensor has been widely applied across many industries and it has been improved to be less sensitive to attitude and

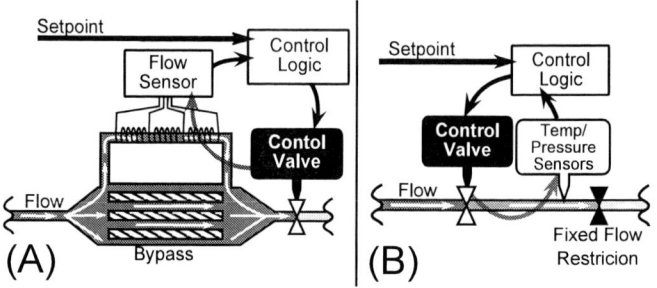

Figure 1. Schematic of flow control schemes for mass flow controllers used in the semiconductor manufacturing industry. A) Thermal Mass Flowmeter: A thermal flow sensor passes data to the control logic which adjusts a control valve to minimize the difference between the flow sensor feedback and an external setpoint. B) Pressure-based flow controller: The relationship between flow rate through the fixed flow restriction and gas pressure is known. The control logic brings the pressure to the required level, based on the external setpoint.

temperature variation. The thermal flow sensor works on the principle of heating and cooling a temperature sensor (usually a wound metallic element), and the flow sensor's response time is a function of both the heat capacity of the temperature sensor and the quality of its thermal communication with the gas. For ultra-high-purity (UHP) applications, where the gas cannot be in direct contact with the temperature sensor, the response time is on the order of 1 sec [4]. This represents an inherent limit to the response time for thermal-based MFCs. To circumvent this limit, some thermal-based MFC designs rely on an open-loop control scheme at turn-on, but since the flow sensor cannot measure quickly enough, overshoot and oscillation in flow can occur without being detected by the flow sensor.

In contrast, the pressure-based flow controller relies on a pressure transducer as its primary feedback sensor. Flow is controlled by varying the pressure upstream of a calibrated flow restriction. The pressure sensor can have a response time of < 1 ms, allowing for fast measurement of flow. Turn-on time for the pressure-based flow controller is governed by how quickly the volume between the control valve and calibrated flow restriction can be pressurized (see Fig. 1-B). Generally, the turn-on time of a pressure-based flow controller is similar to the thermal MFC. Additionally, pressure-based flow controllers are less sensitive to upstream pressure variations, due to the fast response time of the pressure sensor; however, they are more sensitive to downstream pressure fluctuations, as they can possibly change the assumed pressure-to-flow relationship of the calibrated flow restriction. Additionally, since the valve that controls flow is upstream of the pressure-controlled volume, for turn-off or turn-down, the response time of the pressure-based flow controller is limited by the time it takes for the gas in the pressure-controlled volume to be evacuated through the calibrated flow restriction and into the process chamber. This time can be on the order of several seconds for some devices.

To overcome some of the limitations present in the two designs, Pivotal Systems has proposed a new type of gas flow control device which is focused on improving response time and long-term accuracy. In Pivotal's gas flow controller (GFC™), there are two independent control systems for measuring and controlling flow. The primary system is a fast, map-based valve control system that allows the device to respond rapidly to changes in setpoint and upstream pressure. The secondary system, named Gas Flow Monitor (GFM™), measures the flow of the GFC using a pressure-rate-of-change flow measurement. This secondary system does not operate as rapidly as the primary system; it is used to validate and if necessary, update, the control map.

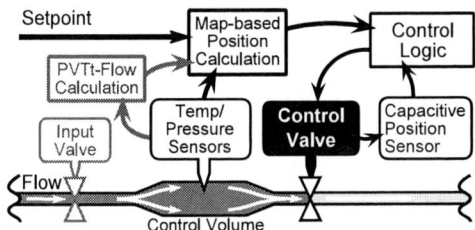

Figure 2. Schematic of flow control schemes for the Pivotal Gas Flow Controller. A capacitive position sensor measures the position of the control valve. Based on the pressure, temperature, and external setpoint input, a position is calculated, and the control logic drives the control valve to move to the desired position. Additionally, an input valve can be closed and the pressure and temperature data can be used to calculate the absolute flow rate through the control valve.

A. Gas Flow Monitor Approach

A real-time in-situ GFM system [5] is incorporated in the GFC to measure flow. The measurement is based on pressure–rate-of-change, in which a total of four independent variables are considered: pressure (P), volume (V), temperature (T), and time (t). The benefit of the pressure-volume-temperature-time (PVTt) type of measurement is that it is a primary standard flow measurement, that is, one that relies only on fundamental state variables and time [6], [7]. The GFM's measurement is performed on the gas in an isolated volume between the control valve and the input valve as shown in Fig. 2. This volume is calibrated and assumed to be constant for all time. GFM monitors pressure and temperature data as a function of time to determine the rate of change of gas density in the volume. A flow rate out of the volume is then established using

an appropriate equation of state—typically either the ideal gas law or the Virial equation of state.

A potential pitfall to the application of a PVTt primary flow measurement such as this, in which volume is held constant and pressure is dynamic, is that adiabatic cooling of the gas must be considered. However, with careful design of the calibrated volume such that the volume primarily consists of passages approximately 0.180" or less in diameter (a typical passage diameter for UHP flow designs), or a geometry with similar sized passages, the deviation of the average temperature of the gas from the stainless steel body is < 0.1% on an absolute scale. This estimation of the gas temperature from that of the stainless steel body, combined with accurate volume and pressure data, is able to yield very accurate flow measurements.

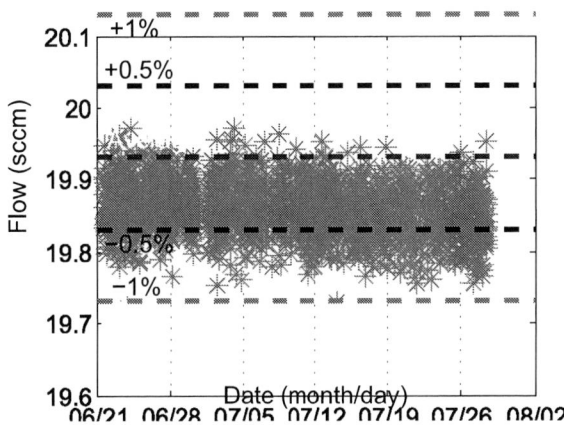

Figure 3. Plot of GFM data on a production plasma oxide etch tool at IBM's Fishkill facility. Each point represents the GFM measurement taken at the beginning of an individual flow run. 1%, and 0.5% deviation bars are shown with respect to the average flow rate observed; the MFC flow setpoint was 20 sccm.

This gas flow measurement method has been tested and implemented in several major chip makers' R&D labs and manufacturing fabs on both thermal-based and pressure-based flow controllers. Figure 3 shows one month's worth of data taken from a GFM running in high-volume production at IBM's Fishkill facility. In this application the GFM measures the gas flow for each process run and shows the variability of the MFC's flow rate over time. Figure 4 shows accuracy tests conducted on the GFM comparing it to a DH Instruments Molbloc™ system, measuring the steady flow from a thermal MFC (100 sccm full scale) at different flow setpoints. The measurements from the GFM show excellent correlation with the Molbloc system.

B. Gas Flow Controller Approach

The control scheme for the GFC is shown in Fig. 2. The primary control system consists of a precision position sensor coupled with a control valve such that when the valve is closed the position sensor output is zero and as the valve is opened the

978-1-61284-408-4/11 $26.00 © 2011 IEEE

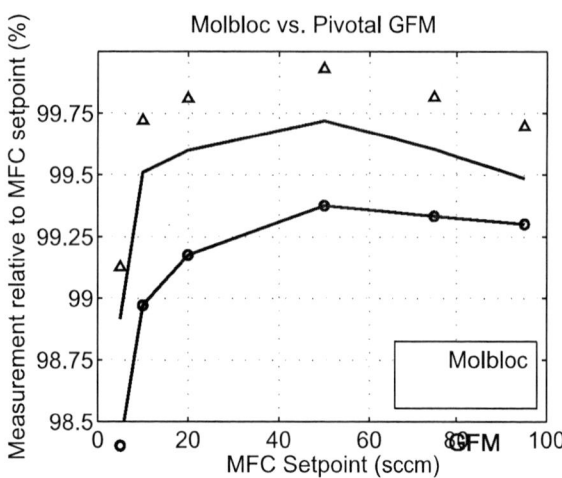

Figure 4. Pivotal GFM measurements compared to Molbloc™ measurements (stated accuracy: 0.2%) on a 100 sccm MFC flowing various flow rates. GFM measurements are within 0.5% of the Molbloc across the range of flows measured.

sensor determines the opening of the valve in units of length. To produce a constant flow rate, a digital signal processor (DSP) receives three inputs: the setpoint, as well as pressure and temperature data from the the gas in the calibrated volume directly upstream of the control valve. The DSP uses this data to calculate a target position that will produce the desired flow rate using a pre-determined control map. This target position is passed to the valve control loop and the drive voltage to the control valve is adjusted so that the output of the position sensor matches the target. A flow rate feedback signal (GFC feedback) is generated in a similar fashion; data from the position sensor, along with pressure and temperature data is passed to the DSP, and using the same control map, a flow rate is calculated that corresponds to the present valve position.

The position sensor is based on a capacitance measurement. This type of capacitive position sensor has a response time $< 1 \times 10^{-3}$ sec; therefore, the control loop can be run at speeds over 1000 Hz. The piezoelectric element that drives the valve also has a response time $\ll 1 \times 10^{-3}$ sec. The desired flow rate can be reached as quickly as the control logic can bring the valve to the desired position; this time can be on the order of milliseconds. This rapid response produces a flow controller that can be turned on or turned down in terms of milliseconds, and is insensitive to sudden changes in upstream pressure. Changes in pressure will simply result in a new target position for the control valve.

This inherent pressure insensitivity allows the GFC to be run with the secondary flow measurement system, GFM, directly upstream of the control valve. In brief, the measurement is taken by closing an input valve to isolate the calibrated volume (V_{cal}) directly upstream of the control valve. This configuration is shown in Fig. 2. As the pressure in V_{cal} drops, the control valve responds by adjusting its position accordingly

per changing pressure, while the GFM uses the pressure data to calculate an independent flow measurement. When the GFM's measurements are complete, the input valve is opened and flow continues normally with the control valve responding to the pressure increase.

The GFM behavior can be configured in several different ways. After the input valve is closed, an accurate GFM measurement can be calculated within a pressure drop of < 1 psi; however, if the pressure is allowed to decrease further, additional GFM measurements are continually calculated until a low pressure limit is reached. These GFM measurements are compared, real-time, with the flow rates calculated by the primary control loop to verify correct operation. The GFM can be configured to run once at the beginning of each flow run, or multiple times throughout the run.

II. TEST SETUP

A test bench, as depicted in Fig. 5, is set up to show both the steady state and transient state of the Pivotal GFC control results. The bench consists of a standard 1-1/8 inch C-seal gas stick, one pressure regulator (set at 27 psig), one Pivotal GFC (200 sccm full scale), one pressure gauge (MKS baratron type; 1000 Torr full scale), one thermal mass flow meter (MFM) and one vacuum pump downstream of the gas stick (pressure < 100 Torr). The pressure transducer is installed between the GFC and MFM to ensure the steady flow. In other words, if the gauge shows a constant pressure during the flow, the flow through the MFM is equal to the flow out of Pivotal GFC. The DSP is running at 1040 Hz for the following tests.

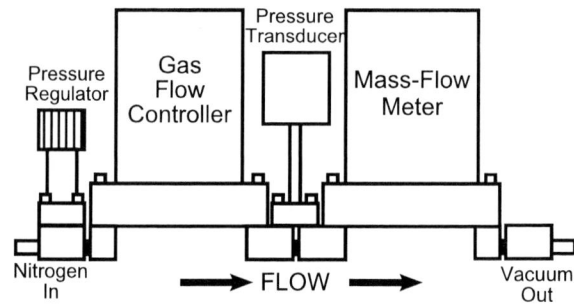

Figure 5. Schematic of the configuration used for GFC testing. The mass flow meter (MFM) has a full scale range of 100 sccm.

The GFC's control valve shown in Fig. 2 is a normally closed flow restriction that is actuated with a piezoelectric element. The capacitive position sensor is mechanically coupled to the flow restriction, such that any movement at the flow restriction results in a linearly proportional response from the position sensor.

A. Response Time

Figure 6 shows how the GFC responds to a setpoint change. Twelve individual runs have been overlaid, with a characteristic response trace highlighted to show the trend; the individual feedback points generated by the DSP are represented in crosses. At time zero, the flow command changed from 0 to 20 sccm; the GFC starts to respond to the command at

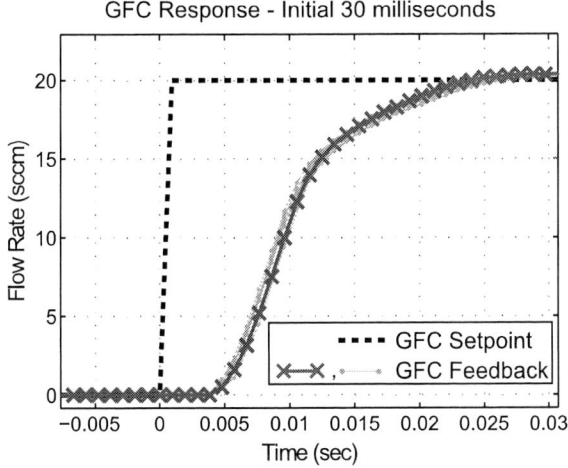

Figure 6. Plot showing the response of the GFC primary control loop to a setpoint change from 0 to 20 sccm. Twelve individual tests are overlaid; a characteristic response has been highlighted.

time 4 msec and its control settling times for all of these 12 trials were about 30 msec. The logic used to drive the valve is a standard proportional-integral (PI), closed loop type. The fast response of the system allows the use of conservative P and I terms that result in over-damped dynamics. As the valve position approaches the target, the changes to the drive signal become smaller and smaller such that overshoot is eliminated. The use of conservative P and I values mean that setpoint is not reached as fast as possible, but because of the speed of the system and its sensors, this can still be accomplished in only tens of milliseconds.

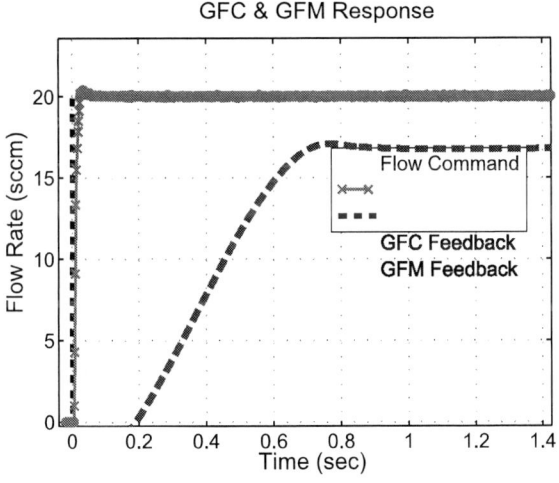

Figure 7. The dynamic response of the gas flow monitor system (GFM), a PVTt, pressure rate-of-change measurement. The response of the primary control valve, GFC (see Fig. 6), is also shown. The response time of the GFM depends on the time over which data is accumulated to calculate a flow rate, in this case, 0.8 second of data is used resulting in a 0.8 sec response time.

B. PVTt Measurement Verification

As described in the previous section, a built-in GFM system is used for the flow verification and calibration. To generate the data shown in Fig. 7, the input valve is closed prior to setpoint change. GFM measurements occur continuously as long as the input valve is closed. In this test, the GFM flow calculation window was set at 0.8 sec; in other words, at each DSP cycle, the most recent 0.8 second of pressure and temperature data are used to calculate the flow through the control valve. This time is not fixed; generally, a GFM calculation uses between 0.05 and 2.0 seconds of data to calculate a flow rate. The amount of data used is primarily a function of flow rate and desired accuracy; higher flow rates can be accurately determined in a shorter time period. As shown in Fig. 7, the GFC was set to flow 20 sccm at time 0, and while the GFC valve took only $\sim$ 30 msec to reach the setpoint, the GFM flow calculation took $\sim$ 0.8 sec to show the true flow change.

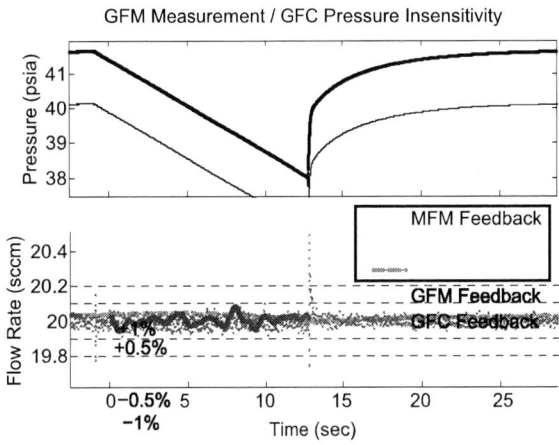

Figure 8. Plot of the combined GFC and GFM feedback signals. The GFM only calculates flow while the input valve is closed. After a measurement is taken, the input valve opens and the pressure increases. The GFC control valve responds to this rapid pressure increase (in this case 10 psi/sec). To verify that flow is not disturbed, a MFM is used to be a third independent measurement; there is no deviation in the MFM output during recharge.

C. Pressure Insensitivity

Figure 8 shows typical steady state behavior of the GFC. At time zero, a GFM measurement is initiated by closing the input valve. The pressure in V_{cal} is shown in the top plot. When the input valve is closed, the pressure starts to decrease from its initial value of 41 psia and continues to drop for $\sim$ 13 sec, to 38 psia. The GFM continually returns flow rate calculations during this time using a rolling window of data points for its calculation. At $t = 13$ sec, the input valve is opened and the pressure in the volume rises back to 41 psia. The initial pressure rise is very fast, approximately 10 psia/sec; however, the control valve is capable of adjusting its position quickly enough to maintain a constant flow rate. For this test, the feedback of the downstream MFM is also shown to corroborate the constancy of flow during this input pressure variation.

III. CONCLUSION

Accurate control of gas flow is critical in semiconductor processing. We have presented a device that incorporates a primary flow standard measurement with a novel control valve that is capable of fast response times. The primary valve control system is map-based and relies on a pressure transducer and capacitance position sensor, both of which have response times on the order of milliseconds. The secondary system verifies the first with a built in primary flow standard measurement systems that uses a PVTt, pressure rate-of-change measurement. The combination of these two systems yeilds a new approach to flow control that offers the semiconductor industry new solutions to difficult gas flow control challenges.

REFERENCES

[1] J. M. Benson, "Benson thermal flowmeter," U.S. Patent 3,181,357, May 4, 1965.

[2] ——, "Thermal flowmeter," U.S. Patent 3,229,522, January 18, 1966.

[3] C. E. Hawk and W. C. Baker, "Measuring small gas flows into vacuum systems," *J. Vac. Sci. Technol.*, vol. 6, no. 1, pp. 255–257, 1969.

[4] L. D. Hinkle and C. F. Mariano, "Toward understanding the fundamental mechanisms and properties of the thermal mass flow controller," *J. Vac. Sci. Technol., A*, vol. 9, no. 3, pp. 2043–2047, 1991.

[5] S. Yedur, A. Sankaran, R. Malone, R. Reed, M. Venkatesh, J. H. Lee, K. Y. Kim, and S. H. Han, "Real-time gas flow monitoring improves mass flow controller performance understanding in wafer fab," *Solid State Technology*, vol. 54, p. 00, March 2011.

[6] R. F. Berg and S. A. Tison, "Two primary standards for low flows of gases," *Journal of Research of the National Institute of Standards and Technology*, vol. 109, no. 4, p. 435, 2004.

[7] J. D. Wright, A. N. Johnson, and M. R. Moldover, "Design and uncertainty analysis for a PVT-t gas flow standard," *Journal of Research of the National Institute of Standards and Technology*, vol. 108, no. 1, pp. 21–47, 2003.

Thermal Budget Reduction and Throughput Enhancement for CMOS Epi Stressors via Wet Clean Interface Contamination Evaluation and Control

Paul Brabant & Keith Chung, Manabu Shinriki,
Scott Hasaka, Dane Scott, Mark Wirzbicki and Terry
Francis
Matheson R&D Albany NanoTechology Center
MathesonGas
Albany NY 12203 pbrabant@matheson-trigas.com

Hong He, Devendra K Sadana
IBM Research at Albany NanoTechnology Center
IBM
Albany NY 12203

Abstract—**In this paper we present characterization, analysis, and methodology for the reduction of surface impurities trapped in the silicon layers at the onset of epitaxial growth. In CVD silicon technology, wet and dry clean of the silicon surface are used to remove native oxide from the surface. However, there are still residual impurities that require desorption via thermal baking to provide a clean interface. This thermal baking leads to unwanted increase of thermal budget. The greater the surface impurities concentration the longer and higher temperature is required for removal of these impurities. In production line environment, long queue times (up to 24 hours) are possible. During these queue times, impurities rebuild up on the surface after the initial wet clean. The combination of ultra-high purity gases and low-pressures during thermal bakes can be used to minimize thermal bake temperatures.** *(Abstract)*

Keywords-H₂ prebake, Low temperature epitaxy, HF last clean,interfacial oxide, moisture, queue time

1. INTRODUCTION

Thermal budget is a major concern for all future generation CMOS 22nm and beyond devices. For Si heteroepitxial films consisting of SiCP, SiGeB for CMOS stressors on SOI the thermal budget for the H_2 pre-bake should be kept at 800°C or below to prevent Si migration and agglomeration. Reducing thermal budgets reliably and repeatably to less than 800°C has proven to be difficult to achieve especially with potentially long queue times in the manufacturing Fab between the HF last wet clean and the onset of the epitaxial deposition.

XPS (X-ray Photoelectron Spectroscopy) has been used extensively to understand the re-growth of native oxides after HF-last wet cleans in the past. It is known that immediately after HF last clean, there is predominate converage of the surface with hydride groups (SiH, SiH₂) and slight coverage with Si-OH, and oxygen related hydrocarbon groups[1]. Over time, depending upon the initial Si surface hydride group bonding and initial O/C(oxygen, carbon) area density, a monolayer has been shown to form in as little as 1 hour up to a period of several days[2].

A non-surface technique evaluating by SIMS(Secondary Ion Mass Spectroscopy) has been found to reliably quantify the remnant interfacial sub-oxide after various wet cleans and dry cleans, and thereby minimize thermal budget in the H_2 pre-bake step. This technique is applied to evaluating the interface after different length queue times that might be encountered in the typical manufacturing area. Hydrogen bake temperature and pressure are recommended to remove the resulting sub-oxide from 2 to 48 hour queue times.

2. EXPERIMENTAL

All epitaxial layers were deposited in a commercially available RPCVD (Reduced Pressure Chemical Vapor Deposition) SWR (Single Wafer Reactor). This reactor was equipped with moisture contamination purification on all house gases (H_2, N_2, He) and also on most specialty gases. House gases and most specialty gases were purified to less than 10 ppb, as verified by TDLS (Tunable Diode Laser Spectroscopy) moisture analyzer. HCl gas was also equipped with a commercially available "cold trap" that reduced moisture to less than 10 ppb by TLDS.

Evaluating the sub-oxide interfacial remnant after a particular wet/dry clean prior to epitaxy is very difficult to characterize without a capping layer of some kind that does not re-introduce interfacial contamination. The use of purification on all gases to chamber is paramount in achieving this. The loading of the wafer into the reactor is done in a way to minimize or eliminate any contribution of contamination from the FEI (Front End Interface) of the deposition system. The wafer is loaded into the reactor chamber "cold" (250°C) and then ultra pure H_2 is flowed for 5 minutes before the temperature is ramped to deposition temperature of 650°C. A silane based Si capping layer is then deposited to "capture" the wet/dry clean interfacial contamination. As the H_2 termination from a wet clean is stable until the temperature reaches approximately 520°C, the technique of wafer loading at 250°C

Identify applicable sponsor/s here. *(sponsors)*

978-1-61284-408-4/11 $26.00 © 2011 IEEE

and purging for 5 minutes, minimizes the interface contamination from any moisture contamination from either a load lock or wafer transfer station. All wafer transfer done at 80 Torr in 3 tiered separation (Load lock to Wafer Transfer to Chamber), from Fab ambient atmosphere.

P- wafers (100) wafers had native oxide removed in both wet/dry methods in 4 different commercial batch and SW(single wafer) systems. All wet cleans were HF followed by IPA. Dry clean chemistry not disclosed. The remnant interfacial O/C from the cleans was measured by SIMS with "no H_2 pre-bake" in the epitaxial reactor after 2 hour queue time (Figure 1). To minimize SIMS only 2 cleans (lowest oxygen 2 hour queue time wet and only dry clean) were chosen for extended queue time subjecting them to 2, 24 and 48 hour queue times in Fab ambient. These 2 cleans were also measured by SIMS with "no H_2 pre-bake" (Figure 2). Again, the 2 cleans were done, and after the various queue times expired, the wafers were loaded into the load lock and pumped down several times before transfer to the chamber. 2 different temperature H_2 pre-bake conditions were used (775°C & 800°C) at 2 different pressures (20 Torr and 600 Torr). The choice of temperature is based on the 775°C-800°C range being the lower limit for thermal H_2 reduction of the sub-oxide in RPCVD reactors. The pressure choice of 20 Torr is due to excessive Si migration on thin recessed Si (SOI) at pressures <20 Torr[3]. The 600 Torr pressure is used due to reports of increase sub-oxide removal efficiency with increased surface coverage of H_2 at higher pressure and this was a commonly used H_2 pre-bake by IBM epitaxy group. The particular commercial RPCVD epi system uses pyrometry for wafer temperature sensing and control. However, this is not a reliable indicator of actual wafer temperature at all conditions since the pyrometer is actually measuring the backside of the SiC coated susceptor. To get accurate wafer temperature at the pressure and H_2 flows utilized for the H_2 pre-bake an in-calibration(NIST) "SenseArray" instrumented wafer(P-300mm) was used. The temperature, as indicated by pyrometry, was offset to obtain the actual wafer temperature. After the H_2 pre-bake step the wafers are cooled to 650°C and received a Si capping layers under conditions described earlier.

3. DISCUSSION

The results in Figure 1 show O/C area density (no H_2 pre-bake) after approximately 2 hour queue time for 4 different commercial clean systems (wet/dry). The range is from 3E13 atom/cm^2 to >1E14 atom/cm^2. This data has been repeated in the past on numerous occasions and have found that the O/C area density to be repeatable between the various cleans. This data represents the actual sub-oxide contamination resulting from the clean with little or no contribution from the vacuum system for the Si capping layer. This is somewhat based on UHCVD data that reports 1/100th of a monolayer(~7E12 atom/cm^2) O immediately after an HF last wet clean[4] and

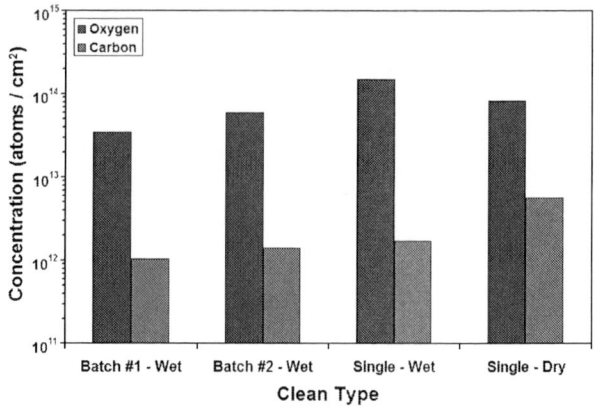

Figure 1: Wet & Dry cleans comparison, no thermal prebake, cold load, ~1500A Si deposition at 650°C. ~2 hour queue time for all cleans

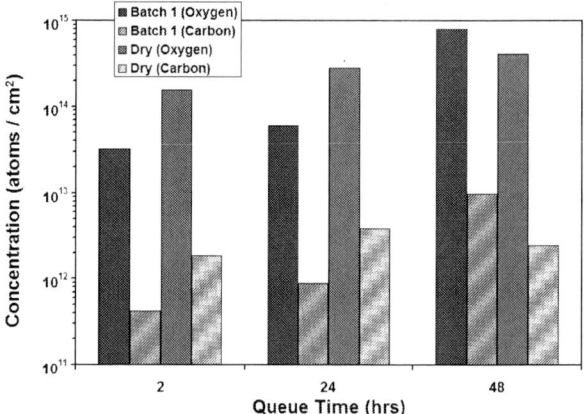

Figure 1: Interfacial oxygen and carbon vs. queue time for Batch 1 and Dry clean, no thermal prebake, cold load, ~1500A Si deposition at 650°C

similar 1/100th of a monolayer reported by RPCVD with no H_2 pre-bake [2]. Prior to this technique of evaluating the interfacial oxide (no H_2 pre-bake) of the various wet/dry cleans, the cleans were used interchangeably by IBM at ANT(Albany NanoTechnology) for epitaxy. Because of this wide variation in O/C area density that is wet/dry clean dependent, even 2 minute 850°C H_2 pre-bakes were sometimes not completely effective at removing the interfacial sub-oxide. This is easy to understand as O area density >1E14 atom/cm^2 take more H_2 pre-bake thermal budget to remove than a low E13 atom/cm^2[2].

Figure 2 shows the O/C area density(no H_2 pre-bake) after 2, 24 and 48 hour queue times in Fab ambient for the wet clean that consistently yielded the lowest initial O/C(Batch 1) and for the "Dry" clean system. For Batch 1, wet clean O area density ranged from 3E13 atom/cm^2 to 8E14 atom/cm^2 over the 48 hour queue time. For the Dry clean, the O ranged from 1.5E14 atom/cm^2 to 4E14 atom/cm^2. It is interesting to note that C is an order magnitude higher at the 2 hour queue for the Dry clean versus the wet clean. The higher C is quite characteristic of the Dry clean.

978-1-61284-408-4/11 $26.00 © 2011 IEEE

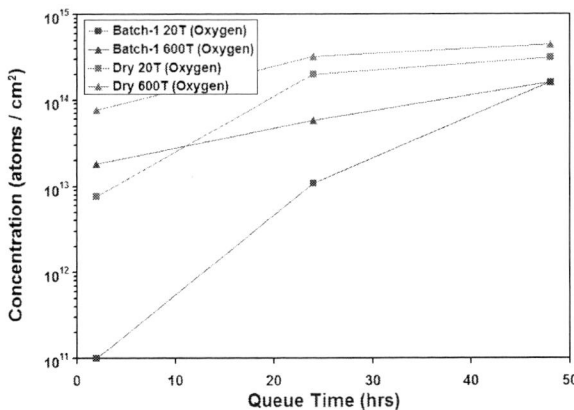

Figure 3: Interfacial oxygen vs. queue time for Batch 1 and Dry clean based on SIMS. Thermal bake at 775°C, 20 torr and 600 torr, with ~1500A Si deposition at 650°C.

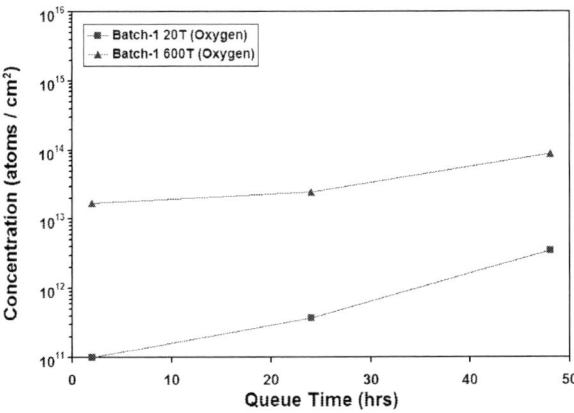

Figure 4: Interfacial oxygen vs. queue time for Batch 1 based on SIMS. Thermal bake at 800°C, 20 torr and 600 torr, with ~1500A Si deposition at 650°C.

Figure 3& 4 demonstrates the combined efficiency of both the choice of H_2 pre-bake temperature and total pressure to remove the wet/dry cleans remnant sub-oxide and oxide regrowth at the 3 queue times. At 775 °C, 20 Torr pre-bakes proved to be much more efficient than the 600 Torr especially at the 2 hour queue time. Some researchers have shown this effect in a different commercial RPCVD reactor[5]. At 48 hours queue, for both cleans there is approximately a monolayer oxide reformed (Figure 2). Once ~monolayer is formed there is not much difference in the sub-oxide removal efficiency between the 20 and 600 Torr pressure H_2 pre-bake (Figure 3). Figure 5 demonstrates the efficiency of a 750°C H_2 pre-bake for 2: duration. The initial area density for this clean at 2 hours queue is 3E13 atoms/cm^2 and after the bake the sub-oxide is reduced approximately an order of magnitude to 2E12 atoms/cm^2.

4. CONCLUSION

We have found a non-surface technique which accurately quantifies the remnant sub-oxide after wet HF last and dry clean methods. By identifying the commercial clean that repeatably delivers the lowest initial O area density and

combining with the 20 Torr pre-bake (nominal 600 Torr) we have been able to lower the epitaxial H_2 pre-bake temperature approximately 75°C at IBM ANT. Queue times up to 24 hours yield acceptable interfacial (<1E12 atoms cm^2) at 800°C

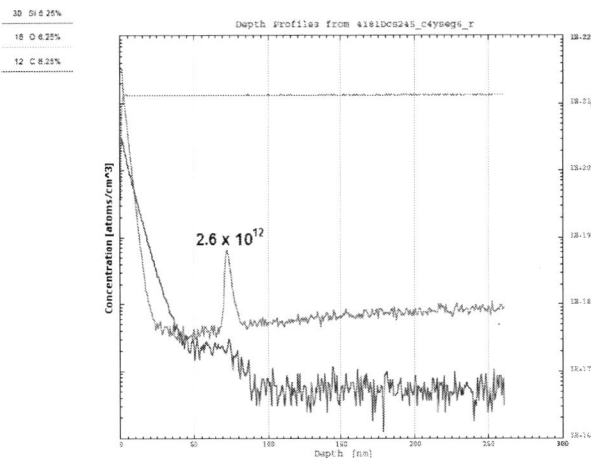

Figure 5: Interfacial oxygen with Batch 1 clean (2 hour queue) based on SIMS. 2: Thermal bake at 750°C at 10 torr,, with ~1500A Si deposition at 650°C.

[1] D. Gräf, M. Grunder, R. shultz and L. Mühlhoff, "Oxidation of HF-treated Si wafer surfaces in air" Journal of Applied Physics vol.68 pp. 5156-5161, 1990

[2] P.Brabant et al. , "Hydrogen termination for extended queue times for Low Temperature Epitaxy, Applied Surface Science vol. 255 pp. 1741-1743 2008

[3] C. Jahan, O. Fagnot, L. Totsi, and J.M. Hartmann, "Agglomeration control during the SEG of Si RSD on Ultra thin SOI substrates" Journal of Crystal Growth vol. 280 pp. 530-538, 2005

[4] B.S. Meyerson, "Low temperature Si and SiGe epitaxy by UHV-CVD:Process Fundamentals", IBM Journal of Research Development vol. 44. No.

[5] F.E.Leys et al., "Low Temperature Epitaxy and the importance of moisture control Thin Solid Films vol. 517 pp.416-418 2008

Optimization of Pitch-Split Double Patterning Photoresist for Applications at the 16nm Node

Steven J. Holmes[1]*, Cherry Tang[3]*, Sean Burns[2], Yunpeng Yin[1], Rex Chen[4],
Chiew-seng Koay[1], Sumanth Kini[5], Hideyuki Tomizawa[6], Shyng-Tsong Chen[1],
Nicolette Fender[3], Brian Osborn[3], Lovejeet Singh[3], Karen Petrillo[1], Guillaume Landie[1],
Scott Halle[1], Sen Liu[4]
John C. Arnold[2], Terry Spooner[2], Rao Varanasi[4] and Mark Slezak[3], Matthew Colburn[2]
Shannon Dunn[7], David Hetzer[8], Shinichiro Kawakami[8], Jason Cantone[7]

[1] IBM Systems &Technology Group at Albany Nanotech, 257 Fuller Rd., Albany, NY 12203; [2] IBM
Research Division at Albany Nanotech, 257 Fuller Rd., Albany, NY 12203; [4] IBM Systems &Technology Group,
2070 Route 52, East Fishkill, NY 12533; [3] JSR Micro, Inc, 1280 North Mathilda Ave., Sunnyvale, CA 94089; [5]KLA-
Tencor, 257 Fuller Rd., STE 134, Albany, NY, USA 12203; [6]Toshiba, 255 Fuller Rd., Albany, NY, USA 12203;
[7]Tokyo Electron America, Inc., 255 Fuller Rd., STE 244, Albany, NY, USA 12203; [8]Tokyo Electron Technology Center,
America, LLC., 255 Fuller Rd., STE 244, Albany, NY, USA 12203

ABSTRACT

Pitch-split resist materials have been
developed for the fabrication of sub-74 nm pitch
semiconductor devices. A thermal cure method is used
to enable patterning of a second layer of resist over the
initially formed layer. Process window, critical
dimension uniformity, defectivity and integration with
fabricator applications have been explored. A tone
inversion process has been developed to enable the
application of pitch split to dark field applications in
addition to standard bright field applications.

Keywords: double patterning, pitch-split, thermal cure,
tone inversion, 16 nm node lithography

1. INTRODUCTION

Pitch-Split lithography techniques have been
identified as a possible method to continue scaling of
semiconductor devices beyond 74 nm pitch applications
[1-4]. We have previously characterized some initial
pitch-split resist materials utilizing both chemical freeze
and thermal cure process options and demonstrated
their use at 64 nm effective pitch [5,6]. Over the past
year, we have further refined the thermal cure materials
to enhance focus/expose process latitude, resist profile,
resist height, lithographic defectivity, and stability of
both first and second layer resist materials. We have
also extended the resolution capability of the materials
to 56 nm effective pitch applications for line/space and
via chain structures. Exceptional defectivity results
have been obtained that are similar to those for
conventional single-layer resist patterning methods. We

have optimized pitch split lithography for wiring
applications by developing and implementing a tone
inversion process, which enables the conductive wire to
be formed from the pitch split resist line. The resist line
has better CD uniformity than the resist space due to
overlay affects on the uniformity of the space. The tone
inversion process has been implemented on a 16 nm
node test site to produce electrical test macros that
include three layers of wiring. A companion paper,
related to the optimization of the lithographic exposure
and track tools for pitch split patterning has also been
recently published[7].

2. EXPERIMENTAL

2.1 Thermal Cure Process Flow

The pitch split process utilizes two separate
resists to form the final pattern. The first resist is coated
on a hard mask stack composed of a silicon-containing
anti-reflective coating (SiARC) at 35 nm thickness over
an organic underlayer at 100 nm thickness. The organic
underlayer is also an ARC, and the dual stack serves to
provide optimum reflectivity control at high numerical
aperture immersion ArF exposure conditions. For 64
nm effective pitch applications, the first resist layer is
exposed at a pitch of 128 nm and a CD target of 32 nm.
After resist develop, the first layer resist pattern is
baked to stabilize the pattern during the resist coat,
expose, and develop processes for the second layer of
the pitch split feature. The second resist layer can be
coated directly over the first layer providing simplified
processing and low cost capability for the dual expose
process needed to form the 64 nm pitch structures. This

978-1-61284-408-4/11 $26.00 © 2011 IEEE

process flow is shown schematically in Figure 1, and a SEM cross-section of the resulting resist pattern is shown in Figure 2.

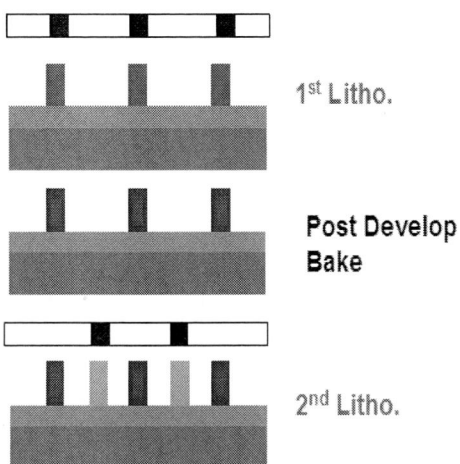

Figure 1. Schematic of pitch split process with thermal cure to stabilize layer 1 during the layer 2 apply process.

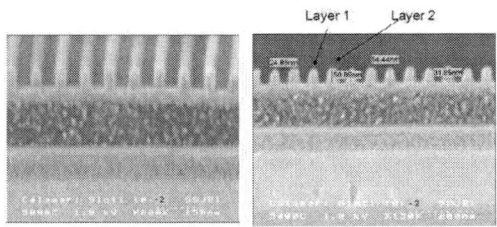

Figure 2. SEM cross-section of pitch split pattern at 64 nm pitch.

The process windows for layer 1 and layer 2 resist materials are shown in Figures 3 and 4 at 64 nm and 56 nm effective pitch. In generating this data, we exposed the full pitch split pattern, while using a single focus/expose condition for layer 2 in order to generate the process window data for layer 1. Similarly, we used a single focus/expose condition for layer 1 for the characterization of the process window of layer 2. With the high NA dipole illumination used for this application, large focus process windows can be achieved, on the order of 300 nm, while expose latitude values of 10-15% can be achieved.

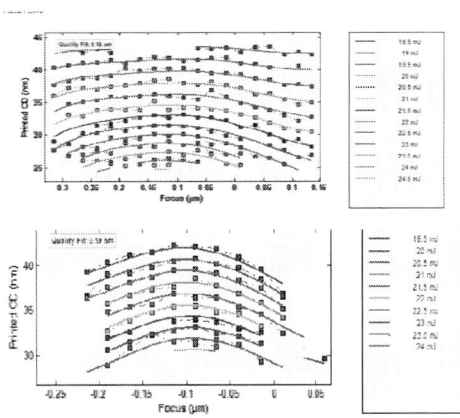

Figure 3. Focus/expose process window for layer 1 (left) and layer 2 (right) at 64 nm effective pitch, measured on wafers with a full pitch split pattern, in which layer 2 is processed at a single focus/expose condition. Dipole illumination and a binary mask were used. At 5% expose latitude, the focus latitude is 0.45 um for layer 1 and 0.29 um for layer 2.

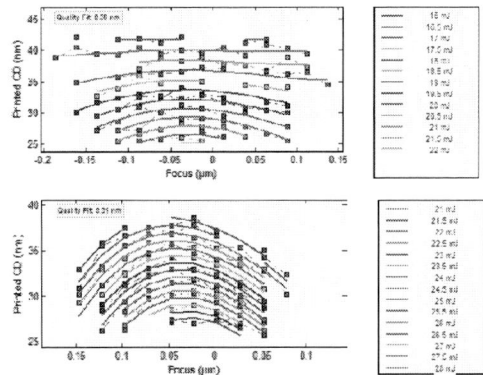

Figure 4. . Focus/expose process window for layer 1 (left) and layer 2 (right) at 56 nm effective pitch, measured on wafers with a full pitch split pattern, in which layer 2 is processed at a single focus/expose condition. Dipole illumination and a binary mask were used. At 5% expose latitude, the focus latitude is 0.27 um for layer 1 and 0.19 um for layer 2.

2.2 CDU Characterization for Multiple Jobs and Track Modules

In order to have confidence in the manufacturability of the pitch split process, it is necessary to evaluate the CD control across multiple wafers, multiple jobs, using a full track configuration that utilizes multiple bake and develop stations. IBM and its partners have made an extensive effort to optimize the exposure tool and track process configuration for the pitch split technique aimed at extending optical lithography to sub-74 nm pitch values [7]. The CDU values have been maintained

978-1-61284-408-4/11 $26.00 © 2011 IEEE

at approximately 1 nm 3 sigma distribution on a 32 nm CD target with consistent results across successive product batches. This enables us to easily achieve our 10% CDU budget for the 16 nm node wiring applications. For this CD characterization, three wafers were processed through each PEB / HB combination for a total of 9 wafers per lot. 18 lots were processed after CDU optimization during the stability check period of close to 2 months. The mean CD after optimization was stable and varied only slightly around 32.00 nm. There was a distinct rise in L1 mean CD of approximately 0.6 nm at lot 10 which continued through lot 13. It was determined that the overlay had shifted on those lots causing the CD shift. The overlay was adjusted for lots 14 through 18. CDU varied only slightly around 1.00 nm. Figure 6 summarizes the CDU monitoring data.

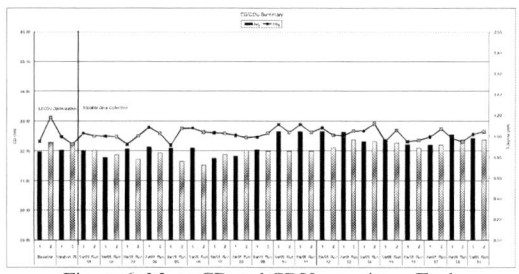

Figure 6: Mean CD and CDU over time. Each bar/point represents the average of nine wafers.

The CDU data up to this point has shown L1 separate from L2. This is practical for the demonstration of the optimization; however, the reality is that the 32 nm 1:1 final structure needs to be treated as one continuous system. When the mean CD of L1 and L2 are sufficiently close, the datasets can be pooled together, to achieve a single mean CD and CDU for the wafer. Pooling of the data was done on a multi-module run of nine wafers (three wafers per PEB plate). The mean CD for all wafers was 32.13 nm and the CDU was 1.03 nm (3 sigma). The variance of the means of those 9 wafers, (i.e., wafer-to-wafer uniformity) was 0.21 nm 3 sigma. Table 1 summarizes the results of that test.

Table 1. Multi-module flow pooled data for nine wafers.

Slot	Mean	3 Sigma
01	32.06	0.93
02	32.10	0.99
03	32.01	0.92
04	32.18	1.14
05	32.18	1.09
06	32.10	0.93
07	32.18	1.10
08	32.22	1.19
09	32.18	0.99
Average	**32.13**	**1.03**
3 Sigma	**0.21**	

2.3 Defect Characterization and Reduction

Remarkable results have been achieved on defectivity with results essentially equivalent to single layer resist processing. Resist line collapse is perhaps the major source for process-related defects for the pitch split lithography process. The occurrence of line collapse is also strongly dependent on the overlay performance of layer 2 to layer 1 resist patterns due to the surface tension forces which can occur during the develop process. Unequal surface tension forces can be caused by variations in the resist space between layer 1 and layer 2 resulting in an increased tendency for line collapse. Optimization of the exposure tool, in combination with optimization of the resist develop process, has enabled consistently low levels of line collapse defects to be achieved.

Process improvements related to defectivity optimization included PEB time and temperature and resist develop for both L1 and L2. It was also determined that the placement of the L2 pattern with respect to the L1 pattern has a significant impact on defectivity. This placement is not the same as the overlay accuracy, which has been shown to be excellent[7]. The process margin for L1 to L2 placement related pattern collapse was determined by manually perturbing the placement until the lines collapsed. Defectivity was reduced from >4500 defects to ~130 defects by optimizing pattern placement.

Another key implementation to reduce defectivity was the addition of surfactinated rinse to the develop process. By adding surfactant to the rinse to the placement based process window (related to pattern collapse) increased significantly from 2-3 nm to over 6 nm. Figure 8 is the pareto for the POR develop process without and with the surfactinated rinse. The rinse reduces collapsed pattern (CP and CP_T) defectivity by 99.9%. The final total defectivity, excluding non-visual (NV) defects, was 15.5 defects (average for two wafers).

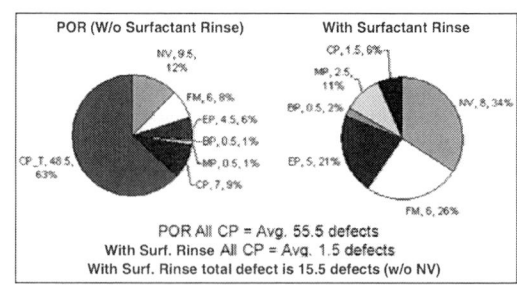

Figure 8: Pareto results showing collapsed pattern reduction when using the surfactant rinse.

After the optimization, the same production-like wafers were used to generate long term CDU data and to collect long term defectivity data. A 9-wafer lot was cycled 20 times through the DETO process, and defectivity data was collected at each cycle, over a duration of three months. After the first nine cycles, it was found that multiple reworks on the same wafers were adding to the defectivity over time. The base level of defectivity on the wafers post rework was confirmed to be increasing. When the lot was replaced with new wafers, the defectivity immediately dropped to similar levels previously observed on new wafers. The defect density for all 20 lots was 0.11 defects/cm^2. The average defect density for new wafers was 0.08 defects/cm^2. Figure 9 summarizes this defectivity results.

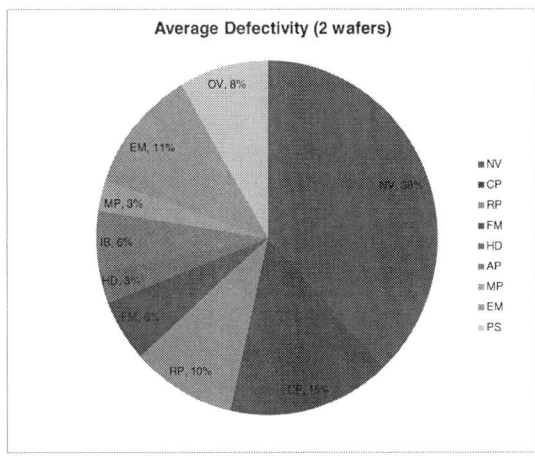

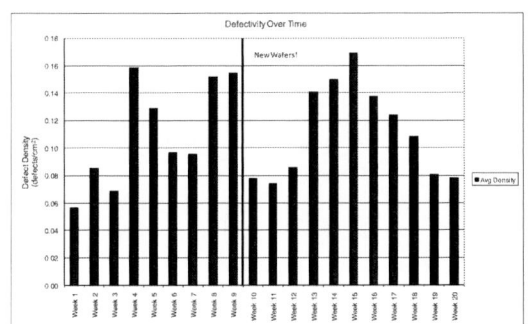

Figure 9: DP defectivity over time. Each line/point represents an average value for a nine wafer lot.

The defectivity level reported here for the 32 nm 1:1 DETO is similar to production defectivity values of a 45 nm 1:1 single layer processes. The next step was to determine exactly what defects were remaining. Figure 10 summarizes the proportions of typical defects remaining on the wafer as well as example images of those defects.

NV : Non Visual
CP : Collapse Pattern
RP : Residue Polymer
FM : Foreign Material
HD : Hole Defect
AP : Attenuated Pattern
MP : Missing Pattern
EM : Embedded Material/FM
PS : Pattern Shift

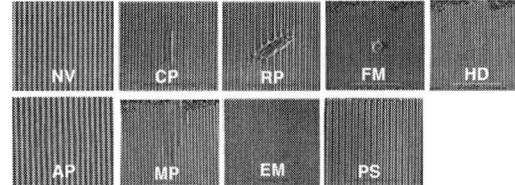

Figure 10: Proportions of defect types

2.4 Tone Inversion Process

For some applications, such as active area and gate structures, the pitch split pattern can be used directly to form the desired features. For other applications, such as wiring structures, it may be desirable to invert the tone of the pattern in order to optimize the yield and performance of the semiconductor device due to the role that overlay between layer 1 and layer 2 can have in the CD distribution of the resist spaces. For the resist spaces, the CDU is affected both by the uniformity of the imaging process as well as by the overlay capability, whereas in the case for the resist line, the CDU is primarily affected by the uniformity of the imaging process alone. This result is shown graphically in Figure 11. For wiring applications, it is believed that the performance of the chip would be optimized if the CDU of the conductive wire is optimized.

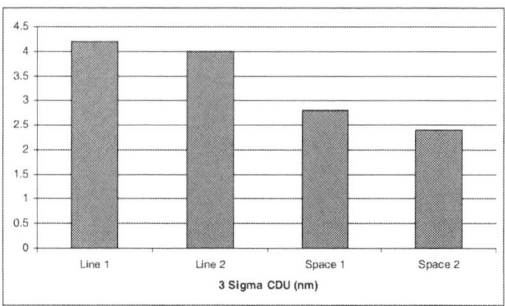

Figure 11. CDU of lithographic line and space after tone inversion for pitch split lithography at 64 nm pitch. Overlay effects degrade the CDU of the line.

Accordingly, we have developed a tone inversion process (shown in Figure 12) that enables us to convert the resist lines into spaces, and the resist spaces into lines. In this process flow, we simplify the hard mask stack at resist apply by omitting the SiARC layer. The resist layers 1 and 2 are coated directly onto the organic underlayer ARC. After the pitch split resist pattern is formed, a silicon-containing hard mask is then coated over the combined resist pattern and etched back with RIE to expose the resist surface. The RIE chemistry is then altered to remove the resist materials and etch down through the organic underlayer with the planarizing silicon-containing layer as a hard mask. We believe that this track-based tone inversion process offers a low cost option for converting the pitch split resist pattern into a mask suitable for wiring applications.

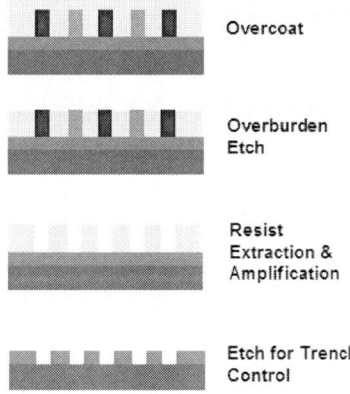

Figure 12. Schematic of tone inversion process on pitch split resist pattern, using an overcoat material which has selective etch properties relative to the resist.

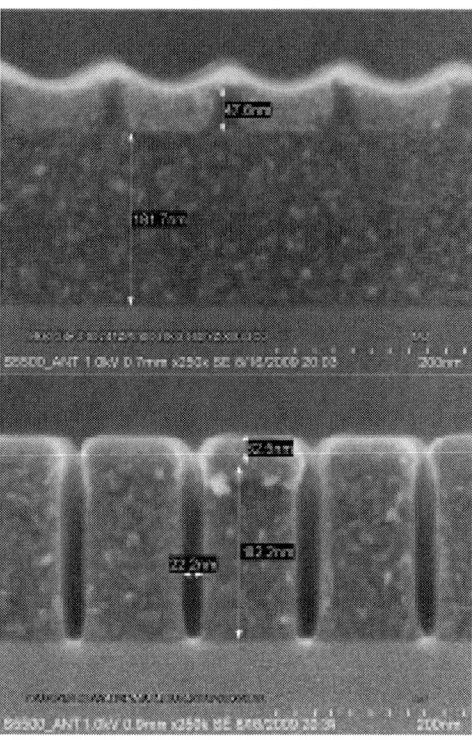

Figure 13. SEM cross-sections of pitch split resist with tone inversion material applied (left), and after RIE image transfer into organic underlayer (right).

Figure 13 shows SEM cross-section data for this tone inversion process: on the left is shown the resist pattern, over an organic planarizing layer (OPL), with the tone inversion material coated over the resist. On the left, the same pattern is shown after RIE etch-back of the tone inversion material to expose the surface of the resist, and then RIE of the resist and OPL to create the final etch mask. The tone inversion process provides an effective method of generating high aspect ratio and high resolution patterns with the desired feature polarity. Figure 14 shows some top-down SEM images of a) the pitch split resist pattern for a via chain structure, and b) the pattern achieved after tone inversion. We have used this 64 nm pitch via chain macro, as well as other electrical macros, to implement and optimize the tone inversion process in a pilot line.

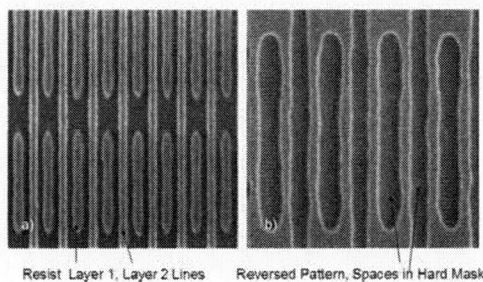

Resist Layer 1, Layer 2 Lines   Reversed Pattern, Spaces in Hard Mask

Figure 14. Top-down SEM view of Via Chain macro a) resist image prior to tone inversion b) hard mask pattern after tone inversion.

2.5 Tone Inversion Materials

In the development of this process, in order to provide an adequate process window for the RIE sequence, it became critical to optimize the planarization properties of the tone inversion hard mask. To quantify the planarization, we measured the thickness of the image reversal material within the resist space and subtracted the thickness of the material over the resist line. Our RIE process must be able to remove the material over the resist line while leaving sufficient material in the resist space to serve as a hard mask for the RIE of the organic underlayer. The design of the tone inversion polymer formulation is also very critical. In our case, a siloxane polymer material is used. The siloxane polymer has sufficient silicon content to provide high etch resistance in an oxygen RIE (which is used for resist and OPL RIE) and be formulated with a solvent which will not cause inter-mixing between the resist layers 1 and 2 during the tone inversion material apply process.

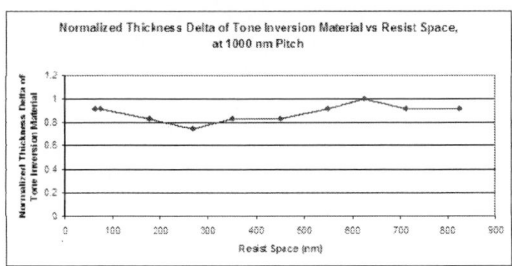

Figure 15. Tone inversion planarization capability after process optimization completed..

The extent of the planarization can be affected by a number of parameters, and after optimization, a uniform planarization can be achieved across a wide range of feature types. Figure 15 shows a consistent difference in thickness of the reversal layer across a resist linewidth varying between 200 and 800 nm for a feature pitch of 1000 nm.

2.6 Tone Inversion Process CDU and Yield Results

An example of the across-wafer CDU achieved after full dielectric etch with the tone inversion pitch split process at 64 nm pitch is shown in Figure 16. As expected, due to the overlay component affecting the resulting dielectric line, the CDU of the dielectric space is superior. The tone inversion process can successfully maintain the CDU of the initial resist line as it is converted into a space and transferred into the product dielectric stack.

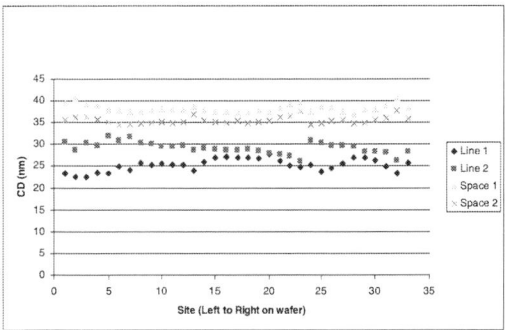

Figure 16. Across-Wafer CDU for pitch split lithography at 64 nm pitch after tone inversion and RIE into dielectric stack. The CDU of the line is degraded by overlay effects, while the CDU of the space is not affected by overlay.

Figures 17 and 18 illustrate the pitch split tone inversion process on a typical electrical wiring macro on our 16 nm node test site. In 17a and 17b we can observe the layer 1 and layer 2 resist patterns separately, while 17c shows the macro structure with both layers in place at the 64 nm effective pitch. Figure 18a shows the pattern after tone inversion RIE into the dielectric hard mask, while 18b shows the pattern after the full dielectric etch. The patterns appear well formed, with low line edge roughness and an absence of shorts and opens. While the array is patterned at narrow pitch, some of the larger features associated with wiring to the probe pads are also visible and can be successfully transformed in the tone inversion process.

978-1-61284-408-4/11 $26.00 © 2011 IEEE

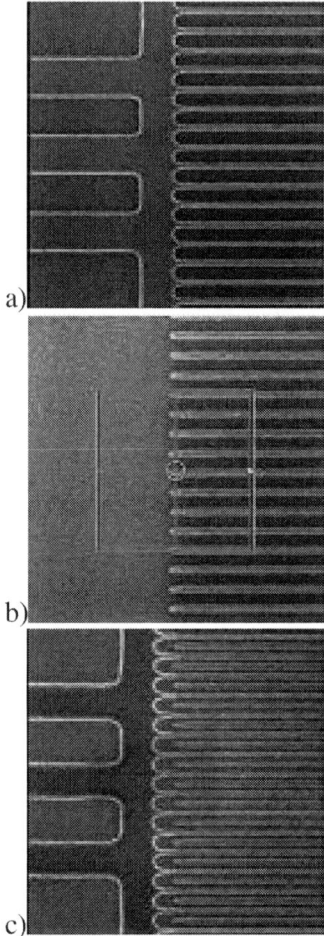

a)

b)

c)

Figure 17. Pitch split lithography for electrical device macro shown at a) layer 1 resist b) layer 2 resist c) layer 1 and layer 2 combined at 64 nm pitch.

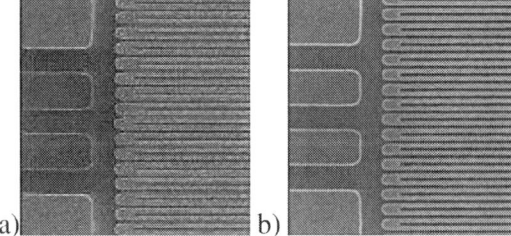

a)   b)

Figure 18. Pitch split lithography for electrical device macro a) after tone inversion and RIE into hard mask b) after tone inversion and full RIE into dielectric.

The electrical shorts and opens for line/space and isolated via structures have been characterized across a range of test site batch jobs in our pilot line application

with consistently high yields at the 64 nm pitch ground rules. Figure 19 summarizes some of this data, with yields typically in excess of 95%.

TEM cross-section data for the electrical macros is also shown in Figure 19. The line/space shorts/opens macro wiring structure for three layers of metallization is shown with each layer being formed by the pitch split tone inversion process.

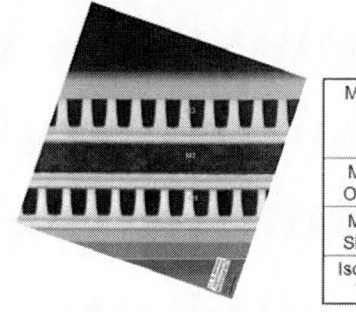

Macro	Electrical Yield
Maze Opens	97%
Maze Shorts	~100%
Isolated Via	~100%

Figure 19. TEM of device macro, showing three layers of wiring, fabricated with a pitch split tone inversion process at 64 nm pitch.

Results indicate that the pitch split process can be extended to 56 nm pitch applications. Figures 20 and 21 compare the pitch split patterns, for 64 nm and 56 nm effective pitch, at lithography and after RIE with tone inversion. The patterns are well-formed at both conditions.

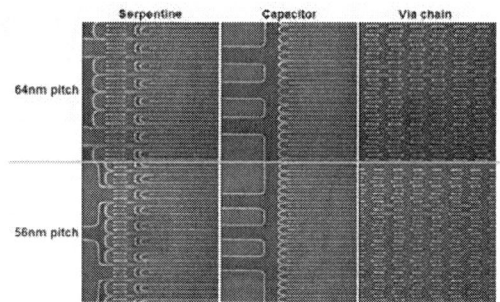

Figure 20. Pitch split resist patterns at 64 and 56 nm pitch for electrical test macros.

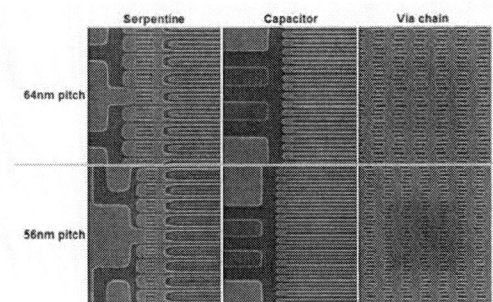

Figure 21. Pitch split patterns for electrical test macros after tone inversion and dielectric RIE at 64 and 56 nm pitch.

A yield comparison for the 56 nm and 64 nm pitch serpentine electrical test macros was performed, with values of 100% yield for electrical shorts for both 64 nm and 56 nm pitch. The opens yield was 87.5 % for the 64 nm pitch, and 85% for the 56 nm pitch. Very little yield degradation is observed for the denser pitch relative to the more relaxed ground rules indicating that the pitch split technique may be extended further. .

3. SUMMARY AND CONCLUSION

Pitch-split resist materials have been developed and implemented for the fabrication of semiconductor devices at 64-nm pitch ground rules. The lithography process has been designed for relative process simplicity utilizing track-only processing for the pitch split process. Integration with the track and exposure tool have been characterized providing excellent defectivity, CDU and overlay capability. While the pitch split process may be directly applicable to bright-field applications, such as active area and gate structures, a track-only tone inversion process has also been developed and implemented for dark field applications with high yield on device macros at 64 nm and 56 nm pitch.

ACKNOWLEDGEMENTS

This work was performed by the Research Alliance Teams at various IBM Research and Development Facilities. The authors would like to acknowledge the TEL Albany team especially Shinichiro Kawakami, Shannon Dunn, Jason Cantone and Dave Hetzer for their excellent work on defect optimization. Also, we would like to thank ASML Albany team, especially Youri Van Dommelen, Mike Manny and Aiqin Jiang, for their prompt support on overlay optimization. We are thankful to Curt Calamari for his great support on cross section sem.

REFERENCES

[1] S. Holmes, C-S. Koay, K. Petrillo, K-J. Chen, M. Colburn, J. Cantone, K. Ueda, A. Metz, S. Dunn, Y. Van Dommelen, M. Crouse, J. Galloway, E. Schmitt-Weaver, A. Jiang, R. Routh, C. Tang, M. Slezak, S. Kini, T. DiBiase, , " Engine for Characterization of defects, overlay, and critical dimension control for double exposure processes for advanced logic nodes" Proc. SPIE, Vol. 7273, 727305 (2009); DOI: 10.1117/12.828483

[2] A. Hazelton, S. Wakamoto, S. Hirukawa, M. McCallum, N. Magome, J. Ishikawa, C. Lapeyre, I. Guilmeau, S. Barnola, S. Gaugiran, " Double patterning requirements for optical lithography and prospects for optical extension without double patterning" Proc. SPIE, Vol. 6924, 69240R (2008); DOI:10.1117/12.771914

[3] G. Wakamatsu, Y. Anno, M. Hori, T. Kakizawa, M. Mita, K. Hoshiko, T. Shioya, K. Fujiwara, S. Kusumoto, Y. Yamaguchi, T. Shimokawa, "Double pattern process with freezing technique" Proc. SPIE, Vol. 7273, 72730B (2009); DOI:10.1117/12.814073

[4] M. Hori, T. Nagai, A. Nakamura, T. Abe, G. Wakamatsu, T. Kakizawa, Y. Anno, M. Sugiura, S. Kusumoto, Y. Yamaguchi, T. Shimokawa, "Sub-40 nm half-pitch double patterning with resist freezing process" Proc. SPIE, Vol. 6923, 69230H (2008); DOI: 10.1117/12.772403

[5] M. S. Holmes, C. Tang, J. Arnold, Y. Yin, R. Chen, N. Fender, B. Osborn, G. Dabbagh, S. Liu, M. Colburn, R. Varanasi, M. Slezak, "Process Characterization of pitch-split resist materials for application at 16nm node" Proc. SPIE, Vol. 7639 (2010); DOI: 10.1117/12.846891

[6] C-K. Koay, S. Holmes, K. Petrillo, M. Colburn, S. Burns, S. Dunn, J. Cantone, D. Hetzer, S. Kawakami, Y. Van Dommelen, A. Jiang, M. Many, R. Routh, L. Huli, B. Martinick, M. Rodgers, H. Tomizawa, S. Kini, "Evaluation of double-patterning techniques for advanced logic nodes" Proc. SPIE, Vol. 7640, (2010); DOI: 10.1117/12.846769

[7] C-K. Koay, S. Halle, S. Holmes, K. Petrillo, M. Colburn, S. Dunn, J. Cantone, D. Hetzer, S. Kawakami, Y. Van Dommelen, A. Jiang, M. Crouse, L. Huli, B. Martinick, M. Rodgers,, "Towards manufacturing of advanced logic devices by double patterning" Proc. SPIE,, (2011), in press.

Investigation of Noise Sources in the Focus Control Process for Immersion Lithography

Jasper Paul Munson, Jay Brown
Applications Engineering
Nikon Precision, INC.
Hillsboro, OR 97124
jmunson@nikon.com

Abstract— Immersion lithography is currently the industry standard for advanced semiconductor imaging. Critical to that imaging capability is adequate focus control across the wafer (or other substrate). The metrology technique called Phase Shift Focus Monitor, PSFM, is used extensively to evaluate across wafer focus control. The Nikon NSR-S610C immersion scanner utilizes an open loop Autofocus control system, meaning each exposure field's topography is measured, or "mapped", prior to exposure. This mapped data is used to move each exposure field into best focus as it is scanned. Focus control is fine-tuned by adding correction maps; for every location on the wafer a fixed focus correction is added. The S610C makes use of two focus correction map types: the first, referred to as a "System Map", optimizes repeatable focus variations unique to the individual scanner, while the second correction map, referred to as a "Recipe Map", optimizes focus variations unique to the product and layer being exposed. Both System and Recipe maps are evaluated using PSFM, and fine tuning is achieved with corrections based on PSFM results. The current method for fine tuning System and Recipe maps consists of PSFM exposure, followed by registration measurement, data evaluation, and correction map generation. This process is repeated until across-wafer focus error meets a target focus (Z) 3σ value. This focus (Z) 3σ target is dependant upon each chip manufacturers operating specifications.

This paper will demonstrate an analytical method developed for identifying and quantifying sources of noise in the Focus Correction Map process. The presence of these noise sources can act as limiting factors to focus control performance. The noise sources investigated in this study are: substrate variation, metrology system noise, and PSFM Rate targeting error. By quantifying the magnitude of noise from each component it is possible to prioritize efforts in order to achieve improvements in focus control. Identifying and minimizing noise sources should allow for improvements in Focus Correction Map accuracy and reductions in the optimization cycle time.

Keywords-immersion; lithograpy; focus; correction map; PSFM; substrate

I.   INTRODUCTION

Nikon immersion scanners make use of a tandem stage system comprised of a wafer exposure (WE) stage and a metrology stage (WM). The WM stage allows subsystem calibrations to be performed during wafer transfer operations. Due to the presence of the water body directly over the exposure field in immersion lithography, Nikon immersion scanners employ an open loop Autofocus (AF) system in which the wafer surface for each exposure field is mapped prior to exposure. To correct focus errors that occur in the open loop focus control system, across wafer focus correction maps are used. These correction maps target focus errors that are both unique to the scanner and unique to each process layout. With Focus Correction Maps applied it is possible to reduce across wafer focus error (Z) 3Sigma levels to below 20nm.

With the implementation of Focus Correction Maps as an optimization knob, additional sources of noise have been introduced to the scanner setup process. This paper will investigate some potential noise sources in the Focus Correction Map creation process and attempt to quantify what impact those noise sources have on the final across wafer focus budget. In the interest of achieving tighter focus control and reducing scanner setup time the noise sources with the largest focus error will be considered for possible improvement.

II.   NOISE EVALUATION IN FOCUS CONTROL

A.   Open-Loop Autofocus (AF)

The transition to immersion lithography required significant changes in how scanner Autofocus systems function. Previous generations of lithography scanners utilized real-time AF measurement; for each shot on a wafer Z focus data was gathered immediately before exposure, in a region very close to the leading edge (front) of the exposure slit. This focus data was used to calculate exposure stage trajectory for the current shot. Focus data was continuously measured and converted to trajectory settings across the wafer, with each region of the wafer measured only a single time.

Immersion necessitated the introduction of an open-loop AF design wherein the AF beams are moved away from the exposure slit, to reside outside of the water body region, and the AF measurement sequence is performed in advance of the exposure sequence. The Nikon NSR-S610C scanner utilizes two rows of AF sensor beams, one on either side of the water body relative to the scanning direction. As a wafer is stepped through exposures the AF system measures, or maps, future exposure areas. Due to AF sensor array size it is common for each exposure field to be measured multiple times prior to its exposure. The mapped focus result for each unit of areas on the wafer becomes a running average of all measurements

978-1-61284-408-4/11 $26.00 © 2011 IEEE

performed at that coordinate. Exposure stage trajectory for each exposure field is calculated from this mapped data.

Across wafer focus control is fine tuned by adding correction maps to the active AF data measured from each wafer. Systematic focus errors are divided into two categories. Across wafer signatures that are unique to an individual scanner are corrected with the "System Map". A signature that is unique to a given product layout is corrected with a "Recipe Map". The accuracy of both System and Recipe Maps is evaluated with Phase Shift Focus Monitor (PSFM) metrology.

The PSFM technique is used to evaluate across wafer focus uniformity in an open-loop AF system. PSFM metrology requires a calibration wafer, called the PSFM Rate, and a test wafer, simply called a PSFM wafer. PSFM metrology works by translating any change in Z during imaging as a printed shift in X and Y. Fig. 1 depicts an example PSFM structure as it is imaged with a near zero focus vs. imaging with a significant focus offset [1]. The PSFM Rate wafer is intentionally stepped thru focus across a reasonably large depth of focus. This wafer is measured on a registration tool, and from the result a rate constant is calculated: for a known focus change the PSFM image shifts by X amount. X and Y image shifts are calibrated separately. Once a PSFM Rate constant is established then across wafer focus uniformity can be evaluated. A traditional PSFM wafer is exposed with a constant focus offset. The wafer is measured on a registration tool, and the results are divided by the PSFM Rate constant to convert into units of focus. Now the PSFM result, in nm of focus, is evaluated for mean and across wafer uniformity, 3Sigma, as a function of all shots on the wafer, complete shots only, and edge shots.

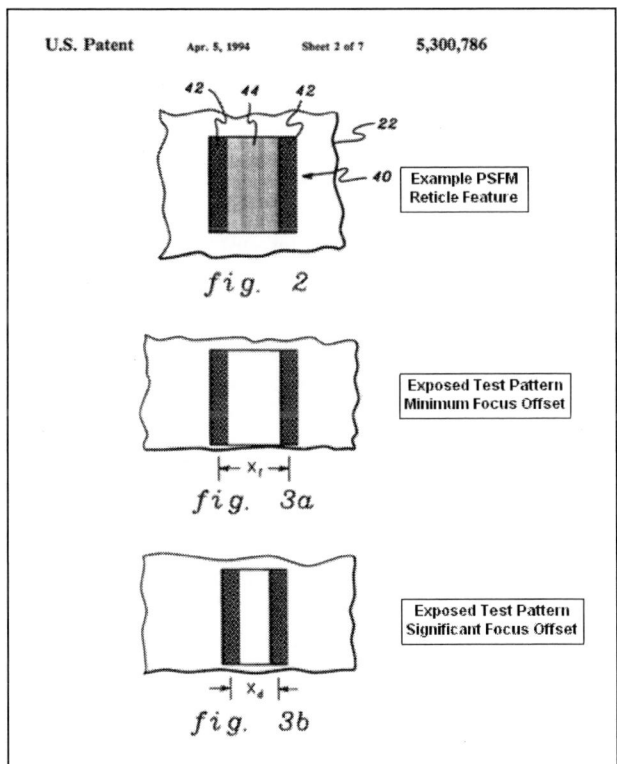

Figure 1. PSFM Principle of Operation

B. Investigation Goal

Tight focus control is necessary for end-users to successfully utilize immersion scanners for advanced imaging. For S610C generation scanners, a 100nm box in box feature is measured and success is evaluated based on across wafer 3Sigma focus variation in the PSFM result. It is possible to establish a baseline of 3Sigma performance, and then to investigate what options exist for reducing this baseline noise. This paper will investigate noise reduction in the Correction Map Process, which is primarily evaluated with PSFM. Analytical methods utilized for the investigation will be described. A second motivation for this investigation targeted the setup time necessary for qualifying new focus correction maps. If noise sources can be identified and minimized, then the iterative map optimization process can be shortened.

C. The Correction Map Process

As stated previously the Nikon S610C scanner utilizes two types of focus correction maps; System Maps and Recipe Maps. Both correction map types use a common creation methodology.

1) Correction Map Creation Process

The initial step in correction map creation, System or Recipe, is to perform a full wafer mapping with the AF system. This AF mapping, called "Foundation Map" for the System and "Auto-Map" for the Recipe, is the base level focus correction information included in the System and Recipe Maps. This AF System mapping is necessary to capture high frequency variations induced by the exposure stage stepping motion. The second step in correction map creation is to expose a PSFM print with the System, or Recipe, map applied. The wafer is measured on a registration tool and then across wafer focus uniformity is evaluated. If necessary an additional correction is created and added to the existing map. The effectiveness of the correction map is then confirmed with another PSFM print. The correct and evaluate sequence is repeated in this manner until a System or Recipe map PSFM result meets end-user focus control specifications. Typically specifications are based on PSFM performance with average dynamic shot error removed, and then itemized for all shots, full shots only, and edge shots only 3Sigma. For comparison purposes in this paper a target value of All Shots 3Sigma less than 20nm will be used.

2) Sources of Noise in Correction Map Creation

Within the Correction Map setup process it is possible to have systematic and non-systematic noise introduced from sources that include: PSFM and PSFM Rate targeting, Registration Metrology non-repeatability, and wafer substrate variation. These noise sources, if not corrected, can limit how low residual focus error can be driven, and the optimized focus correction map will take additional iterations to achieve.

a) PSFM and PSFM Rate Targeting

PSFM as a metrology must be setup correctly for the output results to be accurate and meaningful. For a given substrate and resist process it is necessary to confirm two elements of PSFM: 1) what is the best dose for PSFM, and 2) over what focus range is the PSFM Rate linear?

Best Dose selection for PSFM is evaluated in the following manner: exposure dose is stepped from low to high, and at each dose PSFM shift is measured across a wide range of focus steps. Several criteria exist for defining Best Dose:

- Maximum Depth of Focus, DOF

- Maximum linearity, R^2, across the full DOF

- Maximum slope of PSFM shift per focus change, PSFM Rate

Fig. 2 displays typical results from a PSFM test print for dose targeting. A dose of 25.0mJ yields ideal results for DOF, linearity thru focus, and PSFM Rate. At dose conditions below Best Dose the DOF may be larger, but the linearity thru focus will be degraded. At dose conditions above Best Dose the DOF will be reduced due to pattern collapse at the extreme ends of focus.

How can PSFM dose targeting be evaluated as a source of noise in the Focus Correction Map process? Dose control accuracy on current systems is very good, for most immersion scanners delivered dose has less than 1% error from commanded. Using 25mJ as our Best Dose for PSFM a 1% error in dose would be 0.25mJ. The impact from this variation would be minimal to DOF, linearity thru focus, and Rate. However, if exposure dose is chosen incorrectly, possibly 10% off of Best Dose, the situation is very different. In this case, the PSFM wafer would be exposed at 22.5mJ, or 27.5mJ, and then the 1% dose delivery error is applied. This results in possible delivered dose of 22.275mJ, or 27.775mJ respectively.

The risk of PSFM error from standard dose delivery variation, approximately 1% error, is small. However, if Best Dose selection is done incorrectly, possibly 10% off from ideal, an effect is noticeable in PSFM results. Table I. displays the PSFM Calibration Metric results based on a dose variation of +/- 10%. In an under-dosed condition the PSFM shift response thru focus is not perfectly linear, meaning large spikes in focus will be misinterpreted; they will be falsely diminished due to a flattening in the PSFM Rate. In an over-dosed condition the average PSFM Rate drops making all calculated focus results inflated.

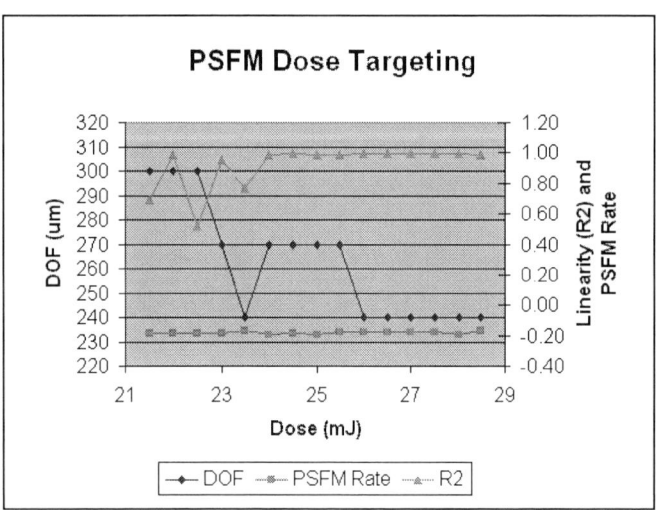

Figure 2.   PSFM dose targetting metrics: DOF, PSFM Rate, & Linearity (R^2)

TABLE I.   PSFM CALIBRATION METRICS VS DOSE

Dose (mJ)	PSFM Calibration Metrics		
	DOF (nm)	PSFM Rate	Linearity R^2
22.5	300	-0.187	0.620
25.0	270	-0.191	0.987
27.5	240	-0.181	0.993

The DOF is also reduced and could cause feature collapse at any locations with large spikes in focus. With gaps in measurement data the accuracy of output focus correction maps is reduced.

Accuracy in PSFM Rate targeting is equally important for preventing noise from being added to the focus uniformity result. A PSFM Rate wafer is exposed with a constant dose while each shot is stepped thru focus. Multiple points per exposure field are measured for evaluation. Each (x,y) field location will have PSFM shift results through focus, thus a PSFM Rate value can be calculated from each measurement point. The following criteria are applied to all intra-field points during evaluation to confirm if the PSFM Rate is reliable:

- Average Z Offset within +/- 30nm

- Linearity $R^2 \geq 0.980$

- PSFM Rate variation between measurement points $\leq$ +/- 0.010

Once a PSFM Rate result is confirmed to meet all of the described criteria all field points are pooled together to create independent average X and average Y Rate constants. The single X Rate constant is used to convert all field points in a PSFM wafer evaluation, with the same being true for the Y Rate constant. The method of using an average X and Y Rate constant was chosen to minimize the impact of any shot to shot variation in the PSFM Rate measurement data.

PSFM Rate inaccuracy can have a direct impact on calculated PSFM across wafer uniformity, and thus impact subsequent Focus Correction Map accuracy. A common procedure during new scanner implementation is to collect fresh PSFM Rate data with every PSFM wafer exposure. With this frequency of PSFM Rate data it is common to see the Rate constant vary +/- 5%. Fig. 3 displays how PSFM 3Sigma results change as PSFM Rate is varied. There is a direct relationship between PSFM Rate error and PSFM results 3Sigma.

If the Rate error is 10% then the calculated PSFM 3Sigma will be in error by 10% as well. For a coarse PSFM result, Sample1, a 10% variation in Rate changes PSFM 3Sigma by >5nm. For an optimized PSFM result, Sample2, a 10% Rate variation changes PSFM 3Sigma by < 2nm. This equates to PSFM Rate error, causing 1.93nm of PSFM All Shot 3Sigma error, consuming 9.65% of the 20nm PSFM target.

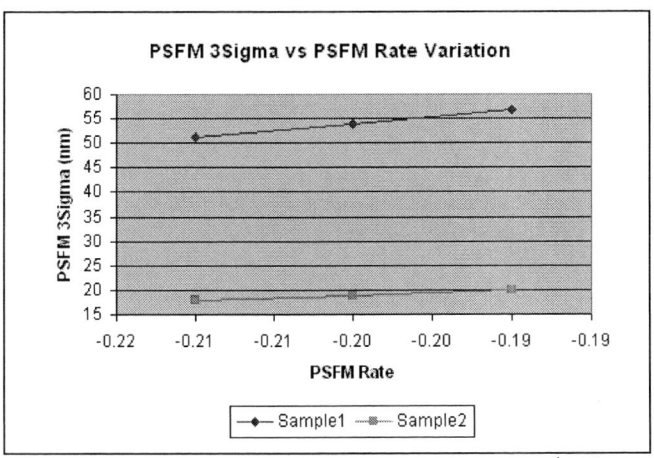

Figure 3. PSFM Rate variation impact on PSFM 3Sigma result

b) Registration Metrology

Using PSFM as a means of evaluating focus across wafer and generating Focus Correction Maps relies upon the accuracy and repeatability of registration metrology tools. Any noise in the registration measurement will be added directly to the PSFM result, assumed to be focus variation, and then corrected for in the Focus Correction Map. This reduces the Focus Correction Map accuracy.

For the purpose of this paper a Registration repeatability investigation was conducted. The goal was to quantify what repeatability was currently present, and how that impacted PSFM results. Two registration measurement tools were evaluated. A System Map PSFM wafer was exposed on a scanner that already had an optimized System Map. This PSFM wafer was measured three times on each Registration system. Initial analysis was performed on raw Registration data with the following manipulations performed on the data: all measurements were converted to units of focus (nm) by dividing X and Y shift results by their respective PSFM Rate constants, for each unique measurement position a max-min, Range, was calculated from the three wafer sample, and from the population of Range data a Distribution and Cumulative Distribution Function, CDF, analysis was performed. Calculating standard deviation from three data points was not statistically reliable, so the Range metric was chosen. Distribution and CDF analysis were chosen to determine whether X & Y results matched and if the Registration systems matched.

Fig. 4 shows the Distribution and CDF analysis from Registration system #2 in our study. The X and Y Focus Range distribution plots are both non-normal in shape. This is reasonable considering the nature of the input data; it is impossible to have a negative measurement error between repeat measurements. This population of Focus Range data closely matches a Gamma distribution. A visual analysis of Fig. 4 reveals that X and Y Focus Range are not matched. Through an evaluation of CDF this can be quantified: For X Range the 98th percentile falls on 2.975nm of focus, for Y Range it falls on 4.075nm.

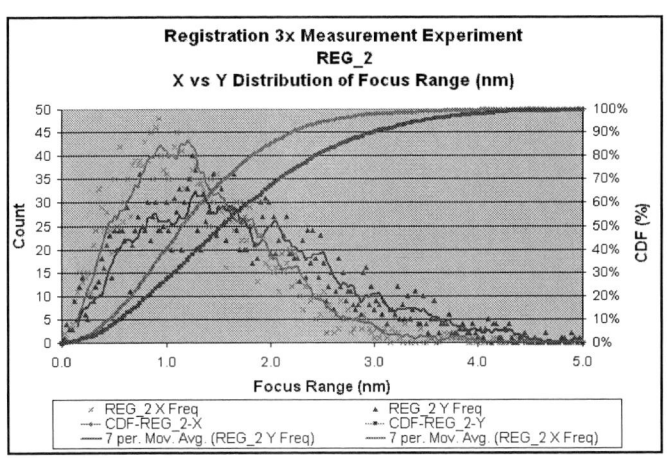

Figure 4. Distribution analysis of registration measurement repeatability converted to Focus (nm)

Putting this into terms of focus measurement error, assuming 98% probability, the X focus measurement will have 2.975nm of error or less, while the Y focus measurement will have 4.075nm of error or less.

Table II shows a comparison of 98th percentile Focus Range for both Registration systems that were evaluated. The results suggest that REG_1 is well matched for X and Y performance, where REG_2 is not. Also, REG_2 has much tighter X performance than REG_1, but REG_2 is worse for Y performance.

In terms of metrology raw data we have quantified the amount of noise present in Registration measurements, but how does that directly impact PSFM measurement results? The three repeat PSFM measurements were analyzed in two different ways to answer this question. The first method was a standard 3Sigma evaluation of PSFM with average dynamic shot shape removed. The three repeat measurements of a single PSFM wafer showed very tight results; after pooling data from both Registration systems the variation of All Shot 3Sigma was 0.15nm. The second analysis method employed was an X-Y difference PSFM 3Sigma evaluation. To do this a -1 sign change was applied to the Y PSFM Rate constant, while the X Rate constant was left positive. The typical PSFM analysis utilizes an average result of X and Y; however with a negative Y Rate constant the PSFM analysis produces a delta of the X and Y results. For the Registration repeatability study the X-Y difference PSFM analysis was used to confirm if the substrate noise was stable from measurement to measurement. After pooling data from both Registration systems together the variation of X-Y difference PSFM All Shot 3Sigma, with average dynamic shot shape removed, was 0.17nm. Table III shows a summary of these results.

TABLE II.   REGISTRATION REPEATABILITY FOCUS RANGE

REG System	Focus Error Range, 98th Percentile Analysis	
	X Focus (nm)	Y Focus (nm)
REG_1	3.775	3.650
REG_2	2.975	4.075

TABLE III. REGISTRATION REPEATABILITY: PSFM RESULTS

Wafer #	PSFM 3σ Analysis, Dynamic Shot Shape Removed	
	XY AVG PSFM *3σ (nm)*	*X-Y Difference PSFM* *3σ (nm)*
REG_1-1	16.59	2.90
REG_1-2	16.56	2.93
REG_1-3	16.65	2.91
REG_2-1	16.63	2.82
REG_2-2	16.71	2.84
REG_2-3	16.69	2.76
Range REG_1	0.09	0.03
Range REG_2	0.08	0.08
Range All REG	0.15	0.17

There is an apparent disconnect between the Focus Range evaluated in the raw Registration data and the 3Sigma range seen the PSFM evaluation. Using REG_2 as an example for comparison, the raw Registration analysis suggests a possible focus error of 2.975nm in X and 4.075nm in Y. An average XY error should be approximately 3.5nm, but the PSFM analysis showed much less with an error range of 0.08nm 3Sigma. This in fact is reasonable due to the nature of data processing used by Nikon for PSFM viewing and analysis. If, at a given measurement location, X and Y results do not match the data is filtered out. An average of X and Y is then used for PSFM evaluation and focus correction map generation. Ultimately the measurement noise from Registration repeatability is small, on the order of 4nm or less in units of focus, and this noise is further minimized by data processing steps down to the level of 0.15nm 3Sigma variation from measurement repeatability. This equates to Registration metrology noise being 0.75% of the defined All Shot 3Sigma target of 20nm.

c) Substrate Variation

Judgment of PSFM results is done primarily with across wafer uniformity 3Sigma. Any variation in the incoming substrate will add directly to the PSFM result. A substrate with an across wafer shape, possibly a bowl or ripple, will be visible in PSFM results. This shape will then be compensated with the focus correction map. If the subsequent substrates do not have the same basic shape then the correction map will apply incorrect compensation. It is also possible, if a substrate has a rough surface or center to edge roughening, that AF measurements of the wafer will have degraded accuracy. Using an X-Y difference PSFM analysis it is possible to quantify what magnitude of noise is coming from the substrate.

PSFM exposure can be performed on many different substrate types; to date BARC, Ultra-Flat (UF), and SLAM substrates have been used [2]. Historically Nikon has recommended evaluating PSFM on BARC substrates. With the introduction of immersion scanners, which increased PSFM measurement frequency, the substrate was changed to reduce

cost per measurement and substrate regeneration time. Currently SLAM is the standard substrate used on Nikon NSR-S610C scanners. Towards the goal of determining which substrate type was best suited for PSFM evaluations the following experiment was conducted. Using a scanner with an optimized System Map three PSFM wafers each were exposed on SLAM and Ultra-Flat substrates. For both substrates two PSFM Rate wafers were also exposed. Analysis of the results included PSFM Rate evaluation, standard XY average PSFM evaluation, and X-Y difference PSFM evaluation.

For the PSFM Rate results a complete analysis was done, including average Z offset, Rate fit linearity R^2, and Rate constant variation between measurement points. For the purpose of this paper the X Rate data will be discussed and compared between SLAM and Ultra-Flat substrates. The PSFM Rate wafer Z offset analysis revealed that the UF substrate performed worse than the SLAM substrate. Both substrate types had measurement points outside the +/- 30nm criteria. Noting that SLAM is the current Process of Record, it was slightly surprising to find SLAM PSFM Rate Z offset did not pass at all measurement points. Fortunately this metric is primarily a gauge of targeting; if the Z offset is significantly outside of target the PSFM Rate wafer can be exposed again with an adjusted center focus. Table IV shows a summary of the PSFM Rate Z Offset evaluation results on SLAM and Ultra-Flat wafers.

A PSFM Rate fit linearity of $R^2 \geq 0.98$ is ideal; the higher the R^2 result the greater confidence can be placed on knowing that PSFM shift is linear thru the full depth of focus. From the investigation of SLAM and Ultra-Flat substrates a clear difference was observed in the PSFM Rate fit linearity performance. On SLAM the average $R^2=0.992$, while Ultra-Flat average $R^2=0.986$; this difference is small, however from the SLAM substrate all measure points had $R^2 > 0.98$, while the Ultra-Flat population had an average of five failing points per wafer. Overall, both SLAM and Ultra-Flat seem adequate in terms of Rate fit linearity, but the SLAM substrate was slightly more robust. Table V shows a summary of the PSFM Rate fit linearity evaluation results on SLAM and Ultra-Flat wafers.

Comparison of the average PSFM X Rate constant showed a 6.9% difference between SLAM and Ultra-Flat substrates. A difference of this magnitude is reasonable; substrate reflectance is a component in the imaging capability of any lithography print. With the resist process held constant, the only variation on imaging was from the substrate. Outside of the average PSFM Rate constant, it is ideal for all measurement points to have variation $\leq$ +/- 0.010.

TABLE IV. SUBSTRATE VARIATION: PSFM RATE Z OFFSET

Wafer #	PSFM Rate: Z Offset Variation (nm)				
	Max	*Min*	*Average*	*Target*	*Fail Count*
SLAM_1	-7.4	-30.2	-18.9	+/- 30	1
SLAM_2	-5.5	-28.6	-17.6	+/- 30	0
UF_1	-13.2	-39.8	-25.4	+/- 30	7
UF_2	-11.1	-32.2	-22.6	+/- 30	2

TABLE V.   SUBSTRATE VARIATION: PSFM RATE FIT LINEARITY

| Wafer # | PSFM Rate: Fit Linearity R^2 | | | | |
	Max	Min	Average	Target	Fail Count
SLAM_1	0.997	0.983	0.993	≥ 0.980	0
SLAM_2	0.997	0.983	0.991	≥ 0.980	0
UF_1	0.993	0.971	0.986	≥ 0.980	2
UF_2	0.995	0.970	0.986	≥ 0.980	8

Neither SLAM nor Ultra-Flat completely passed this metric. The SLAM substrate averaged three failing points, while the Ultra-Flat averaged five. It should be noted that the PSFM Rate constant used for evaluating a standard PSFM result is the average of all 35 points on a PSFM Rate wafer. In practice this makes the Rate constant variation < +/- 0.010 a guideline rather than a strict rule. Table VI shows a summary of the PSFM Rate Constant Variability evaluation results on SLAM and Ultra-Flat wafers.

The second metric of evaluation in the substrate variation experiment was XY average PSFM performance. This test evaluated across wafer focus uniformity 3Sigma on three PSFM wafers exposed on SLAM as well as Ultra-Flat. At the time of exposure all focus corrections were held constant on the scanner. This included system focus offsets, correction maps, and exposure focus steps. A typical PSFM evaluation, using XY average and dynamic shot shape removed, was used to evaluate the data. A wafer to wafer, WTW, variability of Focus 3Sigma was then calculated. Due to the small size of the three wafer sample a max-min, Range, result was calculated to evaluate variability. Fig. 5 displays the difference in WTW range between SLAM and Ultra-Flat when evaluating All Shots, Full Shots only, and Edge Shots only on the wafer.

For the purpose of this paper the All Shots 3Sigma metric is considered; the SLAM substrate has a WTW variability of 2.50nm, while the Ultra-Flat substrate has 1.45nm WTW variability. In consideration of the proposed target PSFM result, All Shot 3Sigma < 20nm, SLAM substrate variability induces 12.5% of the target while Ultra-Flat variability induces 7.35% of the target.

A third evaluation metric in the substrate variation experiment was X-Y difference PSFM performance. This test quantifies the across wafer focus 3Sigma noise induced by substrate variation. As an initial step, for both SLAM and Ultra-Flat, a WTW range was calculated.

TABLE VI.   SUBSTRATE VARIATION: PSFM RATE CONSTANT VARIABILITY

| Wafer # | PSFM Rate: Rate Constant Variability | | | | |
	Max	Min	Average	Target	Fail Count
SLAM_1	-0.206	-0.229	-0.217	+/- 0.010	2
SLAM_2	-0.208	-0.235	-0.221	+/- 0.010	4
UF_1	-0.188	-0.224	-0.207	+/- 0.010	6
UF_2	-0.190	-0.215	-0.201	+/- 0.010	5

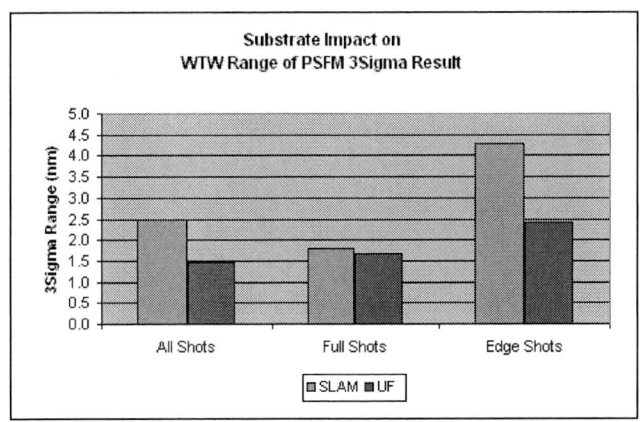

Figure 5.   Substrate Variation: PSFM 3Sigma WTW Variability

With dynamic shot shape removed the All Shot 3Sigma range on SLAM was 0.07nm vs. 0.16nm on Ultra-Flat. Generally speaking this level of repeatability error is small which suggests the absolute magnitude of X-Y difference PSFM 3Sigma is reliable as a performance metric.

Fig. 6 shows a comparison chart of three wafer average X-Y difference PSFM 3Sigma for All Shots, Full Shots only, and Edge Shots only for both SLAM and Ultra-Flat. Considering the All Shots X-Y difference PSFM 3Sigma a few observations can be taken from the data: both SLAM and Ultra-Flat substrates introduce noise into the PSFM 3Sigma results. For SLAM the substrate noise is 2.61nm vs. 1.97nm on Ultra-Flat.

In consideration of the noise evaluation results, where Ultra-Flat had slightly tighter repeatability than SLAM, it could be expected that PSFM 3Sigma results would be equivalent or tighter on Ultra-Flat than on SLAM. In fact the opposite was true. For All Shots, Full Shots, and Edge Shots the SLAM results had tighter across wafer 3Sigma than Ultra-Flat. Fig. 7 shows a comparison of PSFM maps for SLAM and Ultra-Flat.

From a visual analysis of the PSFM maps a distinct difference could be seen, primarily around the wafer edge. In consideration of the experimental conditions this was actually reasonable. As stated previously, all focus corrections were held constant for both SLAM and Ultra-Flat PSFM exposures.

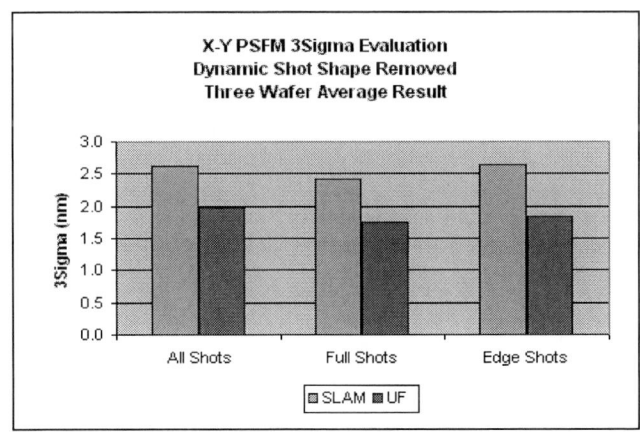

Figure 6.   Substrate Variation: X-Y Difference PSFM 3Sigma

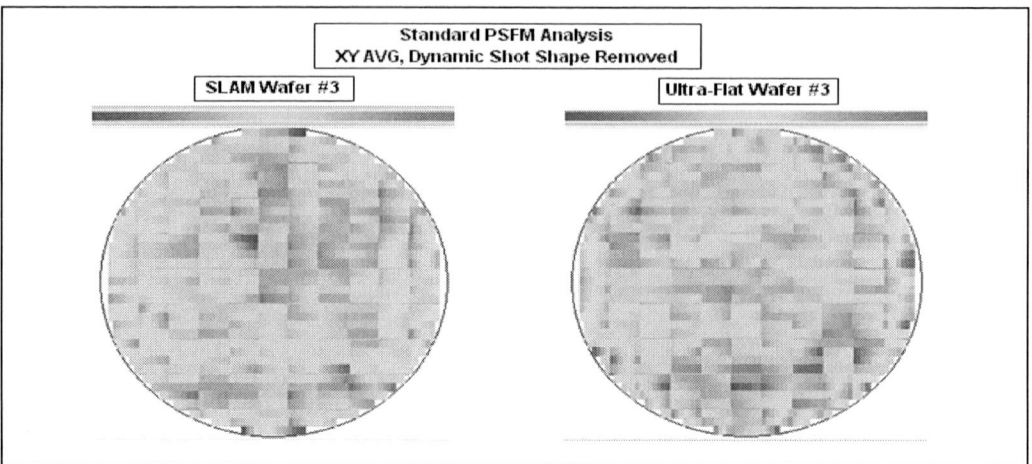

Figure 7. PSFM Map Comparison: SLAM vs Ultra-Flat

This included the System Map. The existing Process of Record for this scanner was to create System Map corrections from PSFM exposed on SLAM. The results from this study suggest that the current System Map is optimized for SLAM, but not ideal for Ultra-Flat with significant differences existing at the wafer edge. A possibility for future work includes re-creation of the System Map using an Ultra-Flat substrate. Once the System Map is optimized the elevated wafer edge signature will be eliminated. The previously discussed WTW stability of the UF substrate suggests that the absolute PSFM 3Sigma magnitude should be lower than what was achievable on SLAM.

III. CONCLUSIONS

Identification of noise sources in the Nikon NSR-S610C focus correction map processes was possible, and for most sources the noise was quantifiable in units of Focus (Z) 3sigma in (nm). However, from the studies performed the potential for improvement was small. Considering the investigation goal, quantifying how much noise a given source introduces into PSFM results, the evaluation of PSFM and PSFM Rate targeting had mixed results. PSFM dose targeting had qualitative success criteria for finding ideal conditions, but there are no quantitative results for impact to PSFM 3Sigma if there is a 10% dose error. PSFM Rate targeting, however, was more successful towards the quantitative goal. The study has shown that if a 10% error exists in the PSFM Rate constant then a 10% shift will occur in the PSFM 3Sigma result. A 10% PSFM Rate error, when evaluating a nearly optimized PSFM result, is equivalent to 9.65% of the defined PSFM target. This amount could influence the decision to assess a focus correction map complete or to perform one more round of optimization.

The investigation of Registration metrology noise was successful in quantifying how much noise was present in the raw Registration results. After converting the shift results to focus (nm), X and Y measurements averaged 3.5nm of error. The investigation also revealed that within a given Registration system the X and Y error may not be matched, and that from system to system there can be a difference in error performance. In terms of PSFM 3Sigma the impact of Registration measurement noise was very small; Registration error is equivalent to 0.75% of the defined PSFM target.

Substrate variation has been evaluated as a source of noise in PSFM 3Sigma results and found to have potential for improvement. The study conducted was limited to the POR substrate, SLAM, and one test case, Ultra-Flat. PSFM Rate evaluations show both SLAM and Ultra-Flat substrates are adequate for PSFM Rate performance, with SLAM being slightly more robust for Z Offset Variation and Rate Fit Linearity R^2. The X-Y difference PSFM evaluation showed both SLAM and Ultra-Flat substrates introduce noise into the PSFM across wafer result, but the Ultra-Flat noise was less. While evaluating all shots, full shots only, and edge shots only the difference between substrates was consistent; Ultra-Flat has 0.70nm X-Y difference PSFM 3Sigma less noise than SLAM. Standard PSFM across wafer 3Sigma evaluations showed wafer to wafer variation for both substrate types. The WTW 3Sigma focus variation was 2.50nm on SLAM and 1.47nm on Ultra-Flat. Substrate variation from SLAM is equivalent to 12.5% of the defined PSFM target, while variation from Ultra-Flat is equivalent to 7.35% of the defined PSFM target. All of these results, comparable PSFM Rate, tighter X-Y difference PSFM 3Sigma, and tighter WTW PSFM 3Sigma suggest that switching to Ultra-Flat substrates could improve the Focus Correction Map process. The iterative correction process would very likely be shortened and possibly the absolute achievable across wafer 3Sigma could be tightened. Future work will be performed to quantify what time savings is possible and how much the absolute magnitude of PSFM 3Sigma can be reduced.

REFERENCES

[1] T. Brunner, M. Hibbs, B. Peck, and C. Spence, "Optical focus phase shift test pattern, monitoring system and process," United States Patent # 5300786, April 1994

[2] E. Andideh, and A. Myers, "Method for making a semiconductor device that has a dual damascene interconnect," United States Patent # 6448185, September 2002

Increase Fab Capacity

With Predictive Short-Interval Scheduling

David Hanny, Applied Materials® AGS, Salt Lake City, USA, david_hanny@amat.com

Abstract

The lithography cell is widely accepted as the most challenging area of a fab to manage cost effectively. Reticle coordination, re-entrant flows and frequent set-up changes create bottlenecks in the Litho module. There is a strong need to use optimization to better match lots and tools, keeping litho tools fully utilized and reducing wafer cycle time.

A new solution, called Applied SmartSched™, for the first time integrates Real-Time Dispatching (APF RTD®) policies with locally optimized short-interval schedules to increase fab throughput and optimize litho tool utilization by converting 'white space' to available capacity. It produces a schedule whose output is executed by the RTD. Its predictive techniques determine precisely where work-in-process (WIP) and reticles will be to efficiently process the WIP. This novel technology is now being deployed in lithography modules of production fabs; initial results show over 2% increase in litho wafers out and litho tool utilization improvements greater than 1%.

This advanced scheduling system looks upstream and downstream to see WIP status in order to better plan and schedule work in the litho area. SmartSched groups WIP better, increasing available capacity resulting in higher throughput. It has demonstrated a throughput improvement of 2.1% in a production litho module. Additionally the software is achieving a >1% improvement in litho tool utilization.

The SmartSched technology uses 3 major components that share the work to better pick the best lot to process.

- Real-time data generation,
- Simulation-based prediction,
- Optimized short-interval scheduling

This advanced scheduling approach is providing manufacturers with dramatically improved results, in critical photolithography areas.

I. INTRODUCTION

Semiconductor fabs have a growing need to use optimization systems in the decision making process to match groups of lots at bottleneck points. The lithography cell is widely accepted as the most challenging area of a fab to manage cost effectively. With new litho equipment sets costing in the $40M to $50M range utilizing these assets is extremely important in today's competitive manufacturing landscape. Reticle coordination, re-entrant flows and frequent set-up changes create bottlenecks in the litho module adding to manufacturing complexity. A solution that has the ability to

better group lots by analyzing the manufacturing flow upstream and downstream can better keep the litho module fully utilized. This will result in increased fab capacity enabling manufacturing to realize a throughput increase and / or reduce lot cycle times.

II. EXPLORING THE PROBLEM

There have been several initiatives in the semiconductor industry attempting to solve this problem. RTD has provided a major benefit to help determine what to process next at a specific step. At first it seemed reasonable to extend RTD rules to build better schedules for the fab. Three limitations in this approach were identified and listed below:

- The inability to look beyond a specific tool's set of rules. It became very difficult to know what was needed to be processed over a time horizon which limited the effectiveness of the option to extend RTD.

- A dispatcher's view for decision making doesn't typically consider other tools in the tool group, therefore decision making was too locally focused to gain a greater benefit than RTD was already delivering.

- As constraints grow and change in the fab a heuristic-based solution will become unmanageable. Developing serialized sets of rules will increase costs and complexity of maintaining the systems. There is also no ability to evaluate different alternative schedules.

For these reasons scheduling requirements for a short-interval solution were created that include an optimization capability, the ability to group equipment, and the element of time.

Some consideration was made towards replacing an existing dispatching system with a scheduler. This was quickly abandoned because data management was such an enormous challenge. The amount of data (which continues to grow) and keeping that data current were requirements that were extremely difficult to meet without the use of RTD. Semiconductor fabs are getting significant value from their RTD deployments and couldn't afford to lose that positive impact to manufacturing. RTD provides the infrastructure necessary to implement and adapt a continuous improvement program. Fabs couldn't afford to take a step backwards, which generated the requirement for a scheduling solution to work with and be seamlessly integrated to the dispatching system.

Mathematical optimization is an important component for a scheduling solution, but is only as good as the data provided to it. It was very apparent early that the infrastructure around optimization was necessary to provide better schedules for the fab. A solution that made maximum use of the data in the fab would far better adapt in the long-run. The idea of the fab data driving the behavior of constraint decisions was born. A best-known-practice was discovered where using the fab data to its maximum would reduce the long-term costs to the fab.

By rigorously testing scheduling in a fab, it was uncovered that there were new opportunities for RTD rules to become more effective by making them aware that a scheduler is providing input to the dispatching system.

III. SOLUTION APPROACH

A new solution, from Applied Materials called SmartSched, for the first time integrates Real-Time Dispatching (RTD) policies with optimized short-interval schedules to increase fab throughput and optimize litho tool utilization by converting 'white space' (lost processing time) into available capacity. It produces an optimized schedule whose output is executed by the RTD. Its predictive techniques determine precisely where work-in-process (WIP) and reticles will be to efficiently process the WIP. This novel technology integrates to fab manufacturing systems quickly and adopts best practices as it seamlessly integrates with RTD. It is now deployed in the lithography module of production fabs where initial results show over 2% increase in litho wafers out and litho tool utilization improvements greater than 1%.

This paper focuses on the use and benefit captured in a mature production environment using this new technique for short-interval scheduling.

IV. SOLUTION COMPONENTS

The SmartSched technology uses 3 major components (Fig 1) that share the work to better group lots to align with factory manufacturing objectives.

- Real-time data generation

- Simulation-based prediction

- Optimized short-interval scheduling

This allows a *Schedule-Aware Dispatching* system to then pick the best lot at a moment which meets the business objectives of the fab.

Applied Short Interval Scheduler Components & Dataflow

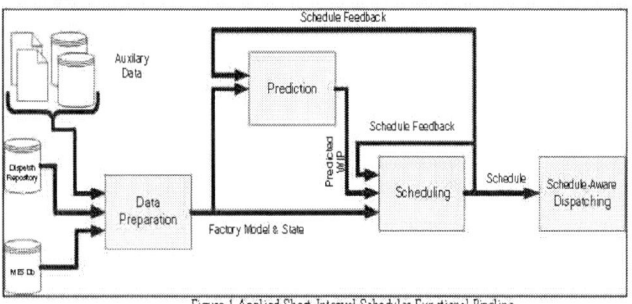

Figure 1 Applied Short-Interval Scheduler Functional Pipeline

(Figure 1)

A. Data Preparation

A prerequisite for any solution is complete and accurate information. Adapting to the dynamics of a complex manufacturing environment requires a very detailed data model. The data generation process imports detailed information from the fab's manufacturing execution system (MES) and other automation data systems (APC, AMHS, etc.). This creates a transformed, detailed data model used to drive the behavior of the prediction and scheduler components.

B. Prediction

The prediction engine determines where WIP and reticles will be located in the future. Prediction provides the accuracy required to create a new good schedule. The visibility into the future enables the scheduling engine to better group lots and helps reduce setups. This generates an expanded fab view for schedule creation. For example, if a high-priority lot will be arriving at a schedule step two hours after the start of the scheduling interval, the best schedule will arrange for the lot to be processed shortly after those two hours by knowing what is happening upstream.

The prediction process is a discrete-event simulation of the fab that forecasts WIP and its expected arrival at each specific step. Prediction execution repeats every few minutes to capture the changing state of the fab. This summarizes information about the load on the fab including the amount of WIP, state of equipment, scheduled PMs, product mix, etc. The simulation requires model details that include factors such as dispatch rules, process times, transport times, and sampling strategies. The prediction engine provides a well-understood framework for doing this type of modeling. The prediction is run over a short interval of time to reduce the inherent error that could build up over long simulation periods. Prediction provides better data input to the scheduling engine's decisions.

C. Scheduler

The scheduling engine is the core of the scheduling process. The scheduler requires three main inputs: a detailed model of the state of the fab, predicted lot arrival times, and the previously generated schedule. It uses the real-time fab data to form litho scenario activities into non-linear mathematical constraints in order to narrow down the schedule choices. Constraint consideration is made for send-ahead wafers, reticle usage, qualification lots, transport delays, recipe qualifications, lot-tool dedication, qtime violations, preventive maintenance, train lengths, tool setups, etc. The scheduling engine also checks input data for data inconsistencies and remedies them before moving on to produce a schedule with correct data. Optimization allows the user to manage the growing number of factory constraints (which have become unmanageable with heuristic or serial solutions in many fabs). The scheduling engine uses constraint programming to generate optimized schedules. Multiple optimized schedules are then created for comparison. The scheduling engine's 'objective function' enables the user to prioritize manufacturing tradeoffs based on business objectives. The key performance indicators (KPIs) are prioritized and weighted. This defines a cost structure allowing the scheduling engine to select the optimal schedule based on the objectives of the business. The resulting schedule includes lot-equipment assignments, lot processing orders, lot and reticle transports and a reticle inspection schedule. The schedule created spans as much as one factory shift and is placed into the dispatching system cache for immediate access and use. The result is also fed back into the scheduling system for the next SmartSched run. This process is repeated every few minutes to account for change and manufacturing fluctuation in the fab.

D. Schedule-Aware Dispatching

The final process in the short interval solution is dispatching the production and transport activities. Initially, the short interval scheduling development teams made minimal changes to existing dispatching rules. They assumed only minor changes would be required because

- The schedule acts as a basic prioritized and time-sequenced work list for each tool

- Processing logic contained in existing dispatch rules must be maintained by each fab that uses the schedule

- Complete dispatching logic is necessary when factory events that occur between schedule generations invalidate specific schedule tasks

Schedule-Aware Dispatching acknowledges that one of the primary inputs to the dispatching rule is a schedule, as opposed to a simple prioritized list of available lots. The schedule is a list of lots for a specific litho tool organized not only by priority but also by temporally designating an earliest and expected start time for each task.

A short-interval scheduling system with "schedule-aware" dispatching incorporates the policies of both conservative and aggressive conformance to a schedule. This is the essence of schedule-aware dispatching, which is a key component in the overall short-interval scheduling solution.

V. RESULTS

A. Throughput and Tool Utilization Increase

This advanced scheduling system looks upstream and downstream to see WIP status and reticle usage activity in order to better schedule work in the litho module. One common source of white space in litho occurs when WIP arrives at a tool, but the required reticle is not yet available. By looking ahead, the scheduling technology helps prevent a reticle from being "locked down" on one tool, when a less flexible tool using the same reticle could soon become idle. SmartSched groups WIP better, increasing available capacity resulting in higher throughput. It has demonstrated a throughput improvement of 2.1% in a production litho module. Additionally the software is achieving a >1% improvement in litho tool utilization with decreased variation

(Fig 2). This provides better distribution of work across the tool group. These KPIs results were further supported by a reduced send-ahead wafers (wafers that need to pass metrology steps to allow litho processing to continue) (Fig 3), More wafers are being tracked out during some of the more down periods of the day (Fig 4).

Litho Equipment Utilization Improvement

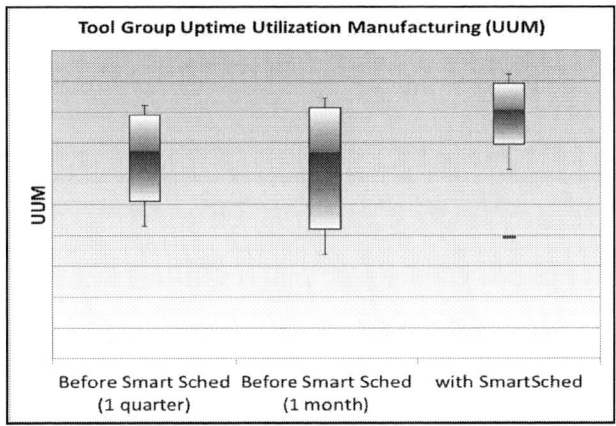

(Figure 2)

Reduced Send-ahead Lots

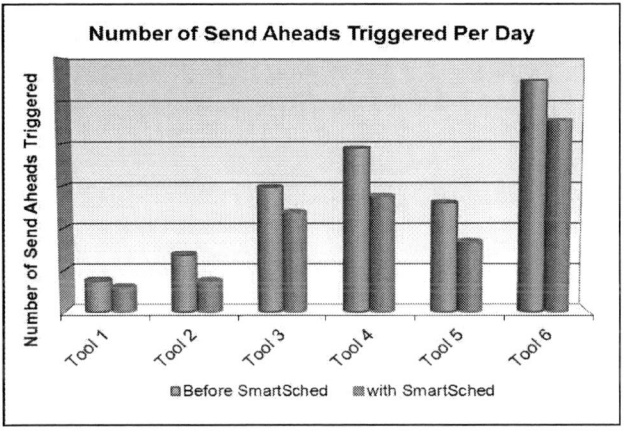

(Figure 3)

Automating Output at Difficult Times of Day

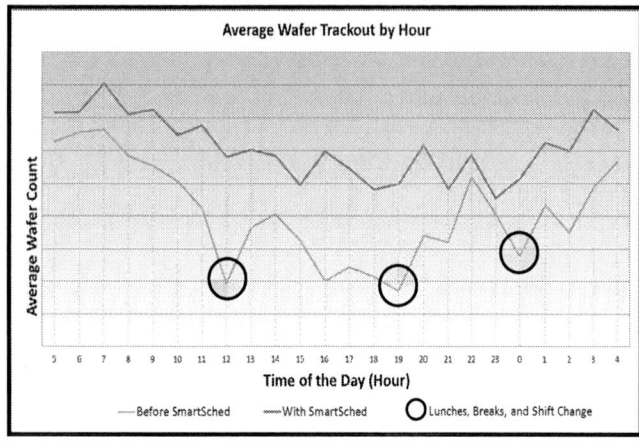

(Figure 4)

Table 1

Economic Analysis of Cycle-Time Improvement

Cycle-time improvement	High Priority Lots	Normal Priority Lots
Lithography	20.83%	-1.67%
Overall	3.04%	-0.04%
Capacity Increase	3.13%	-0.04%
First Year Value	**$6,194,887**	($324,239)

B. Cycle Time Improvement

Opportunities for lot cycle time improvement also may be realized through short-interval scheduling depending on the SmartSched configured fab business objective (table 1). The cycle time analysis was done comparing a pure heuristic system with short-interval scheduling. The results were a significant cycle time reduction in the high priority lots as well as a reduction for other lots. The capacity increase in this case was used to reduce the amount of WIP waiting at litho resulting in reduced lot cycle time.

C. Conclusion

The movement from reactive to predictive operations in the fab represents the next generation in fab productivity and waste reduction. This advanced scheduling approach is providing manufacturers with dramatically improved results, in critical photolithography areas. SmartSched short interval scheduling has also seen positive testing results from the Wet / Diffusion module in the fab. In today's economy, where reducing waste is essential to a company's survival, SmartSched represents an Applied innovation that allow fab's to get more productivity out of their capital assets.

REFERENCES

[1] Dan Muller, Michael Anderson, David Norman and David Hanny, "Solving Complex Fab Challenges with Real-Time, Short Interval Scheduling," *Nanochip Fab Solutions*, Vol. 4 No. 2, (2009).

[2] David Norman and Michael Anderson, "Technological Approach to Short-Interval Scheduling in Photolithography," *Proceedings of the SPIE Metrology, Inspection, and Process Control for Microlithography XXIV Conference*, San Jose, CA, (February 2010).

[3] Robert C. Leachman, Leachman & Associates LLC, (2007). The Economics of Speed, Fab Engineering and Operations (1), 98-102.

CONTRIBUTORS

[4] Steve Marteney, David Norman, Keith Pare, Applied Materials

Extendible Scanner Platforms for Mass Production, Now and In the Future

Hamid Khorram
Advanced Technology
Nikon Precision, Inc.
Belmont, CA
hkhorram@nikon.com

Abstract— Increasing demand in technology requirements from IC manufacturers has necessitated that photolithography equipment suppliers design and build tools with many advanced capabilities. These additional features not only add to the complexity of the tools, but also increase the cost. Therefore, in order to keep lithography affordable, it is essential that equipment suppliers increase equipment productivity and extend tool lifetimes to support multiple process generations.

As a further challenge, the requirements from different IC manufacturers can be quite dissimilar even for the same technology node (as defined according to the ITRS roadmap). For example, for upcoming process nodes logic manufacturers may require enhanced flexibility to enable imaging of smaller and more complicated features, coupled with tighter overlay performance. In contrast, memory manufacturers may put more emphasis on Critical Dimension Uniformity (CDU) control. Therefore, today's photolithography scanners must be equipped with the latest technologies to achieve better resolution and heightened overlay accuracy, while at the same time delivering continuous improvements to throughput.

In order to extend the usable lifetime of the scanner on to next-generation technologies, the system platform must also accommodate further improvements and/or upgrades to achieve the demanding requirements of the future. In addition, the scanners must be flexible enough to integrate Complementary Technologies (CT) that are being developed in parallel to further enhance scanner performance.

Litho scanner suppliers are actively addressing these issues and Nikon has developed the Sx20 platform to provide an extendible platform for dry and immersion scanner applications. Immersion scanners such as the NSR-S620D are currently being used in mass production, and support future upgrades to enable enhancements to accuracy and throughput for next-generation manufacturing.

In this paper some of the key design concepts of an extendible scanner platform will be discussed. These include enhancing the accuracy and maximizing productivity of the manufacturing environment, which is vital in ensuring the affordability of lithography in the future. To show industry progress in these areas, current performance from leading edge scanners that include overlay, auto focus, and Optical Proximity Effect (OPE) matching data will be presented. In addition, it will be shown how these scanners deliver overlay accuracy sufficient for the 32 nm node and beyond, with focus uniformity less than 20nm (3σ), while utilizing a platform capable of throughput up to 200 wafers per hour.

Further, to support future technology nodes, scanners also need to accommodate advanced imaging solutions such as free form illumination sources, Source Mask Optimization (SMO) and Double Pattern Techniques (DPT). Some of these technologies, as well as how they must be integrated with flexible and extendible scanner platforms in order to keep next-generation lithography affordable, will be discussed in this paper [1].

Keywords- CDU, DPT, OPC, S620D, AF, GRP, sPURE, SMO, CDU Master, and DPF

I. INTRODUCTION

The highly competitive IC manufacturing environment is putting increased pressure on lithography equipment suppliers to be more flexible in system design, increase wafer output, and extend tool platform life-time – all while continuously enhancing performance. As a result, equipment suppliers are under heightened pressure to develop tools that are more flexible and have extendible platforms to keep development costs down and offer affordable solutions to customers.

According to the International Technology Roadmap for Semiconductors (ITRS), IC manufacturers must receive their beta tools for development of a given technology node 24 months ahead of the first production tool [2]. In reality this period has been shrinking for the past several years, driven by demand for leading edge technology. This situation has forced IC manufacturers to pull in introduction of next-generation technologies, thereby leaving less time between the start of development and first production of a fully functioning chip. This makes extension of current tool platforms more appealing to equipment suppliers and IC manufacturers.

To satisfy these aggressive timing requirements, especially with the various delays in EUV infrastructure development, it clearly makes sense to extend immersion lithography applications. Most IC manufacturers are only using the initial capabilities of immersion lithography in production, mainly utilizing single exposure technology with a heavy emphasis on OPC and customized illumination sources for enabling low k1 factors, as well as fairly aggressive overlay accuracy. The transition to double patterning technology (DPT) is dependent upon the type of the product (Flash, DRAM, logic…etc) and degree of the IC design complexity. Today some companies

978-1-61284-408-4/11 $26.00 © 2011 IEEE

(i.e. NAND-flash memory manufacturers) are already using various forms of double patterning technology (DPT). For the majority of IC manufacturers, DPT plus immersion lithography will be a likely lithographic solution for their sub 22nm node. This makes it imperative to have an immersion scanner platform that provides comprehensive solutions to extend immersion double patterning.

Innovations and flexibility in design are the keys to success for photolithography equipment suppliers and IC manufacturers. In this paper we take a closer look at current technology requirements, and critical design innovations that are used to satisfy these requirements. We also explore future requirements and present potential solutions.

II. CURRENT SITUATION

A. Challenges

We start with reviewing the current technical requirements and solutions that are offered for each challenge by a scanner supplier. According to the ITRS roadmap, in 2011 the industry should be ramping up 40nm half-pitch technology using single exposure patterning for DRAM and DPT for logic and MPU, with overlay requirements ranging from 8.0nm to 9.5nm (single machine overlay) and CDU control ranging from 2.5nm to 4.2nm [3]. However, some leading edge IC manufacturers, having applied Restricted Design Rules (RDR), are already ramping their 32nm node while developing ~ 22nm node [4]. Based on the ITRS roadmap, overlay and CDU requirements for 32nm node overlay and CDU are as low as 6.4nm (3σ) and 2.1nm (3σ) respectively. Meanwhile these must be achieved while continuously improving defectivity and productivity.

In addition to these requirements, existing technology nodes require reduction of the k1 factor. One important solution to this is use of illumination units (IU) that gives IC makers the ability to optimize printing of individual layers, simultaneously reducing the k1 factor while maintaining a manufacturable Process Window (PW).

B. Meeting the current challenges

In order to meet the current challenges, lithography equipment must be equipped with state of the art Autofocus (AF) systems, highly precise exposure stages, customizable illumination systems and capabilities to achieve high throughput while maintaining tight overlay accuracy.

1) Autofocus: Focus accuracy is one of the vital factors in achieving tight CD uniformity to satisfy manufacturing requirements. Advanced lithographic scanners, such as the S620D, are equipped with an open-loop straight-line AF system. Two main characteristics of this system are dry focus measurement and full wafer coverage. The importance of these characteristics will be discussed in the following sections. Figure 1 is a simple depiction of the S620D AF system and its location with respect to the Projection Lens (PL) and immersion nozzle.

a) Dry measurement: Dry measurement is key to improving accuracy by reducing the impact of humidity caused by the immersion water. Humidity around the AF beam path could change the refractive index of the surrounding air, which in turn can impact AF beam path. Therefore, sensors are positioned ahead of the immersion nozzle and measure focus as the dry wafer moves from the loading position towards the PL/immersion nozzle. Additional air flow and temperature-control measures maintain a stable environment around the AF beam path to further enhance its stability.

b) Full wafer coverage: By covering the entire wafer at once it is possible to measure focus in one constant motion and direction, hence reducing potential focus measurement degradation due to stage movement accuracy. A full wafer, single-directional measurement strategy also reduces the time for focus measurement, increasing throughput and thus productivity. Figure 2 depicts leading-edge scanner across wafer auto-focus performance from two differnet customer sites[5] with results below 13nm including edge shots. The graph demonstrates AF variation to be within 3nm through an entire 25 wafer lot.

2) Exposure stage precision: As overlay requirements become more stringent for the 32nm node and beyond, the need for exposure stages with enhanced accuracy and precision is increasingly critical. To extend immersion DP applications, the scanner wafer stage control system must have good (a) linearity, (b) repeatability and (c) long term stability. Historically, interferometer-based systems were used for tracking wafer stage movements. Interferometers have good linearity and long term stability; however, they lack sufficient repeatability and could be influenced by environmental fluctuations (air fluctuation). Analysis has indicated that >75% of overlay budget components can be affected by air flucuations.

In the past several years, substantial engineering work has gone into reducing the environmental impacts on interferometer performance by means such as shortening the interferometer paths to creating micro environmental chambers surrounding the beam paths to reduce the impact of air fluctuation. However, such solutions are not sufficient to satisfy overlay requirements for current and future technology nodes as outlined by the ITRS roadmap.

In contrast, encoder systems are very repeatable, but may not have optimal linearity due to Grating Plate (GRP) manufacturing errors. Therefore, in order to satisfy all three critical wafer stage control characteristics, it is necessary to combine interferometers with an encoder system to provide a hybrid stage control system. An encoder system can measure both X and Y movement of the exposure stage, as well as Z movements. In the case of the S620D, the stage control system consists of 4 GRP plates surrounding the wafer chuck, with multiple encoder heads that are mounted on top of the exposure stage at a 2mm distance with respect to the GRP plates (Figure 3). The short distance between the encoder head and GRP eliminates the impact of environmental conditions. The hybrid combination of the encoder system and interferometer is an integral part of lithography scanner capability to satisfy overlay requirements below 2nm (3σ). Actual data approaching this requirement are discussed and shown in a later section.

978-1-61284-408-4/11 $26.00 © 2011 IEEE

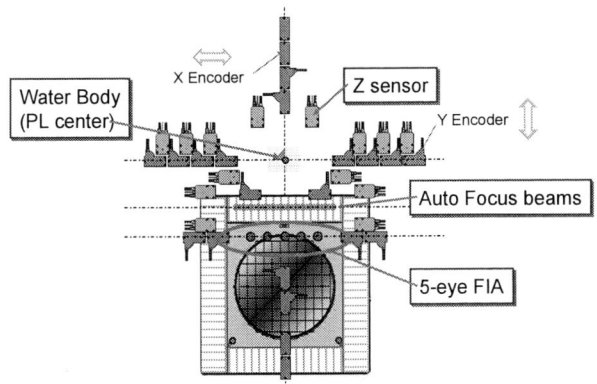

Figure 1. Straightline AF system covering entire wafer diameter at onnce. In this figure PL refers to Projection Lens and FIA refers to Field Image Alignment system.

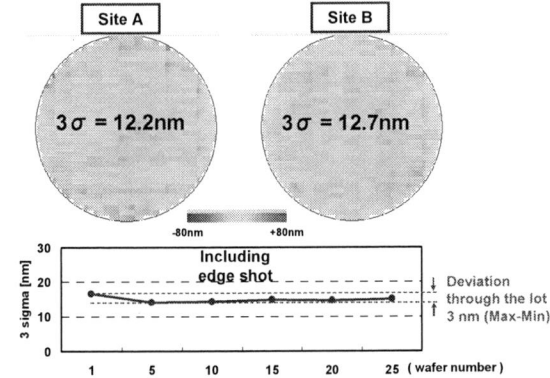

Figure 2. S620D AF stability performance at customer sites A and B using PSFM method.

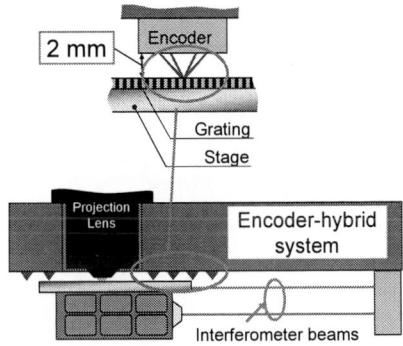

Figure 3. Encoder-interferometer hybrid wafer stage control system. This hybrid system takes advantage of superiority of interferometer and stability of encoder to accuratley control movement of the wafer stage.

3) Illumination unit: As k1 has shrunk, both scanner suppliers and IC manufacturers have taken steps to enable continued printing of complex layers. For instance, IC manufacturers are trying to follow RDR as much as possible and reduce non-essential pattern orientations. Meanwhile, scanner vendors have introduced the ability to use complex illumination patterns as well as illumination polarization to enhance the process window. A state-of-the-art illumination system needs not only the ability to create exotic illumination source shapes, but also appropriate controls to fine-tune errors,

such as illumination and pupil non-uniformity, in the supplied illumination, which can impact CD variation across the the chip.

For example, S620D IU systems can handle up to 14 Super Power Up Resolution Enhancement optics (sPURE), and enable several fine-tuning methods such as pupil-uniformity correction. Upcoming developments in the illuminator will be discussed in a later section.

4) Throughput and overlay: Increasing throughput while maintaining overlay accuracy is another essential characteristic for the current generation of scanners. As a result of mechanical/manufacturing deviations, it is understood two wafer exposure stages will have inherrently diferent stepping accuracy which results in differnet grid signature. In order to improve mix and match overlay, a single exposure stage design will eliminate grid matching errors within the same scanner, but could be a disadvantage in terms of throughput if not accompanied by a fast and accurate global alignment measuremnt scheme.

In the case of the S620D, the Stream Alignment system is used, which consists of five Field Image Alignment (FIA) cameras. This enables dense sampling of alignment marks at a fraction of a time needed for previous generation scanners.

Figure 4 illustrates a simple comparison between a traditional alignment scheme and newly developed Stream Alignment. As depicted by red arrows the zig zag stage motion is reduced to a simple continous motion while sampling more EGA sites with Five-eye FIA. Figure 5 shows current scanner overlay stability results obtained at multiple sites using Stream Alignment. The mean of the data is stable thorughout the five day experiment and |M| + 3σ results satisfy current ITRS roadmap requirements. Finally, by employing Stream Alignment, the straight-line AF system, and increasing the wafer stage scanning speed, a scanner using a single exposure stage has achieved throughput up to 200wph (Figure 6).

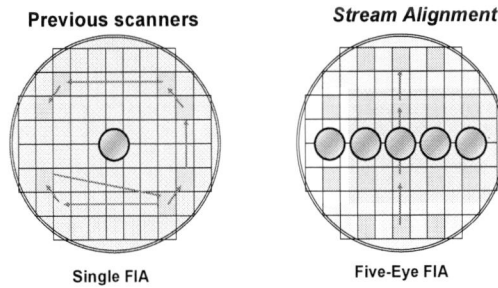

Figure 4. Traditional EGA alignment (left) and Stream Alignment scheme (right). Blue circles represent FIA camera, yellow felds represent alignment sites. In the case of Five-Eye FIA, five cameras are employed to enable increased sampling.

978-1-61284-408-4/11 $26.00 © 2011 IEEE

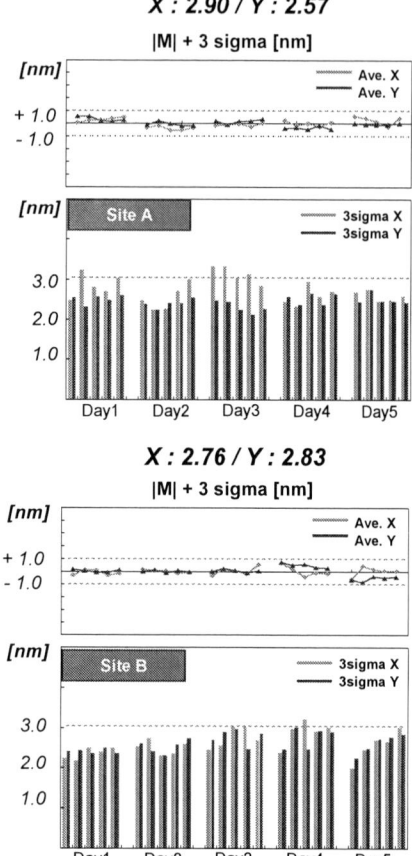

X : 2.90 / Y : 2.57

X : 2.76 / Y : 2.83

Figure 5. Most recent overlay stability data from multiple customer sites.

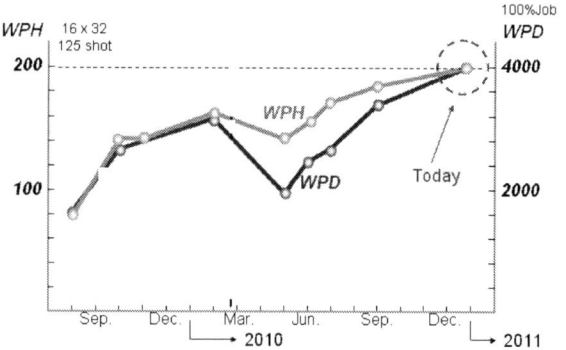

Figure 6. Advanced alignment methods enable 200 wafer per hour immersion scanner throughput.

III. COMPLEMENTARY TECHNOLOGIES

Up to this point, key features of an extendible scanner platform have been discussed. However, as part of the extendibility, the litho scanners must also be flexible enough to take advantage of so-called Complementary Technologies. . CTs can cover a wide range of photolithography aspects. They can contribute to overlay enhancements as well as improve imaging capabilities to enable faster and more accurate OPE matching. This section will concentrate on one of those CTs

that we believe is critical not only for single exposure but also in the future for DP exposure.

CDU optimization: CDU is a critical parameter for gauging photolithography performance now and will be of even more significance when DPT is widely used. It is important to remember that the scanner is not the only contributor to CDU error; other process components such as the lithographic track also contribute. Overall CDU error-mitigation schemes are highly desirable.

Therefore, scanner suppliers have been pursuing various solutions to enhance CDU, and Nikon has developed the CDU Master. This tool corrects intra and inter-exposure field CD uniformity by optimizing focus and dose for each shot. Figure 7 illustrates the entire process flow. This tool requires 3 sets of inputs: focus and dose error files from customer metrology tool (DF map file), wafer exposure/shot information from the scanner, and user offsets (i.e. Z-offset, tilt...etc) from the customer. As depicted in Figure 7, first sample wafers are exposed by the scanner and measured by a metrology tool. Based on this data, focus and dose errors corrections are calculated and used to prepare dose and focus (DF) map files. Then exposure result files, which include wafer exposure maps and other exposure parameters from the scanner, along with DF maps are loaded onto CDU Master. The software then uses the DF file, and the specific scanner's hardware/software constraints, and the exposure result file to prescribe control parameter values for focus and dose per shot. This prescription is then uploaded to the scanner for use during exposures. Dose and focus corrections up to 4th and 6th order respectively can be made using CDU master. The table in Figure 8 shows dose corrections that can be made using offsets in the X and Y directions. A similar table exists for the focus corrections. Figure 9 demonstrates an example of CDU improvement after applying recommended correction parameters by CDU master.

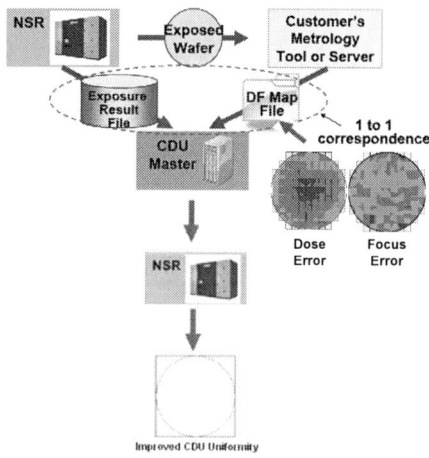

Figure 7. CDU Master process flow

978-1-61284-408-4/11 $26.00 © 2011 IEEE

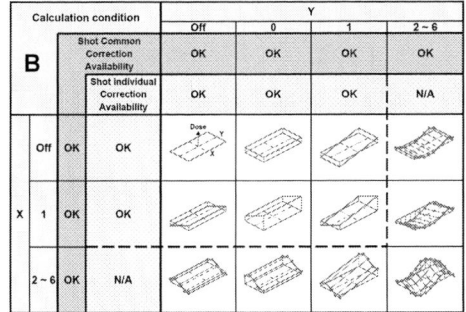

Figure 8. Dose correction capabilities of CDU Master, up to the 6th order corection can be made independently for X and Y.

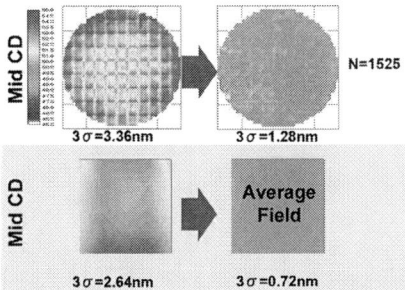

Figure 9. Mid CD uniformity before and after applying focus/dose prescription to scanner. Top two graphs represent entire wafer CDU, bottom two graphs represent mid CD uniformity after exposure shot averaging.

IV. MEETING FUTURE REQUIREMENTS

The previous sections discussed extendible scanner platforms and reviewed one of the main Complementary Technologies that we believe will be essential in meeting the current and future demands of IC manufacturing in terms of imaging and overlay. In this section, further steps that scanner suppliers are taking to extend scanners' useful life to meet future technology nodes will be presented.

In the upcoming years scanner suppliers must stay focused on two main improvement areas: imaging and overlay accuracy (mix and match).

1) Illuminator: As we push the limits of immersion lithography, continuously reducing k1, we are also shrinking the PW due to ever increasing sensitivities to lens apodization, mask errors and illumination pupil errors [6]. This is clearly an untenable situation. Various schemes, including Source-Mask Optimization (SMO), "Inteligent Illuminator" and "illuminator pupil predictor" are used to mitigate this. SMO is beyond the scope of this paper, but clearly any optimization of the illuminator requires an accurate prediction of its output. Pupil predictor software will assess the fidelity of the target pupil as delivered by the illumination system, this means that the predictor can accurately reconstruct the pupil shape at the reticle plane to match the intented pupil shape. The judgment criteria for a good prediction is based on the impact on imaging of the delta between predicted (reconstructed pupil shape by the software) vs. target pupil (intended pupil shape). A sensitive test of this "pupil predictor" is in optical proxity effect (OPE) matching of calculated CDs, since the optical proximity effect itself is sensitive to changes in the pupil

pattern. To demonstrate its accuracy, an OPE simulation was conducted with three pupil inputs: ideal top-hat, predicted pupil and measured illumination pupil. As shown in Figure 10, in this simulation the annular ratio was varied from its nominal value and the impact on OPE curves for different pupil inputs was calculated. Figure 10 compares OPE changes among the three pupils, and clearly the OPE curve for the predicted pupil predictor matches the measured pupil extremely well. This means the predictor can closely mimic the behavior of a measured pupilgram as illumination conditions changes (e.g. annular ratio in this example). These results confirm that the pupil predictor is a powerful tool for an accurate and quick SMO and OPE matching.

The ability to accurately predict the pupilgram by pupil predictor is enhancing the accuracy of SMO and OPE matching. Historical methods of creating customized pupil shapes have been sufficient to satisfy imaging requirements up until now, but additional flexibility in creating illumination source designs is needed to improve PW. To meet this requirement scanner suppliers are equipping their scanners with solutions such as the "Intelligent Illuminator". This solution can generate any source shape from parametric to free form sources.

Freeform illuminators are a crucial part of accurate OPE matching and SMO, and Degree of Pupilgram Freedom (DPF) is a vital part of this. DPF is defined by number of gray scale level multiplied by the total number of usable pupil grid pixels. The conceptual diagram is shown in Figure 11. Although one may be able to judge the quality of the pupilgram by visually comparing the original pupilgram with pupilgrams generated with different DPF values (Figure 12), the ultimate judgment must be made quantitatively, and that can be done by comparing OPE results. In order to determine the minimum DPF value, OPE simulations have been run with 4,000 DPF, 10,000 DPF, 100,000 DPF and were compared to the original pupilgram. Depending on the allowable OPE tolerance one can decide the minimum DPF value. For example, if we set our OPE tolerance to 0.5nm by looking at the OPE curves in Figure 13 one can determine the DPF value must be 10,000 or greater.

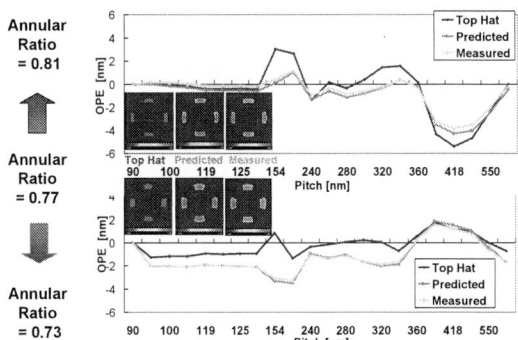

Figure 10. Changes in OPE by modulating annular ratio for 3 different illuminator descriptions.

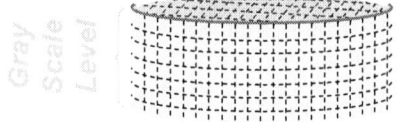

Figure 11. Conceptual diagram of DPF

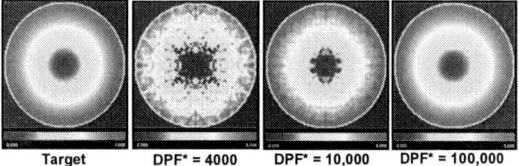

Figure 12. Comparison of generated pupilgrams with different values of PDF

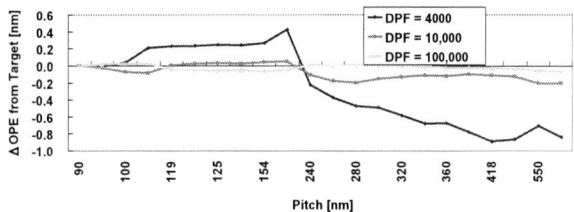

Figure 13. Impact of the value of DPF on OPE delta from target

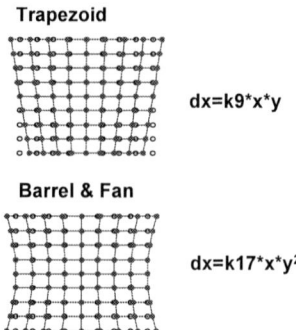

Figure 14. Two distortion models that can be corrected by scanner adaptive lens controller.

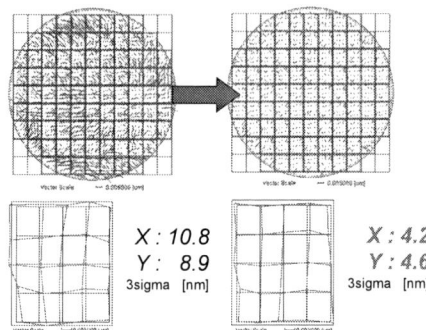

Figure 15. Simulation results demonstrating potential grid matching capabiliy of next generation lens controller systems, with the example here showing grid matching errors cut almost in half.

2)   Dynamic lens controller: The importance of overlay for DPT lithography has already been discussed. When speaking of overlay, naturally wafer exposure stage precision and alignment accuracy come to mind. In reality through, overlay consists of multiple factors as shown in equation 1.

$$\Delta OL_{line} \cong \sqrt{\Delta OL^2_{SO} + \Delta OL^2_{reticle} + \Delta OL^2_{process}} \qquad (1)$$

Overlay errors can be broken down to intra and inter-exposure field errors, and then divided into linear and non-linear errors. Intra-field errors are mainly caused by scanner lens aberration differences and reticle errors; while inter-field errors are caused by wafer thermal expansion, non-linear wafer deformation caused by Chemical Mechanical Polishing (CMP), and other process related items. Scanner suppliers have developed solutions to address each item, such as Grid Compensation Matching (GCM), Grid Factor Feeding (GFF), and Super Distortion Matching (SDM). To further address intra-shot errors, a new lens controller system to dynamically correct for intra-shot grid mismatch caused by lens distortion, wafer chuck, pellicle, reticle, lens heating or the process has been developed. This new adaptive lens controller system can correct for trapezoidal (K9), barrel or c-shape (K17) distortion signatures (Figure 14) dynamically for each shot by adjusting the z position of selected lens elements within the PL. Based on simulation results as shown in Figure 15, using such a method can potentially improve grid matching by over 50%, which will be significant with the transition to DPT.

V.   SUMMARY

In order to keep lithography affordable, it is essential that equipment suppliers increase equipment productivity and extend tool lifetimes to support multiple process generations. Scanner platforms must accommodate further improvements and/or upgrades to achieve the demanding requirements of the future. In addition, they must be flexible enough to integrate Complementary Technologies that are being developed in parallel to further enhance scanner performance.

This paper discussed some of the key design concepts of an extendible scanner platform that include enhancing the accuracy and maximizing productivity of the manufacturing environment. To show industry progress in these areas, current performance from leading edge scanners including overlay, auto focus, and OPE matching data has been presented. It also was shown that such scanners are able to deliver overlay and focus accuracy sufficient for the 32 nm node and beyond, with 200 wph platform capabilities.

Further, to support future technology nodes, scanners also need to accommodate advanced imaging solutions such as free form illumination sources in support of pitch doubling or quadrupling. These technologies, and how they must be integrated in order to keep next-generation lithography affordable, were also discussed.

ACKNOWLEDGMENT

The author would like to thank Ms. Raluca Popescu, Mr. Stephen Renwick, and Ms. Holly Magoon for their critical technical reviewing and feedbacks.

References

[1] M. Preil, "Transition of the Dresden Fab1 from IDM to Foundry in the context of Low k1 Lithography," LithoVision Symposioum 2011, p. 8

[2] ITRS 2009 Executive Summary, p. 16.

[3] ITRS 2010 Table_LITH3.

[4] S. Sivakumar, "Enabling the Future – From 32nm Production to 14/10nm Development," LithoVision Symposium 2011, pp. 1-22.

[5] Y. Shibazaki, "Enabling 32nm Production and beyond,", pp.8.

[6] T. Matsuyama, N. Kita, Y. Mizuno, "Pupilgram adjusting scheme using intelligent illuminator for ArF immersion exposure too," SPIE 2011, Vol. 7973-52 [Optical Microlithogrpahy XXIV]

Strategies for Single Patterning of Contacts for 32nm and 28nm Technology

Bradley Morgenfeld[a], Ian Stobert[a], Henning Haffner[d], Ju j An[a], Hideki Kanai[b], Martin Ostermayr [d], Norman Chen[c], , Massud Aminpur[a], Colin Brodsky[a], Alan Thomas[a],

[a] IBM Semiconductor Research and Development Center, 2070 Route 52, Hopewell Junction, NY, USA 12533;
[b] Toshiba America Electronic Components, Inc, 2070 Route 52, Hopewell Junction, NY, USA 12533;
[c] GLOBALFOUNDRIES Inc, 2070 Route 52, Hopewell Junction, NY, USA 12533;
[d] Infineon Technologies NA Corp, 2070 Route 52, Hopewell Junction, NY, USA 12533;

Abstract - **As 193 nm immersion lithography is extended indefinitely to sustain technology roadmaps, there is increasing pressure to contain escalating lithography costs by identifying patterning solutions that can minimize the use of multiple-pass processes. Contact patterning for the 32/28 nm technology nodes has been greatly facilitated by just-in-time introduction of new process enablers that allow the support of flexible foundry-oriented ground rules alongside high-performance technology, without inhibiting migration to a single-pass patterning process. The incorporation of device based performance metrics along with rigorous patterning and structural variability studies were critical in the evaluation of material innovation for improved resolution and CD shrink. Additionally novel design changes for single patterning along new capability in data preparation were both assessed to leverage minimal impact of implementation of a single patterning contact process into the existing 32nm and 28nm technology programs [1].**

I. INTRODUCTION

Contact patterning for middle-of-line (MOL) has historically been one of the most challenging modules for semiconductor technology development. Process development for this module shows high sensitivities to defectivity, in addition to aggressive critical dimension (CD) and process window targets required to meet shorts, opens and reliability metrics. Between 45nm and 32nm technology nodes the requirements for contact patterning have scaled more aggressively than the standard overall technology node scaling of 70%. These aggressive targets are accompanied by minimal advancements in tooling for both lithography and reactive-ion etch (RIE). Without major tooling enhancements, 32nm contact ground rules necessitated a dual path development strategy for both double and single patterning solutions in case either became infeasible due to cost or fundamental technical hurdles. Challenges include technical constraints as well as logistical implementation of a single patterning process into an existing double patterning design and data preparation flow. Through the use of novel process development and characterization techniques, it has been possible to navigate innumerable issues to enable a final low cost single patterning

contact process for the 32/28nm technology. In addition to tackling technical hurdles, the insertion of a single patterning process into a technology based on the assumption of double patterning is far from trivial. In attempting to insert a single patterning process into the existing technology framework we evaluated data preparation solutions as well as a more single patterning friendly SRAM designs to address these concerns.

II. MATERIALS AND PROCESS IMPROVEMENTS

A. Photo-resist capability

As 32/28nm technologies made the transition from research to development the initial focus was primarily on a double patterning contact solution. One of the major shortcomings in early process development was not only tooling, but also materials performance which gated the migration to a single patterning contact process. One of the primary issues was photo-resist performance. Photo-resist materials used in the 45nm contact patterning process show limited extendibility to single exposure for the 32/28nm node. Materials defectivity and performance were far from target, with significant shortcomings in the areas of resolution, contact edge roughness (CER) and CD uniformity. Early technology development work began using double patterning and pitch splitting as the primary approach. This permitted broader technology learning with the use of existing photo-resists, whose dense resolution were not compatible with a single exposure/single etch solution. Two areas of priority in the photo-resist performance evaluation process were sub-resolution assist feature (SRAF) printing sensitivity and dense feature resolution. These two metrics tend to work against one another in that high sensitivity photo-resists engineered towards dense feature resolution will often have difficulty supporting necessary SRAF sizing for sufficient process window in isolated features. The use of dedicated resources and a standardized work flow for resist material evaluation was critical to identifying new resist candidates, allowing the evaluation of up to thirty materials based on prioritized metrics. Not until well into the development cycle were materials found which adequately balanced performance in a way that would support migration to single patterning process. We took advantage of improvement in circularity, roughness and CD uniformity between the 1st development material also

978-1-61284-408-4/11 $26.00 © 2011 IEEE

used in 45nm technology and the material that was chosen during the 3rd development cycle and will be used for initial manufacturing ramp.

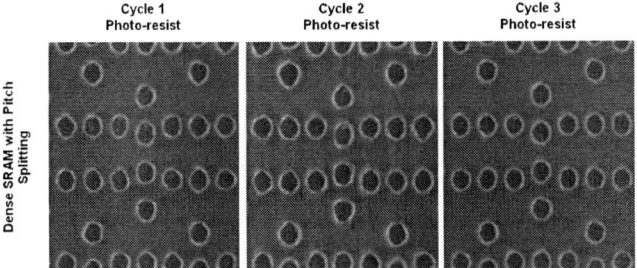

Figure 1: Comparison of photo-resist performance during 1st, 2nd and 3rd development cycles for contact layer. Shown is significant improvements in circularity and CD uniformity between early development cycles and manufacturing resist material.

B. Critical dimension reduction

Even with improved photo-resist performance and better minimum resolution, the technology requirements placed a large burden on RIE to shrink the features to their final target sizes. Rigorous co-optimization of resist CD and RIE process parameters had previously shown that maximum contact opens yield occurs when develop CD was between 70-80nm, while final targets were between 30-40nm. Outside of this range opens yield rapidly degrades as a function of decreasing develop CD. By maintaining litho CD above a critical value of 70nm, a large burden is placed on the etch process to achieve a shrink requirement of nearly 40nm. Two major revisions to the 45nm contact patterning process enabled this large degree of CD reduction. The first was the use of lower temperatures during the hard mask open (HMO) steps. The framework for a double patterning contact process is inherently multi-step and can be leveraged to allow temperature as an additional degree of freedom. By using a two step etch process with low HMO temperatures, more polymerizing by-products are created which will passivate the sidewall and allow for increased taper angles yielding a significant reduction in the final CD. Another important component of this process was the introduction of significant changes in the silicon containing anti-reflective coating (Si-Arc). Initial work was begun in this area in 45nm with the use of a tri-layer patterning stack. In order to extend this methodology, increased selectivity between the organic masking material and photo-resist was required for the purpose of CD shrink. The transition to Si-Arc materials with double the silicon composition enabled significant improvements in CER and shrink capability during the etch process.

C. Integration and materials improvement

Even with the significant improvements outlined above, process capability was still not adequate. Significant changes in integration for both double and single patterning were

needed to achieve the technology requirements for mean CD and uniformity. Initial improvement came paired with a double patterning approach to contact formation. By applying a thin oxide liner we obtained an additional final bottom CD reduction of 5-10nm. This reduced the shrink burden on the RIE process with only a small increase in complexity. The downside of this liner shrink technique is that the chemical vapor deposition (CVD) rates and therefore liner thickness are highly dependent on pattern density, resulting in the introduction of across chip CD non-uniformities.

Post-Etch Contact CD Uniformity (normalized)

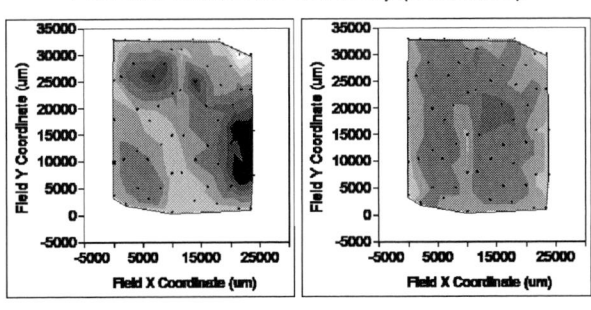

Figure 2: Comparison of across exposure field CD uniformity for oxide shrink process versus shrink process using amorphous carbon hard mask. Reduction of 50% in 3sigma of final contact CD uniformity is observed with amorphous carbon shrink process.

With improvement in resist resolution, more materials and integration schemes became available to remove the oxide liner and eliminate this across chip CD effect. Work with more etch resistant organic hard mask materials such as amorphous carbon began early on in the program but were initially discarded due to its lack of planarizing capability which was crucial for matching surface topography between the first and second steps of a double patterning integration scheme. With the switch to a single patterning integration scheme, the use of amorphous carbon became a powerful tool for CD shrink due to its improved etch resistance compared to most standard organic spin-on hard mask layers used in the industry. This change allowed for removal of the oxide liner shrink process and further improvement of across field CD uniformity, promoting migration to a more cost effective single patterning contact process.

II. SRAF AND OPC SOLUTIONS

A. Sub-resolution assist features

In order to achieve sufficient process window for contact hole imaging, sub-resolution assist feature (SRAF) placements are very important. The design styles encountered in 32/28nm contact levels are sufficiently complex and varied that it is not a trivial matter to define a rule-based SRAF scheme. Optical proximity correction (OPC) software vendors have looked at model-based SRAF placement tools as a way to address this challenge. For newer and more complex nodes after 28nm, it seems that these model-based SRAF placement tools may be

978-1-61284-408-4/11 $26.00 © 2011 IEEE

the only choice for achieving the best possible mask solution. However, with 32/28nm node we wanted to see if we could find an aggressive complex SRAF placement scheme without incurring the high computational cost of a model-based tool. The solution that was ultimately chosen was a hybrid solution which employed a model-based tool to analyze various layouts. This approach resulted in fairly complex rules-based SRAF placement code. The rules for SRAF placement attempt to mimic the placements found by the model-based tool over a wide variety of layouts. Although the resulting SRAF placement is fairly complex, the run time is improved over a model-based tool, shortening the turn around time for mask creation. The quality of the SRAF code depends largely on the careful assembly of good test layouts which span a wide variety of ground-rule clean design styles. These layouts are assembled and then run through model-based SRAF code. Simulations are done varying dose, focus, and mask size to verify that the resulting model-based SRAF placements provide the desired simulated process window. Once satisfied that the model-based SRAF code is adequate, the resulting SRAF placements are studied carefully to try to see what patterns could be turned into rules. SRAF widths, and distances from target features are measured, and then preliminary rules are coded. The coder then iteratively runs both the model-based SRAF placement code and the rules-based code and the simulation results are superimposed though a large variety of test patterns. As configurations are found where the solutions are quite different, the rules-based code is refined, and the process repeated. Over time, the rules-based code grows in complexity, but manages to converge to a solution which is comparable to the model-based solution.

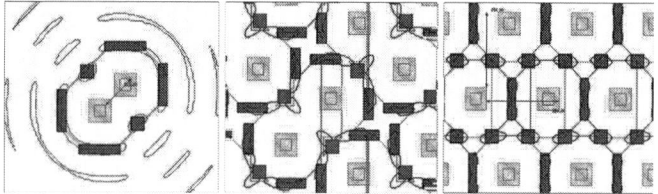

Figure 3: Examples shown of model-based assist feature placement (outline) compared with final rules based placement algorithm (dark bars) for several contact layouts (shaded square).

B. SRAFs and optical rules checking

The other challenge for SRAF placement code is developing SRAF rules that are sufficiently aggressive to help with main feature printing, while not being so aggressive as to result in printing assist features on the wafers. Addressing this problem is done partially through simulation, but also through careful study of performance on wafers. A large degree of fine tuning was required for the simulation settings to be able to predict where SRAF printing starts to occur. This information is then used in optical rules checking (ORC) to identify SRAF printing risks prior to creation of a product mask. Recently, OPC software vendors have been focusing efforts regarding how to deal with printing SRAFs. Several vendors have SRAF

print avoidance tools which leverage a model's ability to detect areas that are likely to have printing assists (as shown in Figure 4.) These tools are an integrated part of the OPC flow. They first perform several iterations of OPC, to get a mask solution which approximates the final mask solution. Simulations are then done to identify assist features which contribute to simulated printing assists. The offending SRAFs are then modified or deleted to address the SRAF printing. Additional iterations of OPC can then be done to converge to an appropriate final mask solution with the updated assist features. This model-based approach to fix printing SRAFs seems to be very efficient, as it targets the worst locations, while not globally reducing the impact of the SRAFs over the entire layout. The success of this approach to dealing with printing SRAFs is highly dependent on having a model that accurately predicts when SRAFs will be resolved in resist.

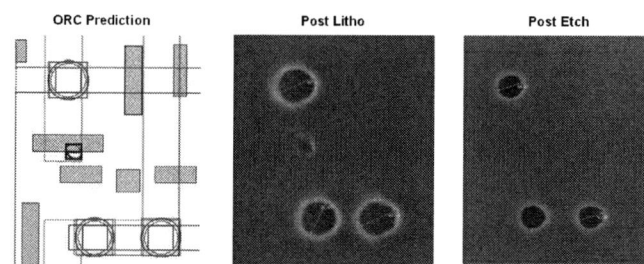

Figure 4: Demonstrate predictive capability for identifying locations susceptible to printing SRAF. Although SRAF printing does appear after lithography it is not transferred into the substrate during etch process.

III. DESIGN MANIPULATION

A. Data preparation re-targeting

The 32/28nm contact data preparation includes significant retargeting code, which manipulates the design before OPC. Current design rules allowed for aggressive placements of redundant contacts with center-to-center measurements below our single exposure lithographic printing capabilities. This was acceptable during early development assuming a technology using double exposure, but became extremely problematic during the later push to enable a single exposure contact patterning process. Rather that creating major disruptions in the design space, an attempt was made to address this problem by employing what we call "contact mapping" code. This code identifies groups of same net source/drain contacts which are closer together than the minimum supported pitch for single exposure lithography. These groups of contacts are then deleted and replaced with larger contact bars. The number of bars is approximately half the count of the replaced contacts, and they are approximately twice the area of the original contacts. The purpose is to match electrical characteristics, resistance in particular. The litho target size of these bars also has to take non-trivial etch biases into account. This code mapped two redundant contacts to one bar, three contacts to two bars, four contacts to two bars, five contacts to three bars and so on. Due to our desire to match resistances, the sizes of the replacement bars have to be varied according to the number of standard contacts they are replacing. For example, the pair of contact bars replacing four contacts need to be larger than the pair of bars replacing a triplet of contacts. Any time significant layout manipulation like this is attempted there are concerns about accidentally affecting the electrical characteristics of the chip. To address this, custom design-rule checking (DRC) was coded and inserted into the tape-out flow. These checks verified that we were not affecting the electrical connectivity within the chip, and that the new replacement bars were not creating risks for shorts to other contacts, gates, or to other metal lines. Utilizing DRC combined with the usual simulation based verification ensures that no harm was being done in these layout manipulations.

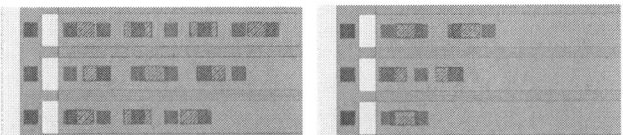

Figure 5: Example of data preparation treatment of redundant contacts. Contacts (dark square) are original design and contact bars (shaded) are post mapping. This results in approximately 50% fewer contacts and mapped contacts have approximately 2:1 aspect ratio.

B. Data preparation for contact redundancy

The contact mapping strategy described above is helpful for moving to a single expose contact strategy, as it removes pairs of redundant contacts which occur at a sub-resolution pitch and replaces them with bars which are easier to print. However, this elimination of redundancy is not ideal, as redundancy has been a powerful tool in maintaining high yield in many design styles. In special cases where there were only two contacts in a given source/drain region, we also employed a technique we call "contact spreading". Contact spreading is really a design modification done within the data prep flow where contacts are spread apart to a distance that is deemed relatively safe for single exposure patterning. The challenge with this spreading code is ensuring that contact locations are only shifted subtly, taking great care to maintain adequate metal coverage over the moved contacts, adequate active area coverage and sufficient dielectric spacing to gates and other contacts. We did this by employing the DRC code used to check ground rules in the OPC run deck. Contacts were spread apart, and when the resulting pair of spread contact passed DRC, they were left in the new spread configuration. When they failed DRC, the spread contacts were deleted, and the layout reverted back to the original drawn configuration, to then be passed into the mapping code. We ran this code on some large microprocessor designs, and found that we were able to spread 60% to 70% of source/drain contact pairs and maintain redundancy in these locations. The remaining pairs were subjected to mapping code converting them into bars.

C. Novel SRAM Design – OPC [2]

In addition to modifications of logic designs during data preparation, we also investigated potential re-designs of dense SRAM cells to make them friendlier for a single patterning process. The criteria of such design changes were to not change the basic footprint of the SRAM cell. A newly proposed SRAM design for contact layer combines contact (long), contact (square) and metal into one shape acting as local interconnect. This is achieved by using a L-shaped contact [4]. Conventional single exposure cells in comparison show foreshortening and large CD variation of the long contact, also leading to degradation of the neighboring square contact's process window.

New Design for Single Patterning SRAM using L-shaped contact (0.120um^2)	Standard Design for Double Pattenring SRAM using pitch splitting (0.120um^2)
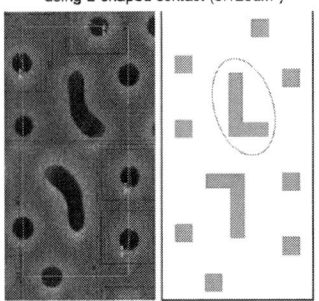	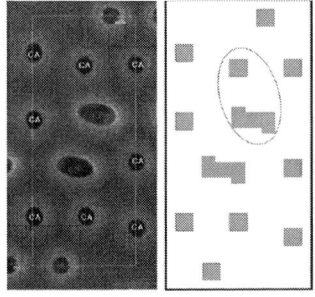

Figure 6: SRAM size of 0.120µm2 single exposed with L-shaped contact on left for NFet and PFet drain connection and connection to X-couple inverter gate (left) [4]; Same SRAM size using double patterning. All contacts split into CA(CA1) & CB(CA2) levels with state of the art inverter connection contact (square) to metal to contact (long) (right).

After the new design concept had been established, single exposure printability had to be proven. Fine tuning of the design layout involved several iterations of manual edge placement and printability assessment. This design optimization cannot be restricted to the contact layer only. Because overlap and electrical isolation are critical, active area and poly-conductor layers need to become part of the optimization process. Fig. 7 indicates qualitatively contact overlap (O) and distance (D) requirements as being critical for the new SRAM design. Fig. 7 also visualizes process variability bands (PV-bands) which were used as a metric to analyze, assess, and compare a variety of layout options.

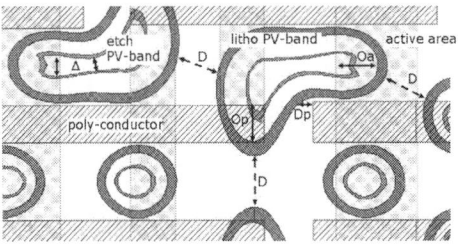

Figure 7: Visualization of parameters taken into account during contact printability assessment as well as target and OPC optimization. "O*" indicates overlap and "D*" minimum (safe) distance requirements which are all measured based on litho PV-bands. "Δ" is defined as the difference between the two inner etch PV-band widths as indicated by the two arrows, effectively representing the necking risk of the final contact acting as a wire.

978-1-61284-408-4/11 $26.00 © 2011 IEEE

Even though some printability assessments were used to define the final target for the SRAM design, the OPC still had to be optimized to achieve the most robust image in resist based on the contact single patterning process conditions. The emphasis for optimization was on the unique L-shaped contact, because any existing biasing in the OPC code was not optimized for such a complex 2D feature. In a first approach, the L-shape was taken as a starting point and attempts were made to meet the inter-level overlap requirements by experimenting with edge placement, fragmentation, and even the strategic introduction of jogs (fig. 8). Specific attention was given to the concave vertex of the L-shape because of its increased shorting risk to the neighboring poly-conductor shape.

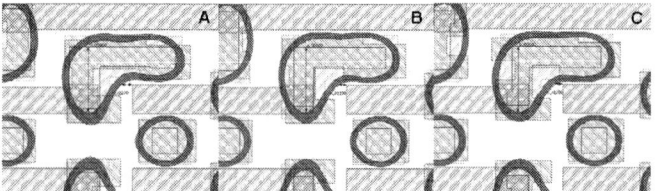

Figure 8: Layout design, litho target, mask layout as well as corresponding PV-bands for various examples of L-shaped contacts constructed from orthogonal edges only.

Out-of-the-box thinking was required to overcome the issues posed by OPC in conjunction with the L-shaped target. In fact, OPC cannot be blamed, because it's an unrealistic assumption that an L-shape can be accurately transferred into resist. In a second attempt 45-degree edges were introduced to the target shape of the "L" with the OPC (fig. 9) to avoid the severe degree of fragmentation which resulted from using target shapes with only 90-degree angles (fig.8). Another attempt at manual targeting and OPC optimization also addressed the high intensity at the corner of the L-shape by splitting the "L" into two pieces leaving a sub-resolution gap in between. This wouldn't work with a standard OPC approach because the gap is not actually intended to be resolved. The OPC code needed to be adapted such that only the edges forming the gap were not part of the target layer anymore. Effectively, the gap forming edges were frozen and not allowed to move during OPC (fig. 9).

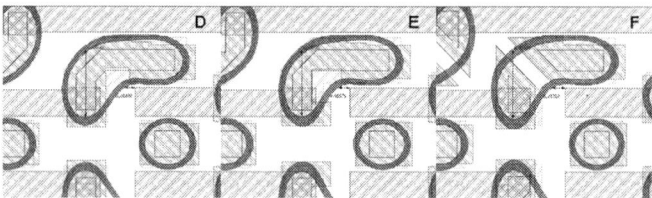

Figure 9: Layout design, litho target, mask layout as well as corresponding PV-bands for various examples of L-shaped contacts using at least one simple 45-degree edge and even separating the "L" into to pieces on the mask with a sub-resolution gap in between them.

As indicated by the contour shape, the corner "rounding" of the convex corner of the "L" shape could be reduced significantly in all three cases. If the orthogonal distance of the nominal contact contour to the poly-conductor is measured, this distance is increased by roughly 10nm. As an added benefit, the OPCed mask shape of the sub-resolution gap case follows relatively closely the given OPC target, which is indication of a healthy process. Not shown but also monitored was the delta of the inner contact opening measured at the final etch PV-band (see fig. 7). Since the final contact CD is in the 20 to 30nm range, even single digit nanometer decreases in local width (necking) increases the risk for yield loss due to opens. Table 1 compares all 6 OPC options A through F based on a simple metrics as shown.

Version	Op	Dp	Δ	O+D−Δ
A	38.4	16.7	9.6	45.5
B	37.5	11.9	8.2	41.2
C	39.6	10.3	7.2	42.7
D	41.2	14.8	6.0	50.0
E	39.7	16.7	6.0	50.4
F	42.2	19.1	4.9	56.4

Table 1: Comparison of printability quality (last column with highest values being best) based on CD measurements of PV-bands (see fig. X2 for measurement locations).

The proposed calculation of a printability assessment sum is certainly somewhat of a simplification but resulted in a quality approximation. All three parameters depend on each other and a maximized sum indicates greatest robustness. The sub-resolution gap construct proves to be superior versus all other shown solutions based on all three parameters (orthogonal distance to neighboring poly-conductor, overlap with poly-conductor, final etch contour necking). It can be further concluded that the sub-resolution gap resist image contour (F) is qualitatively and quantitatively very similar to the examples in fig. 9 above (D & E) with the common denominator of the introduction of 45-degree mask edges.

C. Novel SRAM Design – Patterning Verification

Arrays of the earlier discussed SRAM experiments A – F were placed on a test mask to verify their printing behavior as well as their OPC accuracy. A qualitative comparison of the images in fig. 10, shows that printability and resist contour prediction were close to target.

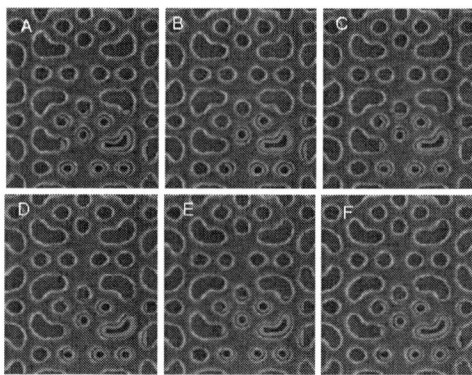

Figure 10: Post litho patterning results of SRAM OPC variations A – F with superimposed simulation results. Upper row (A – C) represents orthogonal targets and lower row (D – F) targets with at least one 45-degree edge.

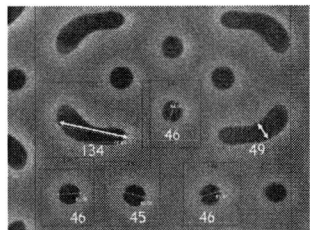

Figure 11: Post etch result of re-optimized OPC for SRAM option E with final CD numbers in nanometers.

An indication of an overall high pattern transfer fidelity can be obtained from the post etch SEM image in fig. 11, showing the final contact dimensions. Both previously explained critical overlap margins (L-contact ends to active area on one end and to poly-conductor on the other) stayed well within the process assumptions. Based on PV-band assessment and on-wafer measurement data, the robustness of this novel SRAM design and OPC solution for a single patterning contact process was comparable if not improved to its double patterning counterpart.

VI. PROCESS LEARNING METHODOLOGIES

Not only was process innovation crucial in the implementation of a single patterning contact process, but meticulous characterization was necessary to maximize process capability. The next portion of the paper will review innovative metrics and characterization techniques used to assess both process, OPC and integration improvements as they are modified throughout the technology development cycle. It is no longer sufficient to use only M inline monitoring structures and electrical metrics to evaluate patterning process capability relative to technology targets. Additionally, it has become even more important to rapidly iterate cycles of learning and ensure "first time right" implementation. Specifically for the 32/28nm technology nodes, three primary techniques have been critical in improving the efficiency and thoroughness of development cycles. These involve the use of patterning verification test masks, voltage contrast short-loop evaluation techniques and contact to gate budget analysis methodology.

A. Patterning verification test mask

Perhaps one of the best success stories in 32/28nm patterning development, specifically for contacts, was the use of a sector level patterning verification test mask prior to the first delivery of fully functional/electrically tested mask sets. This was extremely useful during the half-node development cycle for 28nm which was significantly shortened relative to standard technology development and was heavily dependent upon having a "first time right" solution. This approach utilizes rigorous on-wafer verification of OPC, lithography and etch for a wide range of features prior to creating the first integrated mask set having device performance, yield and reliability objectives. The proposed flow for creating a patterning specific test mask is outlined in Figure 12.

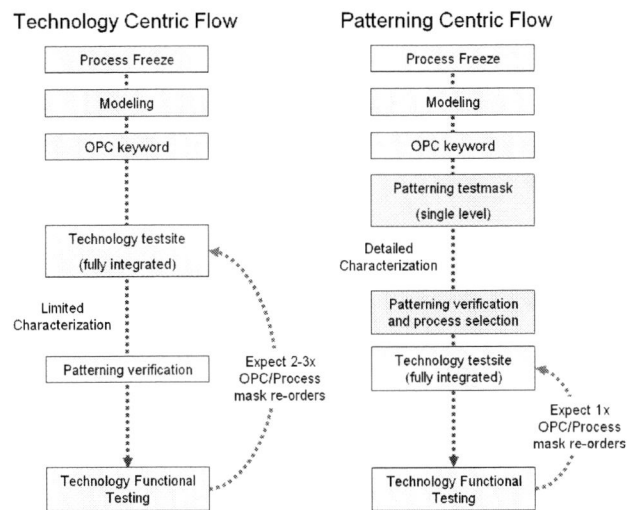

Figure 12: Flow diagram of patterning centric development cycle vs. technology centric development cycle. With patterning capability becoming ever more important in technology success, there is a strong need for more fundamental structural studies at the beginning of technology cycles.

On this test mask a wide array of structures would be placed for modulating spacing between contacts and custom sizing for dense SRAMs as well as assist feature sizing and ORC hotspot verification, that could be fully characterized with sector level methods such optical, CDSEM and voltage contrast inspection.. A specific version of the OPC keyword with the best process window can then be chosen from multiple iterations. Additionally small adjustments can be made in both process and OPC to ensure that there is a high degree of confidence for a following integrated test site. Relative to number of steps and activities of a more traditional development cycle, this methodology is not novel, but instead an improvement to the quality and cycle time of gauging patterning performance. When the goal is a patterning only process assessment we can quickly build simple macros specifically for that purpose, instead of relying on complex fully-integrated testable content intended for device or other broader yield assessments. This enables an improvement in cycle time to address patterning-related issues and increases the thoroughness of the patterning sector assessment prior to committing the process to integrated hardware. Though it may appear to increase the budget and duration of the development cycle with an additional mask order, we dramatically reduce the risk of re-orders on the integrated mask set for minor OPC keyword or process adjustments by performing this detailed preliminary patterning evaluation. This is especially true as technologies become more aggressive and rely heavily upon patterning capability for success. Additionally it makes available the resources and test mask space to evaluate a wide variety of much more high risk/high reward content such as single vs. double patterning contact experiments that may not have otherwise been evaluated on an integrated test site.

978-1-61284-408-4/11 $26.00 © 2011 IEEE

B. Rapid experimental cycles of learning

As hinted above, detailed and thorough process characterization would not be possible without a meaningful evaluation technique to detect all modes of contact patterning failure. For several generations contact patterning has benefited enormously from compatibility with voltage contrast e-beam inspection. This inspection functions by grounding the substrate and scanning the surface to detect differences in partially formed or missing features. Depending on macro design this can be used to detect both contact opens as well as contact shorts to the gate. The enormous benefit of this inspection is that it provides an assessment comparable to contact chain electrical testing but can be performed on low value substrates that have had as little as one lithography level of processing. This allows the capability to run experiments at much higher volume and with much lower cost and cycle time than using fully functional wafers. Additionally this methodology is significantly more effective than traditional optical inspections. Accuracy is improved because the voltage contrast will detect contacts that are fully formed on the surface, but partially etched and not electrically grounded to the substrate. The signal to noise ratio is also superior because any nuisance defects such as surface foreign material or embedded contamination won't be detected unless they directly interfere with the contact connection to the substrate. With this highly sensitive evaluation metric at our disposal, we were able to rapidly iterate cycles of improvement for both lithography and RIE to enable single exposure processes for 32/28nm contact patterning. Specifically, this methodology was extremely useful for resist developer recipe optimization with short cycle times that would not have been possible using optical inspections or functional testing (fig. 13).

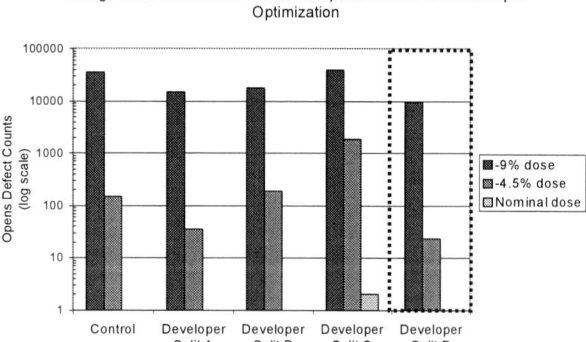

Figure 13: Voltage contrast inspection data for contact opens at nominal dose and two lower dose conditions. No distinction can be made between processes at nominal conditions versus the control cell using the current process of record. With dose and CD conditions at the low end of the process window, developer split "D" demonstrates the best contact opens performance.

By combining voltage contrast inspection with process window modulation, specifically variable across wafer dose or focus conditions, it is possible to detect very subtle improvements between competing process iterations and their impact to process window. For developer optimization, no difference in process splits could be detected at nominal process conditions, however at the lower dose and CD conditions we are able to observe improvements contact opens compared to current (control) process. The ability to detect small improvements at the corners of the opens process window are critical for the development of a manufacturable single patterning solution.

C. Process tolerance budget [3]

In addition to the rapid cycle time methodology above, an additional evaluation method is needed to assess integrated structure and variation relative to process assumptions. This can be a tricky task as structural learning is often derived from physical failure analysis (PFA) at process extremes and is not representative of the mean process center. The procedure outlined below is a combination of both highly detailed top-down metrology combined with inline transmission electron microscopy (TEM) to assess minimum reliability ground rules for contact to gate spacing without the selection bias commonly associated with PFA. At a high level there are five components needed to determine this minimum spacing relative to a fixed design: contact and gate mean CD, contact and gate variability as well as CA-PC overlay variability.

32/28nm Contact – Gate Integrated Scorecard			Values
Mean Line Center		Electrical Gate Width	[mean]
		Contact Bottom CD	[mean]
Variability Components	Contact - Gate	Overlay	[3sigma]
	Contact	CD (shape)	[3sigma]
		LER (edge)	[3sigma]
	Gate	CD (shape)	[3sigma]
		LER (edge)	[3sigma]
Contact – Gate Mean Spacing		Calc/Obs	[mean]
Contact – Gate Minimum Spacing		Calculated	[3sigma]
Process Assumptions Met?			Yes or No

Table 2: Above is tolerance calculator showing components for contact and gate variability assessment.

Overlay variability can readily be determined from standard inline monitoring and metrology. For the purpose of determining CD variability we have devised a high density CD measurement plan using the scanning electron microscope (CD-SEM). This variability component would be further subdivided into lot to lot variability, within wafer variability, line edge roughness and thru pitch variation. The lot to lot variability is determined from standard inline monitoring. For measuring the remaining components we have created a new wafer characterization measurement structure to be placed alongside the mask house product control image (PCI) for high density within wafer measurements distributed equally across the reticle field. We then proceed to measure approximately one hundred of these sites for the wafer characterization PCI (WCPCI) and SRAM across a given exposure field. These same targets would then be measured on approximately twenty exposure fields across the wafer.

This is a fairly robust statistical sampling of several thousand measurement points of data that far exceeds what we can assess with PFA. When choosing target locations and exposure fields it is important to consider sampling an equal distribution of location and pattern density across the reticle as well as radial distribution throughout the wafer. From the measurement data we can summarize the across reticle, across wafer, CER components as well as an approximation for thru pitch contribution. We use the variation between contacts with different local environments in an SRAM as the approximation for thru pitch. This is used instead of the fully allowed design rule space because only a subset of these geometries, typically the SRAM is the best example, are subject to the minimum contact to gate spacing. By combining the variability components from the high density wafer measurements and the inline variation metric we determine a total value for CD variation for both contact and gate that can be directly compared to process assumptions. For CD mean values we use TEM measurements chosen at random from the same minimum spacing structures, typically SRAM, measured in the top-down CDSEM variability assessment. A statistically significant sample size of TEM measurements is needed to determine physical CD for both gate and contact that we can correlate with the high density top-down metrology data. The TEM measurement and top-down CDSEM measurement can then be calibrated based on the offset of the measured sample to the inline mean to determine a mean CD value to be inserted in the tolerance assessment calculation. With all the measured and calculated values above we can produce a tolerance calculator (table 2) to assess current process capability relative to process assumptions.

Viewing the process capability as a tolerance calculator is extremely useful as an evaluation metric for several reasons. It provides an assessment of process assumptions without the selection bias associated with failure analysis sites that are often outside a normal distribution. During this verification procedure for contact patterning, the budget methodology is helpful in identifying critical areas in need of improvement, some of which may otherwise have been overlooked. When inserting measured variability data into a tolerance calculator, we often discover a disparity between the calculated values compared to direct observations from TEM measurement for contact to gate spacing. This is the primary objective of the TEM validation and allows us to identify detractors such as non-linear contact profile, silicide mushrooming and gate undercut that reduce contact to gate spacing but may not be adequately considered in the technology process assumptions. In terms of process development it allows us a robust methodology for assessment of capability with reasonable resource requirements and helps direct development focus towards addressing the largest detractors. This measurement technique has been aggressively utilized to compare process

iteration and gauge degrees of improvement and can be added to the electrical and yield performance as a standard metric for process change review.

IV. CONCLUSION

The implementation of novel process changes and robust characterization methods has been critical to the enablement of a single patterning contact process. Through the introduction of new materials and integration we were able to realize success first using a double patterning process to enable early technology learning and allow the necessary time to overcome the barriers impeding the use of single patterning for contact module. As photo-resist performance improved, unique data preparation solutions including contact mapping and spreading helped minimize potential design conflicts that could have thwarted a single patterning process change from occurring late in the development cycle. It was also demonstrated that novel SRAM designs could additionally reduce single exposure printability issues associated with aggressive technology SRAMs. During the rapid iteration of both double and single patterning process learning, crucial technology decisions were made based on the use of rigorous characterization methods like voltage contrast inspection and tolerance budgeting. As a result of these aforementioned process improvements and characterization methods we have been highly successful in demonstrating a low cost single patterning contact for both 28/32nm technology nodes.

ACKNOWLEDGMENTS

This work has been supported by the independent Bulk CMOS and SOI technology development projects at IBM Microelectronics, Div. Semiconductor Research and Development Center, Hopewell Junction, NY 12533.

REFERENCES

[1] Morgenfeld, B., Stobert, I., An, J., Kanai, H., Chen, N., Aminpur, M., Brodsky, C., Thomas, A., "Contact patterning strategies for 32nm and 28nm technology" in Optical Microlithography XXIV, edited by Mircea V. Dusa, Proceedings of SPIE Vol. 7973 (SPIE, Bellingham, WA 2011) 797319.

[2] Haffner, H., Ostermayr, M., Kanai, H., Chang, C. S., Morgenfeld, B., An, J., Luo, M., Zhuang, H., "Single exposure contacts are dead: long live single exposure contacts" in Design for Manufacturability through Design-Process Integration V, edited by Michael L. Rieger, Proceedings of SPIE Vol. 7974 (SPIE, Bellingham, WA 2011) 79740E.

[3] Brodsky, C., Chu, W., " Lithography budget analysis at the process module level," Proc. SPIE 6154, 61543Y (2006).

[4] H. Kawasaki, M. Khater, M. Guillorn, N. Fuller, J. Chang, S. Kanakasabapathy, L. Chang, R. Muralidhar, K. Babich, Q. Yang, J. Ott, D. Klaus, E. Kratschmer, E. Sikorski, R. Miller, R. Viswanathan, Y. Zhang, J. Silverman, Q. Ouyang, A. Yagishita, M. Takayanagi, W. Haensch, K. Ishimaru: "Demonstration of highly scaled FinFET SRAM cells with high-κ/metal gate and investigation of characteristic variability for the 32 nm node and beyond", Proc. Electron Devices Meeting, 2008. IEDM 2008.

Non-contact Handling and Transportation for Substrates and Microassembly Using Ultrasound-Air-Film-Technology

Gunther Reinhart, Michael Heinz, Johannes Stock
Institute for Machine Tools and Industrial Management
(*iwb*), Technische Universität München
Munich, Germany
michael.heinz@iwb.tum.de

Josef Zimmermann, Michael Schilp, Adolf Zitzmann, Jens Hellwig
Zimmermann & Schilp Handhabungstechnik GmbH
Regensburg, Germany
josef.zimmermann@zs-handling.com

Abstract—The requirements needing to be met by handling technologies and factory automation in semiconductor fabrication and microsystems technology are rising relentlessly. Fragile, surface-sensitive and thinned substrates call for new, innovative approaches in order to tackle numerous material handling tasks. Therefore a new upcoming handling technology based on the ultrasound-air-film-technology and its applications for non-contact handling in PV-Thin-Film and microassembly are presented in this paper.

Keywords-component; Non-contact-Handling; PV-Thin-Film; Microassembly; Ultrasound; Ultrasonic; Air-Film

I. Introduction

In the year 1959 Richard Feynman asked in his famous speech: *Why cannot we write the entire 24 volumes of the Encyclopaedia Britannica on the head of a pin* [1]? What he wanted to talk about in this context is (or was) the problem of manipulating and controlling things on a small scale. Since then, the semiconductor and photovoltaics manufacturing industry as well as the close-by microsystems technology have continuously kept making innovations in circuit design, smaller geometries, larger and thinner wafers, leading to higher circuit density, increased circuit speed, increased functionality and degree of efficiency as well as remarkable price reductions [2]. However, this ongoing success has always been associated with the development of new manufacturing methods, for example the LIGA-technique [3], and massive efforts in fab automation. Especially the material-handling tasks in a fab which are very large nowadays. Therefore automated material handling systems play a key role to prevent human's handling errors, such as dropped wafers, and to reduce manufacturing cycle times. The increase in wafer size in combination with thinning out the wafer continuously, miniaturization and extreme circuit shrinkage require strict quality control and higher-class clean rooms to reduce the increased risk of particle contamination. Thus the number of human operators as a significant source of particles has to be reduced to a minimum. Moreover, these relentlessly rising requirements have to be met by special handling technologies and new approaches in factory automation in semiconductor fabrication and microsystems technology. Fragile, surface-sensitive and thinned substrates call for new, innovative and smooth transportation and handling devices, respectively. Useful applications of smooth handling are in the front-end fabrication as well as in the back-end assembly.

II. Physics of Ultrasonic Levitation

A. Introduction to Ultrasonic Levitation

Basically, ultrasonic levitation can be split into two different phenomena, the standing-wave-levitation (SWL) and the squeeze-film-levitation (SQL), also known as near-field-levitation (NFL).

The first principle (SWL) was first used by the NASA and ESA, respectively, in the nineteen-seventies in an area of research called containerless-processing. For those investigations in the field of micro-gravity it was necessary to place small solid or liquid test specimen contactless in the free space. This can be achieved by positioning a reflector at a distance of multiples of half the wavelength of an ultrasonic sound source. In between the ultrasonic sound source and the reflector small parts can be levitated in the pressure nodes of the generated standing wave pattern. Nowadays those ultrasonic traps are still useful for explorations in experimental physics and chemistry, for example to investigate the chemical dynamics of micro-particles [4].

The second phenomena, the ultrasound-air-bearing that is also known as squeeze-film-levitation in literature [5, 6], is a unique and promising method to handle flat products like wafers, both PV and semiconductors, glasses for thin-film solar cells or flat panel displays, dice in the back-end and microassembly, and foils in a very clean, gentle, fast and secure way without any surface contact. A detailed introduction into the basic principle of squeeze-film-levitation can be found in numerous references [5, 7]. However, the use and the potential of acoustic levitation for contactless handling purposes in production engineering are described for the first time in detail in [8, 9]. Furthermore, its application for handling tasks in microassembly is studied in other papers [10, 11].

The authors would like to express their gratitude towards the German Federal Ministry for Education and Research (BMBF), the Project Management Agency Karlsruhe (PTKA) for the funding of the research project SonicGrip (02PG2324) and towards the Deutsche Forschungsgemeinschaft (DFG-ZA 288/31-1) as well as towards the Bavarian Ministry of Economic Affairs, Infrastructure, Transport and Technology (TP22/07-IBN/p-2277/08) for the financial support.

B. Squeeze-Film-Levitation

The physics of SQL derives more from fluid dynamics than from acoustic principles [5, 6, 7] and is by now the most applied non-contact handling technology in the industry.

The gas pressure in the gap between the levitated part and the vibrating surface of the sound generator rises due to the cyclic compression and decompression of the thin gas film resulting in a mean pressure p_R exceeding the ambient pressure p_m (Figure 1). Any gas - air or process gases - can be used. Typical gap sizes are between 25 μm and 300 μm.

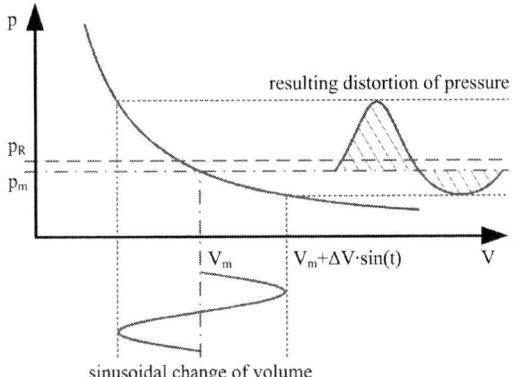

Figure 1. Basic principle of squeeze-film-levitation.

The operational behaviour of an ultrasound air -bearing is comparable to that of a conventional air bearing – except for the losses from generating compressed air as no pumps, pipes and nozzles are required. Furthermore the energy consumption of an ultrasound air -bearing is much lower compared to any other non-contact handling technology. It is to mention, that the damping air film prevents the vibration energy from being transferred to the part. Thus there is basically no vibration of the part that rather levitates at a well defined distance above or underneath the vibrating surface.

For non-contact handling purposes ultrasound is generated by transducers which are supplied by a separate power electronic device. These Langevin-Bolt-Transducers (LBT) are driven at its natural frequency by exciting the piezoelectric discs with a high frequency voltage. By means of special designed horns t he ultrasonic vibrations are amplified to a maximum amplitude at the very right end of the system (Figure 2). This illustration impressively shows the typical shape of the first longitudi nal eigenfrequency and the characteristic distribution of areas with low (light grey to white shade) and high amplitudes (dark grey shade). Typical amplitudes are within the range from 5 to 30 microns.

Depending on the application and the handling task, respectively, special adapted end-effectors are mounted to the end of the amplifying horn. Usually, this has to be done in such a way, so that the natural frequency of the transducer will not be disturbed too much. Therefore, two kinds of handling tools, also called sonotrode, can be distinguished.

For the handling of small parts, for example dice in the back-end, only a small ultrasound -vacuum-pipette is mounted to the end of the amplifying horn (Figure 3 (a)). This can be

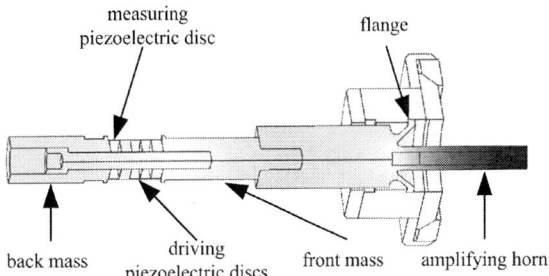

Figure 2. Basic principle for generating ultrasonic vibrations by means of so called Langevin-Bolt-Transducers. Illustration of the characteristic oscillation mode and its corresponding distribution of areas with low (light grey to white shade) and high amplitudes (dark grey shade).

interpreted as an additional extension of the horn, only de - tuning the natural frequency slightly. Furthermore, the extended system is still driven in the oscillating mode as shown in Figure 2 and, if frequency tuning should be necessary, the system can easily be tuned by adjusting the length of the amplifying horn together with the tip.

When handling large parts such as wafers, glasses or flat panels a simple extension of the horn is inadequate to generate a stable and powerful ultrasound-air-film, able to lift the weight of the ob jects. Moreover, large -area sonotrodes are necessary to scale up the lifting forces that are in first approximation proportional to the supported area of the levitated part (Figure 3 (b)). Thus, planar plates are dimensioned for oscillating in a well-defined eigenshape and whose natural frequency corresponds to the working frequency of the transducer as shown in Figure 2. This requirement is essential for the natural frequency as well as for the oscillation mode of the combined system. By now, these planar sonotrodes can be designed and manufactured of diverse materials, for example aluminium or plate-glass for applications in high-class clean rooms.

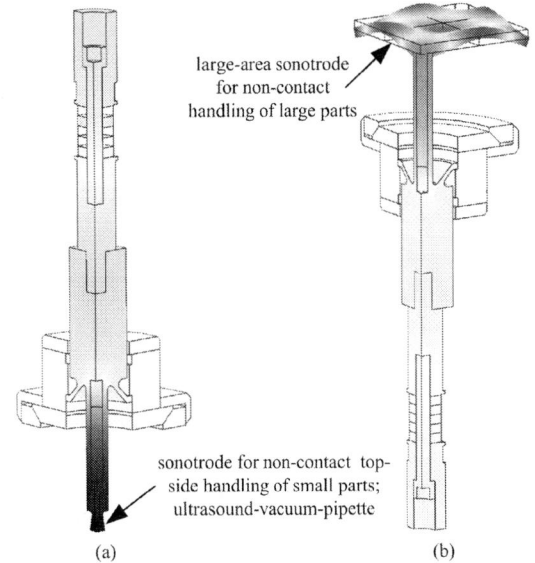

Figure 3. Oscillation mode s for both applications, (a) non-contact top-side handling of small parts with ultrasound-vacuum-pipettes and (b) non-contact handling of large objects with large-area sonotrodes.

978-1-61284-408-4/11 $26.00 © 2011 IEEE

To use squeeze-film-levitation for non-contact top-side gripping tasks, repelling ultrasound forces (F_{US}) are used in combination with attracting low-pressure forces (F_{LP}) in a special sonotrode design (Figure 4). By superimposing the weight of the part (F_G) with those pressure forces, an equilibrium of forces can be established so that the part levitates at a well defined distance underneath the gripping area. Further, the vacuum flow involves low pressure zones at the front surfaces of the part, thus centring the part relative to the gripper without any tactile stoppers. Though, self centring is only operative given geometric congruency of both surface areas and for small, very light parts, for example dice. With increasing size of the part, the ratio of weight to centring forces becomes disadvantageous and tactile stoppers will be necessary to keep the position relative to the gripper. This is the case when handling solar cells or flat panels. However the principle of top-side gripping still remains applicable with increasing size and weight and can be adapted to and applied in many applications and different industrial markets.

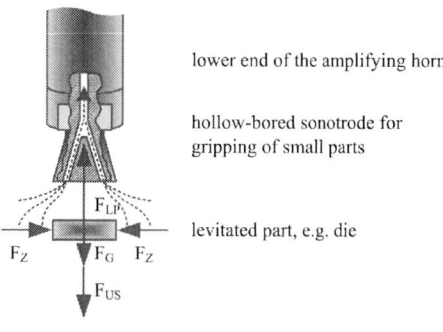

Figure 4. Principle for non-contact top-side gripping, exemplarily shown by a sonotrode for pick-and-place of small parts.

III. APPLICATIONS

As already described versatile products covering a huge range of sizes can be handled contactlessly by using the ultrasound-air-film-technology. Although the basic principle, a combination of repelling ultrasound forces and attracting vacuum forces, remains the same for all part sizes, three different categories of end-effectors can be distinguished (Figure 5). First, ultrasound-vacuum-pipettes (Figure 5(a)) are an innovative alternative to conventional vacuum-pipettes and die-collets, respectively. Especially the absolutely non-contact gripping of surface sensitive parts holds several advantages. In the first place the minimized risk of particle contamination has to be mentioned. For applications in micro assembly the end-effectors are usually mounted to high-precision robots and mounting machines [12]. Non-contact grippers with tactile shoulders for handling tasks in semiconductor and photovoltaics are related to the second category (Figure 5(b)). In contrast to the handling of small parts by means of ultrasound-vaccum-pipettes, tactile shoulders are necessary when handling larger parts in order to keep the position relative to the gripper. But, similar to the ultrasound-vacuum-pipettes these end-effectors are again mounted to robots. Thirdly, non-contact conveyors are slightly different to the first two

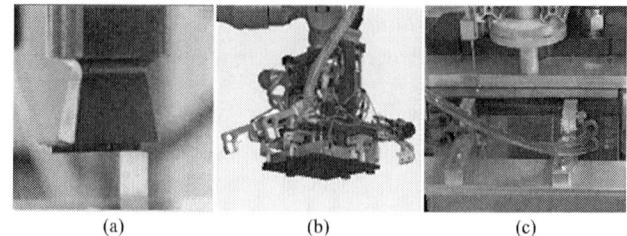

Figure 5. Classification of non-contact end-effectors based on the ultrasound-air-film-technology. Category (a): non-contact ultrasound-vacuum-pipette without tactile shoulders for the handling of small parts, applications in microassembly; category (b): non-contact-grippers with tactile shoulders for applications in semiconductor and photovoltaics; category (c): non-contact conveyors with tactile shoulders or tactile carriers for applications in seminconductor and photovoltaics.

categories. Here, the end-effector itself is fixed and the transported object is moved relative to the supporting area of the conveyor by means of tactile carriers (Figure 5(c)). Typical applications are picker modules for wafer-concentrating and -distributing [13].

A. Non-contact Handling in Microassembly

In [12] an approach was presented that facilitates the dimensioning of the gripping characteristics of ultrasound-vacuum-pipettes (category (a)). However, the focus was set on the design and optimization of the vertical equilibrium of forces. Based on these considerations it is shown in the following how to optimize the horizontal equilibrium of forces and the self-centring of the part relative to the gripper, respectively.

Considering the hollow-bored sonotrode in Figure 4, it is obvious that the ultrasonic amplitude and the pushing force as well as the low pressure and the pulling force are the main parameters for the vertical condition of equilibrium. However, additional factors like diameter, position and number of the vacuum nozzles have indirect but strong influences on the handling performance and the stability of the part. Although it is not possible to calculate an optimized parameter set quantitatively, these factors and its interdependencies have to be optimized during the design at least qualitatively. Thus guidelines for the design of the vacuum nozzles will be defined on the basis of the following considerations.

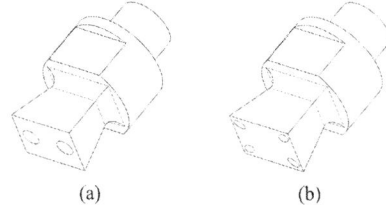

Figure 6. Optimization of the self-centring-affinity and of the handling performance by varying the design of the vacuum-nozzles in the sonotrode.

Obviously the centring of the part relative to the sonotrode has to be due to some forces F_Z that act on the front sides of the part. Basically, two kinds of forces are operative when levitating a part on an ultrasound-vacuum-air-film. On the one side the ultrasonic forces F_{US} due to the cyclic compression and decompression in the air gap could account for the centring. But, on the other side it seems much more auspicious to have a closer look at the flow regime of the low -pressure forces F_U. This issue can be analyzed qualitatively by means of CFD models. Figure 7 (a) visualizes the resulting streamlines due to the vacuum flow in the air gap for the ultrasound -vacuum-pipette as shown in Figure 6 (a). Because of symmetry reasons the considered domain only illustrates a quarter of the air gap. When virtually deflecting the position of the part relative to the sonotrode (Figure 7(b)), the streamlines are accelerated around the affected front edge of the part. Thus the local pressure at the front surface decreases analogue to the Bernoulli equation. Hence, the effective force on the front side tries to pull back the part into position.

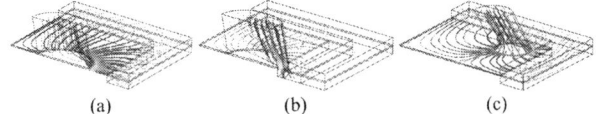

(a)   (b)   (c)

Figure 7. Optimization of the self-centring-affinity and of the handling performance, respectively, by varying the design of the vacuum-nozzles in the sonotrode.

However, it is also obvious for the pipette shown in Figure 6 (a), that the centring forces only act mainly on one front side. This fact is confirmed when identifying the handling performance of the levitated pa rt and illustrating the self -centring-affinity in the corresponding stability card (Figure 8). Here, three different states are distinguished. The first state is denoted as not stable and implies too low as well as too high vacuum force, causing drop of th e part or contact between the part and the gripper. The second state is named unstable. That is, the part is levitating, but oscillating predominantly in the direction perpendicular to the second front side that is only affected slightly by the centring fo rces. Finally, one can observe a relative large area where the part is stable, thus suitable for contactless pick andplace tasks.

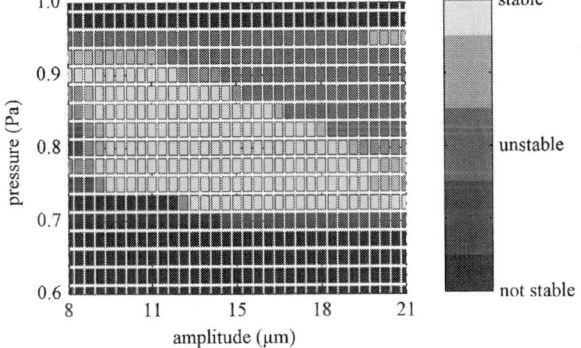

Figure 8. Stability card for part behaviour as a function of the amplitude and the pressure for sonotrode design corresponding to Fig ure 6 (a).

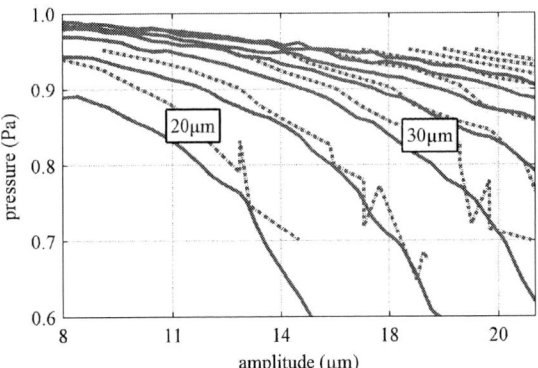

Figure 9. Resulting contour lines for relevant clearances corresponding to equilibrium of forces betweeen weight of the part, pushing ultrasonic force and pulling vacuum force (solid lines) as well as measured contour lines for the clearance (dashed lines).

These considerations allow the design of an optimized sonotrode with rearranged vacuum nozzles. In order to affect both front sides of the part the idea is to position four vacuum nozzles at the corners (Figure 6 (b)). The vertical equilibrium of forc es and the corresponding diameter of the nozzles, respectively, are again dimensioned by following the design method as described in [12]. As shown in Figure 9, the dashed lines, representing the measured data for the clearance, almost match the data gener ated by the force measurement (solid lines).

Additionally, Figure 10 shows the stability card for the optimized sonotrode. In contrast to Figure 8 there's only a very small domain with unstable behavior of the part. Moreover the stable domain is dominating, that is the part is levitating stable without oscillations in any direction now. Again there are two states denoted as not stable implying too low as well as too high vacuum force, causing drop of the part or contactbetween the part and the gripper.

On the basis of these considerations it is finally possible to deduce guidelines for the design of the vacuum nozzles:

- The vacuum nozzles have to be positioned as close to the edges as possible.

- The zone of influence of the vacuum nozzles has to cover an area at the front sides of the part as large as possible. Additionally the vacuum nozzles are best positioned in the corners.

- As the centring forces are increasing with decreasing pressure the vertical state of equilibrium should be moved to lower pressure levels.

- The effective area of the attracting low pressure forces on the surface of the part should concentrate on the boundary areas. The inner area should be provided for the optimal characteristic of the squeeze -film-levitation.

- That is, the vacuum nozzlesare to be designed as small as possible.

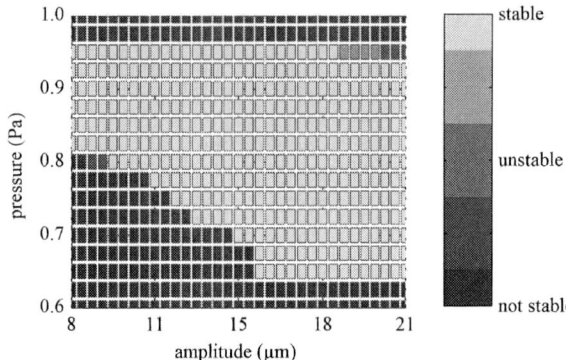

Figure 10. Stability card for part behaviour as a function of the amplitude and the pressure for sonotrode design corresponding to Figure 6 (b).

B. Non-contact Handling and Transportation of Substrates

The system performance of an ultrasound -air-bearing is comparable to that of a state -of-the-art pre-loaded air-bearing. Figure 11 displays the typical payload graph used for backside handling. For handling purposes a distance of approx. 100 μm is the optimized operating point. The system performs quite stiffly, any disturbance is compensated immediately by a growing restoring force. Thus the ultrasound air -bearing can also be used for handling of larger parts up to dimensions of several meters. Applications already implemented in industrial environments are

- semiconductor wafer handlers (end effector),
- handling of thin-film- and FPD-substrates
- as well as photovoltaic wafer handling.

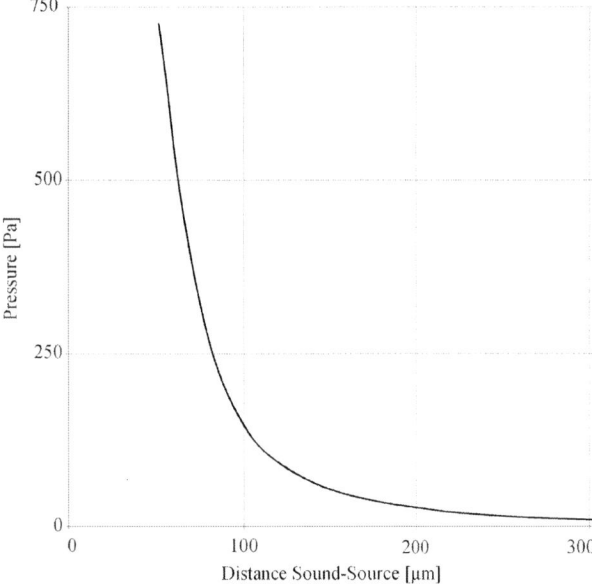

Figure 11. Example of a payload diagram.

The semiconductor end-effector uses a standard ultrasound-air-bearing for backside -handling. The wafer is hovered by squeeze-film-levitation throughout the whole surface. Side -stops prevent the substrate from sliding of the gripper. As this is form closure instead of force closure used in edge -grippers the stress on the edges is reduced tremendously. Single-paddle end-effectors can be realized as well as double-paddles (Figure 12). Both symmetrical as well as asymmetrical versions can be implemented (e. g. 12"/18"-end-effector).

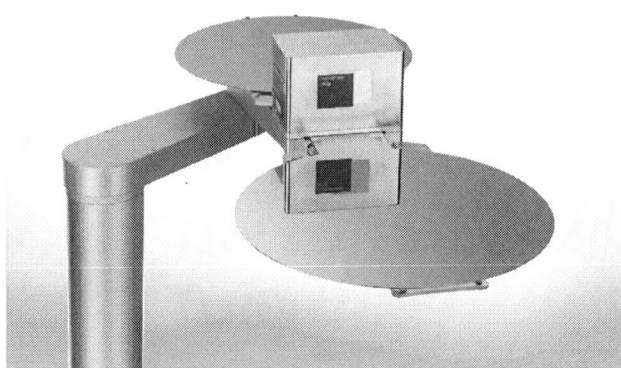

Figure 12. Semiconductor wafer handler (12"/300 mm-end-effector).

In many industrial applications backside handling is inadequate. E. g. thin -film-PV or FPD -substrates (glass) require top-side handling. Also for handling of larger parts the combination of ultrasound and low pressure (Figure 3b) can be used. To achieve a homogenous distribution of handling forces the low pressure nozzles also have to be distributed throughout the whole surface of the handling device. Gripping devices of different sizes have been implemented, for 300mm by 300 mm substrates (Figure 5b) as well as for 1200 mm by 650 mm substrates (Figure 13).

Figure 13. Thin-film-PV top-side-gripper for substrates (1200 mm x 650 mm).

Both semiconductor end -effectors and FPD -grippers are designed to be implemented in existing production environments. For hand ling challenges in photovoltaic manufacturing new machine designs are economically worthwhile. As there is no contact an d therefore no friction between ultrasound air -bearing and substrate it is not mandatory to move the substrate with the handling tool. In fact the substrate can be moved relatively to the ultrasound -air-bearing by means of tactile carriers . Hence those systems can be related to category (c). Using the ultrasound -vacuum-technology overhead handling can be implemented.

Applied to PV -wafers t his machine concept can increase the throughput by factor 2 or more (>7000 wafers/h). Whereas the breakage rate for cSi -substrates is decreased to less than 70 ppm. A n overhead handling system with integrated wafer singulation (pick-up distance up to 40 mm) is sh own below (Figure 14).

Figure 14. Semiconductor wafer handler (end-effector).

IV. SUMMARY

The ultrasound-air-bearing technology and its applications in semiconductors, photovoltaics and flat -panel-display manufacturing show great promise for non -contact handling. Not only stress and damages of the parts that have to be handled can be decreased tremendously but also the energy consumption is the lowest among all known non -contact handling technologies. Consequently this is a further step towards green fabs.

ACKNOWLEDGMENT

The authors would like to express their gratitude towards the German Federal Ministry for Education and Research (BMBF), the Project Management Agency Karlsruhe (PTKA)

for the funding of the research project SonicGrip (02PG2324), towards the Deuts che Forschungsgemeinschaft (DFG -ZA 288/31-1) as well as towards the Bavarian Ministry of Economic Affairs, Inf rastructure, Transport and Technology (TP22/07-IBN/p-2277/08).

REFERENCES

[1] Feynman, R.: There's plenty of room at the bottom. Journal of Microelectromechanical Systems 1 (1992) 1, S. 60 –66.

[2] Lee, T. -E.: Semiconductor Manufacturing Automation. In: Nof, S. Y. (Hrsg.): Springer Handbook of Automation. Berlin: Springer 2009, S. 991-926. ISBN: 9783540788300.

[3] Saile, V.: LIGA and its Applications. Weinheim: Wiley-VCH 2009. ISBN: 9783527316984. (Advanced micro & nanosystems 7).

[4] Mason, N. J.; Drage, E. A.; Webb, S. M.; Dawes, A.; McPheat, R.; Hayes, G.: The spectroscopy and chemical dynamics of microparticles explored using an ultrasonic trap. Faraday Discussio ns 137, (2008), S. 367–376.

[5] Minikes, A.; Bucher, I.: Coupled dynamics of a squeeze -film levitated mass and a vibrating piezoelectric disc: numerical analysis and experimental study. Journal of Sound and Vibration 263 (2003) 2, S. 241–268.

[6] Höppner, J.: Verf ahren zur ber ührungslosen Handhabung mittels leistungsstarker Schallwandler. M ünchen: Herbert Utz Verlag 2002. (Forschungsberichte iwb 164).

[7] Hashimoto, Y.; Ueha, S.; Koike, Y.: A theoretical study of near -field acoustic levitation of planar objects. In: He rbertz, J. (Hrsg.): Proceedings of the 1995 World Congress on Ultrasonics. Berlin 1995, S. 839–842.

[8] Reinhart, G.; H öppner, J.: Non -Contact Handling Using High -Intensity Ultrasonics. Annals of the CIRP Vol. 49 (2000), S. 5 –8.

[9] Reinhart, G.; H öppner, J.: The Use of Acoustic Levitation Technologies for non -contact Handling Purposes. Annals of the German Academic Society for Production Engineering 8 (2001) 1, S. 77 –82.

[10] Reinhart, G.; Heinz, M.; Kirchmeier, T.: Ber ührungslose Greiftechnologien f ür die Halbleiter - und Mikrosystemtechnik. In: Reinhart, G. et al. (Hrsg.): m ünchener kolloquium - Innovation f ür die Produktion. M ünchen: Herbert Utz Verlag GmbH 2008, S. 253 –263. ISBN: 978-3-8316-0844-7.

[11] Schilp, M.: Auslegung und Gestaltung von Werkzeugen zum berührungslosen Greifen kleiner Bauteile in der Mikromontage. München: Herbert Utz Verlag GmbH 2007.

[12] Reinhart, G.; Heinz, M.; Kirchmeier, T.: Integration of the Ultrasonic Handling Technology into Microassembly Systems. In: Lien, T. K. (Hrsg.): 3rd CIRP Conference on A ssembly Technologies and Systems (CATS). Trondheim, Norway: Tapir Uttrykk 2010, S. 91 –96. ISBN: 978-82-519-2616-4.

[13] Schilp, M.; Zimmermann, J.: PV Wafer Distributing. <http://www.zs -handling.com/index.php?option=com_content&view=article&id=95:pv -wafer-verzweigen&catid=44:pv-wafer-handling&Itemid=76&lang=en> - 26.02.2011.

Lithography Cost Savings Through Resist Reduction And Monitoring Program

Terri Couteau[1], Scott Lindauer[1], Chris Stewart[1], Jennifer Braggin[2], Brent Bjornberg[2]

1. Spansion, Inc., 5204 East Ben White Boulevard, Austin, TX 78741
2. Entegris, Inc., 129 Concord Rd., Billerica, MA 01821

Abstract — Photolithography has been one of the primary processes driving semiconductor advances for the past few decades. In order to make faster, more reliable devices, designers drive circuit scaling to its limits. Equipment and material suppliers must create products to meet the throughput and lithographic performance standards required of advanced devices. Track equipment performance and process architecture have improved to meet throughput considerations, and therefore the cost of lithography tracks has remained relatively constant and few equipment changes are remaining which will drastically reduce the cost of lithography. On the other hand, the cost of photolithography materials has drastically increased due to the complexity of the chemistries required to resolve shrinking critical dimensions. In addition, new technology nodes have often required additional processing layers, adding to the high cost of materials in the lithography process. These increased costs are directly delivered to the wafer for processes where high volumes of chemical or multiple layers are needed to properly coat a wafer for further processing.

By fine tuning the coating process, a track equipment engineer can reduce resist cost per wafer by reducing the volume dispensed on the wafer. While reducing the resist volume to extremely low levels is attractive, it can also introduce risks into the coating process. Any interruption in the low volume dispense can cause a poor coating or no-coat situation, thus creating a wafer that requires rework or scrap. If the event is detected at the point of dispense, the wafer can be reworked. If the wafer escapes undetected until a post-lithography metrology step, it must be scrapped.

This paper evaluates the cost benefits of utilizing an advanced dispense system, such as the IntelliGen® Mini, made by Entegris, Inc., combined with an every-wafer point-of-dispense monitoring strategy. This paper will discuss the means to reduce resist dispense volumes by up to 60% and the ability to track every dispense, decreasing overall scrap and the need for some routine metrology. In addition, the authors will show a return-on-investment summary for undertaking such a project.

Keywords – photolithography, return on investment, photo dispense

I. INTRODUCTION

The photolithography sector often has the greatest challenge in reaching yield and cost targets. The photolithography sector is always at the leading edge of technology and is often the first to adopt new materials, process integration strategies, and equipment in order to drive manufacturing. Ultimately, the race to create the newest technologies drives increasing costs.

When a technology and process become stable, process and equipment engineers are tasked with reducing those costs. There are several techniques to reduce the overall cost of lithographic processing[1]. Potential strategies for cost reduction are:

- Choosing less expensive materials

- Adding a solvent pre-wet step

- Adjusting the process equipment to utilize less chemicals

Each reduction strategy comes with a cost. When choosing a new chemical for the process, a new qualification procedure must be followed. This is time consuming, and does not guarantee that the new material, while less costly, will replicate the results of the initial, more costly material.

A pre-wet step is one of the most efficient ways to initially reduce resist consumption. Solvent pre-wet is used to reduce the surface energy of the substrate and allow for a more even spread of less chemical. While this is an attractive strategy, it does introduce a new material which may have its own defect challenges. These materials are often rigorously studied before using them in a processing strategy.

After adding a pre-wet solvent, most engineers focus on fine tuning the process equipment. Dispensing resist on wafers seems like a simple task, but several pieces of equipment exist between the bottle and the wafer. Between the bottle and the wafer are a dispense system, a filter, and a control valve. When the track sends a signal that a wafer is ready for coating, the dispense system draws the chemical from the bottle, pushes the material through the point-of-use filter, and the track or dispense system tells the control valve to open and close to dispense a precise amount of chemical. Coordinating the timing of these events is what can dramatically affect the amount of resist dispensed on the wafer. The IntelliGen Mini's unique two-stage technology design allows for the filtration and dispense functions to operate completely independently, allowing the end user to fine tune the dispense process without

978-1-61284-408-4/11 $26.00 © 2011 IEEE

worrying about the effect of filtration. This also allows much lower filtration rates to be used, with a potentially very positive impact on defectivity. In addition, the dispense system has advanced software-based control features that control the dispense stage and produce accurate, repeatable dispenses within a 3 sigma value of ± 0.02 mL.

Once the process has been tuned to reduce resist waste, it is even more critical to ensure that dispenses are monitored. A high volume dispense has more room for error than a dispense that is using as little resist as possible. That creates a situation where it is even more critical to monitor every dispense. Several monitoring strategies are available for today's lithography sector:

- Macro inspections after coating[2]

- Macro inspections after exposure

- Lithography metrology (defect, critical dimension, etc.) after exposure

- In-line, point of use dispense monitoring

Each monitoring strategy also has its drawbacks. By adding a macro inspection after coating, the wafer must be removed from the lithography cluster. This action alone can introduce defectivity and can greatly reduce productivity. This control is not an option for critical DUV layers, where the time between resist coat and exposure must be limited to control critical dimensions. Macro inspections after exposure are another place to catch issues with dispense. While this is an effective method, it can also reduce productivity and without a full, automated review of the macro image, the technique is not extremely useful.

Allowing the wafers to move through metrology steps directly after lithography is yet another way to detect a dispense issue. Unfortunately, by the time the wafers have reached the metrology stage, it is possible that additional poorly coated wafers have been processed, increasing the number of wafers that will require inspection to determine if they require rework.

The worst possible method to catch a failing wafer is to find an issue at the final sort step. A significant amount of time and money has been spent on a wafer before it arrives at final sort.

The best method to find a coating issue in real time is to monitor at the point of dispense. If an issue is detected within the track, the wafer can be immediately reworked, saving time and cost of further processing. The IntelliGen Mini dispense system is unique in its ability to detect coating issues real-time[3]. Dispense confirmation is a proprietary software solution that compares the current dispense to a defined golden reference. By using algorithms to compare dispenses, the dispense system can determine when a change has occurred and provide a value of merit to the end user. By combining this capability with a Network Platform, the value of merit can be sent to a fab's SPC (statistical process control) chart. When tracking dispense confirmation on an SPC chart, a process engineer can set firm lower limits that will trigger an immediate wafer rework.

The goals of the study undertaken were to reduce resist usage on a common photoresist and also provide more robust monitoring to prevent scrap and rework. Ultimately, the combination of reducing resist consumption and detecting maverick wafer events can show how any fabricator can save a significant amount of money in the lithography sector.

II. RESIST REDUCTION PROGRAM

The ultimate goal of this project was to reduce resist consumption on a commonly used resist. Figure 1 shows the step down effort to reduce resist volume by initially making process and equipment improvements, and further reducing resist volumes by introducing the new dispense system.

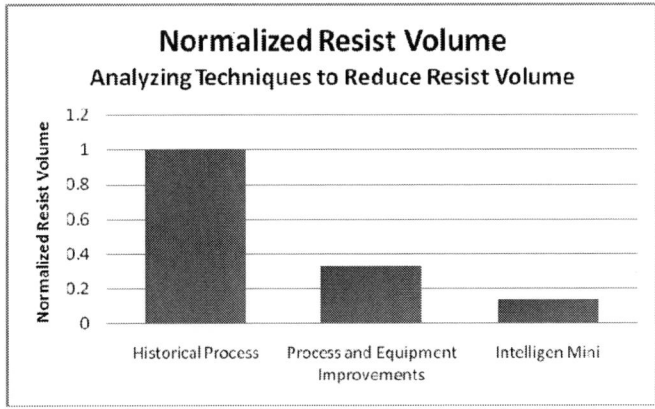

Figure 1. Dispensed volume per wafer when testing new resist consumption reduction methods

A. Process and Equipment Improvements

Spansion engineers were able to further reduce resist consumption by making process and equipment improvements. These improvements included, but were not limited to:

- Adding a pre-wet solvent

- Controlling coater cup exhaust

- Controlling temperature and humidity

- Modification to post apply bake exhaust configuration

- Changes to process recipes

- Changes to dummy dispense recipes

These initial changes ultimately reduced the resist volume usage by 66%.

B. Utilizing a new dispense system

Although initial efforts reduced the resist volume by over 60%, further improvements could be made. Spansion equipment engineers installed the dispense system on the track. The new dispense system allowed the process engineers to have more fine control over the process variables. Small adjustments to the dispense process improved the repeatability of the process, further reducing variation and resist consumption by 17%.

978-1-61284-408-4/11 $26.00 © 2011 IEEE

While achieving the lowest possible dispense volume was the target, the process now ran at the edge of its capability. In order to continue to use the low dispense volume, monitoring became all the more important to make sure the process was repeatable and stable.

III. EVERY WAFER MONITORING STRATEGY

The driver for an every wafer monitoring strategy with the new, low dispense volume was a particular excursion event at Spansion. At final wafer sort, a "key hole" pattern was seen on 25 wafers per quarter (Figure 2). The pattern showed passing die only in the center of the wafer, indicating intermittent coating issues with the wafers at the lithography step. Unfortunately, because these wafers were not caught in the lithography bay at the point of dispense, they were fully processed and the issue was not detected until final sort.

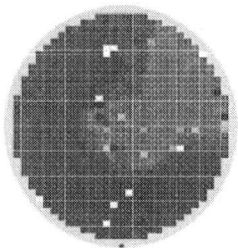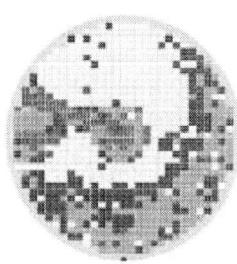

Figure 2.   Keyhole patterns detected at final wafer sort

A.   Monitoring strategy setup

The steps to set up a robust monitoring strategy are:

- Create a golden reference of a low volume dispense that is repeatable

- Collect long term CD (critical dimension), macro inspection, and defect data to ensure a stable process and plot the value of merit to an SPC chart

- Use the long term data to determine statistical limits for rework and wafer scrap

B.   Monitoring strategy implementation

While 100% macro inspection was turned on initially in the fab, the inspection was not highly reliable and frequent escapes occurred. Once it was determined that dispense confirmation was robust enough to catch failures, the macro inspection was eliminated, saving processing time. The value of merit was posted to an SPC chart, as seen in Figure 3, with limits based on the historical data collected during the monitoring phase.

Lots failing the control limits are automatically tagged for rework, saving additional metrology and inspection time, as it is known that these failing wafer will require rework. Additionally, because the wafers are automatically reworked, they will not be fully processed before the problem is detected. Also, this will ultimately reduce the number of wafers failing at final sort This strategy has driven wafers failing from the "key

hole" pattern from 25 wafers per quarter to 6 wafers per quarter.

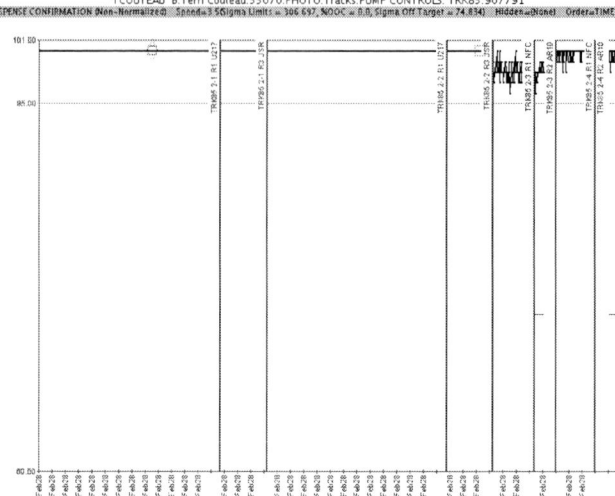

Figure 3.   Example SPC chart with value of merit for one material

IV. ADDITIONAL EFFICIENCY GAINS AND COST SAVINGS

While the ultimate goal of the project was to reduce resist usage and better monitor the sector, additional efficiency gains were made. These gains include:

- Remotely accessing pump data from anywhere to look at the pump data any time there is a fail. This single reduction saved each Spansion engineer 30 minutes per day, specifically reducing the time required to walk to the fab and gown up to enter the cleanroom.

- Triggering filter changes through filter cycle data plotted to an SPC chart.

- Remotely controlling the pump to perform simple tasks, like additional filter venting, when necessary.

- Comparing the same resist process on different tracks. This information is used to standardize dispense recipes, reducing the variation in setup between different pieces of equipment and different operators.

V. RETURN ON INVESTMENT ANALYSIS

There are several factors to consider when evaluating the return on investment in a resist reduction and monitoring scheme. A few include:

- Resist cost

- Resist volume reduction

- Engineer, technician, and operator time savings

- Scrapped wafer reduction

In this particular experiment, the resist volume was reduced by 83.3%. If the resist cost were assumed to be approximately

$1800/gallon, an annual savings of \$185,000 is achieved at a single point of use.

As previously mentioned, scrapped wafers were also reduced from 25 per quarter to 6 per quarter. With an average wafer cost at scrap being \$1000, that results in a \$19,000 per quarter savings on scrapped wafers. More globally, fewer scrapped wafers directly results in increased wafer revenue and opportunities.

Engineering, technician, and operator time is also saved by utilizing the automated SPC charts. A minimum of 40 minutes was saved by technicians when being able to log in remotely to access data for all the pumps in the fabricator during their initial shift checks. This additional free time allowed the technicians to focus on specific errors instead of collecting routine data from equipment that was working properly. The additional time was used to determine the root cause of failures, leading to a reduction in the time to solve track issues.

Considering these figures, as well as additional cost benefits to the user when installing the dispense system, we can calculate the time it would take to receive a return on investing in the installation of one new dispense system. The rough return on investment for purchasing one dispense system, reducing resist volume, and improving processes is within 2 weeks of use of the system.

VI. CONCLUSION

In this study the authors were able to reduce dispensed volume by 83% by making changes to the process and equipment to create a repeatable, small volume dispense. This particular change on one dispense point could save \$185,000 annually on resist.

By being able to detect coating excursions at the point of dispense, the fab can also save time and money on not further processing and testing wafers which have had a known excursion during resist coating.

Ultimately, the changes to a robust process presented in this paper can all be used to significantly save the lithography sector money and time.

ACKNOWLEDGMENT

The authors would like to thank Paul Magoon, Patrick Flores, and Johnathan Vail of Entegirs, Inc., and Lou Schaeffer, Nathan Hein, and Rodger Stagman of Spansion, Inc. for their support of this work.

REFERENCES

[1] Barone, D., et al. "Resist Dispense Volume Reduction Using the Six Sigma Methodology", Proc. CS Mantech Conference, Austin, TX, USA (2007).

[2] V. Menon, et al. "Product and Tool Control Using Integrated Auto-Macro Defect Inspection in the Photolithography Cluster" *Metrology, Inspection, and Process Control for Microlithography XX*, edited by Chas N. Archie, Proc. of SPIE Vol. 6152, 61521R, (2006).

[3] Braggin, J. "Preventing Lithography-Induced Maverick Yield Events With A Dispense System Advanced Equipment Control Method", Proc. Advanced Semiconductor Manufacturing Conference, Boston, MA, USA (2008).

Wafer Placement Repeatibility and Robot Speed Improvements for Bonded Wafer Stacks Used in 3D Integration

Andrew C. Rudack
3D Interconnect Metrology and Standards
SEMATECH
Albany, NY
andy.rudack@sematech.org

Michael Dailey
President
Fabworx Solutions, Inc.
Austin, TX
mdailey@fabworx.com

Abstract— **Robotic wafer handling of bonded wafer stacks (BWS) brings new challenges in placement repeatability and robot speed. Current standard SEMI M1.15 [1] does not contemplate wafer stacks >775 microns thick. Varying BWS thickness of 800-microns (top wafer thinned to 25-microns bonded to a 775 micron carrier wafer) to 1550 microns thick (two full thickness 775 micron wafers bonded together) creates additional mass and thus additional momentum during wafer movement. Without a corresponding increase in holding force, wafer sliding on the end effector is likely. One solution is to reduce robot velocity and acceleration, but this can lead to tool throughput reductions. In this paper we explore wafer handling issues for BWS and the application of a new end effector technology, called Gravity Edge Hold (GEH). Wafer sliding issues will be described in terms of wafer and BWS momentum, lateral holding force, and robot acceleration. It will further describe the wafer placement repeatability issues encountered and the corresponding decrease in robot speeds implemented to counteract this problem. Laboratory experiments were conducted to compare the performance of traditional end effectors used in current 300mm tools to that of the GEH end effector. An evaluation based on lateral holding force is included.**

Keywords: 3D, 3D interconnect, robot end effector

I. INTRODUCTION

SEMI® M1.15 defines many wafer parameters for 300mm crystalline silicon used in semiconductor manufacturing. Bonded wafer stacks (BWS) will not comply with many of these SEMI®-specified wafer parameters (e.g. thickness, diameter, edge bevel, notch and bow), and will compromise wafer handling requirements that expect wafers comply with single silicon wafer specifications. An additional consideration is that the mass of bonded wafer stacks will represent new challenges for traditional robot end effectors used to move silicon wafers within process tools (TABLE I). Accurate placement of BWS within tool process modules and safely returning them to wafer carriers on tool load ports is absolutely essential from a process stability and safe wafer handling perspective.

TABLE I.   SEMI® M1.15 COMPARISON FOR BONDED WAFER STACKS

Parameter	Single Wafer	Bonded Wafer Stack
Thickness	775 microns	1550 microns (bonded) 785– 850 microns (bonded/thinned)
Diameter	300 mm	300 mm (bottom wafer) 294 – 300 mm (top wafer/edge bevel trimmed)
Edge Bevel	T/3, T/4 in M1.15	Modified by edge bevel trim
Notch	Figure 7 in M1.15	Overlay/edge trim modified
Bow/warp	100 microns	30 – 2000 microns observed
Mass - grams	126 g (not specified)	140 – 286 grams

Traditional robot end effectors have successfully handled single silicon wafers for more than 15 years of 300mm wafer processing. End effector contact on the back of the wafer (either through a grooved channel or three points of contact) will seal against the wafer backside when process vacuum is applied. Process vacuum is typically supplied to a tool in the range of 15 - 26 inches of mercury. This resultant holding force allows a wafer to be removed out of a wafer carrier, cycled within a process tool and returned to the wafer carrier with placement precision that enables the process while maintaining safe wafer handing.

Certain process tools are location sensitive, and the process repeatability will suffer if wafer placement repeatability within a process module is compromised (e.g. not centered on a spinning wafer chuck in a photolithography coat and develop module). Safe wafer handling and damage losses are avoided when robot end defectors deliver wafers to locations that avoid potential wafer breakage (e.g. etch chamber door closure) or defect contribution (e.g. scraping wafer carrier sidewalls).

978-1-61284-408-4/11 $26.00 © 2011 IEEE

During the start-up of the SEMATECH 3D interconnect toolset in Albany, NY, one lesson learned [2] was the need to slow down wafer handling speed when placing BWS in various process modules. BWS slippage was observed on robot end effectors, and handling errors (BWS collision with wafer carrier sidewalls) occurred when using default robot setups designed for single wafers. Root cause for the slippage errors was determined to be the increased mass (and subsequently theorized to be associated with thickness) and momentum of the bonded wafer stack that exceeded safe handling speeds. A reduction in BWS handling speed was implemented as a workaround, empirically derived by observations on the velocity changes that minimized BWS slippage.

Wafers are often held in place on an atmospheric robotic end effector by a vacuum holding force. The total vacuum holding force is a product of the process vacuum that is applied to the open area of the end effector contacting the wafer surface, minus vacuum leakage and delivery line losses. Vacuum holding force can increase or decrease based on the size, shape and materials of the portion of the end effector that comes in contact with the wafer. Process vacuum levels in fabs can vary, further impacting this holding force and causing problems for end effectors that do not have significant design margins. Additional consideration should be given to surface roughness of the wafer backside in contact with the robot end effector, as this can induce vacuum leakage, producing reductions in vacuum holding force.

Equation (1) shows that the momentum (P) of a moving wafer is equal to its mass (m) times its velocity (v)

$$P=mv \qquad (1)$$

and the lateral force (F) on the wafer can be expressed as either the mass times acceleration (a) or the time rate of change of the momentum

$$F=ma=dP/dt \qquad (2)$$

In the case of bonded wafer stacks, the mass can increase to more than double that of a single wafer. If wafer handling robot velocities for a BWS are held constant (i.e. not reduced versus velocities used with single wafers), the BWS momentum will double. This increased momentum will significantly increase the need for an end effector that can firmly hold the BWS in place. When momentum exceeds the vacuum holding force for the robotic end effector, wafer slippage occurs. If the end effector design did not anticipate the increased mass associated with bonded wafer stacks, the BWS handling velocity will likely need to be reduced to insure safe BWS handling and repeatable BWS placement. The lateral force on a wafer in motion is its acceleration (g's) times it mass, so it's actually the acceleration (and deceleration) of the bonded wafer stack that creates the force which overcomes an end effector's lateral holding force.

Many atmospheric end effector designs cause a visible wafer "dimpling" when engaged with the wafer. In this experiment the traditional robot end effector was observed to create a dimple measuring 19 microns deep [3]. This may be a concern with regard to device structures (stress) and wafer backside contamination, but in the case of bonded wafer pairs, it is predicted to reduce vacuum holding force. The stiffness of a BWS is greater than that of a single wafer, and the dimpling effect is correspondingly reduced. When the BWS is gripped with the traditional robot end effector, the dimple was not observed. The dimpling is a result of the wafer shape conforming to the applied vacuum holding force. For a bonded wafer pair, the additional stiffness reduces the ability of the BWS to conform to the end effector when vacuum is applied. This can reduce an atmospheric robot end effector's holding force. The BWS is less likely to mold to the end effector's shape, potentially increasing vacuum leakage and reduced vacuum holding force.

II. METHODOLOGY

To make comparisons between robot end effectors and their ability to hold a 300mm wafer without sliding, a metric and test methodology was developed. The test metric is termed lateral holding force, and is defined as the lateral force required to overcome the static friction between a wafer and a robot end effector. Lateral holding force is exceeded when a wafer begins to slide.

A test apparatus was developed to measure this lateral holding force, as shown in Figure 1. An end effector is mounted rigidly and level, with a wafer or bonded wafer stack placed on the robotic end effector. A digital force meter with peak hold measurement capability is mounted on a linear rail on the same plane as the end effector. This force meter is then moved laterally by a low speed electric motor, to drive the meter into the wafer edge until wafer movement occurs. The peak force is read, recorded and averaged for multiple runs on the same wafer or bonded wafer stack being tested. A variety of robotic end effector designs can be evaluated on the test apparatus, including atmospheric and non-atmospheric styles. Wafer movement indicates that the lateral holding force has been exceeded for the robot end effector design that is being evaluated.

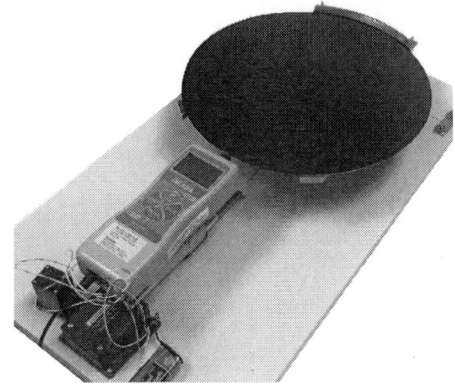

Figure 1. Test apparatus for measuring lateral holding force

978-1-61284-408-4/11 $26.00 © 2011 IEEE

The semiconductor industry is constantly working to reduce defectivity. Excessive contact areas with the wafer backside and wafer sliding on end effectors are known causes of particles [4]. Proper end effector design requires minimal wafer contact area, zero wafer sliding, and adequate lateral holding forces applied to the wafer backside and edges. Ideally, the end effector will enable the robot to perform at high speeds, so that tool throughput is not compromised, regardless of wafer mass, bow, warp, stiffness or backside surface texture. Wafer rubbing or collisions on tool parts and carriers can be considered another source for particles. Wafer placement repeatability is necessary to avoid this contamination source, as wafer slippage can cause particles to be generated when wafers collide.

Five versions of robot end effector styles were tested and compared for lateral holding force including single wafers, bonded and thinned wafer stacks, and double thickness bonded wafer stacks. Test results for lateral holding force are summarized in Table 2.

A. Metallic pocketed end effector

Typically used in vacuum chambers, wafers are held in place by gravity and friction with the metal end effector. There is no external process vacuum applied to add a vacuum holding force, and as expected, this design style demonstrated the lowest lateral holding force. The holding force also scaled with BWS thickness, as would be expected for static friction.

B. Pocketed end effector with embedded o-rings

Many robot end effectors utilize perfluoro-elastomer or similar materials in pad or o-ring shapes to minimize wafer contact and to provide a higher coefficient of friction with the wafer. When designed properly, the lateral holding force of this method is impressive considering there is no additional vacuum holding force. This style end effector can greatly reduce wafer backside contact area and sliding, improving particle performance. Typically, this style end effector is also used in vacuum chambers.

TABLE II.   END EFFECTOR LATERAL HOLDING FORCE

End Effector		Single Wafer Mass = 0.126 Kg	BWP, Thinned Mass = 0.134 Kg	BWP, Full Mass = 0.281 Kg
1. Metallic Pocketed End Effector – No Holding Mechanism		0.19 N	0.19 N	0.36 N
2. Pocketed End Effector with Embedded O-Rings Holding Mechanism		2.92 N	3.22 N	3.89 N
3. OEM/ Traditional Atmospheric End Effector – Vacuum Holding Mechanism		8.82 N	9.00 N	8.82 N
4. Atmospheric End Effector With O-Rings And Vacuum Holding Mechanisms		42.23 N	42.63 N	42.29 N
5. Gravity Edge Hold™ End Effector		>100 N	>100 N	>100 N

C. OEM/Traditional atmospheric end effector

Machined ceramic, stainless steel or aluminum materials are typically used with traditional atmospheric end effectors. Lateral holding force is primarily controlled by a process vacuum holding force created by the end effector's vacuum groove area in contact with the wafer backside. A variation on this type of end effector will use three points of contact with the backside of the wafer, and might include some elastomeric materials to help provide a more conformal grip.

This style of end effector required a reduction in robot speed to control wafer slippage during wafer handling on the 300mm tools used to process bonded wafer stacks. BWS momentum exceeded the lateral holding force of the end effector, and the BWS slipped during acceleration and deceleration of the end effector during BWS placement within the tool.

The data in TABLE II suggests that the dimpling of the wafer is not significant in reducing lateral holding force for this style end effector. The observed wafer slippage is being attributed to the velocity of the end effector during acceleration or deceleration of the BWS.

D. Atmospheric end effector with o-rings

The atmospheric end effector combines perfluoro-elastomer o-rings with a vacuum gripping mechanism, drawing on the best of both wafer holding techniques. Because of the softer o-ring material, this method seals well against rough wafer backsides. Wafer backside contact is minimized, and due to the larger sealing area the dimpling effect on the wafer is also reduced. Significantly higher lateral holding forces were observed. This style end effector is shown to be quite suitable to handling a BWS at normal and even slightly elevated robot accelerations.

E. Gravity Edge Hold™ (GEH) End Effector

When a wafer or bonded wafer stack is lowered onto a GEH end effector, its weight engages four cams. The wafer keeps the cams engaged, eliminating wafer sliding regardless of wafer mass. Wafers are re-centered by this technique. Wafer contact is minimal; end effector contact is only made between the wafer backside and a small bump on the cam bottom, and lightly between the wafer edge and a perfluoro-elastomer post mounted in the cam upright. Overall performance is not sensitive to wafer bow, warp or thickness. The GEH style end effector demonstrated the highest lateral holding forces, and will enable wafer transfers at top robot speeds and accelerations

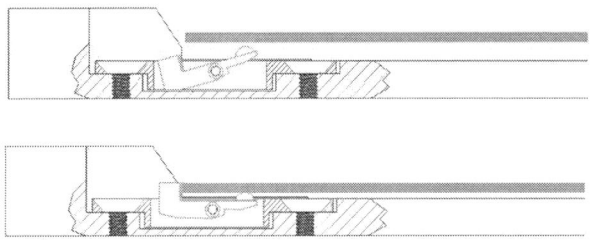

Figure 2. GEH end effector showing cam engagement mechanism on wafer. Wafer weight moves cam to contact wafer edge to hold it in place.

Note that the GEH end effector style is applicable to both vacuum and atmospheric environments.

F. Robot velocity and acceleration

While robot accelerations will vary by manufacturer, typical maximum robot acceleration is 200 inches per second[2] [5]. Based on this acceleration, the lateral wafer force on a single wafer, a thinned BWS, and a full thickness BWS are plotted in Figure 3. These forces can now be compared against the lateral holding force of various end effectors. If the end effector design does not accommodate this increased mass, the acceleration (velocity) of the robot will need to be reduced to insure safe BWS handling.

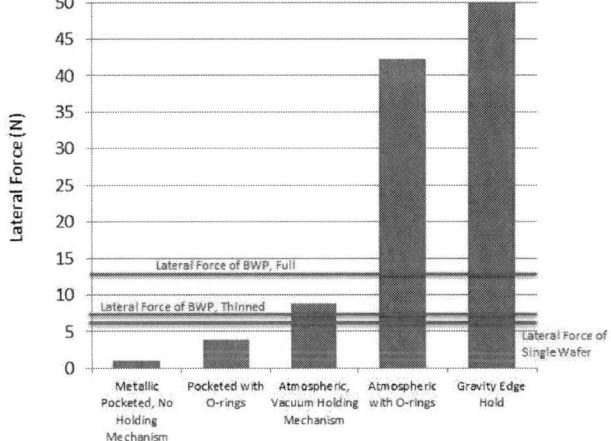

Figure 3. Lateral force on wafer (red line) compared to lateral holding force on end effector (blue bar).

III. Conclusions

Bonded wafer stacks do not comply with many of the parameters specified in SEMI M1.15, including wafer thickness, diameter, edge bevel, notch and bow. BWSs contain additional mass that can create lateral forces which exceed the lateral holding force on robot end effectors. The result is wafer slippage and placement errors on 300mm tools.

Robot end effectors used to process bonded wafer stacks on SEMATECH's 3D interconnect toolset in Albany, NY required a decrease in wafer handling speed to prevent wafer slippage, resulting in safe, repeatable BWS placements.

SEMI standards need to be updated to reflect the specific handling requirements for bonded wafer stacks, including wafer mass as a new concern. Tool design and robot end effectors must reflect the new BWS parameters.

Acknowledgements

The authors would like to thank:

- Traci Miller, Shawn Schmidt and Chris Pokorny (Fabworx Solutions, Austin, Texas) for developing the testing apparatus, collecting the data and helping to organize and write this paper
- Rick Kent (Fabworx Solutions, Concord, New Hampshire) for developing the end effector technologies described in this paper

References

[1] SEMI® M1.15, "Standards for 300mm polished monocrstalline silicon wafers (notched)." Available at www.semi.org.

[2] A. Rudack, R. Caramto, "3D lessons learned," 6th annual ISMI Manufacturing Symposium, October, 2009. Available at http://www.sematech.org/meetings/archives/mfg/index.htm

[3] internal memo from Rick Kent, Fabworx solutions, Concord, NH

[4] Angelo, D., Suh, S.M., Khurana, N., & Sankaranarayanan, K. (2004). "Improving Defect Performance through Better System Design." Nanochip Technology Journal. Issue 2, 65-68.

[5] http://www.ptb-sales.com/semiconductor/robot/genmark/genmark.html

200mm Fab AMHS Improvement During Aggressive Ramp

Bouhnik Sylvain
Manufacturing Enginering Planning Departement
Micron Semiconductor Israel Ltd P.O.Box 1320
Qiryat Gat 82109, Israel
sbouhnik@micron.com

Abstract: **The semiconductor industry experienced increased requirements during the period 2008–2010. These requirements were followed by more wafer starts (loading upside). Micron Israel's 200mm factory capacity increased by more than 40% within one year.**

The automated material handling system (AMHS) was not originally intended to support a 40% load increase. Prior to the load increase, the 200mm AMHS was not considered a factory constraint. However, after the increase, with the AMHS capacity gap at about 40% of the expected loading, a capacity improvement roadmap was implemented for all toolsets, including the AMHS. This roadmap, among other things, included: optimization of the lot destination, stocker capacity increase, stocker cleaning, and leading parameter definition to prioritize manual handling of lots and route changes. Roadmap action plan priorities were set through simulation gain analysis. All of the AMHS improvements resulted in a system capability increase of more than 20% over the original planned capability.

This paper presents a review of the AMHS improvement roadmap actions that led to the system capability increase.

Keywords: AMHS, Simulation, Loading.

I. INTRODUCTION

In the last two years, Micron Israel's Flash factory has experienced an increase in its loading and capacity. A request to increase capacity was initiated for all of the factory capacity limiters. Capacity limiters included the AMHS, which was originally limited to 30% less than the new maximum loading required. Therefore, a joint effort was initiated with the IS, Manufacturing, and Industrial Engineering departments to increase system capability.

The major improvement projects included the following elements:

- Increased stocker capacity

- Steps for traffic jam reduction

- Day-to-day mode for operating projects

- Sustained operation of system components

This article presents a description of the AMHS in use, an analysis of the impact that loading has on AMHS performance, and the improvement project that led to the AMHS capacity increase.

II. AUTOMATED MATERIAL HANDLING SYSTEM

The layout of the AMHS (AeroTrack and stocker) was set during construction of the Fab.

The layout includes:

- **Three independent routes.** These consist of the outer route (black line in Figure 1), the west route (blue line in Figure 1), and the east route (red line in Figure 1).

- **About 70 stockers.** Most of the bays are being served by two stockers.

- **Machine loading robot vehicle (MLRV) systems.** Four of the systems serve diffusion tools; the other four serve lithography tools.

Figure 1 shows the factory AMHS layout and stockers.

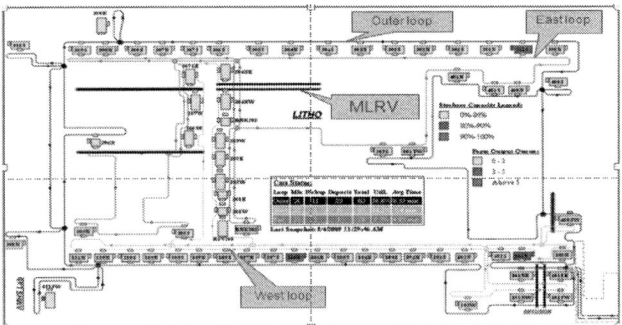

Figure 1. Factory AMHS layout

Minor changes to the AeroTrack system have been implemented during the last 10 years. Most of the changes were made to adapt the AMHS layout to construction requirements.

The AMHS was not adaptable to process flow changes that can occur. In the beginning, the factory ran Intel microprocessors, but it was later converted to run 45nm/60nm Flash memory. As a result of these changes, the three routes became completely unbalanced. TABLE I shows the route loading differences between the three routes, the number of cars per route, and the average delivery time. The table clearly shows how the outer loop is overloaded and the east loop is underutilized.

978-1-61284-408-4/11 $26.00 © 2011 IEEE

TABLE I. LOOP PERFORMANCE

Cars Status:

Loop	Idle	Pickup	Deposit	Total	Util.	Avg Time
Outer	20	14	31	65	69.2%	8.88 min.
West	14	8	17	39	64.1%	6.93 min.
East	15	6	8	29	48.2%	5 min.

III. AMHS LOADING IMPACT

The Fab engineers were approached with a very aggressive goal of increasing Fab capacity loading from 2008 to 2010. The potential impact of increased capacity loading on AMHS performance was then unknown. As a part of factory readiness, the AMHS risk assessment was reviewed by the factory and production manager. At the beginning of the loading increase ramp, analysis shows a linear increase of the transportation time on the three loops as a function of loading increase (see Figure 2). This analysis triggered the improvement working group to define a clear roadmap and execution plan for increasing AMHS loading capacity.

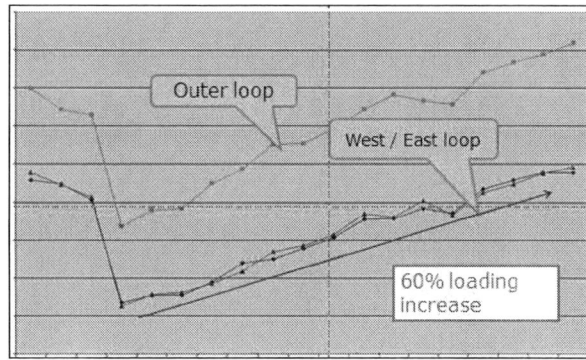

Figure 2. Delivery time increase during loading increase

IV. IMPROVEMENT PROJECTS

The AMHS and equipment were involved in many large improvement projects. Some examples are referenced in [2]. The improvement project for 200mm wafers included increasing equipment performance, changing the operator mode of operation (manual transportation or in-shift changes in operator habits), and routing changes. All these improvements required resource investment; therefore, prioritization of the project was critical. A simulation was used to quantify the project gain. As a result, the following projects were defined and later executed:

- Routing Changes
- Stocker Improvements
- Operation Changes
- Lot Transportation Improvements

A. Simulation Development

The development of simulations helped engineers understand which AMHS parts were causing process bottlenecks (AeroTrack loading, stocker occupation, robot loading, stocker P-port, or turn table). Furthermore, the simulations were key decision tools used to set AMHS improvement roadmap priorities and project gain.

A simulation for the analysis of AMHS bottlenecks was built using the Automod platform. Validation was based on actual AMHS indicators such as transportation time, stocker utilization, and car utilization. "What-if" scenarios were checked through dynamic simulation output analysis. Most of the project gain was evaluated through simulation analysis.

The gap between actual utilization value and the simulation value is 7.5% for the outer loop (see Table II). A good correlation between loop utilization and robot stocker utilization can be observed.

TABLE II. SIMULATION VAILIDATION

		Actual	Simulation
Loop utilization	Outer	73.6%	81.3%
	West	64.0%	61.4%
	East	45.0%	46.6%
Stocker occupancy			113%
			64%
			58.4%
			57%
Robot Utilization		66%	68%
		48%	41%
		46%	47%
		43%	35%
		34%	22%
		40%	27%
		38%	44%

The stocker occupancy was not bounded; therefore occupancy of more than 100% reflects a full stocker. This condition initiated a stocker destination change or the installation of two new stockers.

B. Routing Changes

- **Developing a stocker destination optimization model for the (MLRV) system.** A mixed-integer mathematical model was developed to provide balanced loading among the stockers. The goal of the model was to optimize stocker loading while keeping tool loading balanced. The mathematical model is:

$$\text{Min} \quad \sum_{\text{all stocker}} ABS\left(\text{Total Loading stocker - capacity(stocker)}\right) \quad (1)$$

s.t

$$\text{Loading(operation)} = \text{wafers start per week per operation(include product impact)} \quad (2)$$

For all stocker, Tool_set

$$\sum_{\text{Operation}} \frac{\text{Loading(operation)} \times \text{Processtime(operation)} \times X_{\text{operation,stocker}}}{\text{TotalToolset Loading}} < \frac{\# \text{tool set run operationin stocker}}{\text{Total\# tool\_set run operation}} \quad (3)$$

For all stocker

$$\sum_{\text{Operation}} \text{Loading(operation)} \times \text{CT(operation)} \times X_{\text{operation,stocker}} = \text{Total Loading(stocker)} \quad (4)$$

where $X_{\text{operation,stocker}}$ in (1,0)

The goal of this model is for stocker balancing to maintain load balancing for the tools in a different toolset.

- Equation (1) reflects the sum of the absolute value of the gap between expected loading and stocker capacity.

- Equation (2) reflects product loading of each operation

- Equation (3) ensures that toolset loading is below the toolset capacity in the stocker.

- Equation (4) reflects the loading of the stocker (operation CT * layer loading).

- $X_{\text{operation, stocker}}$ reflects the assignment of the operation to a specific stocker. The goal is to assign each operation to a unique stocker. This reduces consolidation time (time to get a complete batch in the stocker).

Figure 3 shows a graphic representation of the model. Each row reflects an operation; each column is a stocker allocation.

							Goal	Stocker balancing		
								Stocker capacity		Stocker loading constraint
							>=	>=	>= >=	
							Stocker loading= sum(assignment*CT*loading)*%per operation)			
Operation	Tools loading assignment per stocker									
operation /layer	Process loading	Process time	% loading	CT hours	Box inventory	Stocker line1	Stocker line2	Stocker line3	Stocker line4	Stocker line5
1			75%			0%	0%	0%	0%	100%
2			25%			100%	0%	0%	0%	0%
3			31%			0%	0%	100%	0%	0%
4	Process RAMP	Process time per operation	89%	CT per operation =CT*Loading *%loading		0%	0%	0%	0%	100%
5			100%			100%	0%	0%	0%	0%
6			100%			100%	0%	0%	0%	0%
7			100%			100%	0%	0%	0%	0%
8			100%			100%	0%	0%	0%	0%
9			100%			100%	0%	0%	0%	0%
					=assignment * %loading * loading * Process time					Assignment
										Equipment loading constraint
			total tool loading		# tools per line Total number of tool (%loading * Loading/Total Loading					

Figure 3. MLRV diffusion optimization model

- **Optimizing AMHS routing.** AMHS routing was set to minimize transportation distance; however, the east and west routes were underutilized. The routing policy was changed to increase the balancing of loop loading. Transaction routes were changed from the outer loop (the shortest distance) to the west loop, despite the increase in transportation distance.

- **Changing the diffusion batches automation system.** Changing the diffusion batches automation system enabled operators to move diffusion production batches to the next operation without waiting for the test monitor. This change saved approximately 20 minutes waiting for test completion, on top of stocker occupation (shelves) and transportation (about 200 transactions per day).

C. Stocker Improvement

- **Installing additional stockers.** Simulation results showed highly utilized and nonutilized stockers during the ramp increase. Three new stockers were installed to accommodate increases in WIP inventory and stocker loading, while other nonutilized stockers were removed to create free space for process tool installation requirements.

- **Increasing stocker robot and MLRV speeds.** Simulation enabled engineers to determine that stocker robots needed to be highly utilized. The robots were upgraded to provide a 20% increase in speed, although the overall system gain was minor.

- **Restructuring stocker preventive maintenance.** Stocker preventive maintenance policy (through preventive maintenance segmentation) was completely changed to reduce the impact of lengthy system downtime. Unscheduled down time was also reduced by converting a repair task from a failure replacement to a scheduled replacement during preventive maintenance.

- **P_port upgrade.** The AeroTrack section that links the stocker to the route (P_port) was a major limiter in the simulation (see Figure 4).

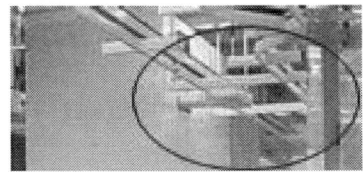

Figure 4. P_port

This was due to the low speed of the P_port. A 56% increase in speed decreased the required outer loop utilization by 6%. TABLE III shows the gain in loop utilization and the gain in the shelf-to-shelf transportation time.

TABLE III.   P_PORT UPGRADE GAIN IN THE SIMULATION

		Shelves to Shelves					Shelves to Shelves			
		Car delivery	Robot waiting	Shef to Shelf			Car delivery	Robot waiting	Shef to Shelf	Gain
73 cars	Outer	5.20	5.59	10.8	73 cars P_port upgrade	Outer	4.81	4.94	9.7	-10%
	West	3.26	2.55	5.8		West	2.69	2.35	5.0	-13%
	East	2.92	2.60	5.5		East	3.26	2.62	5.9	6%
	Loop Utilization					Loop Utilization				
	Outer	81%				Outer	75%	-5.9%		
	West	53%				West	49%	-3.8%		
	East	54%				East	53%	-0.6%		

D. Operation Changes

- **Activating manual lot handling mode.** A leading productivity indicator is the number of lots waiting for car transportation (pending lots). When the number of pending lots exceeded the upper control limit, the selective manual handling mode was activated. Transactions with the highest distance*volume value were prioritized for manual handling. A response flowchart (RFC) was developed for all shifts. The RFC describes which type of manual transportation handling to activate when a loop is down or overloaded and the number of lots pending above the control limits (see Figure 5).

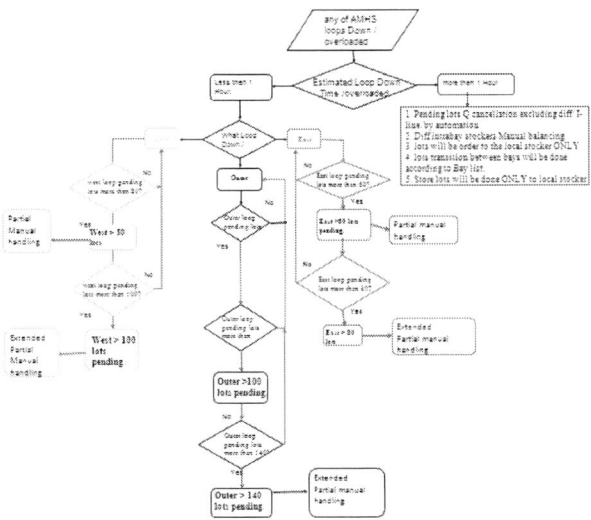

Figure 5. AMHS manual handling RFC

- **Balancing in-shift AMHS capacity loading.** Analysis of the number of lots waiting for car transportation showed a high variability during shifts. Therefore, the action plan for balancing AMHS capacity loading included modification of the start policy and the operator work mode.

The original start policy was to introduce a wafer start at the beginning of the shift. About 20 cars were then ordered at the same time, causing a major impact on the overall AMHS. The new mode of operation (MOO) was to introduce a start smoothly during the shift. This MOO was rapidly changed to the implementation of a manual start.

Figure 6 shows the variability of the AHMS loading during the shift, before and after the MOO change. The graph reflects the impact of the shift start (7:00 and 19:00) and the operation meal time (1:00 and 13:00). As result of this change the standard deviation of the pending lots within the shift was reduced by 2% and the gap between minimum and maximum pending lots was reduced by 6%.

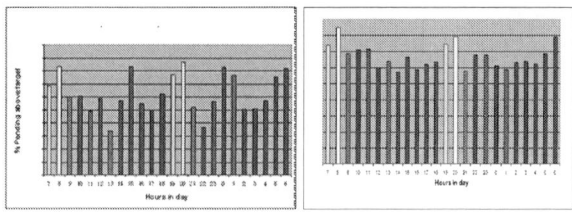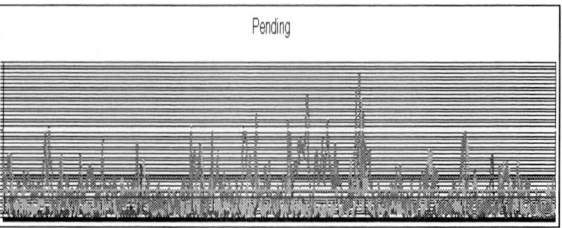

Figure 6. AMHS loading variability during shift

Figure 8. Number of pending lots at full loading capacity

E. Lot Transportation Improvements

- **Increasing the number of cars.** The simulation was used to determine the optimum number of cars required, based on the utilization of loops and shelf-to-shelf transportation time. Figure 7 shows the impact of increasing the number of cars on the AMHS.

The data shows clearly that:

  o Increasing the number of cars increases the delivery time and decreases the number of pending lots. This is due to the increase in traffic jams.

  o Decreasing the number of cars decreases the transportation time but increases the car waiting time (pending lots).

Despite this improvement, and looking forward to future loading increases, the option of using only manual handling is being investigated. The main reasons for considering eliminating the AMHS is the aging equipment and the need for more cleanroom space. Figure 9 shows the number of aging cars removed from the AeroTrack per week.

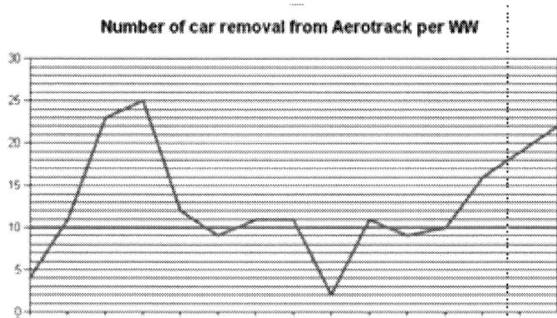

Figure 9. Number of cars removed per week

However, this type of conversion would require a new tracking system such as radio frequency identification (RFID), additional head count to support manual lot handling, and additional lot stocking locations.

67 cars

Shelves to Shelves

	Car delivery	Robot waiting	Shef to Shelf
Outer	5.06	5.94	11.0
West	2.91	2.53	5.4
East	3.26	2.60	5.9

Loop Utilization

Outer	87%
West	53%
East	53%

71 cars

Shelves to Shelves

	Car delivery	Robot waiting	Shef to Shelf
Outer	5.03	5.72	10.7
West	2.92	2.55	5.5
East	3.27	2.63	5.9

Loop Utilization

Outer	83%
West	53%
East	53%

69 cars

Shelves to Shelves

	Car delivery	Robot waiting	Shef to Shelf
Outer	5.09	5.81	10.9
West	2.91	2.52	5.4
East	3.26	2.60	5.9

Loop Utilization

Outer	85%
West	53%
East	52%

73 cars

Shelves to Shelves

	Car delivery	Robot waiting	Shef to Shelf
Outer	5.20	5.59	10.8
West	3.26	2.55	5.8
East	2.92	2.60	5.5

Loop Utilization

Outer	81%
West	53%
East	54%

Figure 7. Impact of increasing the number of cars on the AMHS

- **Increasing car speed.** An increase in car speed was checked on the most utilized loop. The speed gain was estimated to be 20%; however, simulation analysis predicted a gain of only 2% due to increased traffic jams. Therefore, this project was dropped.

V. CONCLUSIONS

The AMHS in Fab 12 was built 12 years ago. However, small improvements have helped to achieve a 20% increase over its original capacity. Figure 8 shows few events with pending lots above the upper control limit while the Fab was at full loading capacity. The overall gain in system improvement is reflected in this reduction of pending lots.

REFERENCES

[1] Watanabe, S.; Wakabayashi, T.; Kobayashi, Y.; Okabe, T.; Koike, A.;Trecenti Technol. Inc., Hitachinaka. "High speed AMHS and its operation method for 300 mm QTAT-fab". Semiconductor Manufacturing, 2003 IEEE International Symposium, Japan

[2] Bo Li Wu, J. Carriker, W. Giddings, R. "Factory throughput improvements through intelligent integrated delivery in semiconductor fabrication facilities" Semiconductor Manufacturing, IEEE Transactions on On page(s): 222 - 231, Volume: 18 Issue: 1, Feb. 2005

[3] Tony Wiethoff; Casey Swearingen "AMHS Software Solutions to Increase Manufacturing System Performance". Advanced Semiconductor Manufacturing Conference, 2006. ASMC 2006. The 17th Annual SEMI/IEEE

BIOGRAPHY

Bouhnik Sylvain – Master's Degree in Industrial Engineering – Technical leader in Manufacturing Engineering, Micron Semiconductor, Israel Ltd.

978-1-61284-408-4/11 $26.00 © 2011 IEEE

Automated SEM Offset Using Programmed Defects

Oliver D. Patterson, Andrew Stamper
IBM Semiconductor Research and Development
Center
2070 Route 52, Mail Stop: 46H
Hopewell Junction, NY 12533 USA

Roland Hahn
KLA-Tencor
20 Corporate Park Drive, Suite C
Hopewell Junction, NY 12533 USA

Abstract - **Defect inspection plays a large role in the development and manufacture of semiconductor technologies. Defects detected in today's inspections tools are generally a fraction of a micron and require SEM review to analyze and justify corrective measures. It is very important that the review SEM drives to the exact location of the defects as a FoV (Field of View) of 2µm is necessary to provide the resolution needed for defect redetection without the inefficiencies associated with repeated 'zooming' of the image. A methodology which allows quick and accurate alignment of the review SEM to the defects in the results file is presented. This methodology uses a special structure containing programmed defects. The methodology is illustrated using the challenging example of PWQ wafers.**

Keywords- SEM Alignment, defect offset, review SEM, deskew

I. INTRODUCTION

Optical defect inspection plays a large role in development and manufacture of semiconductor technologies. Tens of optical inspections are strategically interlaced throughout the process sequence in order to detect, quantify and classify defectivity affecting the wafer. Because of the small feature size of today's technologies, and in turn the small size of critical defects, SEM review is almost always necessary to classify the defects.

Redetection of defects by review SEM has become particularly challenging in recent years, again because of the small size of the typical defect. Robust wafer alignment and a common die corner are two necessary factors for successful defect review. Despite excellent review SEM stage accuracy, a small offset between defects across the wafer still exists. This is because of variability in the calibration wafers, temperature, identification of the center of a defect and other factors. Therefore a third parameter, the defect deskew, is also necessary. The process of calculating the defect deskew, also termed 'defect deskew', may be performed automatically or manually. In addition to correcting offset within a wafer, defect deskew also compensates for a systematic offset between different inspection tools and modes. For example, the coordinate accuracy of darkfield inspection tools gets worse with larger

spot size. Also, the coordinate accuracy of bare wafer inspection degrades with higher throughput. The following offsets can be corrected by using an deskew: translation, scaling, rotation and non-orthogonality.

To perform an efficient defect deskew, a set of reference defects needs to be selected, relocated and marked on the wafer. Defects detected by the inspection tool are sometimes not visible to the review SEM. When they are, the visible ones are not always well distributed across the wafer as required for an ideal deskew. Deskew is especially difficult for Focus Exposure Matrix (FEM), Process Window Qualification (PWQ) and Process Window Centering (PWC) inspections [1]. The nature of these inspections results in very high defect density and large defects in the higher modulations, making it difficult to reliably locate a suitable set of defects for deskew.

In this paper, we propose the use of programmed defects (PD) to assist in the deskew process. This methodology is described in Section II. Application of this methodology to a number of PWQ wafers for comparison with current methods is discussed in Section III.

II. METHODOLOGY

Traditionally, PDs have been used for calibrating the sensitivity of inspection techniques such as e-beam and brightfield inspection [2,3]. This paper introduces a special structure, called the SEM Alignment Structure, which contains PDs at key levels throughout the process. These include active, deep trench, gate-stack, contact and all the metal and via levels. A small area of the structure layout around the PD at the active and contact levels is shown in Fig. 1. A small area of the metal 1 structure layout around the PD and a corresponding wafer image are shown in Fig. 2. This structure is 58um x 58um so that it can easily fit within the scribe line. The structure must have a repetitive pattern so that it can be inspected in array mode. A random mode inspection will not work for 1x1 reticles, which are common in development, because the PD appears on the same location in each reticle field. The PDs for each level are stacked on top of each other so that only a single PD is detected at each level. Since brightfield can sometimes

978-1-61284-408-4/11 $26.00 © 2011 IEEE

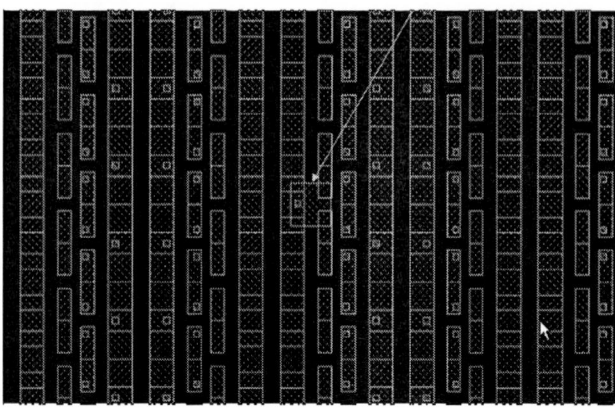

Figure 1: SEM Alignment Structure design showing the programmed defects at the active (yellow) and contact (pink) levels

Figure 2: Metal 1 SEM Alignment Structure design (left) and corresponding wafer image (right)

detect defects at prior levels, this is necessary.

To use the methodology, the inspection must include an array mode test to capture the PDs. All die may be inspected, but it is sufficient to just inspect the die that will be used for deskew. A special class code is assigned to these defects so they may be easily be identified during SEM review for deskew. Currently for the KLA-Tencor eDR-5210 review SEM used for this work, relocation of the PDs and then deskew must be done manually, but a software patch to allow this to be done automatically will soon be available.

III. APPLICATION

A. Case 1: Comparison to manual offset using a metal 1 PWQ wafer

Application of the SEM Offset Methodology to a metal 1 PWQ wafer is described here to demonstrate the usefulness of this structure. The wafer was inspected with a KLA-Tencor 2825 brightfield inspection tool using KLA-Tencor PWQ methodology. An additional array mode test was added to capture the PDs in the SEM Alignment Structure. The result file was sent to the review SEM.

First, manual deskew was performed as accurately as possible using the defects on the wafer other than the programmed defects. Images of all defects were taken. A

special program was used to determine the offset of each defect relative to the center of the FoV of the SEM image. These are plotted in Fig. 3. The defect scatter is 3μm. This inaccuracy is caused by the difficulty of selecting the correct defects within an image for deskew. Figure 4 shows the defect map for this wafer. The wafer can be divided into three zones. In Zone 1, all the defects are non-visual. In Zone 3, each FoV is swamped with defects, so reliable selection of the correct defect is impossible. Only Zone 2 contains discrete defects which are useful for SEM deskew. Unfortunately, this area is a small fraction of the entire wafer and so the deskew is poor.

Figure 5 shows a case from Zone 1. The difference image shows three differences between reference and defect. Unfortunately, a real defect is not visible under the reviewSEM. This defect cannot be used for deskew. SEM non-visual defects can occur on any wafer, not just PWQ wafers.

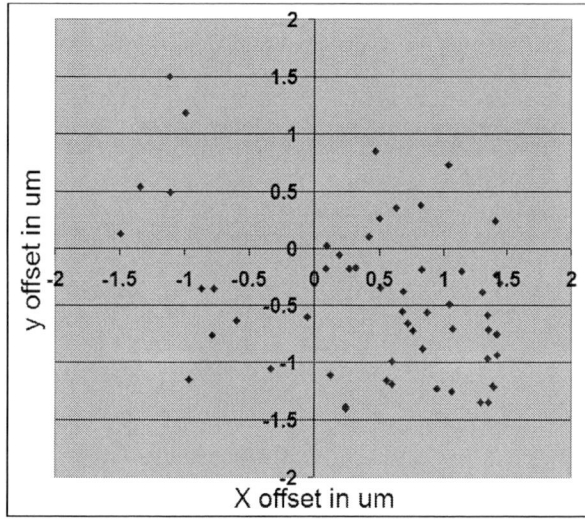

Figure 3: Defect offsets for all SEM visible defects on the metal 1 PWQ wafer

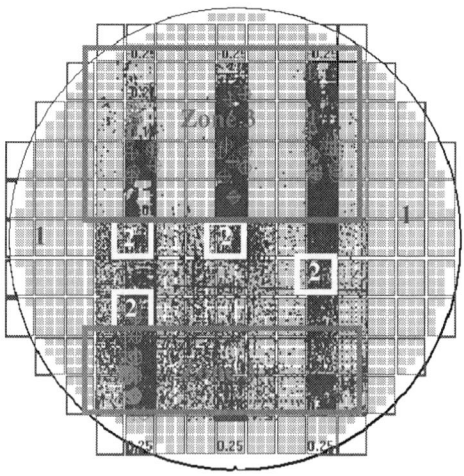

Figure 4: On PWQ wafers, only defects in Zone 2 are generally useful for deskew.

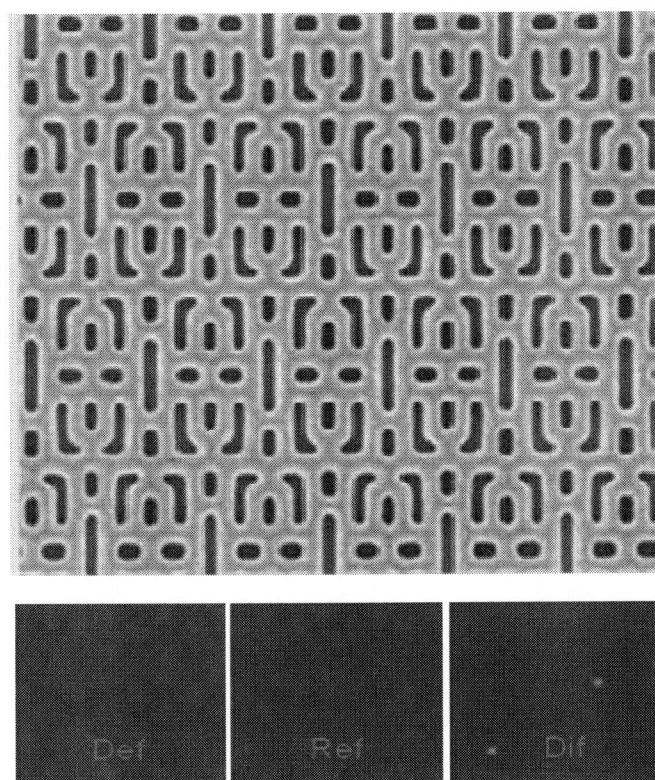

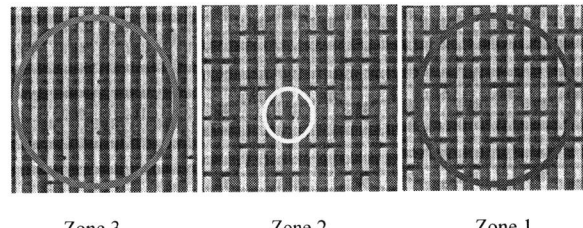

Zone 3   Zone 2   Zone 1

Figure 7: Defects from a gate-level PWQ wafer. Zone 3 is too defective. Zone 2 is good as the defects are discrete. Zone 1 is bad because of no defects or just non-visual defects.

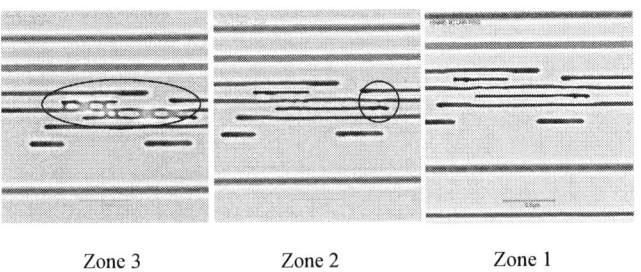

Figure 5: Zone 1: Bottom: optical defect, reference and difference images. Top: review image. The defects in the difference image are just not visible in the SEM image.

Zone 3   Zone 2   Zone 1

Figure 6: Defects from the logic area of a metal wafer. Zone 3 is so defective, the review SEM cannot know which defect to chose.

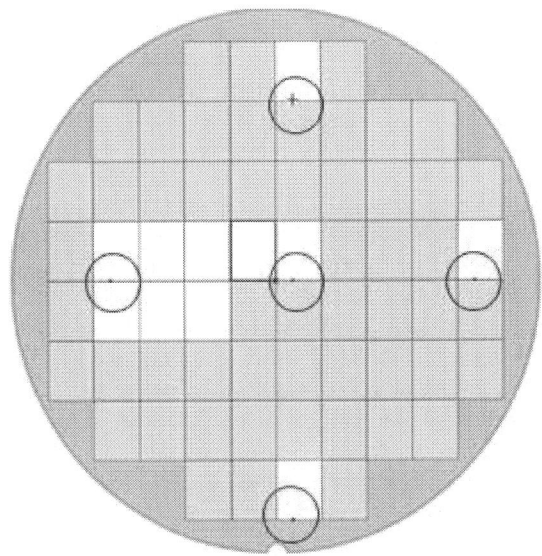

Figure 8: PDs selected for deskew of the M1 PWQ wafer.

Figure 6 shows images for the same site from the logic area of a metal wafer. Multiple defects appear in the Zone 3 (the higher modulation die) review image. It is impossible to reliably select the correct defect for a proper deskew. Figure 7 further illustrates the type of defects seen in the different zones. These images are from within the SRAM for a gate-level PWQ wafer.

Next the programmed defects, at the ideally spaced locations show in Fig. 8, were used to deskew the wafer. After deskew using the SEM Offset Methodology, images of all the defects were taken, and the offsets for these defects were measured again. The results are shown in Fig. 9. The defect scatter is now 0.2um, a very substantial improvement.

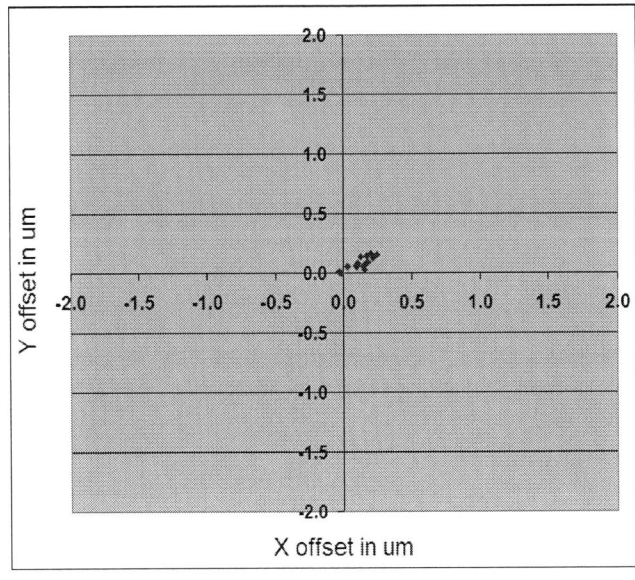

Figure 9: Defect offsets when the SEM Offset Methodology is used

B. Extension to other tools and levels with a deskew file

Once the deskew has been calculated for a particular recipe and inspection tool, it is saved in the review recipe in the form of a "deskew cache file". This file corrects for the systematic offset between the review SEM and the particular inspection tool. It may also be used for different recipes, different modes and even different inspection tools within a device/product. Figure 10 shows the defect scatter when applying the deskew file obtained using the SEM Offset Methodology across multiple recipes from the same inspection tool. The scatter is 1.5um. Figure 11 shows the defect scatter when applying the deskew file obtained using the SEM Offset Methodology across multiple recipes and multiple inspection tools of the same type. The scatter is 2um. While the scatter is better than in Fig. 3, it is not nearly as good as in Fig. 9. Use of the SEM Offset Methodology for each new wafer, would substantially improve the SEM alignment accuracy down to 0.2um error. An advantage of this would be to be able to take higher resolution images, 1um rather 2um FoV, of the defects.

C. Case 2: Comparison to use of a current deskew file with a gate-stack PWQ wafer

The SEM Offset Methodology was also compared to existing methodology, where an existing deskew file is used. For this second study, a gate-stack PWQ wafer was studied. Rather than manual deskew, in this case the existing deskew file loaded in the inspection was used. This deskew file was created using a different level and possibly a different inspection tool. Review images of all defects were captured. Again a special program was used to determine the offset of each defect relative to the center of the FoV of the SEM image. Figure 12 shows the defect scatter, which is 1um. The population is offset -0.5um so the greatest offset is also 1um.

The SEM Offset Methodology was then applied to this same wafer. Figure 13 shows the defect scatter. The scatter is now 0.4um and perfectly centered.

D. Case 3: Review of Voltage Contrast Defects

Redetection of voltage contrast (VC) defects from an e-beam inspection (EBI) tool can be difficult. E-beam inspection tools use high beam currents of 25nA or more for VC inspection. Review tools have maximum beam currents of about 1nA. Review of VC defects can be useful as they may be caused by much smaller physical defects only visible with a well centered, high magnification SEM image.

Figure 14 shows an example at the gate level for a 28nm bulk technology. The EBI image shows a bright gate line. This defect type can be caused by a variety of issues, some visible from the surface and some not. The low magnification review SEM image shows the bright gate line is barely visible to the eye. Rather than detecting the brighter gate line, the review SEM centers on a small unrelated physical defect, merging of spacer. Even if the physical cause for the VC signal is visible, it would be missed in the high resolution image because it is not in the FoV.

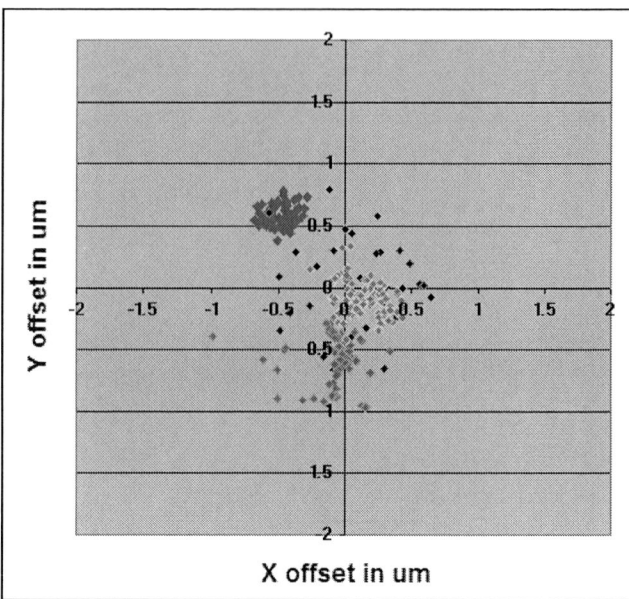

Figure 10: Using the same deskew across multiple layers on the same tool

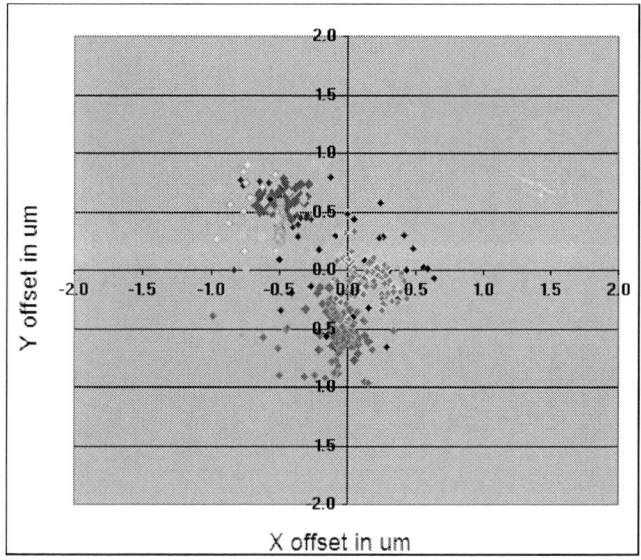

Figure 11: Using the same deskew file across multiple layers and tools

A special beam condition can be used on the review SEM to enhance the VC signal. The problem is this condition will not be nearly as good at imaging physical defects. Therefore, SEM review of EBI defects such as this is an excellent application for the SEM Offset Methodology.

IV. FUTURE PLANS AND SUMMARY

An array mode redetection algorithm is being implemented for the KLA-Tencor eDR review SEM platform. With this improvement, automatic deskew will be possible for every wafer with a SEM Alignment Macro.

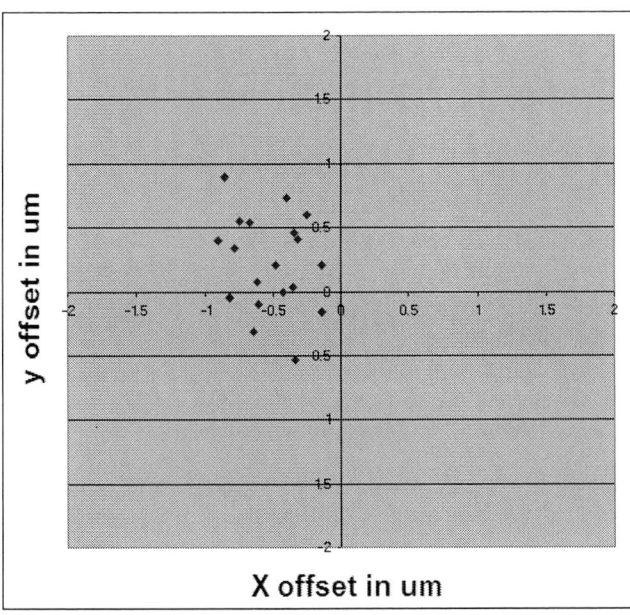

Figure 12: Defect offset using a stored deskew file

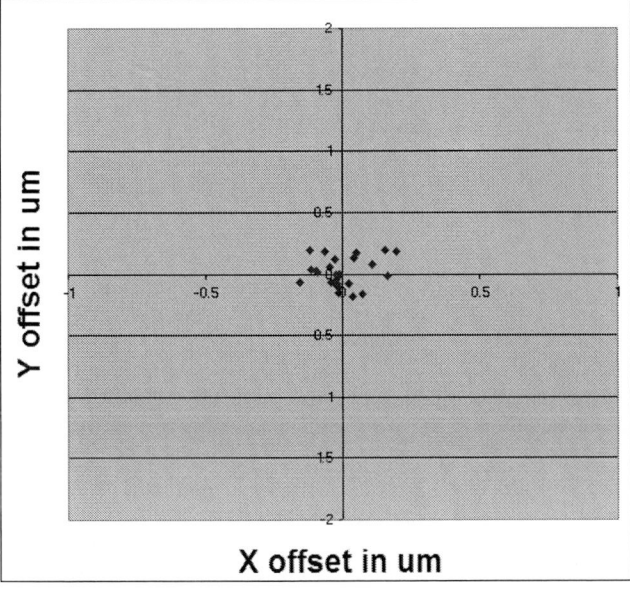

Figure 13: Defect offset using the SEM Offset Methodology

In this paper, a methodology for fast, accurate alignment of the review SEM to the inspection defect map is presented. This methodology utilizes a special macro containing programmed defects. The methodology was demonstrated using several PWQ wafer. PWQ wafers are one good application, but this methodology can be used for any wafer. The key benefits are 1) the time searching for good defects for SEM offset will be eliminated and 2) defects will be centered with an accuracy of better than 0.2um enabling a 1um FoV image for a better image of the defect.

ACKNOWLEDGMENT

This work was performed at the IBM Microelectronics, Semiconductor Research & Development Center, Hopewell Junction, NY 12533. Thanks to Kourosh Nafisi for his help in testing these structures.

REFERENCES

[1] R Buengener, C Boye, B. N. Rhoads, S. Y. Chong, C. Tejwani, S. D. Burns, A. D. Stamper, K. Nafisi, C. J. Brodsky, S. Fan, "Process Window Centering for 22nm Lithography", *Proceedings of ASMC*, pp. 174-178, 2010.
[2] O.D. Patterson, H. Wildman, D. Gal, K. Wu, "Detection of Partial Shorts and Opens using Voltage Contrast Inspection", *Proceedings of ASMC*, pp. 327-333, 2006.
[3] H. Xiao, L. Ma, F. Wang, Y. Zhao, J. Jau, K. Selinidis, E. Thompson, S.V. Sreenivasan, D. J. Resnick, "*Inspection of 32nm imprinted patterns with an advanced e-beam inspection system,*" BACUS, 2009.

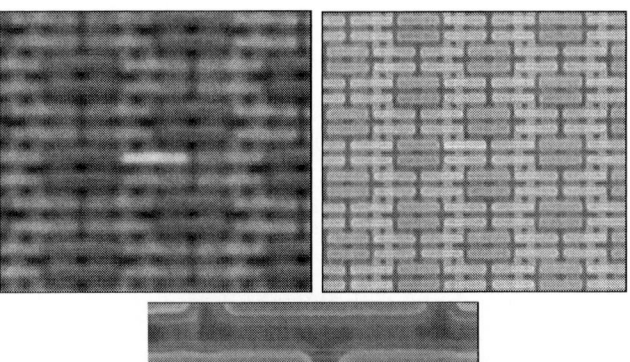

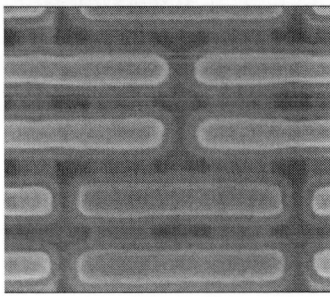

Figure 14: Top left: EBI image showing a bright gate line. Top right: Review SEM image. The bright PC line is barely visible with a standard condition. Bottom: The automatic high magnification review SEM image. If the high magnification image is not centered on the VC defect as in this case, then any physical cause visible at the surface will be missed.

Post Etch Killer Defect Characterization and Reduction in a Self-aligned Double Patterning Technology

Hong-Ji Lee, Sun-Yi Lin, I-Ting Lin, Kuo-Liang Wei, Sheng-Yuan Chang, Nan-Tzu Lian, Tahone Yang,
Kuang-Chao Chen and Chih-Yuan Lu
Macronix International Co., Ltd., Technology Development Center,
Advanced Module Process Development Div.
No.16, Li-Hsin Road, Science-Based Industrial Park, Hsinchu 300, Taiwan ROC.

Abstract – This paper identifies post etch killer defects, e.g., core bridging, small particle and tiny bridging, and investigates the possible solutions in a SADP module. Among the killer defect adders, core bridging and small particle are commonly observed after the oxide core removal by BOE. Core bridging adder is a carbon-containing polymeric by-product during nitride spacer open; by introducing additional diluted HF (DHF) treatment could effectively eliminate such bridging adder. Small particle adder is found to peel from the poly-Si hard mask-1 (HM1) damaged location, where is eroded during the wafer backside cleaning. It is useful for suppressing the formation of small particle by skipping the wafer backside cleaning process. Tiny bridging adder is block of etched poly-Si HM1 causing short between lines. One possibility of blocked etch adder creating is the fine micromasking formed on the opening of BARC during poly-Si HM1 etching. The effective suppression of tiny bridging adder is using high energetic F-/O-radicals to break through the micromasking in the poly-Si HM1 patterning. The reductions of above killer defects successfully boost up the sorting yield in our 45 nm charge-trapping flash memory.

Keywords- SADP, post etch defect, defect characterization, defect reduction, yield improvement.

I. INTRODUCTION

Charge-trapping flash memory technology keeps ever scaling for fabricating high-density nonvolatile flash memory (NVM) [1, 2]. From the trend of scaling of charge-trapping flash memory, the word line patterning becomes a major limitation for cell shrinkage. The set of approaches to overcome the optical lithography limitation is double patterning, where self-aligned double patterning (SADP) is the mainstream technology in the memory manufacturing because it requires only one litho step to define equal word lines on the cell array [3, 4].

With developing the SADP technology, the etch defectivity is regarded as leading edge challenge in the manufacturing line, the major yield killer defects are attributed to the new SADP module. As illustrated in Fig. 1, a list of post etch defect adders of interest, referring to as "core bridging", "small particle" and "tiny bridging", was found worse in the early stage of SADP development. From the observations of scanning electron micrographs (SEMs), these three defect adders pose a tangible and substantial yield risk due to their subtle physical characteristics and high defect density on wafer. Hence, it is important to understand the mechanisms of the killer defect formation in the Oxide Core SADP technology. In this paper, an in-depth study of the adder identifications and the possible solutions are discussed.

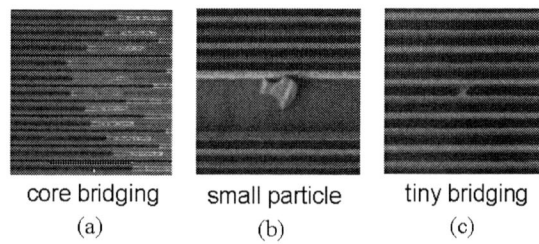

core bridging   small particle   tiny bridging
   (a)   (b)   (c)

Fig. 1. Examples of post etch killer adder in the SADP module.

II. METHODOLOGY

Fig. 2 shows the schematic flow of patterning 45 nm cell array in this work. On the hard mask stack, such as Oxide-Nitride-Oxide, another SADP stack, Oxide-Poly-Oxide, is deposited. After the lithography process, the TEOS core was followed to define in dry plasma etching. Then, a conformal nitride spacer deposition and several etch processes, such as nitride spacer open, TEOS core wet remove, poly-Si HM1 etch and TEOS HM2 etch, were applied to form the 45 nm equal line/space structures.

The defect adders at every process step were inspected with a KLA-Tencor 2351 pattern inspection tool. After inspection, SEM review (JW-7555S) was performed to obtain a defect pareto or defect mode based on the adder map. The chemical composition of the defect adder was also analyzed by using energy dispersive x-ray spectroscopy (EDS).

III. RESULTS AND DISCUSSION

A. Characterization and reduction of core bridging adder

The SADP process flow is much more complicated than the normal lithography only process flow. Any subtle defect from each process step could become a yield killer upon process completion.

978-1-61284-408-4/11 $26.00 © 2011 IEEE

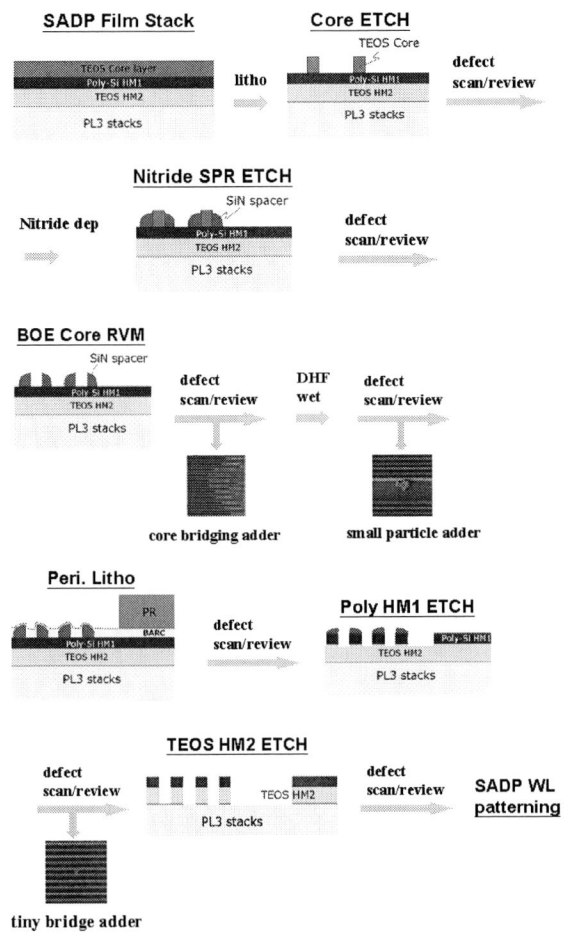

Fig. 2. Schematic process flow of a TEOS core SADP technology.

Core bridging adder is commonly observed after the oxide core removal by BOE wet etching. Fig. 3(a) shows the TEM micrograph of a representative core bridging adder, interestingly, remaining like a film spanned the nitride spacers. The film-like residue was identified mainly containing C, Si and O atoms from the observation of its EDS spectrum (Fig. 3(b)). The data indicate that the bridging adder is related to the $SiOxCy$ polymeric by-product produced during nitride spacer etching back in the CF4/CH2F2/O2 plasmas; followed the conventional O2 ashing and wet stripping (SPM and APM) are not effective in removing such by-product on the etched surface or in the etching damaged layer, as shown in Fig. 4. During the oxide core wet removal by BOE, the $SiOxCy$ polymeric by-product could not be dissolved in BOE completely and form the film-like structure upon patterns. Due to the damaged layer has been oxidized by O2 plasma exposure. Hence, we introduced additional DHF treatment post BOE wet etching. Upon the defect inspection mapping (Fig. 5), the defect density is improved from ~92.76

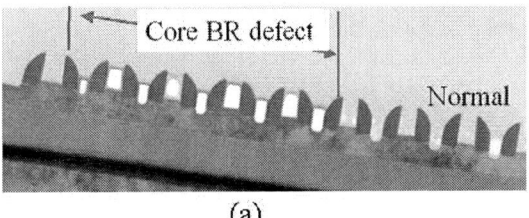

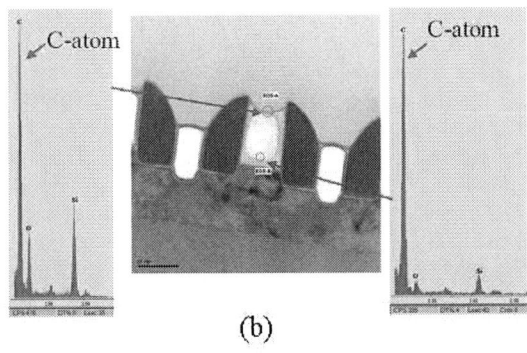

Fig. 3. (a) TEM micrograph of core bridging adder, which is a suspension covered on the array. EDS spectra indicate that the defect mainly contains carbon atom (b).

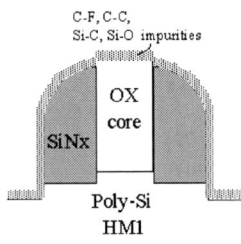

Fig. 4. The impurities containing C atom penetrated into the etched surfaces.

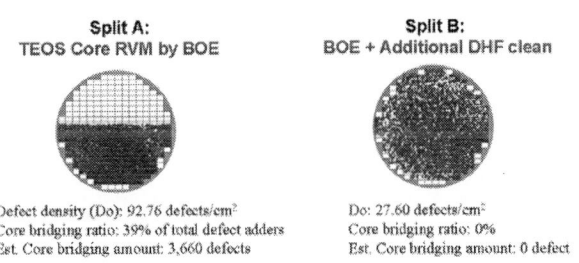

Fig. 5. Defect inspections for core bridging adder in the splits of A: post TEOS core removal by BOE and B: additional DHF clean treatment.

defects/cm^2 to ~27.60 defects/cm^2 in which there is few core bridging adders found by SEM review.

B. Characterization and reduction of small particle adder

Small particle, which we also observed after the oxide core removal process, is found to be concentrated near the wafer edge. With reviewing in previous processes by optical microscopy (OM) and SEM, we found abnormal watermark-like stains around the wafer edge post the wafer backside cleaning, as shown in Fig. 6. The watermark-like stain was identified as the damage of poly-Si HM1 on the front side of the wafer.

A possible mechanism underlying the formation of small particle is proposed in Fig. 7. The mixture of HF/HNO3, which is formulated to clean the wafer backside, caused unexpected damage of poly-Si HM1 on the front side of the wafer. The damaged poly-Si HM1 is difficult in maintaining enough thickness to be the etch mask in subsequent nitride spacer etching back, causing surface pitting. Thus, small particle adder, i.e., peeling defect, is formed as the TEOS HM2 is eroded in the following core removal process with BOE and DHF treatments. Fig. 8 presented the SEM micrographs of the eroded TEOS HM2 at the edge bevel supports our contention. Based on the mechanism, it is useful for suppressing the formation of small particle by skipping the wafer backside cleaning process.

C. Characterization and reduction of tiny bridging adder

Tiny bridging adder is block of etched poly-Si HM1 causing short between lines as shown in Fig. 9. There is a possibility causing tiny bridging such as the micromasking residue formed between nitride spacers on the opening of organic bottom antireflective coating (BARC) in plasma etching; the micromasking would cause blocked poly-Si HM1 etch resulting in bridging. According to the mechanism of the tiny bridging formation, a high energetic break-through (BT) step with F-/O-radicals prior to the poly-Si HM1 main etch was implemented. It is apparent from Fig. 10 that the adoption of the optimized BT step demonstrated effective suppression of tiny bridging (~0.06 adders/die in average) compared to that using conventional CF4 plasma as the BT step (~26.47 adders/die in average).

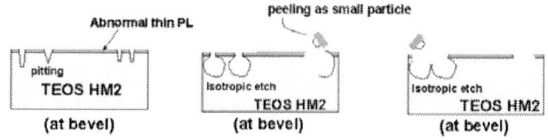

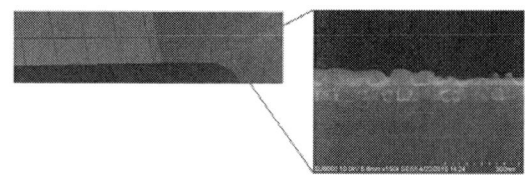

Fig. 7. A possible mechanism underlying the formation of small particle occurred after TEOS core removal process in the SADP module.

Fig. 8. The SEM micrographs of the eroded TEOS HM2 at the edge bevel post core removal with BOE and DHF treatments.

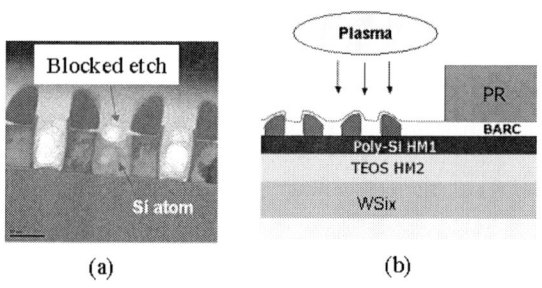

Fig. 9 (a) TEM micrograph of tiny bridging adder, which is block of etched poly-Si HM1. There is a possibility causing tiny bridging such as the micromasking residue formed and blocked HM1 etching in plasmas (b).

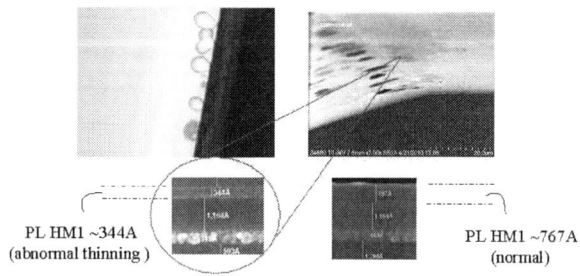

Fig. 6. There are stained marks found in the edge regions post the wafer backside poly-Si cleaning process. The wafer backside cleaning caused unexpected the front side poly-Si HM1 damage, resulting in abnormal HM1 thinning at the edge bevel.

HM1 etching with Conventional CF4 BT step

Do: 54.3 defects/cm²
Tiny bridging defect ratio: 68%
Est. tiny bridging/die: ~26.47 adders/die

HM1 etching with An optimized BT step

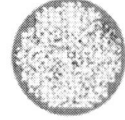

Do: 3.85 defects/cm²
Tiny bridging defect ratio: 2%
Est. tiny bridging/die: ~0.06 adders/die

Fig. 10 Defect inspections post HM etching. An optimized BT step during HM1 etching demonstrated effective suppression of the tiny bridging.

IV. CONCLUSION

Post etch killer defect investigations in this work are summarized as follows:

1) Core bridging adder is the carbon-containing (SiOxCy) polymeric by-product and seems not to be dissolved in BOE; by introducing additional DHF treatment could further eliminate such polymeric bridging adder.

2) Small particle adder is found to peel from the poly-Si HM1 damaged location, where is eroded during the wafer backside cleaning. It is useful for suppressing the formation of small particle by skipping the wafer backside cleaning process.

3) There is a possibility causing tiny bridging such as the fine micromasking residue formed between nitride spacers on the opening of BARC in plasma etching. The effective suppression of tiny bridging adder is using high energetic F-/O-radicals to decompose the micromasking during the poly-Si HM1 patterning.

From the reductions of the aforementioned SADP killer defects, the sorting yield is significantly boosted up in the 45 nm charge-trapping flash memory because less redundant elements are used.

ACKNOWLEDGMENT

The authors would like to thank MD340 & MD450 personnel, Macronix International Co., for their technical assistance and useful discussions during the course of this work.

REFERENCES

[1] A. Chimenton and P.ero Olivo, "Reliability of erasing operation in NOR-Flash memories," Microelectronics Reliability, vol. 45, pp. 1094-1108, 2005.

[2] N. Goel et al., "Erase and retention improvements in charge trap flash through engineered charge storage layer," IEEE Electron Device Letters, vol. 30, pp. 216-218, 2009.

[3] D.Y. Lee, Y. Kang, Y.S. Chae, S.J. Lee, H.K. Cho, and J.T. Moon, "Double patterning technique using plasma treatment of photoresist," Jap. J. Applied Physics, vol. 46, pp. 6135-6139, 2007.

[4] K. Yahashi et al., "Sub-32nm line and space patterning using sidewall transfer process," Dry Process International Symposium, Japan, pp. 279-280, 2008.

978-1-61284-408-4/11 $26.00 © 2011 IEEE

A system to optimize inline defect detection using short loop testchips leading to faster yield learning

Tanya Yang, Hun Chow Lee, Victor Lim, Fang
Hong Gn, Tri Mardiyono, Qionghan Wang, Long
Phan Nguyen
GLOBALFOUNDRIES
60, Woodlands Industrial Park D St 2
Singapore 738406
longphan.nguyen@globalfoundries.com

Fei Li, Sa Zhao, Anand Inani

PDF Solutions, Inc.
333, W. San Carlos St, Ste 700
San Jose, CA 95110, USA
anand.inani@pdf.com

Abstract—**With every new manufacturing node also come new modes of failures. Being able to identify these new fail modes and solve them quickly is the key to bring a manufacturing process to mass production readiness. Inline inspection is typically used for studying defects at critical layers. However, this is often limited by the amount of defects that can be visually inspected and to be able to qualify them between killer and false defects. We describe a powerful methodology combining electrical measurements from CV® testchips and inline inspection to make efficient usage of limited inline inspection resources and be able to identify new defect types that will eventually cause yield loss. This methodology can also be used to optimize inline inspection recipes and apply to production wafers.**

Keywords-CV® testchips; inline inspection; defect pareto; short loop; array mode inspection

I. INTRODUCTION

Defectivity is one of the top yield killers in any semiconductor manufacturing process. During the development phase at every generation, several new types of defects or fail modes occur and accelerated ways to understand and solve these are of a great help in bring the process to the manufacturing phase. During the mass production phase, some new fail modes may be uncovered depending upon specific product layout styles. In order to accelerate yield learning, it is of utmost importance to reduce such new types of defects quickly. However, as the critical dimension reduces and manufacturing processes become more complex, it is getting difficult to capture such small killer defects with new fail modes. PDF Solutions' Characterization Vehicles (CV®) testchips provide a large critical area spread over a large variety of design of experiments to be able to identify such new fail modes faster thereby helping the process integration and inspection teams identify and debug issues so as to bring the process into mass production. Here we describe a methodology developed to combine inline inspection and CV testchips at an advanced 45nm technology to help identify new fail modes and optimize inspection recipes that eventually lead to a faster yield learning.

II. USE OF CHARACTERIZATION VEHICLES AND INLINE INSPECTION IN PROCESS DEVELOPMENT

As part of the 45nm process development at Globalfoundries' Fab 7 in Singapore, PDF Solutions' Stackable CV testchips are run on a regular basis in short flow mode [1]. Electrical data obtained from these testchips is used to quantify fail rate performance by layer, study process and design systematics and evaluate split results. From the early data from the testchips, the electrical fail rate for metal (Mx) opens was estimated to be quite high. On the other hand, Mx opens performance from inline inspections, using KLA-Tencor 28xx tools [2], performed on the SRAM vehicle that was used for process development prior to introduction of the CVs was far lower. As a result, Mx opens was not considered to be a top yield loss mechanism and there was no significant activity to improve performance of this layer. Inline inspections followed by SEM review performed on randomly picked samples on the CV also did not reveal a high level killer defect density. The inspection on the CVs was also done using the same inspection recipe that was optimized by a combination of experience from past nodes and the SRAM vehicle. This was of great concern as the electrical fail rate predicted from the CV, if true, would bring Mx opens to the top of the yield loss pareto.

In order to further study this problem further, a CV lot was commissioned. Inline inspection was performed at M1 layer followed by random SEM review for 50 defects. The CV wafers were then measured using PDF's pdFasTest tester [3]. The electrical signature from analysis of the defect size distribution from the NEST structures [4] showed that single line opens were the dominant fail mechanism. An overlay analysis was performed between the CV electrical data and the inline inspection data using PDF's pdCV™ analysis software [3] as shown in Fig. 1. Typical size of electrical DUT (and hence localization accuracy of electrical defect) is of the order of 0.1mm^2 and tolerance used for overlay with inline defect data is 50µm. This analysis showed a low capture rate, in particular for the single line opens that were the top yield killer as seen in the electrical data. Moreover, the SEM review of the randomly selected sites also did not find many single line opens. Given all this, many in the team started to believe that

978-1-61284-408-4/11 $26.00 © 2011 IEEE

these electrical fails, that could not be caught by inline inspection, were "false electrical defects" just like one would often observe false defects in inline inspection reviews. This could be possible if there is something wrong with the reticle or the tester used for testing the CV. This mystery needed to be solved quickly.

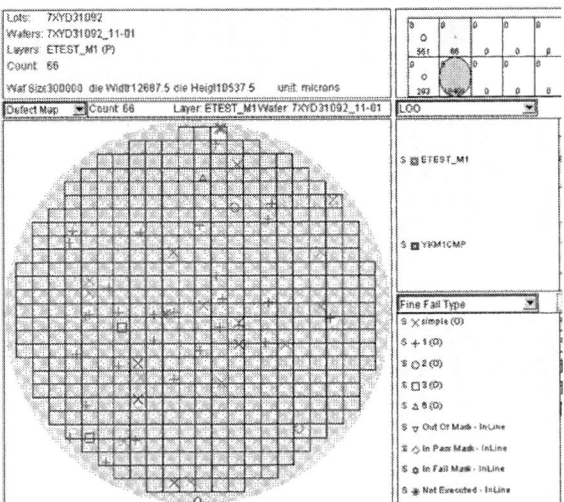

- RED fails are electrical opens
- BLUE fails are electrical opens with corresponding "hit"
- Lucky-Charms correspond to #line opens

Figure 1. Overlay of electrical fails seen on CV® testchips and inline inspection. Number of defects overlaid between electrical data and inline inspection is very low.

In order to study the electrical fails further, a few sites on a wafer that did not overlay with inline inspection were selected for failure analysis. The result of failure analysis using voltage contrast showed very fine metal open for every site selected as shown in Fig. 2. A cross section on one of the sites showed a very clear metal open which was immediate input into the BEOL team for process improvement.

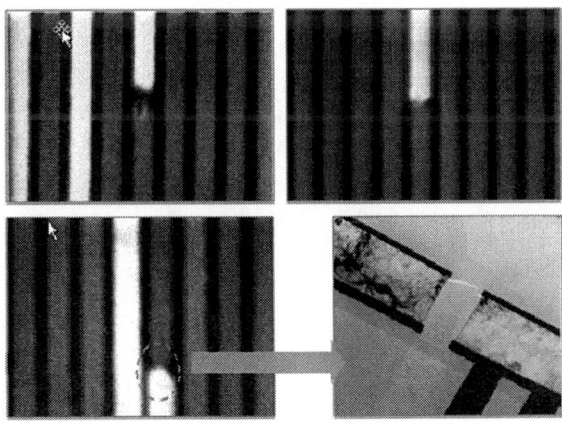

Figure 2. Failure analysis results showing single line metal open at several sites selected on Stackable CV.

This confirmed that the electrical fails were indeed actual fails and hence needed to be caught in inline inspections as well. But the problem the inspection team faced was that if inline inspection did not provide coordinates for a defect, it was almost impossible to manually hit upon such small opens by manual review in the large area DUT implemented on the Stackable CV for higher critical area. Moreover, the inline recipe was already optimized with several factors into consideration and it was not clear what more could be done. A new method had to be developed that will help them solve this problem.

III. SMART OVERLAY METHODLOGY

A. Typical Inline Inspection Methodology

In typical inline inspections at any layer, the location of all the defects are captured in a file, such as a KLARF file [2]. Following this, a random review of a fixed number of defects is performed to generate a defect pareto. The problem with this methodology is that many of the reviewed defects may not be killer, and hence the effort involved is not very efficient for yield learning. This problem is aggravated further at advanced technology nodes when the size of defects of interest are very small, newer defect modes may exist, and the inspection recipes developed using past experience may not be optimized to detect new types of defects.

B. Smart Overlay Methodology

After several rounds of discussion with the inline inspection team, we came up with a flow we called "Smart Overlay" that will help them not only make their SEM review effort more targeted, but also can help optimize recipes for newer kinds of defects. The underlying methodology is to get rid of random review and do a directed review on known electrical fails [5]. This can be achieved with a short flow vehicle that is testable inline at the layer of interest. The flow of the methodology is shown in Fig. 3 and the steps involved are as follows:

1. Inspection at Mx layer and generation of a KLARF file. No review at this point.

2. Electrical measurement at Mx layer, quantify electrical fails and perform overlay with the data from step 1. This gives two sets of data – a list of all defects overlaid and a list of known electrical fails, but not captured by inline inspection.

3. Review of overlaid defects to generate a quick directed defect pareto with minimal effort.

4. Manual review of the electrical fails that could not be overlaid. This can help find new defect modes as well as optimize the inspection recipe further. If the DUT size is large (typically ~ 0.1mm^2) such that manual review may not localize small defects, an optimization of various parameters of inspection recipe combined with an array mode of inspection can help find the defect.

5. If all else fails, submit for failure analysis (FA). This filtered set of fails saves FA resources by doing FA only on the new types of defects. If needed for inline inspection recipe optimization, the coordinates of the fail location can be

identified during this step without doing a destructive FA. Following this the wafer can be re-introduced in the fab to do inline SEM review and recipe optimization.

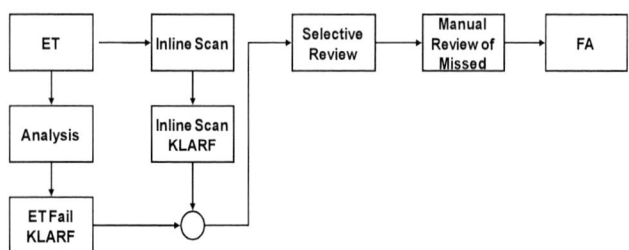

Figure 3. Smart Overlay flowchart

Following this flow achieves the following goals:

- Generates a quick defect pareto and saves resources by avoiding doing review on "non-killer" defects.

- Finds newer defect modes that may not have been captured by inline inspection so far.

- Optimizes inline inspection recipe further.

- Saves FA resources

IV. APPLICATION

Smart Overlay discussed above was applied to study M1 and Mx opens. Using steps 1-3 discussed above, the directed SEM review led to identification of "Missing Trench" as the key defect type causing yield loss as shown in Fig. 4. Steps 4-5 were then applied to optimize inline inspection recipe further, leading to more than a 2x increase in the number of "Missing Trench" defects captured as shown in Fig. 5. This was then used to run process splits that led to almost a 6x reduction in M1 opens in a very short amount of time. The same inspection recipe was then also used to inspect product wafers and even to wafers from the previous node. The result was an accurate capture, monitoring and fixing of this large yield loss mechanism.

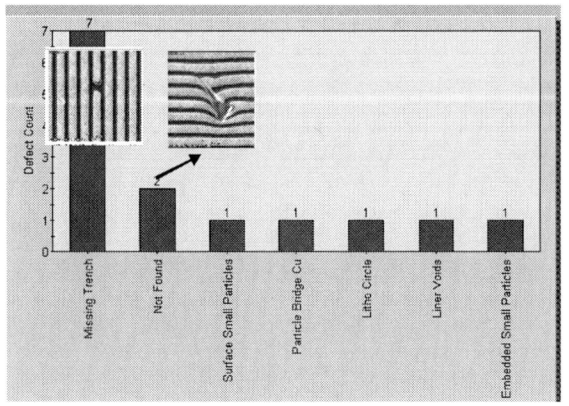

Figure 4. Defect pareto generated from smart overlay before reciepe optimization

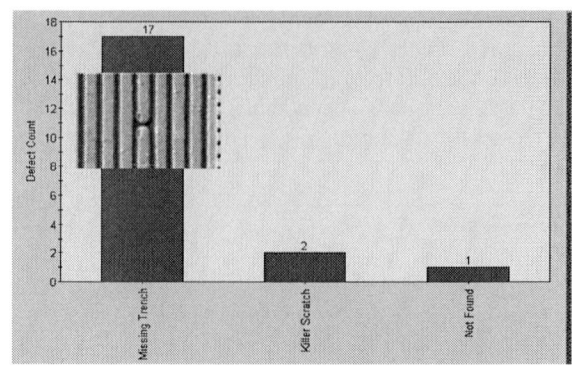

Figure 5. Defect pareto generated from smart overlay after inline inspection recipe optimization. More than two times more single line opens could be detected

Smart Overlay was then used for front end of line (FEOL) layers - active area and poly. This could be achieved by enabling probing of CV wafers inline after silicidation.

In one case, there were massive poly shorts on the wafer edge during early process development as seen from the electrical results from CV wafers in Fig. 6 (a). After applying a new clean split, many of these gross systematic shorts were removed as shown in Fig. 6(b).

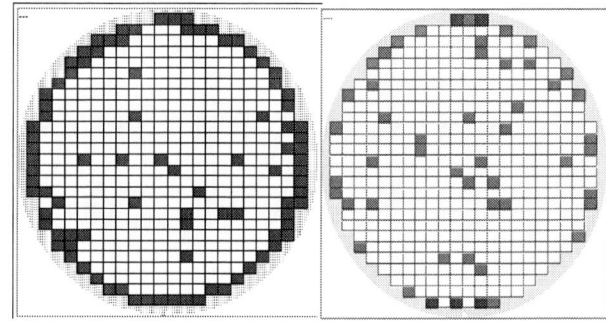

Figure 6. (a) Poly shorts with massive failures on wafer edge, (b) Poly shorts after applying clean split showing a major reduction in wafer edge fails

Now that the systematic issue was gone, there was need to study the remaining defects. Smart Overlay was commissioned on these wafers and a comparison of clean reference dies vs. the electrically failing dies showed the remaining types of defects on the wafer edge as shown in Fig. 7. During this process, poly opens could also be studied quickly even though that was not a top killer mechanism. The results of the directed SEM review are shown in Fig. 8. This quick defect pareto generation was extremely useful for the FEOL integration team and they could design splits to study and solve these issues quickly thereby enabling an acceleration of process integration and yield learning.

978-1-61284-408-4/11 $26.00 © 2011 IEEE   260

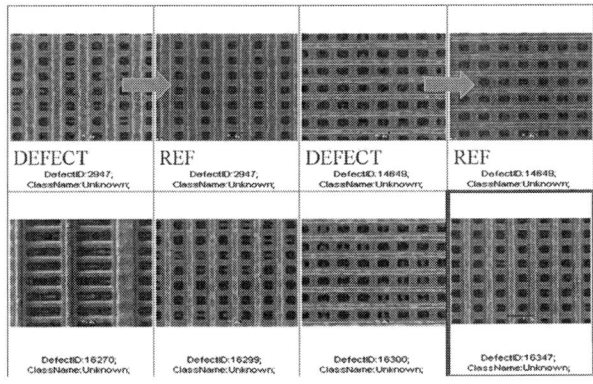

Figure 7. Result of directed SEM review of poly shorts from Smart Overlay

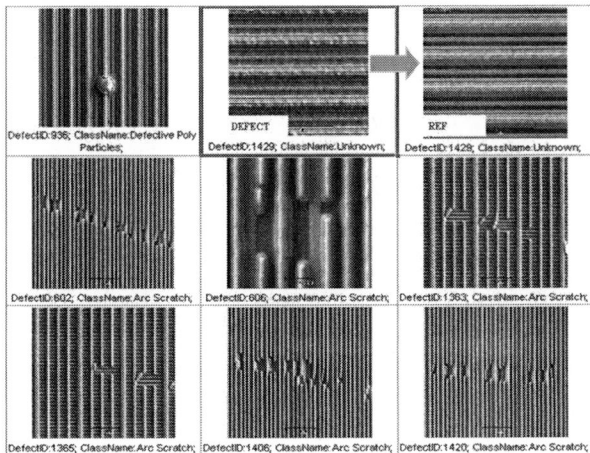

Figure 8. Result of directed SEM review of poly opens from Smart Overlay

The next step as in the BEOL case was to apply steps 4-5 of Smart Overlay to optimize the inspection recipe. During this phase, we found that while Smart Overlay helped to quickly generate pareto of systematic and random defects, optimizing inline recipes for FEOL has been more challenging due to higher topography of embedded particles, and defect sources from several layers. This work is still ongoing and the inline inspection team is working to overcome these and optimize a recipe that can be applied not only for process development, but also for production monitoring.

Note that although Smart Overlay can be used effectively at planar layers, it may not necessarily be applicable to generating paretos for contact and via layers.

V. FUTURE APPLICATIONS

Smart Overlay has proven to be an effective methodology to generate yield relevant defect pareto quickly. It shows the effectiveness of combining electrical characterization using sophisticated characterization vehicles and inline inspection tools. We envision that in the future this methodology can be used very effectively for quick technology and yield bring up. Additional applications would include benchmarking various inline inspection tools and new tool bring up.

VI. CONCLUSIONS

Smart Overlay methodology combining inline inspections and CV® testchips has proven to be extremely valuable in accelerating yield and making effective usage of resources for defect pareto generation. The methodology was applied for studying defects for both BEOL and FEOL planar layers. It was also applied for optimizing inspection recipes that were eventually applied to product wafers and even ported over to older technology nodes. It continues to be used to drive down the defectivity further as the process enters the random defectivity phase after having solved several process integration related systematic.

REFERENCES

[1] C. Hess, A. Inani, A. Joag, S. Zhao, M. Spinelli, M. Zaragoza, L Nguyen, B. Kumar, "Stackable short flow characterization vehicle test chip to reduce test chip designs, mask cost and engineering wafers," Advanced Semiconductor Manufacturing Conference, 2010

[2] KLA Tencor, Inc., http://www.kla-tencor.com

[3] PDF Solutions, Inc., http://www.pdf.com

[4] C. Hess, D. Stashower, B. E. Stine, L. H. Weiland, G. Verma, K. Miyamoto, K. Inoue, "Fast Extraction of Defect Size Distribution Using a Single Layer Short Flow NEST Structure," IEEE Transactions on Semiconductor Manufacturing, pp. 330-337, Vol. 14, No. 4, 2001

[5] V. Wei, E. Fuhrmann, A. Junge, R. Lutz, M. Rochel, M. Seider-Schmidt, R-S. Unger, C. Wallace, and J. Kuei, "Fast characterization of electrical fails overlaying to inline defect inspection during 90 nm copper logic technology development," International Symposium on Semiconductor Manufacturing, 2007

A Quality Metric for Defect Inspection Recipes

Ralf Buengener
GLOBALFOUNDRIES
Yield & Characterization, FCO TDV
Hopewell Junction, USA
ralf.buengener@globalfoundries.com

Julie L. Lee, Brian M. Trapp, John A. Rudy
IBM
Systems &Technology Group, Microelectronics Division
Hopewell Junction, USA

Abstract—**This article describes a metric that measures the quality of defect inspection recipes. It takes into account several factors including non visual (NV) rate, defect of interest (DOI) rate, defect count per wafer, and inspection time. The calculation runs automatically and only minimal user input is necessary. Different weighting models allow giving each factor more or less importance, thus making the metric flexible for different applications in development and manufacturing. The result is a final score for each recipe. A list of inspection recipes can be sorted by scores to show which recipe needs optimization. Several examples show the use of the quality metric in IBM's East Fishkill development facility.**

Keywords: Defect inspection, recipe, quality, defect of interest, non visual

I. INTRODUCTION

Defect inspections are an important part of modern semiconductor manufacturing. Inspections can be inserted after most process steps. They are non-destructive and give quick inline feedback about problems in the manufacturing flow. Comparison of inspection results from several steps can usually locate the root cause of a problem.

There are two main tasks for defect inspections: For tool monitoring, the defect inspection is located right after the step in which the tool to be monitored is used. So the inspection can track defect densities and defect types and highlight excursions as soon as they happen. The tool can be shut down and troubleshooted. For process monitoring, the defect inspection is located at the end of a process module. This allows gauging how well the process is working and is very important in development, when different processes are tested to choose the best one for manufacturing.

In both cases, to fulfill its task, the inspection must be sensitive enough to pick up all defects of interest (DOI) but at the same time suppress non visual (NV) or nuisance defects. Nuisance defects are real defects that are not of interest at this point. A typical example is prior level defects that have been caused by previous processes but are still detectable. Another example is defects that are known not to cause electrical problems or fall into areas where no active circuits are located. A good inspection would pick up most DOI and no or very few NV or nuisance defects.

II. CHARACTERISTICS OF A RECIPE QUALITY METRIC

If a defect inspection recipe is insensitive, excursions might be missed because the DOI are not detected. If the recipe picks up too many NV or nuisance defects, false alarms may occur or the DOI may be missed among the high number of other defects. In process development, critical decisions may not be possible because the inspection misses the difference between different processes. And valuable inspection tool time is wasted for inspections that yield no useful results. These recipes should be improved as soon as possible.

But recipe improvements need resources: A trained technician or engineer has to do the work, usually a wafer is needed which has to be held and in most cases the work has to be done on the inspection tool which has to be taken out of production and cannot inspect while the recipe work is being done. Therefore it is important to use these resources wisely and improve the recipes that have most need of improvement. In order to make this decision and find the worst recipes, a metric is needed that is able to gauge the quality of inspection recipes.

Traditionally, the NV Rate is a metric for the quality of inspection recipes. NV Rate is the ratio of the number of defects that are NV in the review (NV Count) over all reviewed defects (Review Count). A recipe with low NV Rate is considered a good recipe.

$$NV\ Rate = NV\ Count/Review\ Count \qquad (1)$$

Other suggested metrics concentrate on recipe stability by measuring the run-to-run variation [1] or the sensitivity distribution across the chip using statistical methods [2]. But we think that one factor alone is not enough to measure the quality of an inspection recipe. E. g. a low NV Rate can easily be achieved by making the recipe insensitive but this does not mean a good quality if DOI are missed.

The DOI Rate is the fraction of defects that are considered DOI in the review (DOI Count) over all reviewed defects (Review Count). This is an important factor, but not the only one.

$$DOI\ Rate = DOI\ Count/Review\ Count \qquad (2)$$

978-1-61284-408-4/11 $26.00 © 2011 IEEE

Another factor that should be taken into account in a manufacturing environment is the throughput. If an inspection takes a long time per wafer, only few wafers can be inspected with a given tool capacity. In some cases it might be worthwhile to take the time to run a sensitive but slow recipe. In other cases, inspecting more wafers with a faster, less sensitive recipe might be beneficial. Throughput is particularly important for tool monitoring recipes used in high volume manufacturing.

The number of defects detected on a wafer should also play a role. Very high defect counts can lead to scan aborts, so data from parts of the wafer may be lost. If this happens repeatedly it is usually an indication that the recipe is too sensitive. In addition, this puts a strain on the yield management database and servers and may slow down data processing. A very low number of defects, on the other hand, may indicate that the recipe is not sensitive enough and DOI may be undetected.

Besides these technical factors, the human factor plays a role, too. If a lot of manual work is necessary to compute a value for the quality of a recipe, it will not be used widely. A system that runs fully automatically and presents scores for all available recipes without extra work will be accepted more easily.

A. Factors

In a brainstorming meeting with inspection tool owners, recipe builders, level owners (defect engineers who use the inspection results), and data administrators we agreed on six factors that should play a role in the recipe quality metric. Each factor would be evaluated individually and get a score. The range for a score is from 0 % (worst) to 100 % (best). These are the factors:

1) NV Rate

This is the classical metric, the ratio of NV over all reviewed defects. The score is calculated by subtracting the NV Rate from 100 %. A good recipe that captures no NV would get a score of 100 %, a recipe that detects only NV would get 0 %.

$$\text{NV Rate Score} = 100\% - \text{NV Rate} \qquad (3)$$

2) DOI Rate

This is the most important factor because it shows how well the recipe performs its main task: Detecting DOI. This is the only factor where user input is needed. The level owner has to enter the defect types considered DOI for this particular recipe. This has to be done only once when the recipe is set up or when the purpose and therefore the DOI list for this recipe changes. Generic lists of defect types can be used, e. g. the same list for every post etch inspection. The score is the ratio of DOI over all reviewed defects. A good recipe that detects only DOI would get 100 %, a recipe that detects only NV and nuisance would get 0 %. More details about the DOI list are given in a later chapter.

$$\text{DOI Rate Score} = \text{DOI Rate} \qquad (4)$$

3) DOI Coverage

As a sanity check for the DOI list, an additional factor is introduced. It checks if every defect type in the DOI list is detected at least once in a while. If some defect types in the list are not detected, either the recipe is not sensitive enough or the list should be updated. The score is the fraction of DOI types that have been detected at least once in the observed time frame. If all DOI types are detected, the score is 100 %, if only half the defect types in the DOI list are detected, the score is 50 %, etc.

$$\text{DOI Coverage Score} = \frac{\text{number of DOI types detected}}{\text{number of DOI types in list}} \qquad (5)$$

4) Adders (Maximum Defect Count/MaxDef)

The upper threshold for the number of defects per wafer (Max Threshold) is set to reduce the impact on the yield management system. The more defects are detected, the more storage and calculating power is needed. As long as the average number of defects per wafer is below this threshold, the score is 100 %. Above the threshold the score decreases linearly until it reaches 0 % at twice the threshold or above. Only adder defects are taken into account for this calculation (Adders Count). This means defects at coordinates where no defect has been detected at a previous inspection.

$$\text{MaxDef Score} = 100\% - \frac{\text{Adders Count} - \text{Max Threshold}}{\text{Max Threshold}} \qquad (6)$$

Values below 0 % and above 100 % are corrected to 0 % and 100 %, respectively.

5) Randoms (Minimum Defect Count/MinDef)

This lower threshold (Min Threshold) prevents very insensitive recipes. If the average number of defects per wafer is above the threshold, the score is 100 %. If the defect count is below the threshold, the score is the ratio of the defect count over the threshold. Thus, if no defects are detected the score is 0 %. Defects in clusters are not taken into account for this calculation because even a small cluster can contain many defects. Only random defects are used (Randoms Count).

$$\text{MinDef Score} = \text{Randoms Count/Min Threshold} \qquad (7)$$

Values above 100 % are corrected to 100 %.

6) Scan Time

There is a target value for the Scan Time that depends on the type of scan tool (Scan Target). If the average scan time per wafer (Scan Time) is below the target, the score is 100 %. For longer scan times, the score decreases and reaches 0 % at twice the target time and above.

$$\text{Scan Time Score} = 100\% - \frac{\text{Scan Time} - \text{Scan Target}}{\text{Scan Target}} \qquad (8)$$

978-1-61284-408-4/11 $26.00 © 2011 IEEE

Scores

Recipe	Weighting Model	# Wafers	Last Update	NV Rate	DOI Rate	DOI Coverage	Adders (MaxDef)	Randoms (MinDef)	ScanTime	Score (%)
	POR	8		85.7%	10.2%	58.3%	100.0%	100.0%	100.0%	51.8%
	POR	6		61.0%	19.1%	66.7%	100.0%	17.8%	100.0%	52.0%
	POR	20		6.0%	5.9%	58.3%	100.0%	41.8%	100.0%	60.6%
	POR	8		3.9%	54.8%	16.7%	100.0%	26.5%	100.0%	70.0%
	POR	12		20.6%	68.3%	66.7%	100.0%	35.3%	100.0%	76.6%
	POR	3		4.3%	74.0%	8.3%	100.0%	100.0%	100.0%	82.2%
	POR	13		12.5%	73.1%	75.0%	100.0%	64.5%	100.0%	83.4%
	POR	42		8.9%	62.1%	75.0%	100.0%	100.0%	100.0%	84.4%

Figure 1. Example list of recipes for one technology in development. Recipe names are hidden. The list is ordered by final score (right column) to show the recipe with the lowest score first. The recipes at the top of the list probably need improvement.

Values below 0 % and above 100 % are corrected to 0 % and 100 %, respectively.

B. Score Calculation

For the final recipe score, all six scores are taken into account. To allow flexibility, each score is weighted with a weighting factor. The reason for the weighting factor is that different recipes may have different purposes. E. g. a standard production monitoring recipe should run fast and stable because many wafers are inspected every day. So Scan Time would get a relatively high weighting. On the other hand, for an engineering recipe designed to find a single defect type that runs only on a few wafers, Scan Time might not play a role so the weighting factor for Scan Time could be set to zero. Thus, for different applications, different weighting models can be used. In practice we have one weighting model for a mature technology in manufacturing, another for a new technology that is still in development, and a third for engineering recipes. If necessary, more weighting models can be used for special applications. They can be varied to fit the needs for each application. The only condition is that the sum of all weighting factors must be equal to 100 %. To calculate the final score, the score for each factor is multiplied by the appropriate weighting factor yielding a weighted score. Eventually the weighted scores are added up to yield the final score for the recipe which will also be a value between 0 % for a bad recipe and 100 % for a perfect recipe.

$$\begin{aligned} \text{Score} = \ &\text{NV Rate Score} * \text{NV Weight} + \\ &\text{DOI Rate Score} * \text{DOI Weight} + \\ &\text{DOI Coverage Score} * \text{DOI Count Weight} + \\ &\text{MaxDef Score} * \text{MaxDef Weight} + \\ &\text{MinDef Score} * \text{MinDef Weight} + \\ &\text{Scan Time Score} * \text{Scan Time Weight} \end{aligned} \quad (9)$$

C. Recipe Quality List

Fig. 1 shows a list of recipes with the corresponding scores. The first column displays the recipe name, the second column the weighting model used for this recipe. The third column shows the number of wafers used for the calculation. Our program usually uses the wafers of the last 30 days. Wafers that have been manually excluded from the trend charts due to known excursions of misprocessing are excluded from the recipe score calculation. The next column shows the date of that last change of this recipe (if known). This is helpful to compare the recipe scores before and after optimization. The next six columns show the scores for each factor. The first column shows the NV Rate. This column is inconsistent with the others as it displays the actual NV Rate and not the score for the NV Rate (which is 100 % - NV Rate). The other columns show the scores for DOI Rate, DOI Coverage, Adders (MaxDef), Randoms (MinDef), and Scan Time. The displayed scores are calculated as outlined above but not yet weighted. The last column shows the final score which is the sum of the six weighted scores.

#RecipePattern	DOIs (Comma or space separated)	WeightingModel	LastUpdate	NV Weight	DOI Weight	DOI Count Weight	MinDef Weight	MaxDef Weight	Scan Time Weight
	EC, FM, DD, MP, IP, HO	POR	5/28/2010	20	30	10	10	10	20
	DD, DS, EC, FM, RL, SS, MP, IP, CO	POR	6/6/2010	20	30	10	10	10	20
	FM, DD, DS, RL, EC, MP, IP, CO	POR	6/8/2010	20	30	10	10	10	20
	MM,BP,EC,FM,RR	POR	5/11/2010	20	30	10	10	10	20
	MS,PS,RN,FM,OP,EC	POR	9/23/2010	20	30	10	10	10	20
	BP,MM,AP,EC,RL,MP,FP	POR	9/24/2010	20	30	10	10	10	20
	CO,FM,BG,CF,NO,OF	POR	9/27/2010	20	30	10	10	10	20
	CO,FM,BG,CF,NO,OF	POR	9/20/2010	20	30	10	10	10	20
	FM,CO	POR	8/9/2010	20	30	10	10	10	20
	AS,IP,EC,FM	POR	8/30/2010	20	30	10	10	10	20
	DR,RL,SV	POR	9/28/2010	20	30	10	10	10	20
	DR,RL,RP,SV	POR	8/6/2010	20	30	10	10	10	20
	MV,RA,SV	POR	10/7/2010	20	30	10	10	10	20
	HO,BP,DE,DC,MP,EC,FM	POR	8/30/2010	20	30	10	10	10	20

Figure 2. Definition Table. Recipe names in the left column are hidden. For each recipe there is a list of DOI in the second column and the name of the weighting model in the third. The date of the last recipe update can be entered in the fourth column. The remaining columns are automatically filled based on the weighting model in the third column

Category	Value	Weight	Score	Wtd. Score
NV Rate	85.7	20.0	14.3	02.9
DOI Rate	10.2	30.0	10.2	03.1
DOI Coverage	58.3	10.0	58.3	05.8
Adders	6973.5	10.0	100.0	10.0
Randoms	1094.0	10.0	100.0	10.0
ScanTime	13.0	20.0	100.0	20.0
Total				51.8

Figure 3. Recipe score calculation for the first line in Fig. 1

D. DOI List and Weighting Models

As mentioned, the calculation should run automatically with as little user input as possible. Some input, however, is necessary. For this input there is a table that is shown in Fig. 2: In the first column, the recipe name is entered. The second column is for the list of DOI types as comma separated list. This list is the only input needed from the level owner, who knows the purpose of the recipe and the relevant defects at this inspection. Wildcards are allowed in the recipe column. So it is possible to create a generic list, e. g. for all post CMP inspections. Then no more input is needed for a new post CMP inspection. If a particular post CMP inspection needs a DOI list or weighting model that is different from the generic list, it has to be entered in a separate row.

The third column contains the name of the weighting model. POR is the model for standard inspections. The weighting models are defined in a separate table. Each user can create new weighting models if necessary. In the fourth column the date of the last recipe change can be entered. The following columns show the weighting factors (in %) for the six factors defined above. The sum of all weighting factors in each weighting model has to be 100 %. These columns are filled automatically once the name of the weighting model is entered in the third column. Thus, the user input is limited to recipe name, DOI list, and name of the weighting model.

E. Example Calculation

An example for the calculation of a recipe score is given in Fig. 3. These are the values for the recipe in the first line of Fig. 1. The first column of Fig. 3 shows the names of the six factors. For each factor, the value is given in the second column. The NV Rate is 85.7 %, the DOI Rate 10.2 %, DOI Coverage 58.3 %, this means 7 out of 12 DOI types were detected. The average number of adder defects per wafer is 6973.5, the average number of random defects is 1094.0, and the Scan Time is 13.0 min. The next column displays the weighting factors in %. The fourth column shows the scores in %, calculated according to the rules explained in chapter II A. The score for NV Rate is 100 % - NV Rate=14.3 %. The scores for DOI Rate and DOI Coverage equal their respective values. Scores for Adders (MaxDef) and Randoms (MinDef) are 100 % because the number of adder defects is under Max Threshold and the number of random defects is above Min Threshold. The Scan Time is below the threshold for this tool type, so the score for Scan Time is 100 % as well. For the last column each score is multiplied by its respective weighting factor. At the bottom right is the sum of all weighted scores, the total score for this recipe.

III. APPLICATION

The recipe quality list has first been implemented in IBM's development facility to monitor the quality of brightfield inspection recipes. A brightfield expert checks the list and decides which recipes need improvement. He sets priorities and coordinates his activities with the recipe build team depending on available tool time and other resources. After a trial period of several months the system has been established in the manufacturing department as well. Meanwhile a second version has been implemented with additional features such as drilldown reports for individual recipes and trend charts of recipe quality.

A. Drilldown Report

The second version of the recipe quality tracker allows

a) **PLY Recipe Quality Drill Down Report**

Recipe	I
Start Date	2010-09-25
Change Date	2010-11-11
End Date	2011-02-24
Tests	(e.g.: 1,2 or leave blank for all tests)
By Eqpid	☐ Check this if you would like the results summarized by eqpid
NV Weight	20.0
DOI Weight	30.0
DOI Count Weight	10.0
MaxDef Weight	10.0
MinDef Weight	10.0
Scan Time Weight	20.0
DOIs	AD,DE,HM,IP,IS,LC,LE,LP,MP,OP,PU,UP

[Submit]

b) **Scores**

Recipe	Start	End	Test	Eqpid	Weighting Model	# Wafers	NV Rate	DOI Rate	DOI Coverage	Adders (MaxDef)	Randoms (MinDef)	ScanTime	Score (%)
I	2010-09-25	2010-11-10	All	All	Custom	23	54.9%	12.0%	50.0%	85.0%	100.0%	100.0%	56.1%
I	2010-11-12	2011-02-24	All	All	Custom	14	35.0%	46.1%	75.0%	100.0%	100.0%	100.0%	74.3%

Figure 4. Drilldown report for a recently changed recipe. Recipe score increased from 56.1 % to 74.3 %.

a)

							Scores						
Recipe	Start	End	Test	Eqpid	Weighting Model	# Wafers	NV Rate	DOI Rate	DOI Coverage	Adders (MaxDef)	Randoms (MinDef)	ScanTime	Score (%)
	2010-12-01	2011-01-09	All	All	Custom	9	28.8%	36.5%	57.1%	80.4%	100.0%	35.0%	55.9%
	2011-01-11	2011-01-30	All	All	Custom	12	35.6%	18.3%	71.4%	100.0%	100.0%	30.0%	51.5%

b)

							Scores						
Recipe	Start	End	Test	Eqpid	Weighting Model	# Wafers	NV Rate	DOI Rate	DOI Coverage	Adders (MaxDef)	Randoms (MinDef)	ScanTime	Score (%)
	2010-12-01	2011-01-09	1	All	Custom	9	66.7%	15.4%	57.1%	100.0%	49.2%	35.0%	38.9%
	2011-01-11	2011-01-30	1	All	Custom	12	65.1%	12.1%	57.1%	100.0%	65.8%	30.0%	38.9%
	2010-12-01	2011-01-09	2	All	Custom	9	26.7%	25.0%	42.9%	100.0%	16.9%	35.0%	45.1%
	2011-01-11	2011-01-30	2	All	Custom	12	60.0%	9.5%	57.1%	100.0%	5.2%	30.0%	33.1%
	2010-12-01	2011-01-09	3	All	Custom	9	12.5%	25.0%	57.1%	100.0%	100.0%	35.0%	57.7%
	2011-01-11	2011-01-30	3	All	Custom	12	4.3%	23.7%	71.4%	100.0%	28.5%	30.0%	52.3%

Figure 5. Unsuccessful recipe improvement. Fig. 5 b shows the data by test. Note that the score for Adders (MaxDef) for each individual test is 100 %. Only if all three tests are combined (Fig. 5 a), the number of adders is over the threshold and the score drops below 100 % for the old recipe.

taking a closer look at a particular recipe. This is useful for example to see the effect of a recipe modification. The drilldown is opened by clicking on the recipe name. A new window opens that is shown in Fig. 4.

Fig. 4 a contains fields that are populated automatically and can be edited. The first field contains the recipe name. The next three fields are for dates. Start and end date can be changed. In the normal table, data for the last 30 days is used. In the drilldown window longer or shorter times can be investigated. The change date field allows checking the effect of a recipe change. If a date is entered here, two lines will appear in the lower part of the window (Fig. 4 b). One with data between start date and change date (old recipe), the second with data between change date and end date (new recipe). If a recipe consists of several tests, each test can be analyzed separately by entering the test number in the next field. In a similar way, different tools that run the same recipe can be compared by checking the box "By Eqpid". One line will be shown for each tool. The following six fields contain the weighting factors. The default is the weighting model used for this recipe. Values can be changed to see the effect of a different weighting model on the total score. The final field is the list of DOI for this recipe. Defect types can be added or removed to see the effect on the score.

Fig. 4 b shows the scores for the recipe with the values entered above. These lines contain the same data as the lines in the normal recipe quality table with the addition of columns for test and Eqpid (tool). If a change date is entered or display by test or tool is selected then several lines will be displayed, one for each test or tool, respectively.

B. Examples

1) Recipe Improvement
Fig. 4 b shows the drilldown report for a recipe that has

been optimized. The first line shows the dates and scores before the recipe optimization, the second line after the optimization. The total score in the last column increased from 56.1 % to 74.3 %. Main drivers for this improvement are the reduction of the NV Rate from 54.9 % to 35.0 % and the increase of the DOI Rate from 12.0 % to 46.1 %. DOI Coverage and Adders (MaxDef) scores increased as well but have less influence because the weighting factors for these scores are lower. Weighting factors can be seen in Fig. 4 a.

2) Unsuccessful Recipe Improvement
The recipe quality table can also highlight that a recipe optimization was not successful. Fig. 5 a shows the drilldown report for another recipe optimization. This time the score decreased. Main reasons are an increase of the NV Rate and a decrease of the DOI Rate.

Fig. 5 b shows a more detailed view of this recipe. It consists of three tests; each test is shown separately here before and after the recipe change. Scores for test 1 are unchanged, scores for test 3 show little change either. But the score for test 2 decreased from 45.1 % to 33.1 %. Reason is the increase of the NV Rate and decrease of the DOI Rate. It should also be noted that the score for MinDef decreased from 16.9 % to 5.2 %. This does not have a strong influence on the total score but it means that fewer defects were detected. Test 2 was changed back to the old settings and the score returned to its original value. Another approach is needed to improve the quality of this recipe.

3) Tool Migration
Fig. 6 shows the effect of migration to a new tool platform. One reason for the migration was the hope that the new platform would allow to scan with a larger pixel size thus reducing the scan time without loosing sensitivity.

Fig. 6 a shows data for the old recipe. The NV Rate is good

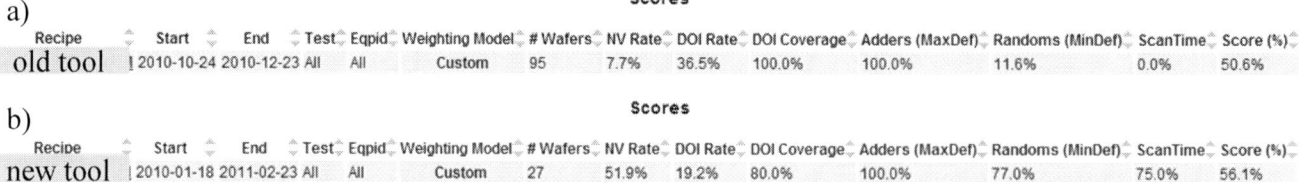

a)

							Scores						
Recipe	Start	End	Test	Eqpid	Weighting Model	# Wafers	NV Rate	DOI Rate	DOI Coverage	Adders (MaxDef)	Randoms (MinDef)	ScanTime	Score (%)
old tool	2010-10-24	2010-12-23	All	All	Custom	95	7.7%	36.5%	100.0%	100.0%	11.6%	0.0%	50.6%

b)

							Scores						
Recipe	Start	End	Test	Eqpid	Weighting Model	# Wafers	NV Rate	DOI Rate	DOI Coverage	Adders (MaxDef)	Randoms (MinDef)	ScanTime	Score (%)
new tool	2010-01-18	2011-02-23	All	All	Custom	27	51.9%	19.2%	80.0%	100.0%	77.0%	75.0%	56.1%

Figure 6. Tool migration. The recipe score for the new tool (b) is higher than for the old tool (a) with the POR weighting model.

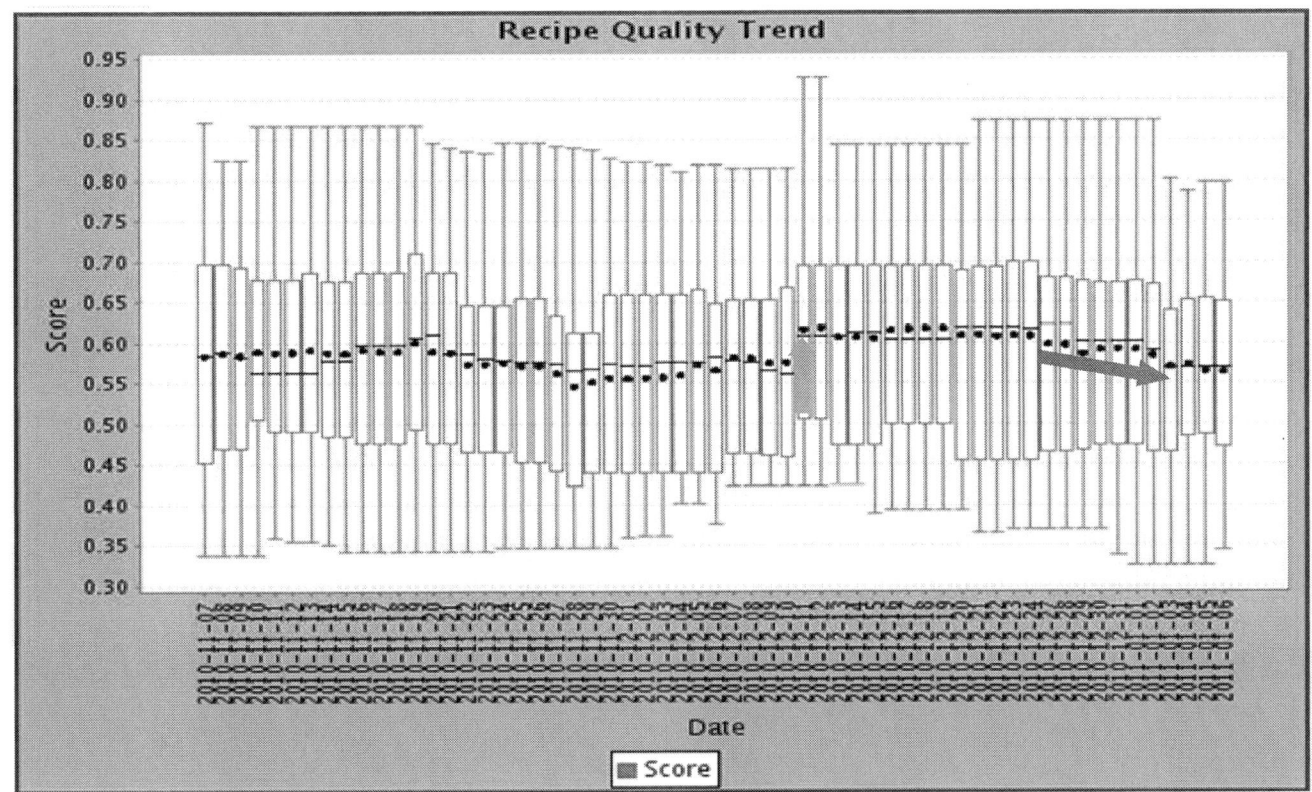

Figure 7. Recipe quality trend. Average recipe score for one technology. The first arrow shows the effect of recipe improvement. The second arrow shows the degradation due to negligence.

with 7.7 %, DOI Rate is 36.5 %. DOI Coverage and Adders (MaxDef) are good, Randoms (MinDef) score is low which means low defect count and the score for Scan Time is 0.0 % because the scan takes so long. Total score is 50.6 %.

Fig. 6 b shows data for the new recipe on the new tool. The NV Rate is increased to 51.9 % and the DOI Rate decreased to 19.2 %, as well as the DOI Coverage which decreased to 80.0 %. Randoms (MinDef) score is increased to 77.0 % which means that more defects per wafer are detected. The Scan Time is reduced; the score is now up to 75.0 %. The overall score is 56.1 % as compared to 50.6 % on the old tool platform.

This shows the importance of the weighting factors. The weighting model here is the same as in Fig. 4 a. Weighting for Scan Time and NV Rate is 20.0 % each. Had the weighting for NV Rate been 30.0 % and for Scan Time 10.0 %, the total score would have decreased from 59.8 % for the old recipe to 53.4 % for the new recipe. But in this case, the reduced scan time and therefore improved throughput of the tool were considered more important than the loss of sensitivity and the recipe now runs on the new tool platform.

4) Recipe Quality Trend

Fig. 7 shows the trend of the average recipe score for one technology in development. The average score fluctuates between 55 % and 60 % most of the time. There is an uptick on December 11 (2010-12-11) because some recipes were optimized this day. Between December 24 (2010-12-24) and January 3 (2011-01-03) there is a steady decrease in average recipe quality. Most engineers were on vacation during the

holidays and recipe quality was not monitored during this time. This shows that constant work is necessary to keep the recipe quality at a certain level.

IV. SUMMARY

A recipe quality metric has been invented and implemented in IBM's East Fishkill development and manufacturing plant. The metric consists of six factors: NV Rate, DOI Rate, DOI Coverage, Adders (maximum defect count), Randoms (minimum defect count), and Scan Time. For each factor a score is calculated and each score is weighted with a weighting factor. The sum of the weighted factors is the recipe score, a value between 0 % for a bad recipe and 100 % for a good recipe.

Different sets of weighting factors (weighting models) can be used to account for different types of recipes, e. g. engineering vs. production recipes. The calculation runs automatically and requires only minimum user input: List of DOI and weighting model. Generic entries can be used to further reduce the need for manual input.

A list of recipe scores is calculated based on recent inspection data and published every night. The list allows the equipment engineers and technicians to track recipe quality and identify recipes that need improvement. A drilldown reports helps to find the cause for a bad score and shows if an improvement has been successful.

ACKNOWLEDGMENT

R. B. thanks John Lupi for helpful discussion and use cases.

REFERENCES

[1] Henry Huang, "Using statistical median to check sensitivity of defect inspection," 2002 Semiconductor Manufacturing Technology Workshop, 2002, pp. 181-182

[2] Makoto Ono, Yohei Asakawa, Takao Sato, "Inspection recipe management based on captured defect distribution," IEEE International Symposium on Semiconductor Manufacturing, 2003, pp. 145-148.

Managing Data for a Zero Defect Production

The contribution of Manufacturing Automation to a corporate strategy

Gottfried Schmid
Manufacturing IT Director
Engineering Data Analysis & Reporting
Infineon Technologies AG
Regensburg. Germany
Gottfried.Schmid@infineon.com

Tilmann Hanitzsch
Senior Management Consultant
Leipzig, Germany
Tilmann.Hanitzsch@me.com

Most chips used for automotive applications do not tolerate failure. Next to design considerations, the effective use of all available production data is key to a successful "Zero Defect" strategy.

The challenge to the Manufacturing Information Technology & Automation department begins with a massive amount of raw data to compute and extends to aggregation, categorization and evaluation of this data to trigger automated online reactions and support efficient engineering analysis.

Analysis, aggregation, automated, automotive, categorization, configurability, detection, equipment, functional, inline, measurement, metrology, monitoring, production, results, statistical, test, tracequality, visualization, yield

I. INTRODUCTION

Chips produced for AUTOMOTIVE SAFETY APPLICATIONS insist on an absolute reliability standard: "Zero Defect". Such circuits in control of automotive brake systems, power trains or restraint devices could inflict a security hazard on its own. This translates upfront into a design challenge, which we call the first aspect.

Embedded self-monitoring functionality needs to guarantee the early detection of any irregularity and trigger appropriate measures to avert disaster.

This paper focuses on the second aspect of the "Zero Defect" mission: The data management dimension of Quality Assurance (QA) measures during production and test.

II. DATA SOURCES

The available data sources for QA can be categorized into three areas:

- "Advanced Process Control" / "Failure Detection and Classification" (APC/FDC)

- "Inline Measurements" (Metrology Data, Statistical Process Control [SPC])

- Functional Test Results

A. "Advanced Process Control" / "Failure Detection and Classification" (APC/FDC)

APC/FDC data directly reflects the processing conditions at any equipment that touches productive silicon. The key to innovation in this dimension is the configurability of raw data aggregation. Monitoring temperature and gas flow during a furnace process yields to thousands of data points. While every single one needs to stay within defined limits, a meaningful condensation into a significant Key Number reflects the quality and stability of the entire process.

The challenge is to support continuous learning and innovation in this sector is to provide the utmost configuration flexibility with a mission critical production support system in charge of an extremely high data throughput. Multiple aggregation levels with sophisticated inclusion/exclusion rule sets need to provide the fuel for trend analysis algorithms during the process.

Along with the trace data from the various process and metrology equipment, environmental data and calendar events may be used to enrich processing stability transparency. Examples are clean room and external climate data. As such data is collected from separate sensors, accurate time stamping is essential, as well as proper mapping tables to reference measurements from multiple sensors to the related processing equipment.

Notable events to enrich the APC/FDC data context:

- Power outages, drops or spikes

- Unusual weather conditions (e.g. lightning)

- Public holidays and events like major sports competitions (may impact workforce fitness)

- Failed Process Tool Qualification runs (PTQ).

- Scheduled and unscheduled equipment down times

- Extended equipment idle times

The combination of trace data and the environmental factors listed above may be used to compute and categorize individual process stability confidence levels. In addition to limit monitoring and corresponding online reactions to limit violations, the confidence factor can be suitable to control metrology sampling factors, i.e. when the overall confidence level for any process drops, a higher percentage of wafers is submitted to physical measurements.

B. "Inline Measurements" (Metrology Data, Statistical Process Control [SPC])

While the data source of any problematic process data typically also identifies the source of the problem, this is not the case with Inline Measurements.

A metrology step to validate layer thickness may follow a series of etch and deposition processes. When a measurement indicates a significant deviation from targets or an increased particle count, any of the previous process steps since the last measurement may be at fault.

To maintain the highest process stability, not only spec limits are subject to SPC supervision, but much narrower "Control Limits" are defined and observed for any "Out of Control" (OoC) conditions. Sampling and batch building add to the challenge of identifying tools and material associated with a conspicuous measurement.

Intelligent interpretation of Single Wafer Tracking data (Front End) and Single Device Traceability (Back End) becomes vital to trigger automated online reactions. Again, the automation group is challenged to support OoC intelligence with configurable rule sets in the continuing innovation effort to automatically separate out just the right equipment and material.

C. Functional Test Results

The most important dimension of QA data stems from Functional Testing. The given examples refer to such functional test results.

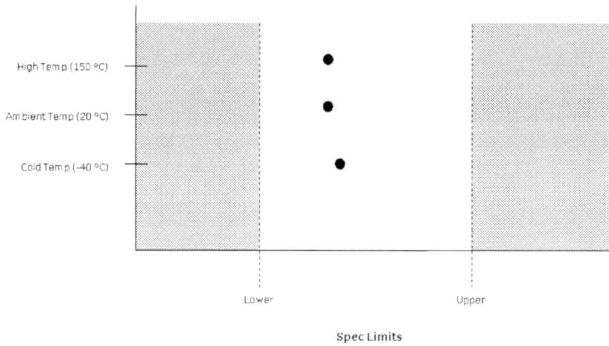

Figure 1. "Healthy" looking test result

Figures 1 and 2 show a common sense example applied to some simplified functional test results. In this case, an electric parameter is tested at three temperatures.

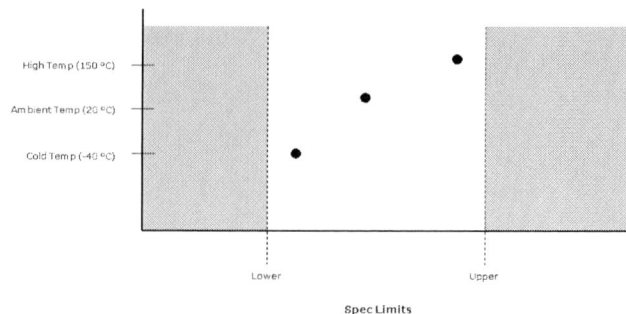

Figure 2. "Suspicious" looking test result

While all measurements fall within spec limits, Figure 2 presents a strong suggestion, that the device becomes unreliable shortly outside of the tested temperature range.

The second example reflects an evaluation by a process called "Part Average Testing" (PAT, see Figure 3). We assume a lower spec limit of 90 and an upper limit of 100. In the depicted case, 998 out of 1000 chips show results between 97 and 99 (symbolized by the green curve).

Only two chips return a value of 93.

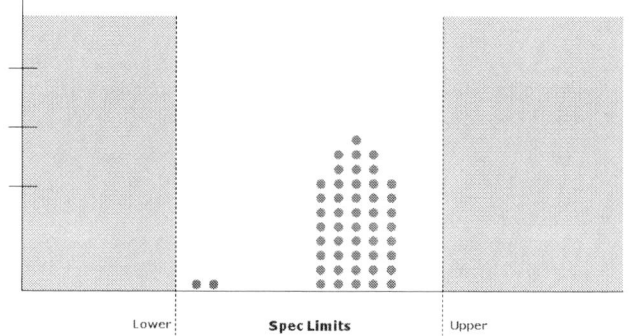

Figure 3. Simulated PAT result

With an isolated view (as applied by the tester equipment) there is nothing wrong with the two devices producing results of 93. When seen in the context of 1000 tests, those two measurements clearly stand out.

While the aforementioned analysis methods concentrate on dies which *passed* functional testing, the YIELD LEARNING WORKFLOW aims to recognize patterns amongst dies which *failed* the test. By analysis of test results in failed bins, Yield Learning first eliminates particle induced faults, which are random by nature and subject to a separate Defect Density (DD) screening. The workflow then applies a host of pattern recognition methods to identify common denominators amongst the remaining failures and suggest potential root causes and mitigation measures. These might address chip design and/or process parameters.

978-1-61284-408-4/11 $26.00 © 2011 IEEE

III. CONTEXTUAL LOGISTICS INFORMATION

Single Wafer Tracking in the FE has become a standard element of lot history protocols in the industry. This is not only a prerequisite to comprehend the exact path of any wafer down to single process chambers, it also allows identifying neighboring wafers in batch processes. Fault conditions with a certain geographical extent can be identified with this proximity factor in view.

As functional testing produces single die related results, die level proximity becomes an important factor, too.

With data originating from Wafer Test departments, FE Wafer IDs and chip coordinates are readily at hand. In the past, these identifiers were commonly lost in the back end process and consequently the identification of devices that had been neighbors on the wafer became almost impossible.

Today, "FRONT END STYLE BACK END" describes the goal of complete traceability of devices throughout the entire production cycle (see Figure 5).

Where microcontrollers are part of the circuitry, unique Chip IDs can be embedded. For other devices, Lead Frame IDs may be used to serve the same purpose. Our systems need to provide the adequate mapping tables to trace back the exact position of the device in question on the FE wafer.

Using color codes for the investigated failure classes and chip coordinates to reconstruct a visual image of the wafer, regional effects can easily be recognized and validated.

As illustrated before, it's not any one ingenious algorithm that ensures a zero defect production to be successful, but the consequent utilization of all available production data in an integrated approach.

IV. AUTOMATION SYSTEMS SETUP

Most tester equipment delivers its data in STDF V4 (Standard Tester Data Format), with the occasional need for format conversion, i.e. from an older STDF V3 to the current STDF V4 format.

Infineon utilizes Solaris clusters to host a modular set of preprocessing components, which deal with the tester data before it is loaded into the Engineering Base System (EBS, see Figure 4).

The major components of the EBS PreProcessing chain are STEPP (Standalone Test Data Post Processing) and PAT (Part Average Testing). All of these components allow very customized settings and adaptations, for example to detect any prober card failure at the earliest point in time.

This example both illustrates the need for a flexible setup as well as an inherent performance requirement. Testing with a defective prober card is a waste of time and worst case a reason to scrap wafers. Turnaround time for such an alert should be within a few minutes to abort such a measurement process at the earliest point in time.

Similarly, the overall performance of the entire EBS PreProcessing chain is a relevant success factor. The Automated Lot Release (ALR) is a very efficient method to pass lots through a quality gate without the need for a manual review. The required evaluation of test results is dependent on the availability of Lot level aggregations within the EBS database.

Within the EBS databases, several months of chip level data and even more than one year of lot level data is retained for on-line analysis. Data which expired from the online retention period may be retrieved from the archive directly by the users with the help of a special de-archiving GUI.

V. SUMMARY

It was shown that Zero Defects for components for the automotive industry can be reached by a twofold approach: Design for Zero Defects and Manufacturing with a Zero Defect mentality. Zero Defects in manufacturing can be achieved with Advanced Process Control, Inline Measurement and using functional Test Results. The key to implementing Zero Defects in Manufacturing is an advanced Manufacturing IT and Automation System which is able to handle huge amounts of data and transform it into analytical information.

VI. RELATED LITERATURE

[1] AEC - Q001 Rev - C: Guidelines for Part Average Testing (provides guidelines for using statistical techniques and extended operating conditions to establish part test limits; this approach could be used to provide "Known Good Die.")

[2] AEC - Q002 Rev - A: Guidelines for Statistical Yield Analysis (provides guidelines for using statistical techniques to detect and remove abnormal lots of integrated circuits)

[3] AEC - Q004 Proposed DRAFT: Zero Defects Guideline (describes a set of tools and processes which suppliers and users of integrated circuits can use to approach or achieve the goal of zero defects during a product's lifetime)

All documents above published by: Automotive Electronics Council, Component Technical Committee, Downloadable from: http://www.aecouncil.com/AECDocuments.html

[4] Douglas C. Montgomery, Introduction to Statistical Quality Control, Publisher: John Wiley & Sons Inc, 2000, ISBN 10: 0471316482 / 0-471-31648-2, ISBN 13: 9780471316480

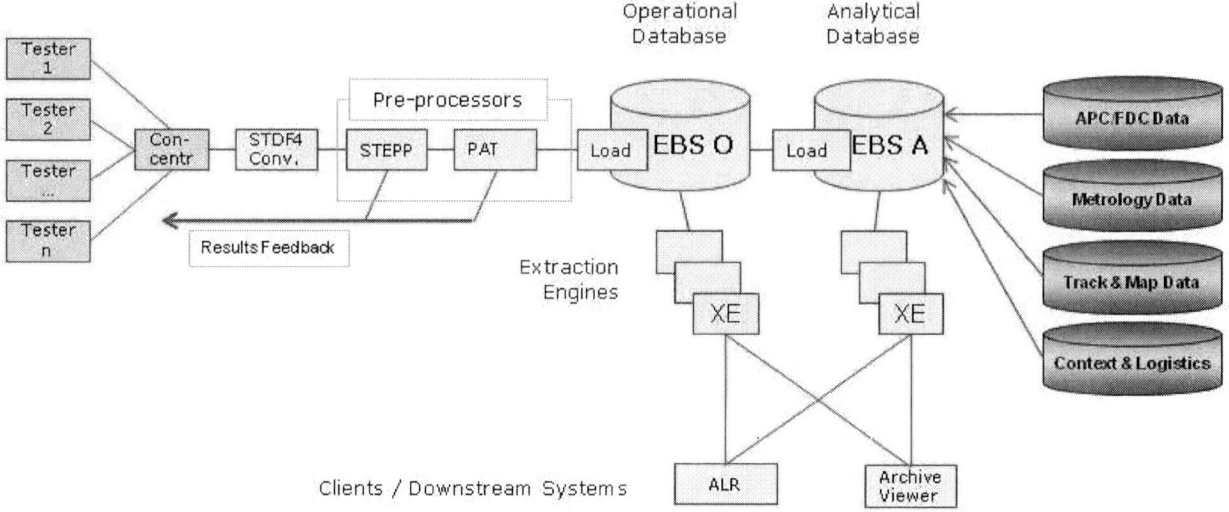

Figure 4. Data Flow Diagram of Infineon's Engineering Base System

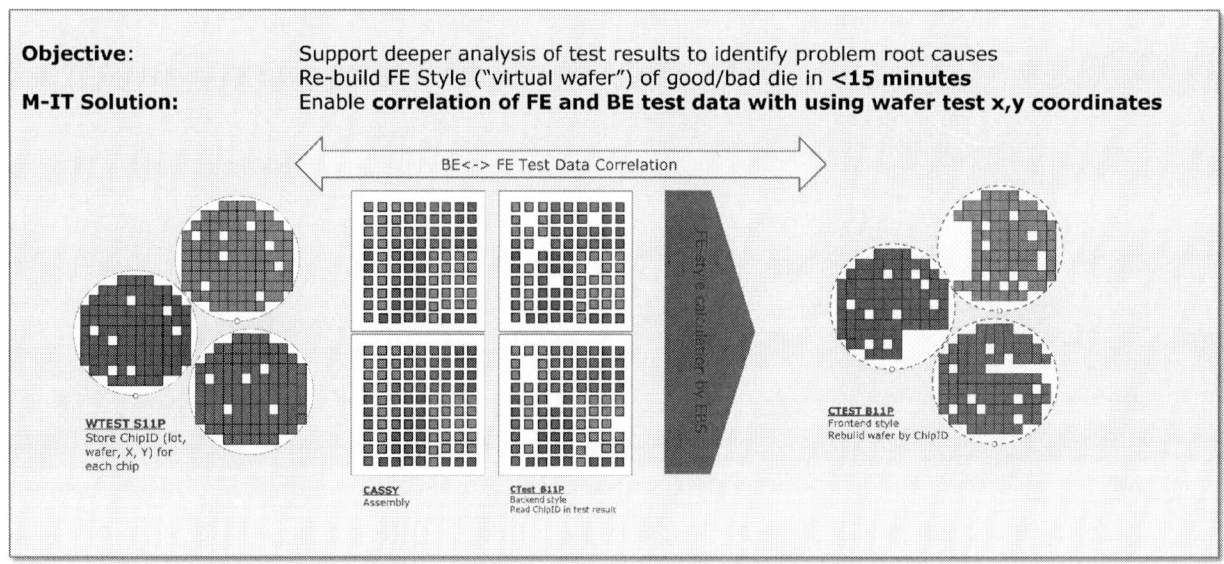

Figure 5. Tester Data Front End / Back End Correlation

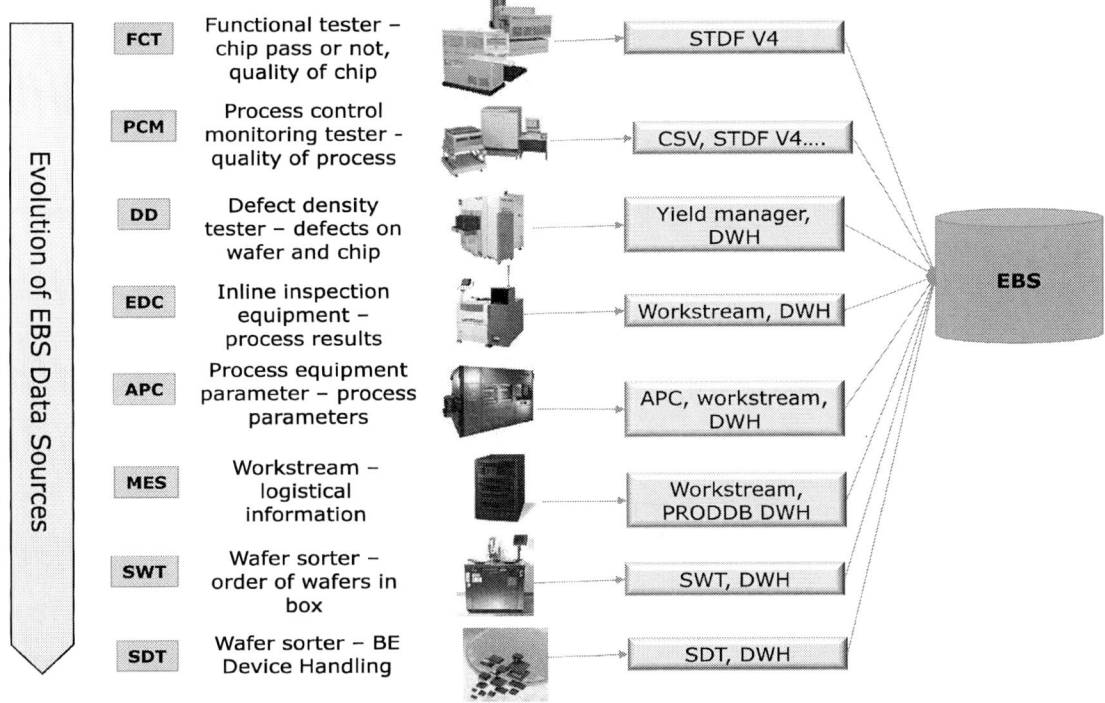

Figure 6. Evolution of EBS Data Sources

The Deployment Page:
Integrated Real Time Views of Tools, Operations, and Lots

ASMC 2011 Data Management

Henry Antonovich

IBM Systems and Technology Group, Microelectronics
1000 River Road, Essex Junction, VT 05452 USA

Abstract – **Running a semiconductor manufacturing facility with a large product portfolio creates unique challenges for knowledge management. Complex process controls, tool configurations, and wide ranges of roles and responsibility can quickly lead to increasing information technology (IT) support requirements with decreasing effectiveness. The IBM Burlington 200mm manufacturing facility has successfully met some of these challenges with an effective and efficient knowledge management system known simply as "The Deployment Page" (TDP). Through integration of key pieces of information, this system has successfully built a foundation to enable standardized work for deploying resources, both people and equipment, across a wide range of roles and responsibilities. At the heart of this system is the integrated view of tools, operations and lots. This paper will cover some of the basics of TDP, its evolution in use, planned future enhancements and some examples of how it is used.**

I. INTRODUCTION

As the complexity of the product mix for wafer fabricators increases, the old "simple" ways of managing the flow no longer work. These complexities include the number of unique technologies and parts, rapid changes in the starts mix, and lots with different committed velocities, among other demands. "Managing the flow" refers to:

1. Tactical capacity planning
   Establishes limits and trade-offs between technologies or parts used by central or supply planning in its starts decisions (quantity and lead time);
2. Near-term deployment decisions
   Which tools can handle which manufacturing operations, with what preference;
3. Manufacturing engineering decisions
   Temporary modifications or restrictions to deployment decisions – e.g., precluding all wafers of a certain part from exclusively being processed by one tool, or temporarily inhibiting a certain lot or operation from a certain tool;
4. FAB-exit demand management
   Involves maintaining an accurate and updated demand statement (part, date, quantity, and importance) and linking this to the lots in the FAB;

5. Scheduling /Dispatching
   Creating an anticipated short-term sequence of lot assignments to tool to balance competing objectives; and
6. Operational staffing deployment
   Prioritizing maintenance and manufacturing resources..

The IBM Microelectronics Division has, and continues to have, an ongoing set of initiatives that integrate information technology, analytics, and optimization to improve decisions in each of the areas listed above. One of the lynchpins in this application suite is the integrated real time view of tools, operations, and lots (ITOL).

The heart of all FAB flow challenges and decisions is the complex interaction between tools, operations, and lots. This interaction covers such diverse items as expedited lots; variations in tool deployments; monitor lots; tool qualifications; load balance between photo levels; near term engineering requirements, etc. The IBM Burlington (Vermont) FAB decided that one key to its continued success was to consolidate efforts in developing real-time decision support applications. This consolidation would enable a diverse set of users (from third level managers to manufacturing technicians) to have common access to all key information about tools, operations, and lots in a manner that captured the complexity of their relationships and supported substantial variations in interest, ease of visualization, and the ability to drill down to the support details if needed. This "one stop shopping" application would serve as the base to support the critical, but never ending, ad hoc reviews and analyses required to successfully manage a FAB.

History

Process deployment (what operations can run on what tools) was historically tracked using spread sheets. Although this was a time-consuming task, it provided a visual to help understand tool set restrictions. A web-based automation (2001) of some of these basic reports was quickly followed by the addition of WIP and tool status, turning this report into a new tool for understanding current cycle time detractors (i.e., which tools are actually causing the issue).

978-1-61284-408-4/11 $26.00 © 2011 IEEE

As this methodology for tracking process deployment was implemented across the FAB, and new features continued to be added, this integrated view of tools, operations and lots has grown into what is now the core of a manufacturing knowledge management system known simply as "The Deployment Page" (TDP). This system is currently used throughout the IBM Burlington 200mm FAB by a diverse set of users (from third level managers to manufacturing technicians, engineering, maintenance, and manufacturing) to understand daily operations and to help deploy resources (both people and equipment).

The Thought Process

The primary goal of TDP is to eliminate wasted time by better organizing the information. Think top down (from FAB level issues, be able to drill down to root cause), but build the application from the bottom up (root causes to high-level summaries). As various questions get asked, or root causes get identified, capture ("go see") the re-occurring manual actions needed to identify and quantify the problem and integrate selected features into the TDP system. This will reduce the time needed to do the repetitive actions and let the customers of the application stay focused on the issues at hand. Building from the bottom up naturally allows for effective drill-down capability. Effective drill-down capabilities make people's jobs easier, structure the system, and standardize the way information is presented.

Application Performance Measures

To build a successful application, one must know what the measurement of success is. The application must be balanced in terms of Quality, Delivery, Cost, Team (QDCT).

Quality – Fast, reliable, easy to learn and use; leads people in the right direction. TDP has spread to hundreds of users, by choice, with minimal to virtually no formal training required.

Delivery – Adaptability of new features to support the ongoing change in needs; provides the base content required to support the users QDCT objectives. TDP has balanced content relating to product quality, delivery and cost.

Cost – Easy to support and develop; Minimal hardware requirements. To date, TDP has only required one person to design, develop, implement and support. Because it uses existing hardware resources, it has not required any new capital.

Team – Enables standardized work across organizational boundaries. The focus is on the customer base, alternative applications and different points of view.

II. CORE CONTENT

Status information is the most widely and regularly used type of information. Spending extra time understanding the status information needs of the various users is vital to enabling standardization across organizational boundaries.

The Individual Group Status Page (IGSP)

The individual group status page (Fig.1) is the core of the status information visualization (many reports drill down into this view and many reports are available from this view). A group is generally defined based on tools that run similar operations or "cascades" of tool plans. This view provides many customized features for each group using a consistent look and feel across the FAB. This makes it much easier for someone to quickly learn about an area that they may not be that familiar with. The main content of this detailed matrix of tools and operations can be categorized into restrictions, status and performance.

At a high level, areas of interest in the FAB are often identified because of high cycle time/WIP levels or a stop-in-production at particular process operations. This leads to questions of "where (exactly) does this process run?" (Restrictions) This then leads to questions of "where can this process run now?" (Status) This then leads to questions of "how did it get to this point?" (Performance) Without an efficient knowledge management system, seemingly simple questions can require a lot of data analysis to obtain the answers. The goal of the IGSP is to turn data already in the system into easily accessible information in order to help answer these types of common questions.

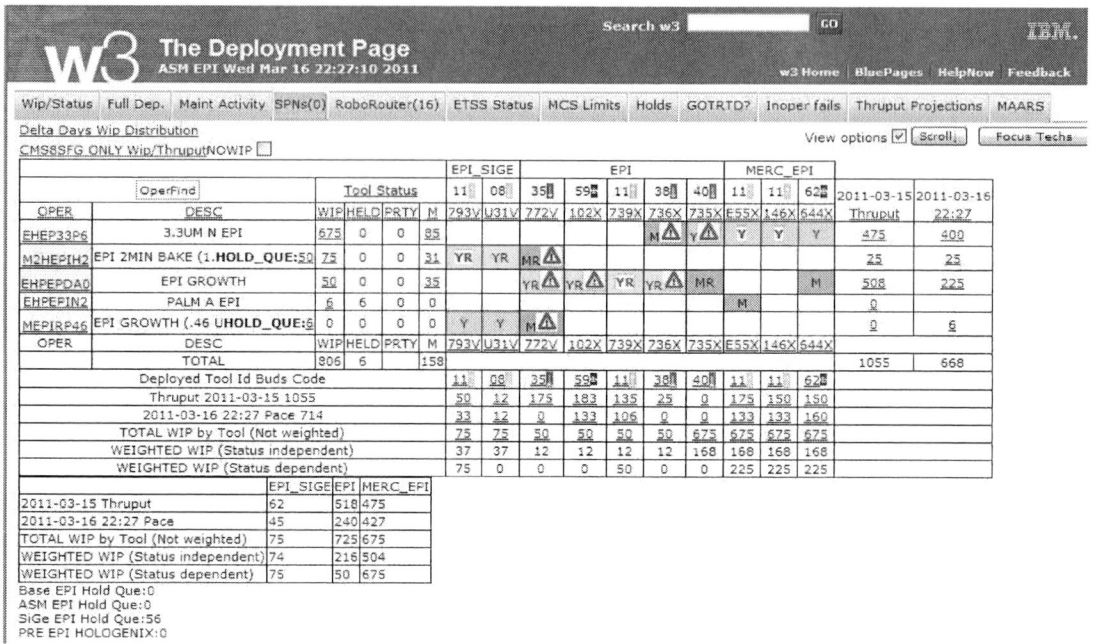

Fig. 1. An 'Individual Group Status Page' (IGSP). This view allows users to determine which tools are down and causing the greatest impact to production. Easy navigation and targeted features then help determine why it is down, how long it has been a problem and even what is being done.

Integrated Chart/Details

Some additional key information can be obtained without leaving the IGSP. A single click will cause the information to appear in a targeted location near the point of interest. This type of integration allows the user to keep the focus on the reason they wanted to learn more (which tools are up/down, operation in focus, etc)

- WIP details (Click on operation or tool WIP)
  - Allow the user to quickly answer common questions relating to:
    - lot technology;
    - elapsed time in the operation;
    - lot priority and
    - lot-specific restrictions.

- Throughput charts (Click on operation or tool throughput)
  - Help the user to identify gaps or changes in production rates. Initially "yesterday and today" by hour is presented, and then the ability to zoom out to a daily trend is available.

- Hourly Equipment Availability charts (Click on the Tool Status)

  - Allows the user to quantify recent equipment availability in real time, and then zoom out to understand a longer term history.

- Statistical process control inhibits and required chamber status details, (Click on the tool/operation),
  - Allows the user to understand very complex process deployment situations.
  - Although an equipment mainframe may be in a ready state, a required chamber may be unavailable, leaving the equipment unable to service the operation or only in a reduce throughput state.
  - Tool status and detailed information about these types of inhibits allow the user to identify the point of inhibition, such as foreign material (FM) or uniformity issues, or to determine that a particular qualification needs to be run before production processing can continue.

TABS

Tabs provide easy navigation to select standard features, allowing the user to get a much more in-depth knowledge of the area.

- Full Deployment tab (Full Dep.)
  Provides further information about the process deployment for all of the operations in the group, such as long-term throughput by tool/operation.

- Maintenance Activity tab (Maint Activity)
  Provides real-time access into detailed maintenance activity logs, allowing the user to get a much more in-depth knowledge of particular situations.

- Restriction tabs (SPNs and RoboRouter)
  Provide detailed information about Stop Production Notices (SPN) and WIP balancing restrictions (RoboRouter). The tabs indicate how many rules/restrictions exist that are related to that group, creating a visual to get a feel for what to expect without even clicking.

- Qualification tabs (ETSS Status and MCS Limits)
  Provide a tool to understand the current state of planned qualifications. This feature allows users to answer questions such as "what qualification timers are coming due on my shift?"

- In operation Fails tab (Inoper Fails)
  Provides tools to identify and quantify failed attempts at loading lots.

- Throughput Projections tab (Throughput Projections)
  Provides tools to analyze both historical and future workload (from short term to several days), as well as historical performance with that workload. Drill-down capability allows the user to better understand how other areas affect the area in focus. Tools are provided to help quickly estimate WIP work-off plans.

III. HIGHER LEVEL SUMMARIES

The real power of the IGSP can be leveraged when it can be easily navigated at the most effective time. Many types of higher level summaries are available that link into the IGSP. Different perspectives, based on roles, all lead to the same location.

Deployment Group Summaries (Fig.2)

The Deployment Group Summary (DGS) format is the most widely used summary format, favored mostly by people responsible for large sub-sections of the FAB. This is designed as a mash-up of group summary information. Using a mash-up technique allows for the ability to efficiently make many variations of groups-of-groups in order to satisfy different spans of control (equipment support, manufacturing, process

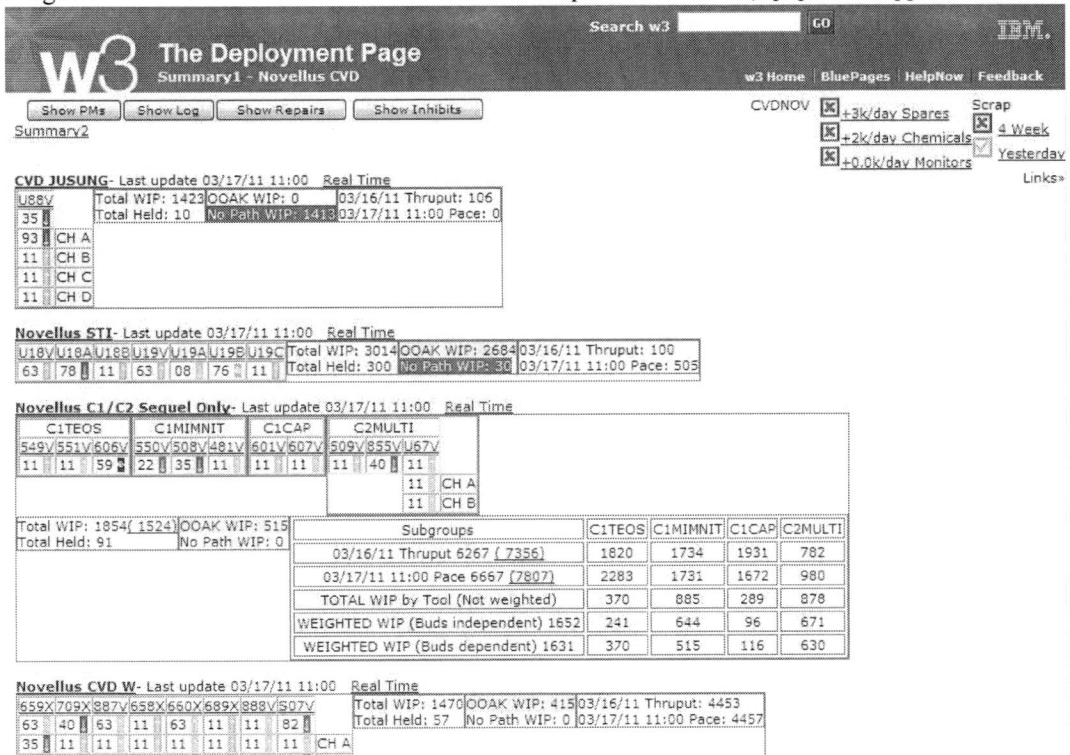

Fig. 2. A 'Deployment Group Summary' (DGS) page: one of many high-level summaries utilizing the IGSP as the core. Highlighting focus areas, providing key information and easy access into more detailed information, these views allow a wide range of users to find out what is going on over a much wider range then a single group.

groups/tech centers, etc) and increase page load speed. This format has been extremely effective at enabling a wide range of users to identify and understand key issues.

The groups in a DGS page are presented in an order that is based on many different factors (e.g., the amount of No Path WIP, number of tools in a production state), bringing the most relevant areas toward the top. Clicking the group title will drill into the IGSP view. Buttons at the upper left of the page allow the user to toggle the display (show/hide) of additional short term information including: upcoming planned maintenance (PMs); maintenance activity logs and SPC inhibits. Each option has drill-down capability into longer term/more detailed formats of the associated data. Clicking on an individual tool status will display an integrated (without leaving the page) real time status trend and any information in the options section related to the particular tool. Color-coded indicators for spending and scrap, located in the upper right, are based on current performance and drill into the respective trending and detailed information.

Days Out viewer (Fig.3)

The days out viewer is used to graphically display WIP relative to the end of the production line. Both the legend and bar stack are ordered by total quantity to improve identification of key areas. Filters based on elapsed time in the operation allow for very quick identification of "trapped" WIP. Fig. 3 shows an example of how the user can drill down to the point of cause. In this case, two out of three deployed tools are down, currently having foreign material (FM) issues.

Deployment on Demand look ahead (Fig. 4)

This feature allows users to easily monitor the current and future state of a population of lots in real time. The application will retrieve the routing of the population of WIP and resolve the deployment status of each future step, highlighting any potential no-path, one-of-a-kind or time-window situations. Each icon represents a single process operation. Clicking the icons immediately displays the details for that lot/operation, including the status of every tool capable of running the process.

Originally developed to simply better understand potential

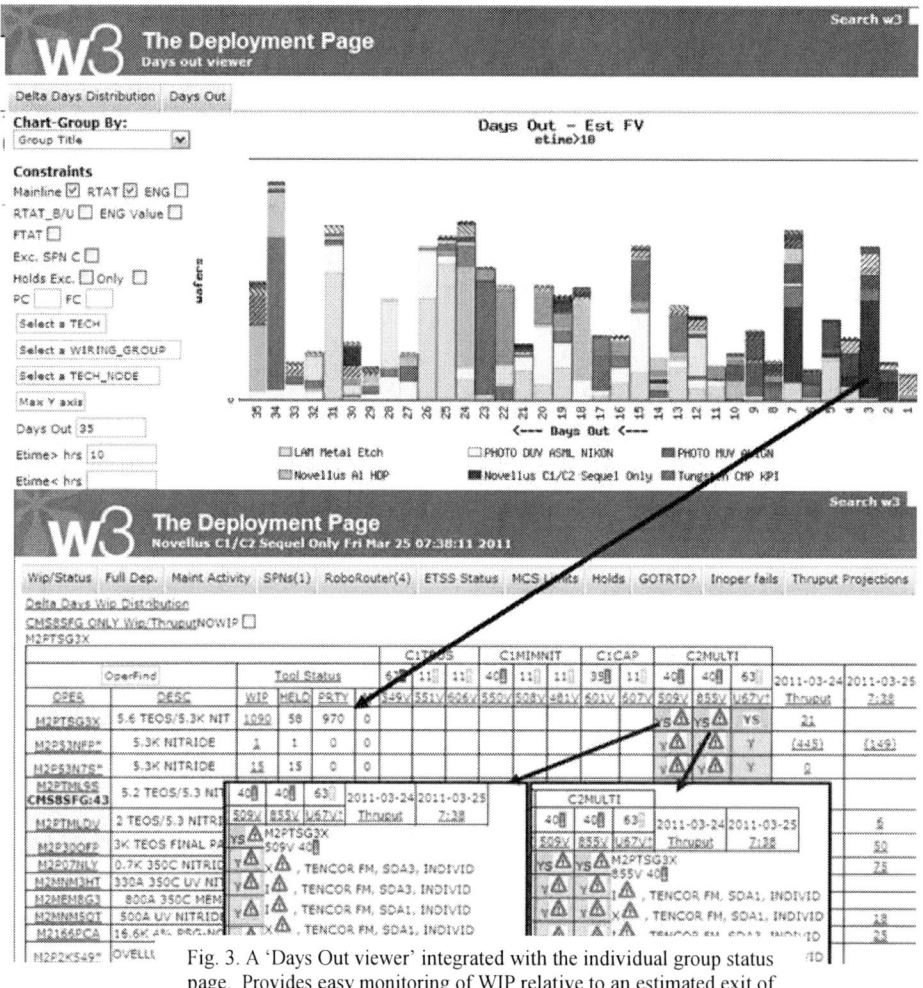

Fig. 3. A 'Days Out viewer' integrated with the individual group status page. Provides easy monitoring of WIP relative to an estimated exit of the FAB. Options allow a diverse range of users to look at many different populations of WIP and quickly determine and understand key pinch points in real time.

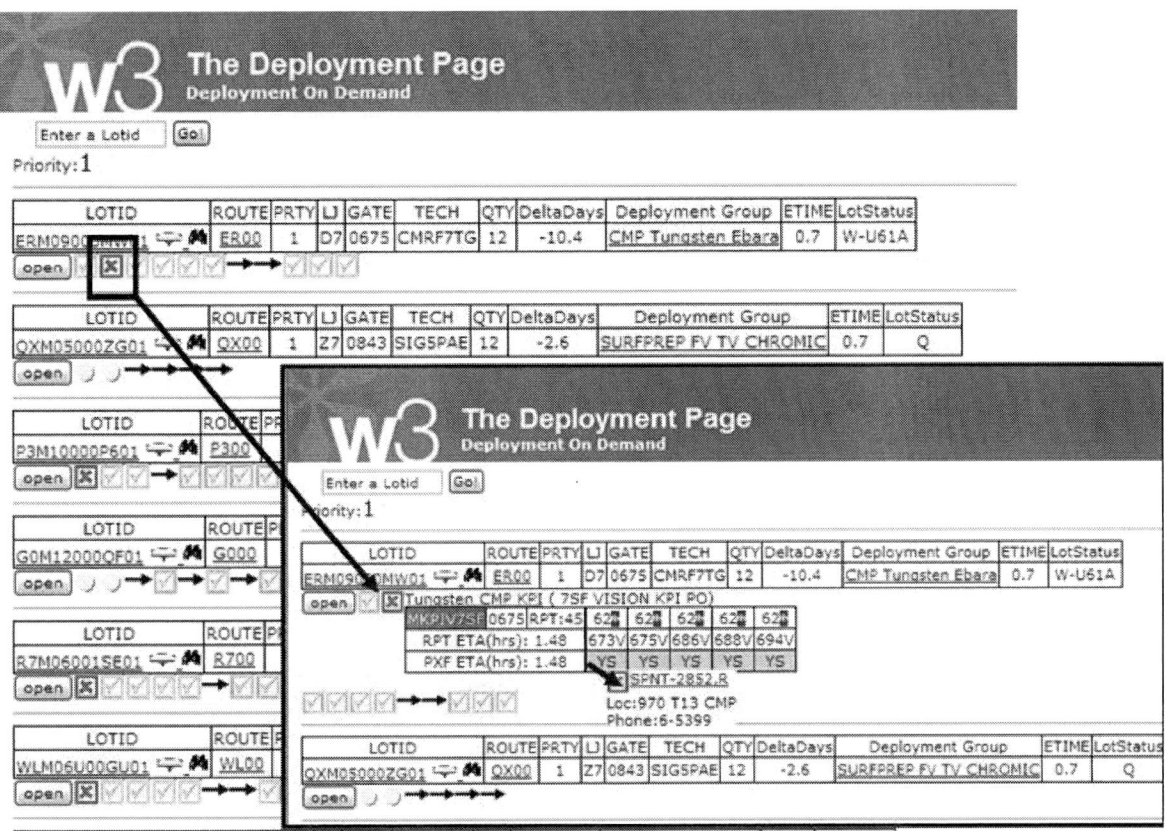

Fig. 4. 'Deployment on Demand Lot Look Ahead' vastly improves the ability of users to monitor the ready state of current and future process operations. Color and symbol coded icons provide an easy visual of any no-path or one-of-a –kind situations, each icon representing a single process operation. This application has proven to be useful for monitoring high focus / rapid turnaround time (RTAT) lots.

impacts to upstream operations, it was quickly utilized by operators responsible for keeping track of rapid turnaround time (RTAT) lots.

IV. CONCLUSION

Efficient and effective knowledge management in a semiconductor manufacturing fabricator can be very challenging due to the quantity and diversity of information needed to make daily decisions. IBM has found that the integrated view of tools, operations and lots is not only important as a foundation to ad hoc analysis and ongoing management, but serves as an incubator for the development of new modules that can be quickly implemented through The Deployment Page.

ACKNOWLEDGMENT

Thank you to everyone who has contributed ideas, feedback and support during the development of The Deployment Page. A special thanks to Ken Fordyce for being the "enzyme" to get this paper going.

Cycle Time Prediction in Wafer Fabrication Line by Applying Data Mining Methods

Israel Tirkel

Industrial Engineering and Management Department
Ben-Gurion University of the Negev
Beer-Sheva, Israel
tirkel@bgu.ac.il

Abstract — **Wafer fabrication is considered the most complex and costly challenge in the semiconductors industry. Cycle Time (CT), which denotes flow time, is one of its key performance measures. This work develops CT prediction models by applying Machine Learning (ML) and Data Mining (DM) methods. The models can assist in improving manufacturing and supply chain efficiency. They rely on historical production line data taken from the fab's Manufacturing Execution System (MES), and include wafer lot processing details of various operations. The prediction is done for an average CT of a single lot, processed through a single operation step. Two types of classification techniques are used. The best fitted Decision Trees (DT) model achieves 76.5% accuracy, and the best Neural Network (NN) model (two hidden layers) achieves 87.6% accuracy. The significance of this study is in establishing dynamic CT prediction models, which can be used to predict CT of a single operation step, a line segment or a complete production line.**

Keywords – Cycle Time prediction; Data Mining; Machine Learning; semiconductor wafer fabrication

I. INTRODUCTION

A. Background

The field of Machine Learning (ML) refers to the study of algorithms and techniques which allow computers to learn from experience [1, 2, 3]. Learning denotes change in a system that enables to do the task more efficiently with time. Data Mining (DM) is known as the extraction of implicit and potentially useful information from data [4, 5]. It is a process of discovering meaningful new correlations, patterns, and trends by sifting through large amounts of data stored in repositories and using pattern recognition methods as well as statistical and mathematical techniques.

Mitchell [2] describes a broad range of ML algorithms used for DM. Many organizations routinely capture huge volumes of historical data describing operations, products, and customers. Scientists and engineers in many organizations are capturing increasingly complex experimental datasets. The field of DM addresses the question of how to best use this historical data to discover general regularities and improve the process of making decisions [6].

MLDM methods have also been used in manufacturing. Harding et al. [7] present a comprehensive overview of applications in various areas, including: production process and control, maintenance, customer relationship, decision support systems, quality improvement, fault detection, and engineering design review. Semiconductors manufacturing is mentioned as one of the potential areas of research. The wafer fab we worked with did not have any known experience using MLDM, although literature offers such specific studies.

B. Semiconductor Manufacturing

MLDM methods have been applied in semiconductors manufacturing for product, equipment and process technology [8, 9, 10, 11]. Following are some of the MLDM studies performed, most of which were in the area of process technology.

Gardner et al. [8] describe the work of researchers from Motorola who solved wafer fabrication problems using DM. They applied a combination of self-organizing Neural Networks (NN) and induction rules on poor Yield factors taken from standard manufacturing data. This illustrated that problems can be solved ten times faster than in using standard statistical process control approach. Yield was shown to increase in the range of 3% to 15%.

Two Intel employees reviewed the benefit of applying ML on enormous amounts of data, described in Ultaut et al. [9]. They presented results of two internal case studies using Decision Tree (DT) models. The first was an equipment commonality done for defects generated by RF performance. The second was a chip performance level detected by early in-line electrical test results. This indicated that ML is powerful and can replace traditional statistical methods.

Industrial Engineering researchers from Taiwan universities investigated the quality of the bump soldering process, described in Chien et al. [10]. An empirical study was conducted using DT induction for Yield improvement. This showed that the relationships among controllable input factors and the target class can be effectively derived.

Six Intel employees from various organizations (process technology, manufacturing, sales & marketing, information systems) exhibit a comprehensive research in Goodwin et al. [11]. Studies using enormous data, numerous characteristics and high dimensionality are presented:

- Process tools signals are identified and studied for variation in Yield between/within production lots.

978-1-61284-408-4/11 $26.00 © 2011 IEEE

Gradient Boosted Tree (GBT) is used on a unique Intel platform, named Interactive Data Exploration and Learning (IDEAL).

- Unit-level speed prediction is done based on individual microprocessors final testing results (Bin Speed, Yield) and fab & sort data. Again, GBT is used on IDEAL.

- Cycle Time (CT) prediction was also studied, and is discussed in the following paragraph.

C. Wafer Fabrication CT

MLDM methods have been specifically applied for wafer fabrication CT [11, 12 and 13], although existing literature is less common. Following are some of the studies performed for CT prediction and classification.

CT prediction study using DM, performed by Arizona State University researchers and an Intel's senior process control person, is described in Backus et al. [12]. The prediction was done for wafer lots while running in the production line. It suggested that Regression Trees modeling is preferred over Nearest Neighbor predictors and NN. This is explained by the ability of Regression Trees to handle both categorical and continuous variables. Wafer lots processing data was collected in the first part of the production line, and then used to predict CT in the last one-third of the line.

In a different research, Chien et al. [13] presented a CT prediction model using DM, given the production line status (e.g. WIP, Throughput). The prediction model was constructed by dividing the fab into a few clusters, based on their utilization performance. Several models were used in sequence: Self-Organizing-Map (SOM) which is an unsupervised NN, DT, Polynomial Regression (PR), and Back Propagation NN (BPN). In a case study presented, prediction was performed for a fab divided into three clusters, using historical data. The results showed prediction accuracy of several percents per each of the clusters.

Finally, Goodwin et al. [11], mentioned above, have used a Clustering and DT model named Classification and Regression Trees (CART) developed by Intel and Arizona State University team. This tool categorized current flow of lots based on past performance and predicted lots CT within two days accuracy.

Concluding the literature above, it was shown that CT prediction were bounded by: (a) reliance on current lot processing through the production line, (b) prediction for only the last part of the line, (c) prediction for large parts (clusters) of fab, or (d) achieving relatively low accuracy measured in days. Our study attempts to reduce the reliance on current lot processing, enable CT prediction throughout the production line, apply prediction single operations, and increase the prediction accuracy to a few hours.

This work follows the phases of the Cross Industry Standard Process for Data Mining (CRISP-DM). Thus, the process of data understanding and preparation is described in Section II, the model construction in Section III, and the classification results and evaluation presented in Section IV. Finally, the conclusions drawn and future research opportunities are discussed in Section V.

II. DATA UNDERSTANDING AND PREPARATION

A. Data Extraction and Description

This phase includes the data collection process, the dataset organization and structure, and the definition of attributes.

The data were taken from the fab Manufacturing Execution System (MES), which continuously records all operational information and transactions of wafer lots processed in the production line. The dataset extracted contains wafer fabrication data sampled in a period of two years, from 37 operation steps which reflect generic fabrication process types, including: Diffusion, Thin-Films, Lithography, Etch, and Planarization. The data are organized in records, each including information of a single wafer lot processed in a single operation step. This information consists of the following: (i) manufacturing fab id, (ii) silicon wafers vendor id, (iii) process tools type, (iv) operation step name and serial number, (v) product allocation of the lot, (vi) lot type allocation: production (intended for sale), engineering (used for process or tool experiments), development (of future technologies), or short-loop (mostly for in-line monitoring), (vii) lot transaction times and wafers quantity, (viii) handling technicians id, and (ix) additional lot processing information (e.g. rework, hold, scrap).

The raw data were extracted onto MS Excel spreadsheet for initial analysis, and then uploaded to SPSS Clementine. It is composed of 8,000 records (rows), each with 43 attributes (columns), containing the information detailed above. The attributes were defined according to SPSS Clementine definitions for type and content, and named as follows: 22 quantitative scale ('range'), 11 qualitative nominal ('set'), and 10 qualitative two-categorical ('flag').

B. Data Understanding and Verification

This phase includes data analysis for understanding the records content, and data audit for verification of the records integrity and quality.

Understanding of the data was achieved by analysis of descriptive statistics. Quantitative attributes were analyzed for their range of values, outliers and extremes. Qualitative data were analyzed for their unique values and range of types. This process further enabled meaningful data verification, preparation, and features selection.

Verification of the data applied techniques of auditing the records for completeness and quality. This was done using both MS Excel and SPSS Clementine for analysis. Records with incomplete data (e.g. missing values) and inadequate content (e.g. unfit values) were first eliminated. Attribute columns completely empty, were verified with the fab's personnel as not in use, and then eliminated. Other attributes that were identified as redundant (e.g. duplicate) were also eliminated. By the end of this phase, the dataset was reduced to 6865 records each with 27 attributes.

C. Data Preparation

This phase includes the physical dataset preparation for adequate and effective application in the model, using further statistical analysis.

978-1-61284-408-4/11 $26.00 © 2011 IEEE

Relying on the data understanding, we concluded that some of the existing attributes required separation into newly formed attributes, as follows:

- All date & time attributes were each split to create three new attributes, including: sequential day number, day in week, and time of day.

- Wafer lot number attribute was split to create five new attributes, including: fab id, silicon vendor id, lot start year, lot start work-week, and serial lot number.

This finalized the creation of all explanatory (independent) attributes.

Some of the date and time attributes were now used to extract (calculate) target attributes directly needed to measure the individual wafer lots CT, as follows:

- Queue CT and Process CT attributes were created by using the date & time of a wafer lot move-in and move-out attributes.

- Total CT attribute was then composed following a similar path (Queue and Process CT's make up the total CT of a wafer lot in a single operation).

As a result of the separation and extraction, ten new attributes were added to the dataset. Of those, Queue CT, Process CT and Total CT, were defined as target attributes of the prediction model.

Further statistical analysis was then performed considering all attributes. Pearson correlations were calculated among quantitative attributes, and a few high correlations were identified, but only for expected pairs (e.g. between move-out and move-in time). Scatter plots were drawn for pairs of qualitative nominal attributes, and a few associations were identified, but only where expected (e.g. between operation steps and tool types). No additional filtering was performed as a result of this analysis.

The goal of the study is to establish a CT prediction model for production lots that run throughout the production line, and are intended for sale when fabrication is complete. Thus, records of wafer lots designated for other purposes (i.e. engineering, development, short-loops) were eliminated. At the end of this phase the dataset was resized to 5927 records by 37 attributes.

In finalizing the preparation, the dataset formed was now split into two parts: (i) a training dataset with 70% of the records – for the model's training, and (ii) a testing dataset with 30% of the records – for evaluating the prediction model results. The random split of records into the datasets was performed per SPSS Clementine standard recommendation.

III. THE MODEL

A. Feature Sets Selection

This phase includes the selection process of attributes for their assignment as features in the prediction model.

Initial selection of attributes was performed by relying on the data analysis and understanding acquired in Section II.

Total of 19 features were selected out of 37, while considering attributes elimination as follows:

- A few original attributes were previously left in the dataset for reference, and were now redundant (e.g. date & time).

- Target attributes, other than the Total CT, were not required for the model.

Table I presents the initial 'Full' feature set selected, detailing the features titles, contents, and format.

In order to further reduce the size of feature set, additional elimination of attributes was manually performed. Thus, 11 features were selected out of 19, while considering attributes elimination as follows:

- Some attributes were highly correlated with others, and thus indicated redundancy.

- Other attributes were considered to have minor contribution to the prediction of Total CT, per the author's knowledge and experience.

This concluded a second optional feature set 'Manually' selected for constructing the prediction model.

Finally, SPSS Clementine statistical procedures were applied to select a minimum size feature set (with test significance level of 0.1). The third optional 'Auto' set consisted of 7 features.

By the end of the feature selection process, three optional sets were generated for the model's training and classification:

1) 'Full' set with 19 features.

2) 'Manual' selected set with 11 features.

3) 'Auto' selected set with 7 features.

TABLE I.   ALL SELECTED FEATURES

Feature	Type	Content	Format(interval)
Year	Scale	Last digit of year	Integer (0,8)
Week	Scale	Week no.	Integer (1,52)
Serial_N	Scale	Serial no.	Integer (1,999)
Pr_MO_D	Scale	Prev. Move-Out Day	Integer (1,99999)
In_Q	Scale	move-In Quantity	Integer (1,25)
Pr_MO_Tm	Scale	Prev. Move-Out day Time	Decimal (0,1)
Load_Tm	Scale	Load day Time	Decimal (0,1)
SRC_Snt_Tm	Scale	SRC Sent day Time	Decimal (0,1)
Sil_Vndr	Nominal	Silicon Vendor	Digit (1,9)
Pr_Op_N	Nominal	Previous Operation no.	Digit (1000,9999)
Op_N	Nominal	Operation no.	Digit (1000,9999)
Route	Nominal	Route name	Name
Prod_Name	Nominal	Product Name	Name
Tool_Code	Nominal	Tool Code	Name
MO_Action	Nominal	Move-Out Action name	Name
Controller_F	Flag	station Controller Flag	Y or N
Hold_F	Flag	Hold Flag	Y or N
Rw_Sts	Flag	Rework Status flag	Y or N
Scrap_F	Flag	Scrap Flag	Y or N

Figure 1 presents the optional feature sets selected, and illustrates the relations among them.

B. Target Function

This phase includes the analysis of CT density, and the definition of the CT prediction target function.

Total CT data were analyzed for the complete dataset: CT Mean 0.369 days, and CT SD 0.838 days. Density distribution was plotted for values less than 2 days (only 2.8% fall beyond 2 days). See Figure 2.

Total CT prediction target function was defined relying on the above analysis. Furthermore, the fab's operations personnel were consulted for their ability of managing work-in-process. They considered the structure of the working shift (e.g. shift tasks, shift duration and breaks) relative to the CT durations. It was concluded that shift operation management ability should be considered for intervals of 3-4 hours each. Thus, the target function was then defined by classification of CT durations into ten intervals (bins), as follows:

- 8 equal intervals of CT of the first day (3 hours each)

- 9th interval for CT between 1 and 2 days

- 10th interval for CT more than 2 days

This CT intervals structure is used in the target function of the prediction model.

C. Model Selection

Selecting the model should consider the type of data, target function, and the learning process. In our case, the feature types are both continuous and categorical. The target variable is continuous but was constructed to fit nominal variables. The learning type is supervised.

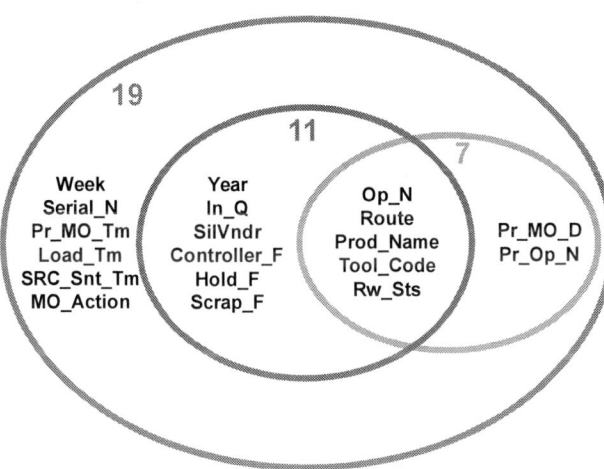

Figure 1. Three optional selected feature sets

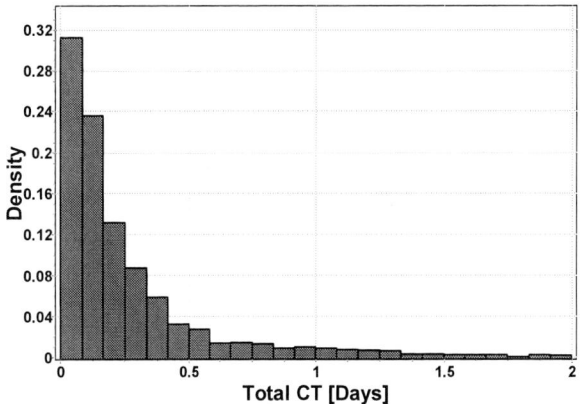

Figure 2. Total CT density distribution

Considering the above, both DT and NN are applicable and can handle the classification model. Other models may not be applicable (e.g. K-Means) since they fit un-supervised training model. Other models are less preferred (e.g. Bayesian) since they rely on statistical assumptions that are less adequate for our case. Thus, the general types of models selected were:

- Decision Trees – inspired by information models, where observations of items are mapped to conclude on target values; branches represent combination of features, and leaves represent classifications; considered a linear model.

- Neural Networks – inspired by the biological models (interconnected neurons), where the model structure changes based on learning; considered non-linear statistical based model, applied for complex relationships.

The models selection here is also in accordance with most models mentioned in literature for CT prediction.

IV. RESULTS AND EVALUATION

In order to maximize the prediction accuracy, a vast variety of model parameters (permutations) was tested for each of the two models, with each of the three feature sets.

The DT model achieved the highest accuracy at 76.5% with the 'Full' features set, and the following parameters:

- Boosting – no. of trials: 50

- Cross validate – no. of folds: 10

- Punning severity [%]: 75

- Minimum records per child branch: 2

The NN model achieved the highest accuracy at 87.6% with the 'Full' feature set, and the following parameters:

- Method: Prune Sample

- Prevent Overtraining Sample [%]: 95

Network topology at best accuracy: Input layer - 20 neurons, Hidden layer 1 – 23 neurons, Hidden layer 2 – 15 neurons, Output layer – 10 neurons.

V. CONCLUSIONS

Overall, both best fitted models achieved high level of accuracy. This is also in comparison to the accuracy detailed in published literature. The best results of both models were obtained using the 'Full' 19 features set, tested with 30% of the dataset. The DT model achieved 76.5% accuracy, as illustrated by Table II. The NN model (two hidden layers) achieved 87.6% accuracy, as illustrated by Table III. The following are the major conclusions drawn:

- The choice of NN with two hidden layers indicates that a linear discriminator (single layer) is insufficient for this prediction model. It may also hint on the inferiority of the DT model, which is structured as a linear function.

- The NN model consistently achieves higher or equal results compared with DT, when tested with all three optional features sets. This strengthens the NN model capability.

- The 'Full' feature set consistently achieves higher prediction accuracy compared with the other two features sets, in DT and NN models. This strengthens the selection of this feature set.

- Though distinguished in prediction accuracy, both models results are characterized by little prediction confusion between distinctly different CT intervals. Clearly, the NN results are characterized with smaller confusion (see Tables II and III). This strengthens the quality of the models applied.

The significance of this study is in establishing practical and accurate models for predicting CT, which rely on commonly available actual data. Fab specific applications can use the models to predict wafer lots CT in a single operation step or in accumulated line segments, with high accuracy. Future studies can include additional types of data, alter the models development (e.g. techniques).

TABLE II.   DT MODEL RESULTS AND COINCIDENCE MATRIX

Correct	1340	76.5%
Wrong	411	23.5%
Total	1751	100%

		Prediction Level									
		1	2	3	4	5	6	7	8	9	10
Actual Level	1	750	16	1	0	0	0	0	0	4	4
	2	89	265	17	0	0	0	1	0	3	6
	3	14	40	135	13	1	1	0	0	4	3
	4	5	1	23	51	7	0	0	0	2	4
	5	8	0	8	8	35	0	0	2	3	0
	6	2	2	4	0	5	10	1	1	2	2
	7	6	0	1	1	2	5	8	7	2	3
	8	8	0	1	0	0	2	3	10	5	0
	9	9	9	5	2	2	0	1	3	48	8
	10	6	3	3	0	2	0	0	0	5	28

TABLE III.   NN MODEL RESULTS AND COINCIDENCE MATRIX

Correct	1533	87.6%
Wrong	218	12.5%
Total	1751	100%

		Prediction Level									
		1	2	3	4	5	6	7	8	9	10
Actual Level	1	759	8	1	1	0	0	1	0	4	1
	2	9	362	6	1	0	0	0	0	3	0
	3	2	3	202	3	0	0	0	0	0	1
	4	0	1	4	86	1	0	0	0	0	1
	5	0	0	1	6	41	4	0	0	2	10
	6	0	0	0	0	2	21	1	0	0	5
	7	1	0	1	1	0	1	27	0	4	0
	8	1	0	1	0	0	0	4	19	4	0
	9	14	22	12	7	2	3	2	6	11	8
	10	3	3	8	3	6	3	6	3	7	5

ACKNOWLEDGMENT

The author thanks Sylvain Bouhnik, of Micron Fab12, for supporting this study with data and consultation.

REFERENCES

[1] J. R. Anderson, R. S. Michalski, J. G. Carbonell, T. M. Mitchell, Machine Learning: An Artificial Intelligence Approach, Morgan Kaufmann, 1986

[2] T. M. Mitchell, Machine Learning, McGraw Hill, 1997

[3] P. Langley, Elements of Machine Learning, Morgan Kaufmann, 1998

[4] J. Błażewicz, W. Kubiak, T. Morzy, M. Rusinkiewicz, Handbook on Data Management in Information Systems, Birkhäuser, 2003

[5] M. P. Singh, The Practical Handbook of Internet Computing, CRC Press, 2005

[6] T. M. Mitchell, "Machine Learning and Data Mining, Communications of the ACM", vol. 42(11), pp. 30-36, 1999

[7] J. A. Harding, M. Shahbaz, S. Srinivas, A. Kusiak, "Data Mining in Manufacturing: A Review", Journal of Manufacturing Science and Engineering, vol. 128, pp. 969-976, 2006

[8] M. Gardner, J. Bieker, "Data Mining Solves Tough Semiconductor Manufacturing Problems", Conference on Knowledge Discovery in Data, pp. 376-383, 2000

[9] T. Utlaut, K. Anderson, "On the Use of Machine Learning in the Semiconductor Industry Examples and Case Studies", Intel Joint Statistical Meeting, work paper, 2004

[10] C. F. Chien, H. C. Li, A. Jeang, "Data Mining for Improving the Solder Bumping Process in the Semiconductor Packaging Industry", International Journal of Intelligent Systems in Accounting and Finance Management, vol.14 , pp. 43-57, 2006

[11] R. Goodwin, R. Miller, E. Tuv, A. Borisov, M. Janakiram, S. Louchheim, "Advancements and Applications of Statistical Learning Data Mining in Semiconductor Manufacturing", Intel Technology Journal, vol.8(4), pp. 325-336, 2004

[12] P. Backus, M. Janakiram, S. Mowzoon, G. C. Runger, A. Bhargava, "Factory Cycle-Time Prediction With a Data-Mining Approach", IEEE Transactions on Semiconductor Manufacturing, vol.19(2), pp. 252-258, 2006

[13] C. F. Chien, C. W. Hsiao, C. Meng, K. T. Hong, S. T. Wang, "Cycle Time Prediction and Control Based on Production Line Status and Manufacturing Data Mining", IEEE International Symposium on Semiconductor Manufacturing, pp. 327-330, 2005

[14] R. O. Duda, P. E. Hart, D. G. Stork, Pattern Classification, Wiley and Sons, 2nd Edition, 2001

Recent Innovations in CMOS Image Sensors

Ray Fontaine
Technology Analysis Group
Chipworks Inc.
3685 Richmond Road
Ottawa, Ontario, Canada K2H 5B7
rfontaine@chipworks.com

Abstract - **The trend in semiconductor manufacturing over the last decade has been an accelerated rate for both materials integration and wafer fabrication process development. While the cutting edge of semiconductor technology is driven by digital logic and memory applications, several other technology sectors benefit from innovation by the leaders. Of these technology sectors, image sensor manufacturers have realized many benefits from the selective use of developments within advanced technology node manufacturing.**

The motivations for the imaging industry to pursue Moore's Law type of scaling are comparable to that of the broader semiconductor industry. Additionally, image sensor companies seek a reduction of camera module form factor, an increase in pixel resolution, and an increase in pixel array performance.

Today, semi-professional grade digital single-lens reflex (DSLR) pixels have scaled down to the size of what were state-of-the-art "small pixel" consumer grade camera phone sensors just a few years ago. The pixel size of recent camera phones has shrunk to 1.12 μm. The resolution for recent camera phones has reached 16.4 Mp. Beyond silicon foundry processes, imaging companies must also concern themselves with the optical systems and packaging solutions required to integrate their silicon devices with the consumer electronics supply chain.

Chipworks, as a supplier of competitive intelligence to the semiconductor and electronics industries, monitors the evolution of image sensor technologies as they come into production. Chipworks has obtained charge-coupled devices (CCD) and CMOS image sensor (CIS) chips from leading manufacturers and performed structural, compositional, and design analyses to benchmark the technology of the market leaders.

Keywords: CMOS Image Sensors; Advanced Processes; Advanced Materials

I. INTRODUCTION

A. Background of CMOS Image Sensor Pixels

Consumer grade CMOS image sensors (CIS) for mobile applications, including camera phones, laptops, tablets, etc., have gone from novelty to ubiquitous status in about a decade. As the successor to passive pixel charge-coupled devices (CCD), current system-on-chip (SoC) CIS devices comprise an array of active pixels and supporting circuitry which enables both still and motion photography [1].

Current CIS devices are a result of quickly evolving pixel architectures, which resulted in the now common four-transistor (4T) pinned photodiode structure [2]. 4T pixels include a photon collection region (pinned photodiode), charge transfer transistor, reset transistor, source follower transistor

(in-pixel amplifier), and column bus transistor. Essentially, photon generated charge is collected and converted to the voltage domain within each pixel, and ultimately read out to an on-chip analog-to-digital (ADC) to be transformed to the digital domain for image processing.

The fundamental motivations for shrinking pixel size, discussed elsewhere [2, 3], include decreasing the overall chip size and camera module form factor for a given resolution, or increasing the resolution for a given camera module size. In either case, there is both push from industry and pull from consumers for CIS technologists to shrink pixel size and improve pixel performance. Therein lies the fundamental challenge of the image sensor community: how to do more with less available light at each collection node.

II. THE EVOLUTION OF PIXEL STRUCTURES

The conventional planar CMOS fabrication process lends itself quite well to the task of digital imaging. A silicon substrate, optically transparent dielectric stack, and the ability to produce fine pitch interconnect lines are all complementary to CIS device requirements. Fig. 1 shows an 8.0 μm pixel pitch front illuminated (FI) CIS produced by Hynix in 2002 (analyzed in 2004). The structure comprises, from top to bottom, a microlens, planarizing lens buffer layer, color filter, dielectric film stack, pixel interconnect, tungsten silicided transistor gates, and LOCOS isolation. The device was fabricated in a 0.5 μm CMOS process and used spin-on-glass (SOG) to partially planarize the dielectric stack. The color filters, deposited by an iterative spin-on process, and lens buffer film have provided a nearly planar surface for the microlenses. Microlenses significantly increase the light-gathering capabilities of small pixels, effectively increasing the

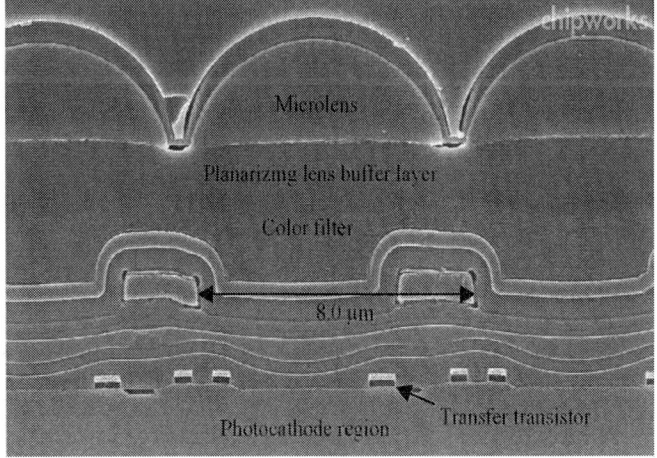

Figure 1. Hynix 8.0 μm Pixel, 0.5 μm Process Technology Generation

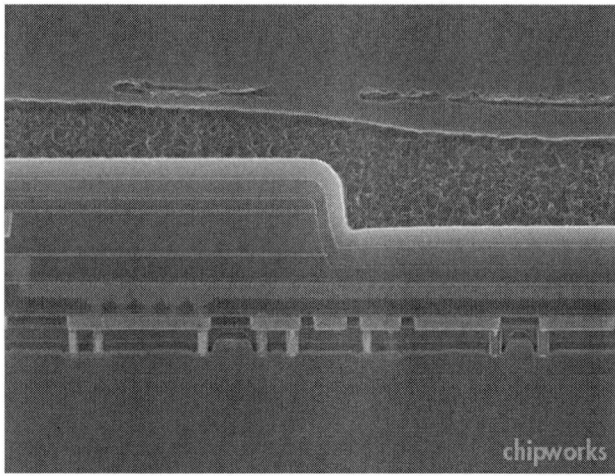

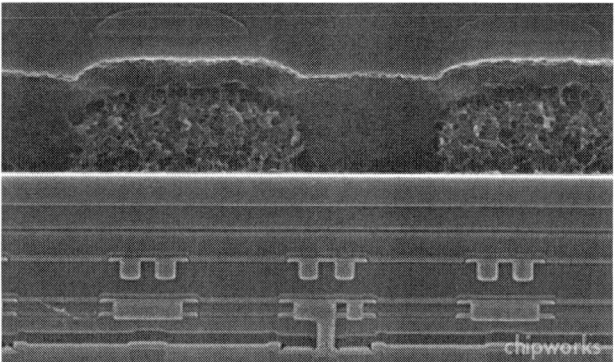

Figure 2. STMicroelectronics 2.2 μm Pixel, 0.13 μm Process Technology (Peripheral and Pixel Regions)

pixel fill factor. While fabricated in a mature process by today's standards, modern FI CIS devices represent refinements of this type of structure.

The 3.0 μm to 3.5 μm pixel generation CIS devices circa 2005 generally featured 0.18 μm process technology with aluminum metallization, chemical mechanical planarization (CMP) of the dielectrics, and shallow trench isolation (STI). Most manufacturers optimized general logic processes for CIS production, and the four to five levels of interconnect enabled sophisticated SoC features to be integrated with the pixel process flow.

A disadvantage of the pixel structures of this era was the shadowing caused by the thick back-end-of-line (BEOL) stack for pixels at the outer edges of increasingly larger pixel arrays. The effects of these comparatively tall structures were somewhat mitigated for outer pixels, where progressively shifting the microlenses toward the array center partially compensated for the chief ray angle (CRA) variance between the outer and central pixels.

A. Shared Pixels

Early implementations of the 4T architecture featured all four transistors occupying valuable real estate which could otherwise be used to collect incident photons. When pixels scaled to the 2.2 μm generation circa 2006, designers evolved

to a pixel sharing scheme whereby neighboring pixels, each with a dedicated transfer gate, shared common readout transistors (source follower, reset, row select). Both two-shared and four-shared pixel architectures are currently in production, netting an effective 2.5 transistors (2.5T) or 1.75 transistors (1.75T), respectively, per pixel. These designs, and other pixel sharing configurations, continue to be significant enablers for pixel scaling.

B. BEOL Stack Reduction and Optical Symmetry

Coinciding with the 2.2 μm pixel generation was the broad adoption of 0.13 μm process technology for production CIS devices. In turn, CIS companies utilized divergent approaches to fabrication. For example, STMicroelectonics, an independent device manufacturer (IDM), utilized a copper BEOL scheme with relatively narrow interconnect lines and a four-shared pixel layout that allowed the implementation of the pixel wiring within only two levels of metallization. Fig. 2 shows a cross section view of STMicroelectronics' pixel and logic BEOL structures, illustrating the partial thickness reduction of the pixel BEOL. The reduced metal interconnect in the pixel BEOL greatly reduced the metal shadowing effects.

A new challenge coincident with the use of damascene copper processes was the presence of silicon nitride trench etch stops and copper diffusion barriers in the pixel BEOL. If left in the optical path, the presence of these films could potentially introduce optical interference effects. STMicroelectronics chose to include a mask and etch step to open windows in the nitride films over the photocathodes.

Another approach was taken by Micron (now Aptina), also using a 0.13 μm process, who converted its depreciated DRAM fabs to CIS production. While some IDMs and foundries switched to (then) more expensive copper processes, the narrow aluminum metallization afforded by a DRAM process allowed Micron to produce competitive devices at the 1.75 μm pixel generation. In addition to the advantageous BEOL design rules, the low-leakage transistor requirements for DRAM wordlines were a natural fit for the low noise performance of CIS pixels.

Fig. 3 shows an overview of a Micron 1.75 μm pixel generation CIS, which featured a pixel array dielectric etch back resulting in a pixel BEOL (including filters and lenses), which was essentially the same thickness as the BEOL stack in the logic regions.

Another approach to reduce the pixel BEOL thickness was employed by Samsung, who used a tungsten "metal 0" local pixel interconnect to reduce metal obstruction. Both approaches served to aggressively move the microlenses closer to the silicon surface, thereby improving the angular response of the sensors.

Out of necessity, image sensor technologists focused intently on the pixel BEOL for the 1.75 μm pixel generation. After implementing shared pixel architectures and reducing the thickness of the pixel BEOL stack, there was an awareness by technologists that performance gains could be obtained by optimizing the pixel metallization for optical symmetry.

978-1-61284-408-4/11 $26.00 © 2011 IEEE

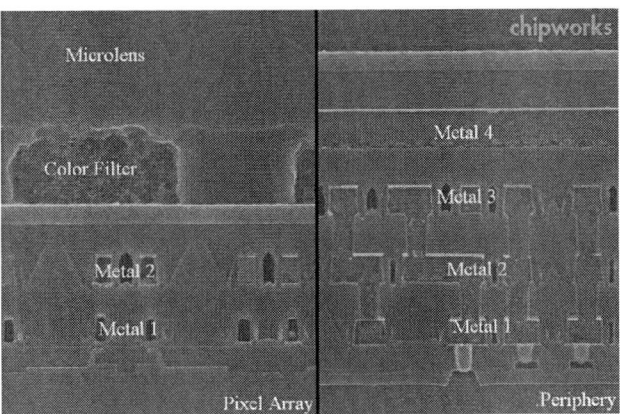

Figure 3. Micron (Aptina) 1.75 µm Pixel, 0.13 µm DRAM Process Technology

Prior to this, the pixel wiring was primarily routed for electrical interconnect requirements, with less emphasis on lateral symmetry through the optical path. The 1.75 µm pixel generation saw the prevalence of balanced interconnect patterning, particularly the use of dummy metal structures to reduce optical cross talk. A high degree of symmetry is considered critical for the optical performance of small pixel FI CIS devices.

Fig. 4 shows two examples of the symmetrical top metal layout of 1.75 µm generation pixels from Samsung and OmniVision. Samsung, an IDM, used a 90 nm process for this generation, while OmniVision's foundry partner TSMC utilized a 0.11 µm aluminum process. Interestingly, the 1.75 µm pixel generation was found to be implemented using 180 nm through 90 nm design rules.

C. Light Pipes

Material integration for CIS production began to accelerate at the 1.4 µm pixel generation with the introduction of so-called "light pipes," or optical waveguides, in the pixels. The addition of light pipes serves to couple the upper region of the pixel BEOL to the photocathode region with minimal photon loss or cross talk.

The processes in production generally comprise: a deep trench etched into the dielectric stack, a silicon nitride trench liner, and a spin-on light pipe fill having an index of refraction

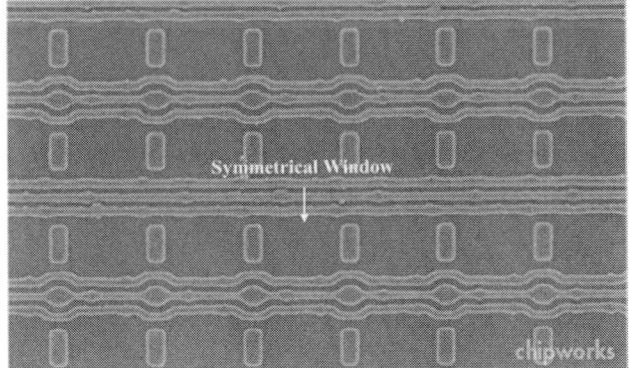

Figure 4a. Samsung 1.75 µm Pixels with Symmetric Interconnect Layout

Figure 4b. OmniVision 0.11 µm Process

higher than the surrounding dielectrics [4]. With this type of structure, the microlenses are optimized to shift the focal point to the upper region of the light pipe, enabling the collection of more off-axis light into the light pipe.

The results are improved quantum efficiency, angular response, and optical cross talk performance. Fig. 5 shows an example each of Aptina and Sony light pipes in production for their 1.4 µm pixel CIS devices. Aptina's light pipe fill is an organic film. The shape of the light pipe is reported to effectively collimate the incoming light, thereby, reducing electrical cross talk in the substrate [5]. This device was fabricated using hybrid 130 nm FEOL/90 nm aluminum BEOL design rules.

Aptina chose to employ pixel dielectric stack thinning and two levels of metal interconnect, while Sony's light pipe implementation included the full dielectric stack and four levels of metal interconnect. Sony's light pipe fill was found to be titanium-based, likely titanium dioxide (TiO_2). Sony's light pipe process was implemented in a 90 nm copper process.

III. BACK ILLUMINATED CIS DEVICES

Despite the cleverness of sharing pixels, the use of advanced node wafer fabs, and innovation in the pixel BEOL, there exists within a FI CIS device an intrinsic competition between the light sensitive area and the pixel control and readout circuitry.

Amongst the several analyzed commercial FI 1.75 µm pixel and smaller generations, the fill factor is typically less than 50% (excluding the effects of microlenses). In 2009, Sony and OmniVision each began mass production of first generation small pixel back illuminated, or back side illuminated (BI, BSI) CIS devices.

Originally used in specialty CCD applications, BI technology had previously been cost prohibitive for introduction into the consumer electronics supply chain. BI devices potentially enable fill factors approaching 100%.

978-1-61284-408-4/11 $26.00 © 2011 IEEE

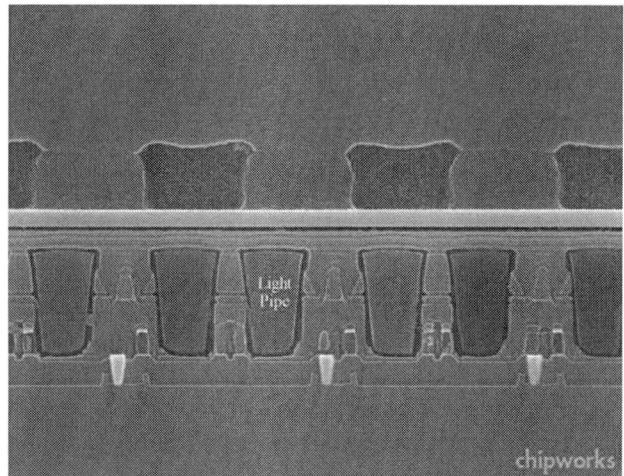

Figure 5. Aptina and Sony 1.4 µm Pixels with Light Pipes

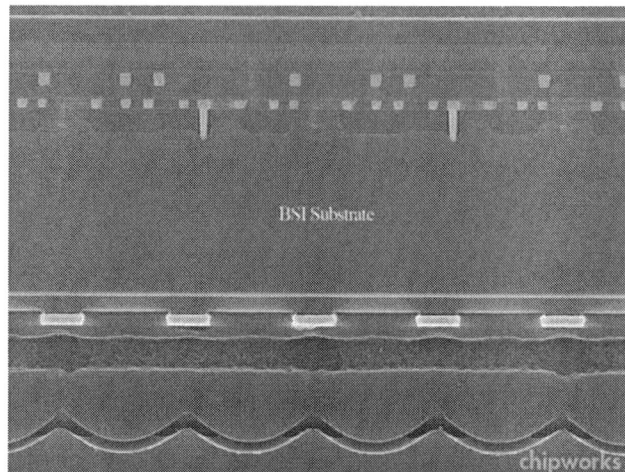

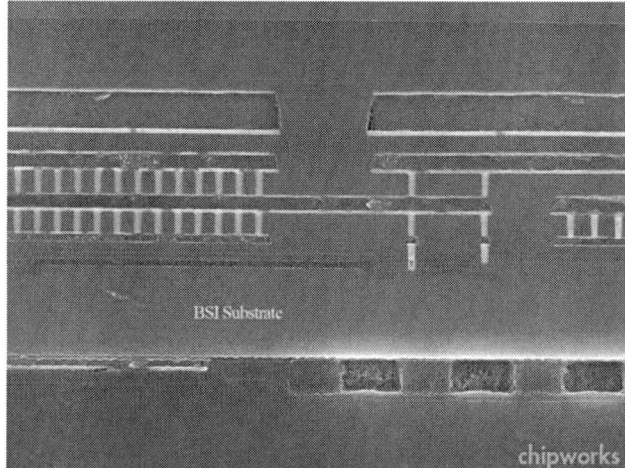

Figure 6. Sony and OmniVision First Generation Back Illuminated Devices

Fig. 6 shows examples of the Sony and OmniVision first generation BI devices. Sony's device was fabricated using a silicon-on-insulator (SOI) process, in which a conventional CMOS process flow was run on SOI wafers with 3 µm thick active silicon.

After completion of the FEOL and BEOL structures, the planarized dielectric stack served as a bonding surface for the joining of the SOI wafer to a silicon chip carrier wafer using an adhesive bonding technique. Next, the sacrificial SOI handle wafer and buried (BOX) layers were removed. The unobstructed back of the substrate functions as the light receiving surface.

For many FI devices, the reflectivity of planar silicon had been addressed through the use of silicon nitride films deposited over the photocathode regions to serve as anti-reflection (AR) layers. For its first BI device, Sony chose to use a hafnium oxide (HfO_2) AR layer blanket deposited over the back of the die. It also included tungsten metallization on the back which optically shielded the peripheral regions, and was patterned to form an aperture grid improving color separation in the pixel array.

While Sony's device had a 1.77 µm pixel size and relatively large form factor module for camcorder applications, OmniVision's first BI device featured a 1.4 µm pixel size, and was deployed in a small form factor camera module. OmniVision and TSMC developed a bulk BI process for their 1.4 µm pixel size sensor, fabricated using 0.11 µm generation process technology [6, 7].

The device used oxide bonding to join the planarized finished wafer to a silicon carrier wafer. The bulk P-type substrate with P-epitaxial layer was mechanically back ground and wet etched, with the epi serving as an etch stop. The resulting substrate thickness was 2.1 µm and displayed some back surface roughness, an artifact of the wafer thinning process. A back surface implant and laser anneal process were used to reduce the crystal defects induced by the wafer thinning process. Due to the small form factor application, OmniVision also adapted its wafer level chip scale (WL-CSP) packaging, first used for its FI devices, to the new BI device structure.

IV. PRESENT AND NEAR FUTURE TRENDS

Many novel CMOS process technologies have been adopted in recent front and back illuminated CIS devices.

STMicroelectronics has incorporated deep trench isolation (DTI) as pixel isolation in FI devices, while Sony has used DTI to isolate the bond pads in BI devices. The emergence of BI devices in mass production has driven the evolution of advanced packaging for CIS devices. Through silicon vias (TSV) were first used by Toshiba and STMicroelectonics in FI devices, and have been used by Samsung in BI devices. Sony has incorporated embedded passive components and semiconductor die in the printed wiring board (PWB) substrates.

Today, CIS devices using 65 nm design rules and 300 mm wafer production are in production. The trend of advanced technology generation pixel manufacturing will continue, likely to below sub-diffraction limit pixels [1]. Beyond silicon, InVisage Technologies, Inc. introduced its quantum dot based image sensors in 2010. Progress in these and other areas ensure that the acceleration of pixel innovation will continue for the near future.

V. REFERENCES

[1] E.R. Fossum, "CMOS Active Pixel Image Sensors: Past, Present, and Future," January, 2008

[2] A. Theuwissen, "CMOS Image Sensors: State-of-The-Art and Future Perspectives," Solid-State Electronics, Vol. 52, 2008, pp. 1401-1406

[3] M. Okincha, "OmniBSI™ Technology Backgrounder," OmniVision Technologies, June, 2009

[4] J. Gambino, et al., "CMOS image sensor with high refractive index lightpipe," 2009 International Image Sensors Workshop (IISW), June 2009

[5] G. Agranov, et al., "Pixel continues to shrink . . . Pixel Development for Novel CMOS Image Sensors," 2009 International Image Sensors Workshop (IISW), June 2009

[6] H. Rhodes, "The Mass Production of BSI CMOS Image Sensors," 2009 International Image Sensors Workshop (IISW), June 2009

[7] S. G. Wuu, "BSI Technology with Bulk Si Wafer," 2009 International Image Sensors Workshop (IISW), June 2009

Parametric Composite Limited Yield Index For Functional Circuits Yield Prediction

Jiun-Hsin Liao[†], Ishtiaq Ahsan[†], Ronald Logan*,
George Rudgers[†]

†Semiconductor Research and Development Center
*Advanced Microelectronics Solutions
Microelectronics, IBM
Hopewell Junction, NY 12533, USA

Fred Towler
Silicon Solutions Engineering
Technology Development, IBM
Essex Junction, VT 05452, USA

Abstract—**In this paper we present an early detection mechanism for semiconductor circuit yield prediction and tracking. Several discrete devices used as components of functional circuits have been examined by their first-metal level test data and correlated to the higher metal level functional yield. A concept of Device Health Composite Yield is also introduced in this paper.**

Keywords-Yield prediction; Parametric limited yield

I. INTRODUCTION

In CMOS IC manufacturing, functional test-structures (e.g. SRAM functional test-structures or scan-chains) are always tested at metal levels much higher than the first metal level. It is desirable to establish a matrix able to predict the yield of such test-structures at an earlier test level, for instance, the first metal level so the reactions to faulty processes or yield improvement can take place at the earliest possible stage. In CMOS technology development, yield of functional circuits is often limited by process induced defects and parametric centering [1]. These parametric factors could be device parameters which are measured at the first metal level. It is therefore desirable to have a matrix that predicts limited yield of the functional test structure driven by parametric centering (from here on called "parametric limited yield").

In this paper, we demonstrate a methodology to predict and track parametric-limited yield of functional test structures using parametric data collected at the first metal level. The SRAM functional test structure will be used as an example.

II. METHODOLOGY

The SRAM functional test structure is composed of the SRAM array and the peripheral circuit. Both include component devices, such as logic FETs, SRAM FETs, decoupling capacitors, precision resistors, eFuse blow FETs, ESD diode, etc. Each of these component devices have a discrete test structure electrically tested at the first metal level.

A. Parametric Composite Limited Yield (ParCLY)

We identified these discrete devices and defined the test conditions and output parameters for each. The upper and lower specification limits of these parameters were carefully specified to be consistent with the maximum variation window allowed for the functionality of the SRAM provided by the circuit designer. A parametric-composite yield was created to compute the yield of all parameters from the SRAM component devices mentioned above. It was called the "Parametric Composite Limited Yield (ParCLY) Index".

The composite yield can be expressed as

$$Y_{Composite} = Y|_{Parm1} \cdot Y|_{Parm2} \cdots\cdots\cdot Y|_{ParmN} \qquad (1)$$

In general, one defines target, upper specification (USL) and lower specification (LSL) limits to specify the yield Y of a parameter. Y = 1 when the data is located between USL and LSL and Y = 0 when the data is located outside of USL-LSL (Fig. 1). Therefore the chip-level ParCLY value is either "1" or "0". However, when utilizing the ParCLY index to predict the functional yield of a SRAM array, the yield dependency varies among component devices. For instance, the SRAM array will probably still yield when some less critical component parameters fall outside of USL-LSL window.

A real case is shown in the Fig. 2. The SRAM array peripheral-circuit leakage in the y-axis is measured at the same level as the SRAM functional yield test is taken, and directly correlates to the functional yield (i.e. the peripheral circuit leakage has to be below a certain level to ensure the SRAM array will be functional). The x-axis is the NFET gate leakage converted dielectric thickness measured at the first metal level. This is a parameter to evaluate the gate oxide thickness by performing the through-gate leakage current measurement and then convert the value of the leakage current to the oxide thickness according to the direct tunneling model. The correlation chart shows that for some chips where the gate dielectric thickness (gate leakage) is lower than the LSL, the SRAM array might still yield, indicating using conventional component yield at the first metal level to predict the SRAM array yield is not fully realistic.

978-1-61284-408-4/11 $26.00 © 2011 IEEE

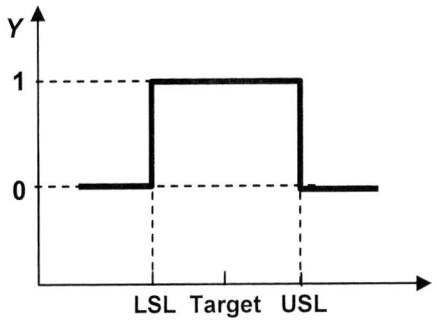

Figure 1. Conventional yield definition.

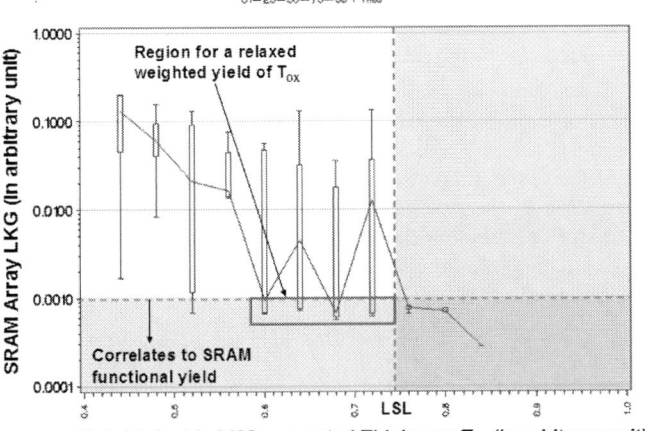

Figure 2. SRAM peripheral circuit leakage current tested at a higher metal level vs. gate leakage converted dielectric thickness. Boxed area indicates possibility of SRAM functional yields with the gate thickness outside the LSL.

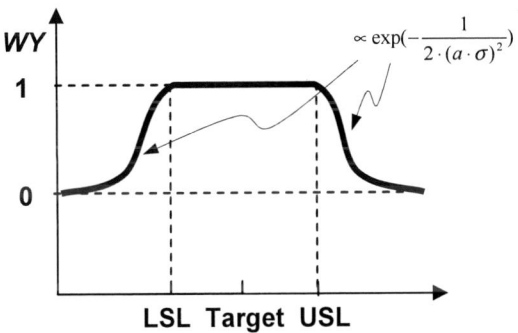

Figure 3. Weighted yield defined by relaxing the LSL and USL edges.

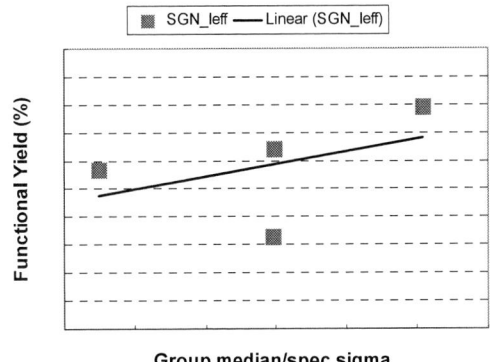

Figure 4. Four-group yield vs. nFET Ieff median. The slope determines the sentativity of the functional yield to the parameter.

B. Parametric Composite Limited Weighted Yield (ParCLWY)

The weighted yield (*WY*) is hence introduced with relaxed yield definition at the edges of USL and LSL in distinction to the conventional yield waveform (Fig. 3).

The determination of the weighted yield for each component parameter is a critical process. It can be based on the data correlation to the SRAM circuit yield or on the model extractions. For demonstration, we utilized the nature of data distribution – Gaussian function to express the waveform at the specification limit edges. The *WY* function for each parametric parameter therefore becomes:

$$WY = \begin{cases} \exp\left(-\frac{1}{2}\cdot\left(\frac{x-LSL}{a\cdot\sigma}\right)^2\right) & \text{Where } x < LSL \\ 1 & \text{Where } LSL < x < USL \qquad (2) \\ \exp\left(-\frac{1}{2}\cdot\left(\frac{x-USL}{a\cdot\sigma}\right)^2\right) & \text{Where } x > USL \end{cases}$$

In this simple approximation the Gaussian distribution waveform is governed by the parameter specification sigma σ, (USL-LSL)/6, and an arbitrary coefficient *a* (also shown in Fig. 3). The coefficient *a* is introduced here to control the order of the relaxation of the edge waveforms.

The parametric composite weighted yield (ParCLWY) index therefore can be expressed as

$$WY_{Composite} = WY\,|_{Parm1} \cdot WY\,|_{Parm2} \cdots\cdots \cdot WY\,|_{ParmN} \qquad (3)$$

In addition, three discrete parameters were selected from the set of component devices comprising the SRAM functional circuit. They are standard gate oxide nFET I_{eff}, thick gate oxide pFET I_{eff}, and SRAM Passgate FET I_{on}. We exercised statistical correlation analysis [2] to identify three critical parameters that have significant functional yield impacts. The statistical correlation analysis first separates the functional yield into four groups based on the yield value. Each group represents a quartile of the data. The median of functional yield and each parametric component are used for correlation analysis. Ranging the functional yield data into four groups can help to filter the average yield impacts of other parameters during the single parameter correlation, since they affect all four groups the same amount. Therefore, the critical parameters having significant yield impacts are identified. Fig. 4 illustrates the statistical correlation analysis on one of the critical parametric parameters – Standard gate oxide nFET I_{eff}. It shows that the higher the drive current, the higher the yield. The linear curve fit slope is the sensitivity of the functional yield to the drive current. We used the same process to identify the other two critical parametric parameters thick gate oxide pFET I_{eff}, and SRAM Passgate FET I_{on}.

III. FAB DATA CASE STUDY

The in-line test data from several wafers processed with the mainstream technology have been analyzed for demonstration. The defect related failed chips were first removed by screening the continuity and scan-chain fails at the same metal level as the circuit functionality is tested. The three most sensitive parameters are tested at the first-metal level. Their weighted yield (WY) were computed by the specification limits and sigma and coefficient a at $a = 1$ for this exercise. The composite weighted yield was also computed according to expression (3). The functional yield data tested at the higher metal level were also retrieved for these wafers for analysis.

A. Parametric histogram and yield charts

Fig. 5 shows the histogram of the ParCLY index and corresponding functional yield. Needless to say, there are only two modes of ParCLY due to its step-function nature. However, the SRAM circuit is still functional for some chips reported as "0" in the parametric composite yield. Note that for this example we see about 90% of the chips failed parametric composite yield due to not only its step-function nature but also on purposely picked wafers with one off-centered parameter for this exercise. The CDF (Cumulative Distribution Function) plot is also presented for the SRAM array circuit fail count in Fig. 6 sub-grouped by the ParCLY index = 0 and 1. It is clear that ~34% chips failed ParCLY (dotted curve) but exhibit null fail count in the functional circuit. On the other hand, the ParCLY =1 group (square – curve) shows about 76% chips null fail count.

Fig. 7 is plotted with the ParCLWY index and the function yield. The weighted yield index was plotted from 0 to 1 with increment of 0.1. The functional yield reading ranges about 90% to 80% when ParCLWY index greater than 0.3 and drops to 60% to 70% when ParCLWY index below 0.3.

Fig. 8 is the CDF plot for the SRAM circuit fail count grouped by the ParCLWY index from 0 to 1 with increment of 0.1. It is roughly seen that higher ParCLWY index value tends to have more chips of perfect yield in the circuit.

Another example is the electrical gate length measurement L_{gate}. L_{gate} is not a parameter from a circuit component device. However, it strongly relates to product yield and performance. Fig. 9 shows the histogram distribution of L_{gate} and the circuit functional yield plotted with dotted curve. The parametric specification limits for the L_{gate} is also illustrated in the chart. It is clear that for this particular data set, the gate length is off center toward to the shorter end. There is about 18% (from

cumulative distribution plot, not shown) of data falls outside the specification limits. However, the chip-level corresponding SRAM functionality shows lower yield with the shorter gate length but not completely non-functional at the range where the L_{gate} is outside the specification limits. The weighted yield index for the L_{gate} data is therefore computed and plotted with the functional yield in Fig. 10 and Fig. 11.

There is a concern about how well the ParCLWY index correlates to the real circuit functional yield. In Figs. 7, 8, 10, and 11 we do not see much yield difference between high and low ParCLWY index values. One main reason for that is that we did not include parameters from all component devices comprisng the functional circuit into the composite yield computation. The yield loss in the plots can be also limited by these other factors.

B. Determination of the coeficient a

It seems to be challenging how to determine how relaxed the specification limit edges should be. In our demonstration in Section A, we utilized $a = 1$. Practically, one should determine this limit edge relaxation coefficient based on the order of correlation of the parameter to the circuit functional yield and the maturity of technology development. A large sample size of fab test data and modeling exercises can help the analysis. Both will be targets of future work.

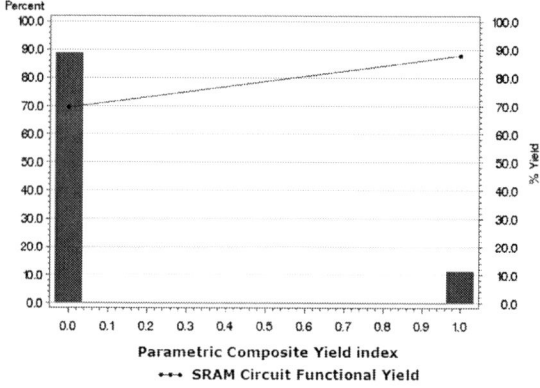

Figure 5. Parametric composite yield index and the function yield curve

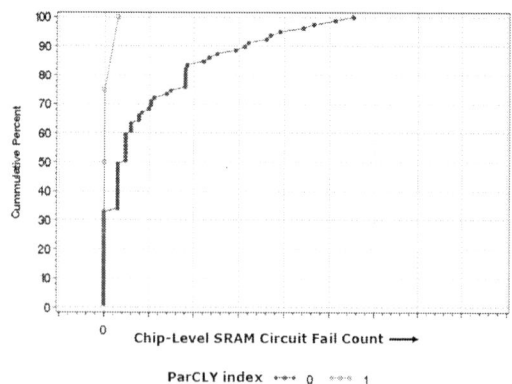

Figure 6. CDF plot for circuit fail count group by ParCLY index 0 and 1

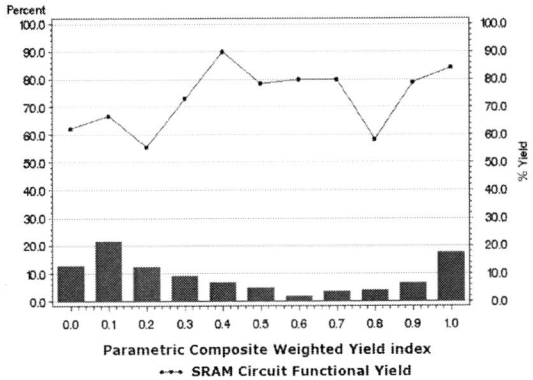

Figure 7. ParCLWY index and the function yield curve

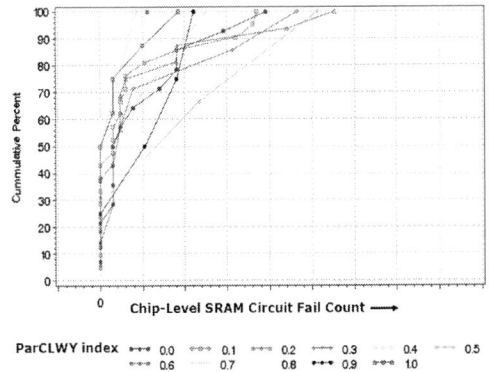

Figure 8. CDF plot for circuit fail count group by ParCLWY index

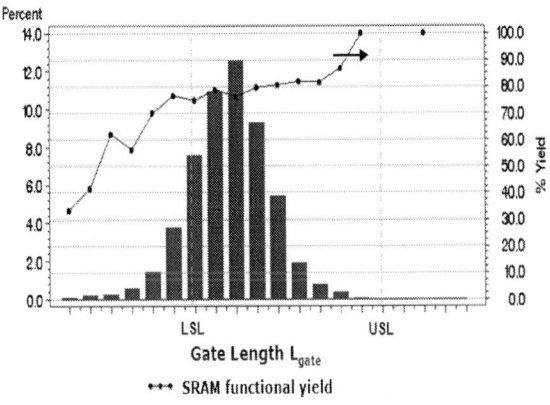

Figure 9. L_{gate} histogram distribution and SRAM functional yield curve

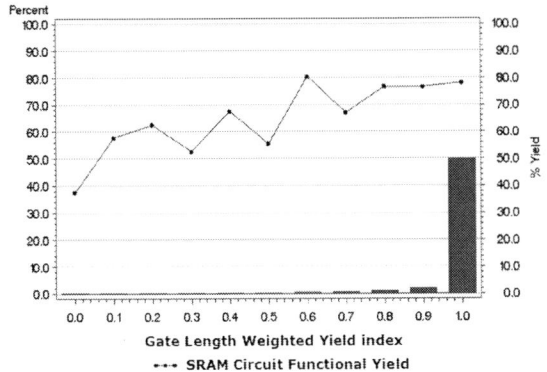

Figure 10. L_{gate} WY index and the function yield curve

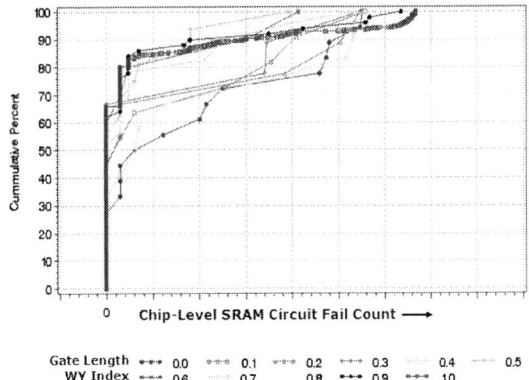

Figure 11. CDF plot for circuit fail count group by L_{gate} WY index

IV. CONCLUSION AND FUTURE WORK

In this paper, we have demonstrated Parametric Composite Limited Yield (ParCLY) index and Parametric Composite Limited Weighted Yield (ParCLWY) index tested at the first metal level for functional yield prediction. We have compared both approaches and pointed out the lack of confidence with the ParCLY due to its strict specification limit edges for the parametric yield. The concept of ParCLWY introduced in this paper has put the prediction into a more realistic stage.

We have demonstrated the weighted yield approach can be implemented for early yield estimation. We have also exercised the parametric-functional yield correlation analysis on several wafers using selected input parameters.

For future work, we plan the items below.

1. A larger sample size of data is desired for the correlation.

2. A composite weighted yield index consisting of a complete set of parametric parameters is underway.

3. Further analysis into the mechanism of determining the relaxation of specification limit edge (coefficient a) for each parametric parameter.

[1] T. S. Kim, S. H. Ahn, Y. G. Jang, J. I. Lee, K. J. Lee, B. Y. Kim, and C. H. Cho, "Yield prediction model for optimization of high-speed microprocessor manufacturing processes," 2000 IEEE/CPMT Int'l Electronics Manufacturing Technology Symposium, pp. 368-373, 2000.

[2] A. Y. Wong, "A systematic approach to identify critical yield sensative parametric parameters," 1997 2nd International Workshop on Statistical Metrology, pp. 56-61, 1997.

Embedded Memory Fail Analysis for Production Yield Enhancement

Youssef Baltagi [1], Daniele Li Rosi [1], Vincenzo Tancorre [1],
Christophe Garagnon [1], Eric Faehn [1], Mario Barone [2], Davide Appello [2]
[1] STMicroelectronics ZI Rousset 13106 Rousset cedex
[2] STMicroelectronics, Via C. Olivetti, 2, I-20041 AGRATE BRIANZA – Italy
youssef.baltagi@st.com

Christophe Suzor
Synopsys, Inc. 700 E. Middlefield Road, Mountain View CA 94043 USA
csuzor@synopsys.com

Abstract :

The traditional approach for memory fail bitmap analysis is to identify the topological signatures and perform a Failure Analysis investigation on the most frequent signatures, based on the (x,y) coordinates of the fails. This approach is inappropriate when a large portion of the fails are single bits, because too many investigations are required to statistically identify the major repetitive failure mechanisms. This becomes a problem for fast product development and production yield ramp. This paper presents a methodology to classify single fail bits by their unique fault signature, based on the sequence of failing march element read operations from multiple data backgrounds, in a standard Memory BIST flow. These classifications allow investigations to focus on the most important failure mechanisms with greatest yield impact. The methodology is demonstrated in an industrial environment, with identification of critical yield detractors. Starting from a yield problem associated to MBIST failures at high operating temperature, the fault signatures were used to identify a static noise margin parametric problem and a dislocation fault physical problem.

Keywords :

Bitmap, MBIST, Yield, March Algorithm, Fault Signature.

1. INTRODUCTION

With the very deep sub-micrometer technology node complex System on Chip (SoC) are able to integrate tens of embedded memories which are tested using Memory Built-In Self Test (MBIST) methodologies. Generally each memory is wrapped by a dedicated logic circuit that can apply the test algorithm and shift out (on dedicated POs) the test results, typically using the standard protocol 1149.1 (JTAG).

In a production environment these results are stored in a central data base for analysis during the Yield ramp phase, and later for process monitoring. During the Yield ramp phase, the statistical analysis based on volume data analysis is increasingly relevant for detecting criticality of the design or the manufacturing flow. Yield engineers usually cross-correlate results from different sources: inline inspection, parametric tests, Static Timing Analysis, ATPG diagnosis, memory BIST, etc.

For memory BIST, the traditional approach is to use a failed bitmap, but this is limited to topological signatures analysis,

and to understand the physical failure mechanism it is necessary to perform Physical and Electrical Failure Analysis. Often the failure mechanism is not a simple short/open issue, and Failure Analysis directly on the failing bit(s) is not able to explain the real cause of the fault. In current SoCs, failing memories are often the major yield detractor, and it is necessary to reduce the time required for root cause analysis. This paper presents an industrial methodology to recognize the top signatures in a production MBIST flow, and presents results using this methodology to solve a critical yield issue at high operating temperature. The conclusion will highlight the benefits and discuss next steps.

2. EXPERIMENTAL DATA

During the yield investigations for a 90nm (SoC) product with multiple embedded memories, some wafers from several lots have been selected for extensive data collection and analysis, including some wafers for failure analysis.

3. CURRENT METHODS

A common Design For Test (DFT) technique to verify embedded memory functionality is to use a MBIST, a technique which embeds a test routine controller into the memory circuit. An alternative approach is to use Direct Memory Test (DMT), if the array can be directly accessed externally by a memory tester, but this mode is impractical with the large number of small memories found in modern SoC.

Traditionally, the memory test results (from MBIST or DMT) are collected and analyzed by a bitmap tool. The final result is a memory and wafermap representation where failing bits are colored, with the color range determined by the number of failed bits in each memory, Fig.1.

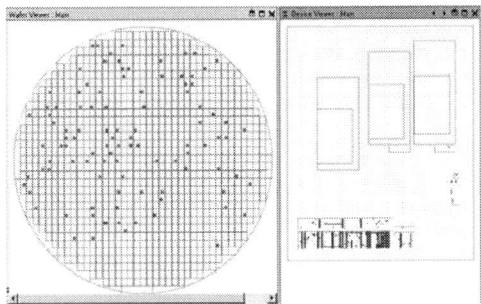

Fig.1: Wafermap and stacked memory representation of bitmap results showing a stacked die with multiple embedded memories and their failures

The user can validate the coordinates using an overlay with the layout to provide precise topological information to the Failure Analysis engineers. Fig.2 shows a Single Bit (SB) fail near the edge of an SRAM memory.

Fig.2: Layout overlay of a Single Bit fail. The green box marks the location of the bad bit.

Failure pattern recognition techniques aim to categorize a failing memory from the topological point of view; several papers describe methodologies adopted in a volume production environment [5].

In our flow based on Yield Explorer ™ by Synopsys, the failing bitmaps are automatically classified in topological signature categories such as single bit, double bit vertical, cluster in die, solid full column, double bit horizontal etc. Fig.3 shows the Pareto from several wafers of the 90 nm product. The Y axis indicates the number of die affected by each signature.

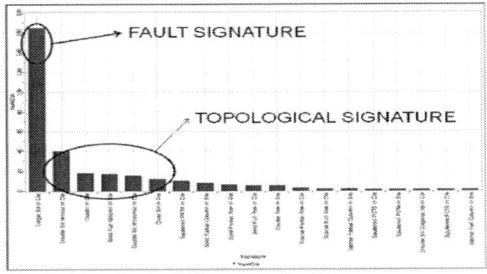

Fig.3: Topological Signature by Die count

The (x,y) coordinates of the most important (most frequent and with greatest yield impact) topological signatures are exported for Failure Analysis, Fig.4.

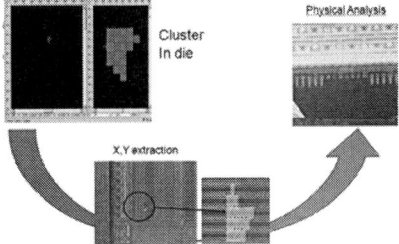

Fig 4. From topological extraction to Physical Analysis

A topological signature does not provide any precise information on the failure mechanism, but there is a certain experience that can be gained with specific signatures often associated to some known recurring root causes. The failure mechanisms for the main topological signatures were identified, as shown in Fig.5.

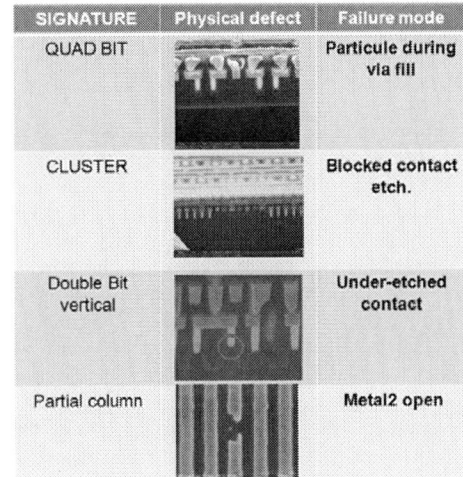

Fig.5: Physical defects on topological signatures

4. PROBLEM STATEMENT

Although useful, the current methods described above present a significant problem: It is inappropriate for situations with many Single Bit failures.

In Fig.3 the most important topological signature was a Single Bit, which has many potential failure mechanisms. It is very costly and time-consuming to investigate a sufficiently large population of Single Bit failures to determine the pareto of failure mechanisms in order to focus on the highest priority yield problems.

By identifying additional characteristics of the Single Bit failures, in addition to the (x,y) coordinates, the Failure Analysis Lab can choose the most appropriate physical or electrical analysis technique to identify the failure mechanism.

5. FAULT SIGNATURE METHODOLOGY

Here we present a methodology for analysing Single Bit failures based on the signature analysis of failing March tests by Yield Explorer from Synopsys.

A March Algorithm

The MBIST controller is able to launch several at-speed test algorithms optimized to rapidly detect a large range of potential faults. These algorithms involve a long series of March Elements, each of which is a short series of write and read Operations, to be performed sequentially, called the March Algorithm. Each March Element is performed on the entire memory array, in ascending ($\Uparrow$) or descending ($\Downarrow$) or undetermined ($\Updownarrow$) address order *[6]*.

In addition, the test can be launched with a number of stress conditions, such as *the addressing scheme: row fast or column fast, or* by varying temperature and voltage.

A typical March Algorithm is a 14N algorithm, for detecting realistic linked faults *[6]*. This algorithm is described as follows:

$\Updownarrow$(wx)	$\Downarrow$(rx,wx*)	$\Uparrow$(rx*,wx, rx,wx*)	$\Uparrow$(rx*,wx)		$\Uparrow$(rx,wx*, rx*,wx)	$\Uparrow$(rx)
M0	M1	M2	M3	M4	M5	
	R0	R1R2	R3		R4R5	R6

This algorithm consists of 6 March Elements (M0 to M5), each composed of up to 4 write and/or read operations, with a total of 7 Read Operations (R0 to R6). The 1st March Element M0 is a write operation to set the initial memory state, called the Background. Each additional wx write operation corresponds to the Background and wx* writes the opposite of the Background. The actual value that is written to each bit depends on the memory architecture, and is typically a function of the address, the memory bank, the physical row, and the IO. Each March element is performed in a specific increasing or decreasing address order, except the initial M0 which is undetermined. The MBIST controller checks the result of the read operations, compares it to the expected value, and reports any failed addresses. The march element and operation and actual and expected values may be logged for further analysis.

In advanced MBIST algorithms, to increase the fault coverage, the March Algorithm is repeated several times, with different Backgrounds, such as solid, column stripe, row stripe, checkerboard, etc. The preferred implementation depends on the memory type and the targeted fault types, and is the subject of theoretical models *[1,3]*.

The MBIST datalog from the tester is parsed into 2 results files, by using the memory configuration, as shown in Fig.6.
1. The Bitmap Results contain all failed addresses and bits. They are used for Topological Signature analysis.
2. The MBIST Fail Results contains all failed addresses and the associated March Element, Operation, plus the Read Data, and Expected Data, for Fault Signature analysis

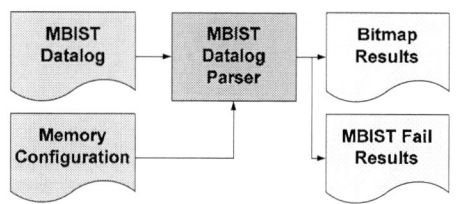

Fig.6: Memory BIST data loading flow

B Fault Signatures definition

The MBIST Fail Results are analyzed to create Fault Signatures, each of which is a binary word composed of the failures at Read Operations captured during the execution of a complete MBIST Algorithm composed of 1 or more March Algorithms. Each memory bit with a failure may be detected by 1 or more read operation, and this is used to create the Fault Signature specific to that Address and Fail Bit.

To illustrate the Fault Signature, let's consider an MBIST algorithm using the 14N March Algorithm , repeated 4 times with different Backgrounds, and a logical "Stuck-at 1" on an Address and Bit where x* is 0. Thus, all read operations where a 0 is expected would fail. The Fault Signature has 7x4=28 bits, where 0 is a correct value and 1 is an incorrect value at each read operation. The Fault Signature could be 0110110110110110110110110000

The Fault Signature for each failed bit is determined by considering the complete March Algorithm: the Expected Value at each Read Operation, which depends on the Background and which may be different by Address. Therefore, 2 different failed bit cells with the same physical root cause may have 2 different Fault Signatures, depending on their Address. In addition, the relationship between a logical fault and the associated physical failure is not exclusive, so a Fault Signature may be linked to 1 or more logical faults, which may have 1 or more physical failure root causes.

Examples of logical faults are *[6]*:
- Stuck-At-Fault: the cell is always 0 (SA0) or 1 (SA1).
- Stuck-Open-Fault: the cell cannot be accessed (SOF).
- Transition Fault: the cell fails to change logical value within the required time period (TF).
- Retention Fault, where the cell fails to retain its logical value, to become 0 (RF0) or 1 (RF1), after some time.
- Coupling Faults, where the cell is sensitized to the logical value in another cell, usually a near-neighbour. The subcategories are Inversion couling, Transition coupling, State coupling, Disturb coupling.

Each of these logical faults will lead to a Fault Signature. With a more complex MBIST algorithm, by using longer March algorithms and multiple Backgrounds, the probability increases that 1 Fault Signature is uniquely linked to 1 logical fault and 1 physical failure root cause.

C Fault Signatures by Simulation

Several papers have previously demonstrated the concept of generating fault signatures for memory diagnosis, by associating logical faults to hypothesized physical failure mechanisms through simulation *[1,2,3,4]*. These papers are based on automatic fault injection (static and dynamic) into a fault simulator based on the memory design and the March algorithm, to generate expected Fault signatures.

This approach through simulation has several complications:
- The fault simulator is complex to implement, requiring multiple inputs from memory design and test configuration.
- Not all fault types are targeted, so the analysis is prone to "unsimulated failure" results.

D Fault Signature Analysis

The objective of memory diagnosis is to identify the physical root cause of the failures to help with yield ramp.

The analysis starts with Bitmap Results, which allows the user to identify the most important groups of similar fails by Topological Signatures, which are exported for Failure Analysis, using the logical-vs-physical memory definitions.

For Bitmap Fails without a Topological signature, such as isolated single bit fails, the MBIST Fail Results are used to create Fault Signatures for each bit fail. Automated and interactive analysis using charts and statistics, guide the engineer to identify the most important (most frequent and with greatest yield impact) Fault Signatures.

The Fault Signatures are compared with the Fault Dictionary, to identify previously known and new failures. For the new Fault Signatures, the coordinates of the failures are exported for Failure Analysis, using the logical-vs-physical memory definitions. The results from Failure Analysis are used to update the Fault Dictionary, for future analyses, thereby building an "expert" system to increase productivity. The flow is shown in Fig.7.

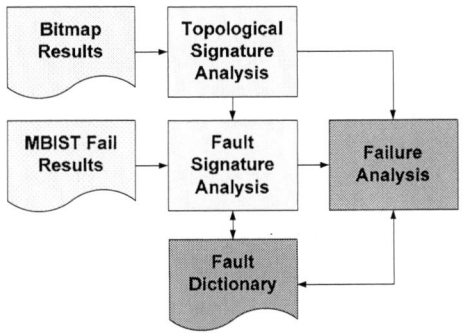

Fig.7: Memory BIST data analysis flow

By focusing on the most important Fault Signatures, the physical failures which are identified can be solved by manufacturing or design changes, bringing significant yield ramp benefits.

6. EXPERIMENTAL VALIDATION

A yield problem associated to MBIST failures at high operating temperature was analyzed with the methodology.

The flows and methodology described above, with the support from Failure Analysis techniques such as nanoprobing (to find mismatched transistors) and transmission electron microscope (to find dislocation faults), were used to identify the yield problems.

A Fault Signatures

With the most important Topological signatures being a Single Bit (Fig.3), Yield Explorer was used to identify the Fault signatures and to select the main SRAM yield loss signatures (Fig.8). Several die were selected for Failure Analysis. The affected bit cells were located on the layout to provide the (x,y) coordinates for rapid localization. The March algorithm was applied. The following signatures were recorded.

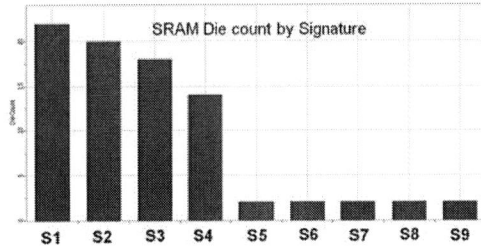

Fig.8: Fault Signature pareto by Die count

Signature	Signature type
S1	001110001110001110001110000
S2	010000000000000000000000000
S3	001110011011001110011011000
S4	011011011011011011011011000
S5	000100000000000000000000000
S6	001100000000000000000000000
S7	011100110110001110011000000
S8	001000000000000000000000000
S9	011000000000000000000000000

By careful expert review, the failures represented by S1 S3 S4 all follow a write operation of a previous March element, they show a dependency on specific backgrounds, they are similar but their difference depends on whether the specific address and row are odd or even. The failures represented by S2 occur on a solid background only, indicating a dependency on adjacent cells at state 1.

B Static Noise Margin (SNM) problem

The 1st series of Fault signatures (S1,S3,S4) were analysed using transistor nanoprobing, performed on an Atomic Force Prober (AFP). This equipment lands tips on the surface of the sample with nanometer precision allowing electrical characterization of elementary structures at the contact or via levels.

Four probing heads based on Atomic Force Microscope (AFM) technology are available allowing the full electrical characterization of transistors. From the topographical image obtained it is possible to land the tips on the nanostructures present on the sample like contacts for example. Coupled with a parametric analyzer it enables the realization of full electrical characterization.

Micro-machining of the sample with Focused Ion Beam (FIB), to build microscopic probing point is not required to implement this technique. The only sample preparation steps are the parallel polishing of the sample with specific procedure to reach the layer to probe. It allows the probing of different transistors on the sample even in dense area like the six transistors composing a RAM bit cell.

For comparison, nanoprobing was performed on the failed bit cell and an adjacent cell (Fig.9).

Fig.9: Areas of design investigated by nanoprobing

For each measurement, the well has been grounded (Nwell for PMOS & Pwell for NMOS) with the 4th probe. For transistors T1 to T6, the following characterization has been performed:

- Id(Vd) Vg=0v, 0.6v, 1.2v @Vb=0V
- Id(Vg) Vd=0.025v Vg 0 to 1.4v @Vb=0V
- Id(Vg) Vd=1.2v Vg 0 to 1.4v @Vb=0V
- Is(Vg) Vs=1.2v Vg 0 to 1.4 @Vb=0V

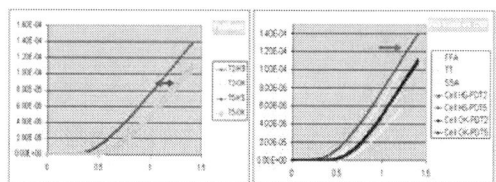

Fig.10: Electrical results from nanoprobing

A Vt shift (Fig.10) was observed between T2 & T5 (Pull Down) on the bad bit cell. Transistor NMOS T5 drives more current than transistor T2. Transistor T5 is not matched Technology Computer Aided Design (TCAD) values. This was observed on 15 different die.

Feedback to the design team helped to re-simulate the bit cell and understand that it was affected by a low SNM band gap. A new process was implemented to tune the electrical parameters and improve the yield at high operating temperature.

C Dislocation problem

The 2nd series of Fault signatures were obtained with expected 0 but are the same as S1,S3,S4 and in this case the nanoprobing showed a different transistor's behavior (Fig.11a). The suspected physical root cause was a short between the Source and Drain. To validate this hypothesis a transmission electron microscopy (TEM) analysis was used.

The TEM analysis on three die affected by the same signature confirmed the short between Source and Drain as a dislocation issue (Fig.11b). Dislocations are typically associated with preferential paths for diffusion of dopant species within silicon. A dislocation across the channel provides a path for source / drain dopants to get into the channel, thus providing a low resistance path between the two regions.

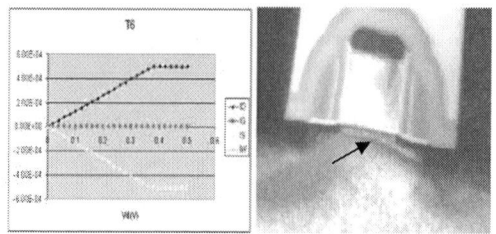

Fig.11: TEM analysis of a dislocation fault

7.CONCLUSIONS AND NEXT STEPS

Memories are useful Intellectual Property blocks (IPs) to use for technology qualification, production yield ramp, and process monitoring, due to their unique characteristics (percentage of silicon area, repetitive structures, localization). In this scenario statistical tools like Yield Explorer by Synopsys are used for selecting the major failure mechanisms, through their unique signatures, with the highest impact on yield.

The main target is to reduce the time needed to understand the physical root cause that affects the failing memories. For this purpose Yield Explorer provides the (x,y) coordinates of bit cell, and is able to suggest a possible physical root cause by comparing the electrical signature with the Fault Dictionary based on historical expertise gained on current devices, and previous devices with similar embedded memory designs. This paper presents a methodology to classify single fail bits by their unique fault signature. These classifications allow investigations to focus on the most important failure mechanisms with greatest yield impact. The methodology is demonstrated in an industrial environment, with identification of critical yield detractors.

The target of a future work will be to extend this methodology by creating the Fault Dictionary for all logical failures from any March Algorithm through simulation. This would enable to identify the logical failure from the electrical signature of the test results, and facilitate the identification of the physical root cause.

REFERENCES

[1] C.-F Wu, C.-T Huang, C.-W Wang, K.-L Cheng and C.-W Wu *"Error catch and analysis for semi-conductor memories using March tests"*, in Proc IEEE/ACM Int. Conf. Computer-Aided Design (IC-CAD), San Jose, pp. 468-471, Nov. 2000.

[2] J.Segal et al. *"Using electrical bitmap results from embedded memory to enhance yield"*, IEEE Design & Test of Computers, vol.15, n°3, pp 28-39, May 2001.

[3] C.W Wang, K.-L Cheng, J.-N Lee, Y.-F Chou, C.-T Huang, C.-W Wu *"Fault Pattern Oriented Defect Diagnosis for Memories"*, ITC Oct 2003 pp29-38

[4] A.Ney, A.Bosio, L.Dilillo, P.Girard, S.Pravossoudovitch, A.Virazel, M. Bastian. *"A History-Based Diagnosis Technique for Static and Dynamic Faults in SRAMs"*

[5] D.Appello, M.Sonza Reorda, A. Fudoli, V.Tancorre, F.Corno, M.Rebaudengo, *"A BIST-based Solution for the Diagnosis of Embedded Memories Adopting Image Processing Techniques"*, IOLTW2002

[6] A.J. van de Goor, *"Testing Semiconductor memories: Theory and Practice"*, ComTex Publishing, Gouda, The Netherlands, 1998

[7] M. Bushnell, V. Agrawal, *"Essentials of Electronic Testing for Digital, Memory, and Mixed-Signal VLSI Circuits (Frontiers in Electronic Testing Volume 17)"*, Springer, Nov 2000

45nm Yield Model Optimization

Brian L. Walsh, John Colt Jr., Daniel Poindexter, Thomas Joseph
IBM Systems & Technology Group
Hopewell Junction, NY 12533
bwalsh1@us.ibm.com

Abstract—**Elements of a yield model combining multiple input metrics will be reviewed. This model has been applied to multiple products across 65nm and 45nm SOI technology nodes. It provides long term yield metrics as well as yield diagnostics. Focus will be on the addition of After Develop Inspection (ADI) yield metrics into an existing framework which incorporates high resolution defect scans (PLY) and scribe kerf electrical test data.**

Keywords: Yield, Yield Modeling

I. INTRODUCTION

Driven by increasing die size, process complexity and wafer cost, an enhanced yield learning application has been developed incorporating multiple datastreams to provide timely and relevant yield diagnostics, feedback to the fabricator, and wafer by wafer yield predictions. The ability to quickly yield ramp the Power 7[TM][1] processor and other part numbers fabricated using IBM's 45nm SOI process [2] has been enabled by the augmentation of existing yield models through the use of the After Develop Inspection (ADI) datatype and a hierarchical correlation engine. This paper outlines the motivation behind the need to upgrade existing yield modeling methodology, correlation engine enhancements to leverage the ADI datatype, and lessons learned from applying this correlation engine to several products fabricated in IBM's 45nm SOI process.

II. YIELD MODEL INPUTS

There are four primary inputs into the correlation engine:

- High Resolution Defect Scans (HRDS / PLY): At select points in the process flow, a subset of wafers are routed into a patterned wafer defect detection / classification system. The resulting datastream consists of a classified set of defects with resolution from <1um to >5um. The primary focus of these inspection steps is to provide in situ feedback to tool and process owners.

- After Develop Inspection (ADI): At each photolithography step a low resolution (>50 um defect sensitivity) full wafer inspection is done after photo resist develop. This step is primarily to allow rework of wafers with high levels of large resist defects. This inspection is built into the resist apply / develop track and thus adds little delay to processing. A sample ADI defect wafermap and corresponding defect taken from a

PLY inspection after metal deposition is shown in Fig 1.

- Scribe Line (Kerf) Parametric / Functional Testing: At multiple points during the wafer process flow, subsets of wafers are routed to parametric or functional testing of electrically active structures in the product Scribe Line or Kerf. Parametric kerf structures consist of comb and serpentine structures of silicon or metal. The macros are designed to easily identify defects in a specific structure. Kerf functional macros consist of small product-like SRAM, eDRAM, or logic arrays which are testable either through a broadside or BIST interface. Historically, kerf data has been the basis for yield models due to ease of use, relatively high sampling rate when compared to PLY, and ease of tracking the failure mode back to a tool or process root cause.

III. 65NM AND 45NM YIELD LEARNING EXPERIENCE

The yield learning of Power 6[TM][3] in the IBM 65nm SOI process [4] and Power7[TM] in the IBM 45nm SOI process provided an invaluable dataset for development and vetting of multiple yield models

The ramp of the Power 6[TM] processor in the 65nm technology node provided a unique opportunity to explore different yield model methodologies applied to a technology at varying stages of maturity. The 65nm SOI process [4] was developed using the Power6[TM] chip as the early yield learning vehicle. During the early yield learning timeframe, a simple yield model based on a process grade which captured planned process changes and empirical correlations to the yield of a large kerf SRAM was able to model roughly 50% of the observed lot to lot variability. As the process matured and the defect density reached ultimate targets, the variability in the kerf SRAM became too small to adequately reflect product yield variation. A second yield model based on classified fails of the kerf SRAM was developed to increase the sensitivity of this monitor. This model employed a partial least squares engine [5] operating on the output of a fail classification system, and provided a two month outlook to expected product yields as well as a yield diagnostic in terms of key SRAM fail type drivers. The additional complexity of the yield model based on classified bit fail map was able to recover the sensitivity of the yield model to be roughly equivalent to the original kerf yield and process grade based model.

978-1-61284-408-4/11 $26.00 © 2011 IEEE

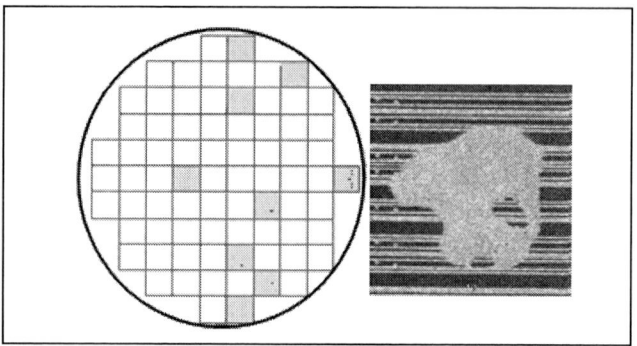

Figure 1. Left - ADI wafer map showing low but elevated defect count at a BEOL lithography level. Right - SEM image from standard defect scan which happened to be available for the same wafer at the previous process level. The defect is a metal puddle caused by an insulator defect.

This methodology was applied to Power 7[TM] with limited success. The 45nm process was relatively mature by the time Power 7[TM] started in the fabrication line. The kerf electrical defect monitors were already nearing their ultimate yields given their limited critical area. In addition, the 25% larger die size of Power 7[TM] vs Power 6[TM] reduced the number of electrically good kerf sites by 40% reducing effective kerf critical area.

The ADI datatype has proven to be a valuable yield diagnostic; as described in the previous section it provides quality metrics in terms of defect count and size and low resolution images of defects. These yield metrics are available on every wafer at every lithography level. The large amount of data provided by the ADI datatype drives the need for a data reduction scheme to identify yield drivers.

Defects detected by these inspections are typically those that cause gross electrical faults, such as probe or power supply shorts. The low resolution, and hence low defect counts detected are increased in value due to the fact that defects which are detected have a high probability of failure. Examples of defects detected are large particles, macro-scratches, and foreign material embedded in insulator films. Since inspections are done on every wafer at every lithography level, point of cause for many defects can be determined down to the process tool/chamber. These ADI lithography data are also valuable in correlating with other in-line defect metrics when debugging problems.

IV. POWER 7[TM] YIELD MODEL

With a reduction in the effectiveness of an SRAM based yield model for Power 7[TM], a new scheme has been developed to incorporate all four of the datatypes described in Section II at the cost of increased model complexity.

The two primary areas of concern with the Power 7[TM] yield model are model over-specification and non-orthogonal yield metrics. These concerns have been addressed by developing a PLS based hierarchical model. The combination of electrical Kerf, functional Kerf, PLY, and ADI can result in greater than 1000 metrics describing yield and parametric centering. Care needs to be taken in model development to avoid model over-

specification, especially on low volume products where extending the timeframe of the model training set to incorporate a large number of wafers dilutes the model sensitivity to current yield signals. Each of the above datatypes

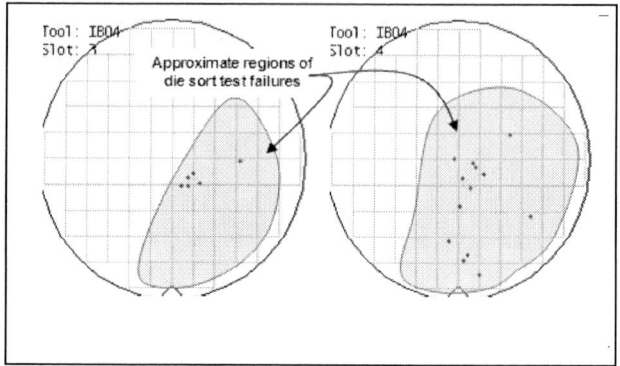

Figure 2. ADI wafer maps from wafers in adjacent slot positions show high defect counts at a well implant level lithography step. In this case, the wafer final test failure mode was predominantly logic failures; the approximate failing regions are indicated by shading. The defect type and cause are unknown but by inference it can be determined that there were many defects of the same type on these wafers, with most below the detection limit of the ADI tool.

have the ability to quantify the yield impact from each defect mode. An example would be the sensitivity of SRAM to gate level defects detectable by PLY.

Eigenvector methods are used to provide a solution to the problem of input independence, and a hierarchical combination of models is used to achieve a reasonable parameter count for the final yield model. Partial Least Squares (PLS), also known as Projection to Latent Structures, is an extension of the familiar Principal Component Analysis method of extracting variability mechanisms from multivariate data sets. The operative difference is that the combinations of variables identified and ranked by PLS are optimized for the best correlation to a desired set of output variables instead of the variability mechanisms identified by PCA.

Each combination of variables can be thought of as a new measurement vector penetrating the multivariate data cloud, that is, a virtual measurement stick that is not aligned with any of the physical dimensions defined by the actual measurements, but include, in different proportions, influence from each of the original inputs. These new virtual measurement vectors are extracted in rank order, starting with the combination of variables having the best possible correlation to the desired output, and followed by the next best combination (or vector) that is selected from that set of virtual measurements that are orthogonal to the previous vector or vectors. In this way, each new virtual measurement is defined to be independent of the others.

When applying PLS against our various data types, models generated with datasets having ~10 times the number of observations as input parameters are accepted as being generally free from over specification. A hierarchical PLS based model is created from a small set of the most significant vectors taken from PLS models created from each datatype.

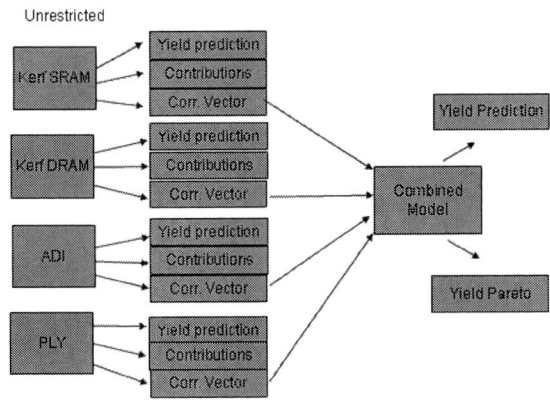

Figure 3. Data flow diagram illistruating the building of the combined yield model from the outputs of unconstrained models built from four key datatypes. An unconstrained yield model is built against Kerf functional data, PLY and ADI The top four correlation vectors are then used as inputs to a combined model. From the combined model, predicted yields and information on the most significant inputs are availble.

Each "sub-model" is then generating its own set of new virtual measurements (vectors), and ranking them in order of importance. In this specific application, the first four vectors in each of the models capture most (provide some percentage here) of the information available from that model. Thus, by utilizing these eigenvector- like "PLS scores", we can describe the information from any model using four original variables, instead of the 50 to 100 in the original data set. These reduced data sets can then be combined into a high level model using, for example, an application with four in-line data sources, or 16 parameters to predict the final yield instead of the many hundreds that would result if one model were used.

By making these sub-models, the full suite of inputs can be analyzed for any sample of wafers for any of the individual models. A decomposition of the highest level model will indicate which of the virtual measurements coming out of the lower level models is responsible for the strongest signal. The information from the specific lower level model indicated can then be decomposed into individual or combinations of the original physical measurements to provide insight into the mechanisms driving yield.

There is no need to start at the highest level. Each sub model and the final model are all "trained" using the same learning data and yield parameter(s). Thus, each low level model, while also generating a reduced set of orthogonal virtual parameters to pass up the hierarchy, is also producing a predicted yield output that is sensitive to any variability in that model's inputs that correlates to yield. Interesting yield modulation seen in the predicted values can be investigated directly, and with more resolution at the lower levels than from the combined model.

Large 45nm Product	Goodness of Fit
Kerf Model Only	1 (baseline)
PLY Model Only	1
ADI Model Only	1.2
Kerf + PLY	1.7
Combined Model	2.2

Table 1: Decomposition of model inputs for large 45nm product indicating the relative goodness of fit metrics based from the correlation engine based on the input metrics.

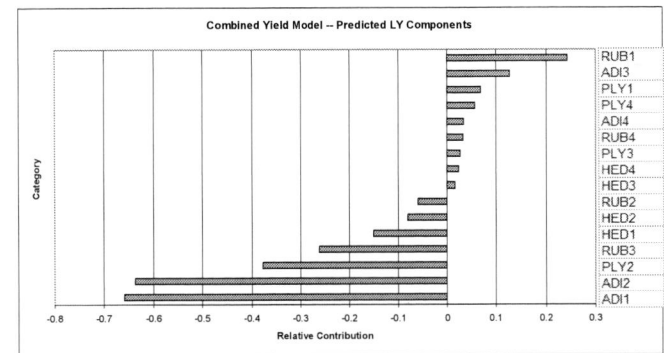

Figure 4. Typical Very Important Parameter (VIP) chart generated from the combined yield model. Each bar indicates the relative contributtion of the most significatnt yield vectors producted by the unconstrained models. RUB and HED compoents are derived fom kerf eDRAM and SRAM kerf yield monitors.The vectors that postively impact yields are toward the top of the chart, those which capture negative impacts ot the yields are toward the bottom of the chart

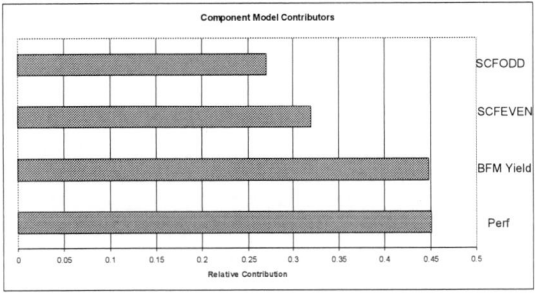

Figure 5. A VIP chart exploring the most significant contributors to the RUB1 component of the VIP chart in Fig. 4. In this instance several classified bitfail map signatures. are highlighted as the most important components in describing the yield.

IV. RESULTS

Table 1. demonstrates the added value of the ADI datastream over the PLY and kerf-only yield models on a large 45nm product chip. Relative goodness of fix metrics are shown for each component of the combined yield model. In the case of Power7[TM], the addition of the ADI data to the kerf and PLY model increased the goodness of fit metrics by 30%, while providing physical defect based yield paretos.

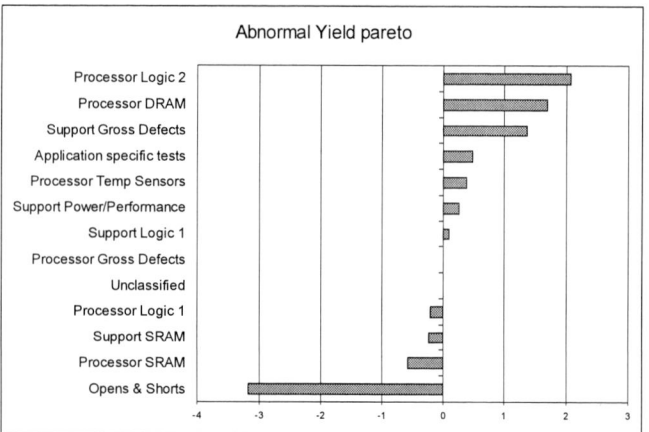

Figure 6. Normalized abmormal limited yield pareto on a lower than average yielding lot indicating higher than normal fallout for probe opens and shorts.

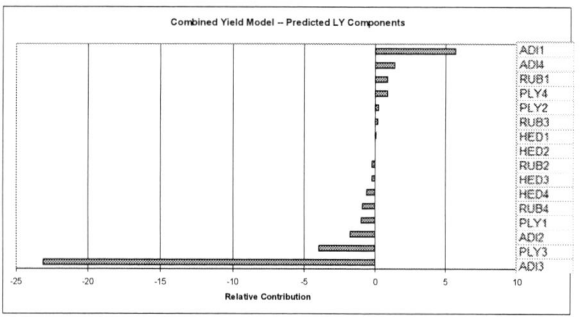

Figure 7. The VIP chart output of the combined model from a lot with an abnormal yield pareto as described in Fig. 6, indicating a strong deviation from the model coming from the "ADI1" yield vector producted from the output of the unconstrained ADI yield model.

The yield diagnostics enabled by the hierarchical correlation engine have been integrated into the process yield learning activity. PLS operates on a centered and scaled (or standard normalized) dataset. It is sensitive to the correlations between parameters and how they deviate from a center value. The predicted values at each level are referenced to the center value of the learning data, but the virtual measurements passed along to the higher level model are only indicators of the sense and magnitude of the variability any given wafer is exhibiting for the yield mechanism captured by that vector. The yield diagnostics are represented in terms of VIP charts as shown in Figs. 4 and 5. Figure 4 is a typical result from the combined model indicating that wafers yielding below model baseline tended to have a strong input on the functional kerf macro "RUB1". Figure 5 is a drill down into that yield vector indicating strongest components, in this case an even/odd SCF fail signature. This is a fail signature with a limited set of

possible root cause drivers, which is fed back to the appropriate process owners for resolution.

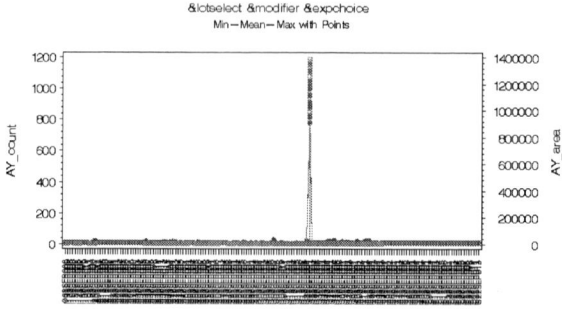

Figure 8. A second example works backwards from an abnormal product fail pareto as showing in Fig. 6 to process root cause Trend of the BEOLVia found to be the primary contributior to the ADI1 component highlighted in the VIP chart in Fig. 7

Fig. 6 contains the relative yield difference between a suspect lot and the baseline, indicating abnormal product scan string fallout. Drill down of the model results on this lot indicate a strong ADI signal as shown in Fig. 7. Further drill-down into the unconstrained ADI base model clearly indicates this lot was a flyer at a Back End Of Line (BEOL) via level (Fig. 8).

V.   SUMMARY

With the increase in product die size and additional process complexity, enablement of yield models using more diverse and larger datasets is a key to learning and prognosticating yields. The Power 7^{TM} experience has led to the development of a two-stage PLS based correlation engine acting of four distinct data types. Focus has been placed on and successes observed with leveraging the ADI datatype as a basis and supplement to existing PLY, and electrical kerf yield metrics.

[1] R. Kalla, B. Sinharoy, W. Starke, and M. Floyd. "Power7TM: IBM's Next Generation Server Processor" IEEE Micro, 16 Feb 2010 Vol 30, Issue 2, pp 7-15.

[2] S. Narasimha, "High Performance 45-nm SOI Technology with Enhanced Strain, Porous Low-k BEOL, and Immersion Lithography", IEDM 2006, pp 1-4.

[3] B. Stolt, Y. Mittlefehldt, S. Dubye, G. Mittal, M. Lee, J. Friedrich, E. Fluhr, "Design and Implementation of the Power 6 Microprocessor" IEEE Journal of Solid-State circuits, Jan 2008 Vol 43 pp 21-28

[4] E. Leobandung, "High performance 65 nm SOI technology with dual stress liner and low capacitance SRAM cell", VLSI Technology, 2005. Digest of Technical Papers. 2005 Symposium, pp 126-127.

[5] Multi- and Megavariate Data Analysis: Principles and Applications L. Eriksson, E. Johansson, N. Kettaneh-Wold, and S. Wold, ISBN 91-973730-1-X)..

Yield Optimization for Third Party Library Elements

Jeanne Paulette Bickford, Francis Chan, Mark
Styduhar, Lee Wang
IBM Corporation Systems and Technology Group
1000 River Street
Essex Junction, VT 05452 USA

Robert Arelt, Ioana Graur, Steven Parker, Deborah
Ryan, Tina Wagner
IBM Corporation Systems and Technology Group
2070 Route 52
Hopewell Junction, NY 12533 USA

Anand Kumaraswamy
IBM Corporation, ASIC Circuit Enablement Team
Bangalore, India

Abstract— **Optimization of semiconductor product yield requires control of systematic defects. A variety of industry tools are available to check designs for systematic layouts that will be difficult to manufacture. Because of the cost associated with setting up the rules for a checking tool and the cost of licenses needed to evaluate designs, manufacturing process lines typically enable a limited set of tools to evaluate designs for systematic yield sensitivity. Semiconductor product design systems typically incorporate library elements designed by third party design companies. When third party library suppliers do not have access to the licenses or expertise to use the tools selected by the target manufacturing line, a barrier is created to having all library elements in a semiconductor design system checked to the same level . This results in a yield exposure for library elements that are not evaluated and fixed. This paper describes a method used to enable systematic yield evaluation for 32nm library elements procured from third party library suppliers. The third party library supplier provides layout data to the contracting library owner for yield sensitivity analysis. Changes are prioritized and fed back to the third party library supplier. This method interlocks design practices with the third party library supplier and provides a means for the contracting library owner to evaluate third party library elements and provide feedback to optimize the design.**

Keywords-Design for Manufacturing (DfM), Yield, Yield Checking Deck(YCD), Litho Friendly Design (LFD), Manufacturability

I. INTRODUCTION

Successful semiconductor product yield requires managing systematic and random yield components. Random yield improvement can be predicted using Critical Area Analysis (CAA) techniques for both products and library elements used to build products [1,2,3,4]. Sensitivity to random defects requires completed layout for the library element or product and the planned manufacturing line defect density. Systematic yield depends heavily on the technology node elements and patterning solution used to manufacture a product. As products move to denser technologies with smaller dimensions, systematic yield has a larger impact on total yield. While some elements of systematic yield loss can be addressed by applying techniques that add design margin or manufacturing screens,

higher cost results because of lower yield and/or larger product area [5,6,7]. Identifying and minimizing the source of systematic yield loss in the library elements that will be used to create products avoids this problem and optimizes product yield.

A variety of yield analysis tools have been developed to analyze library element layout for sensitivity to systematic defects so that the source of these potential defects can be removed before products are built using these library elements. Yield analysis tools consider layout sensitivity to process and tool limitations in a target manufacturing line. Technology development teams select yield checking tools and develop rules for the selected tools. The tools and rules are used by library element and product designers to optimize designs for yield [8].

Semiconductor design systems typically incorporate library elements designed by third party design teams. Because third party library suppliers have expertise with complex IP and often control intellectual property needed to design such IP, tools and rules specific to the design system and the target manufacturing line need to be shared with the third party library suppliers.

Because of tool license availability and expertise with particular tools at different companies, third party library design teams often use different yield optimization tools than the owner of the design system and/or technology.

Integrating library elements from different suppliers creates challenges like those faced by pharmaceutical companies where the supplier of a material may not have access to the same analysis tools as the contracting company, but needs to ensure that the quality of the end product meets the quality standards for the end product. [9]

This paper describes a process implemented by the IBM 32nm SOI ASIC design team and IBM technology development teams to optimize yield for library elements so that products built with those library elements realize the yield benefit from design optimization and have improved yield. Details of process implementation with two different third party IP suppliers will be discussed including the optimization of design choices, routing impacts, reconciliation of design styles, and manufacturing line specific impacts.

II. YIELD OPTIMIZATION PROCESS

A. Semiconductor Product Design Overview

In past technologies, recommended rules were used to provide guidance to design teams on how to optimize for yield. In some cases, design teams used Design Rule Check (DRC) decks that included some or all recommended rules, so that designs could be checked to the recommended rules. In newer technologies, following all the recommended rules would significantly impact product area utilization (increasing die area and product cost), power and performance. Since products need to meet required specifications, recommended rules that impact product power or performance cannot be fully implemented unless product specifications are changed. While DRC checks are standardized to a set of constraints and a limited number of widely accepted software is used for DRC, tools and rules used to optimize yield by checking designs for manufacturing (DfM) compliance do not have the same level of standardization.

Semiconductor products in current technologies use base library elements and predesigned combinations of base library elements called cores. The base library elements and cores are characterized for timing and power. Base library and core timing and power are used to design products. Changing layout in the base library or core can impact timing and power so changes in layout within the predesigned unit cannot be allowed after timing and power characterization is complete. If yield optimization is not achieved in the library elements, it will not be possible to make changes within the library element during product level design. Opportunities for yield optimization at product level are limited to placement of library elements and die level interconnect wiring optimization.

B. Yield Optimization Tool Overview

Compliance to all recommended rules would result in substantial size growth for library elements and the products that use them. Key recommended rules are selected and prioritized to minimize area growth and design impact, and to maximize yield improvement.

Possible opportunities for yield optimization need to be flagged for library element designers so that possible changes to improve yield can be considered. DfM recommended yield checking deck (YCD) rules are binned based on the severity of a fail type. As shown in Fig. 1, recommended rules are prioritized using the severity of product yield impact. Must fix changes are assigned a priority of 0 (must fix). Higher priority numbers indicate less impact on yield. Library elements use very different layout types, but within a particular type of library element, layouts are usually similar. Setting up a DfM recommended YCD

YCD Recommended Rules Implementation

YCD checks Recommended Rules
    Meet R rules where opportunity
    Set Pass/Fail compliance threshold

YCD rules are prioritized (0-2 required)

Rule	Priority
Vx redundancy	0
Mx small metal islands	0
Performance / maximize PFET stress	1

Criteria for Each Library Element Type

Figure 1. Implementation of YCD Recommended Rules

rule based on the severity of a fail for a particular library element type allows yield to be maximized considering the layout that is most likely to be used in a particular type of library element. DfM recommended YCD rules are implemented in an YCD tool and provided to library element designers. Many YCD tools are available in the industry. IBM uses the Magma Quartz tool for YCD analysis. An example of a problem flagged by YCD that can be fixed without growing area or impacting power or performance is shown in Fig. 2.

Litho Friendly Design (LFD) evaluation flags layout areas that present patterning problems in manufacturing by applying pattern simulation to layout prior to tape-out. Pattern simulation identifies layout changes that need to be fixed in the design. If changes cannot be made, technology development and manufacturing need to modify the process to support the layouts. Experience from past technologies and early test sites is used to identify problem layouts. In 32nm, IBM uses the Mentor Graphics Calibre tool to

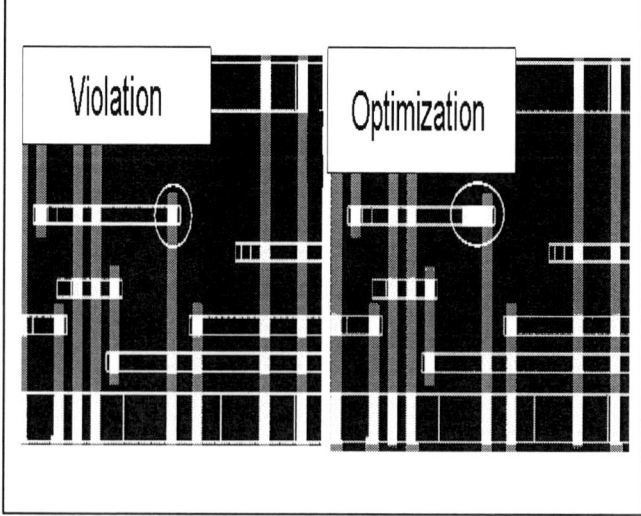

Figure 2. Optimization with no area, performance or power impact

978-1-61284-408-4/11 $26.00 © 2011 IEEE

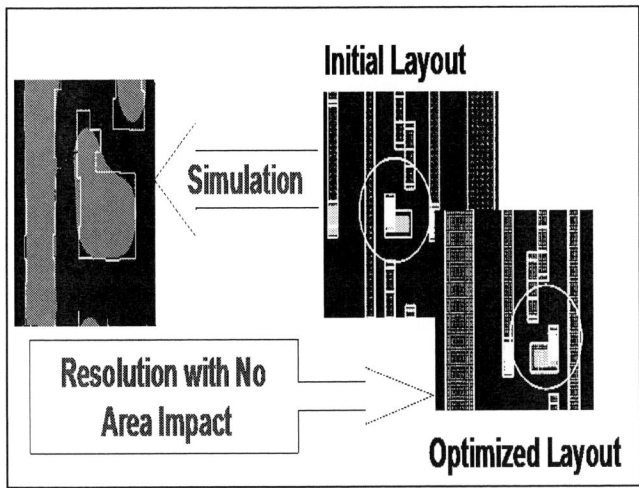

Figure 3. LFD uses pattern simulation to identify hot spots.

evaluate library elements. Fig. 3 shows layout changes flagged by LFD that can be fixed without growing area.

C. Internally Designed Library Element Optimization

Since many different designers are involved in creating an ASIC design system, a system that enforces YCD and LFD use at specific points in the development cycle is required for successful library element design optimization.

The IBM ASIC design team has embedded links to the YCD and LFD tools in the library element workflow. The workflow tool specifies particular points in the design process where running both YCD and LFD is required. Each library element must meet the specified criteria for that library element type. If the library element does not meet those requirements, it must be updated to pass the check, or manufacturing must support a waiver and accept the library element so that yield learning can be planned and prioritized appropriately.

Fig. 4 provides a diagram of this process [10]. The use of embedded links provides version control for YCD and LFD tools.

D. Third Party Yield Optimization Process

Many third party design teams have expertise in design of complicated cores that make these teams excellent choices to design related cores for an ASIC library, therefore use of third party design teams for core designs is increasing. Third party design teams use industry recognized tools for DRC checking, but often do not use yield checking tools to evaluate compliance with recommended rules as part of their design flow. Where third party design teams use yield checking tools, the available tools may not be the tools used by a particular manufacturing line. Licenses that are used for yield checking are expensive and the knowledge on how to effectively use the tool may not be present at a particular third party supplier.

The YCD and LFD flow shown in Fig. 4 for internally designed library elements is not workable for core library elements designed at many third party suppliers. Because these cores occupy a large amount of die area, it is particularly important that these cores be optimized for yield. To enable yield checking for third party cores, the process shown in Fig. 5 was developed. This process establishes a partnership to optimize yield with yield checking done in IBM and results provided to the third party suppliers so that the third party library element provider can optimize the library element for yield.

Third Party Library Element suppliers design macros (subcomponent blocks) that will be used to assemble the completed core and use the DRC deck to ensure that the design is compliant with all required design rules. Each macro is supplied to an IBM team and the macro is evaluated with both YCD and LFD. Results are fed back to the third party library supplier and possible changes are discussed. Since third party engagements are governed by the contract between the two parties, changes need to be agreed to by both parties. If changes cannot be made to remove the identified yield exposure, the part of the design that is not compliant needs to be identified and manufacturing needs to

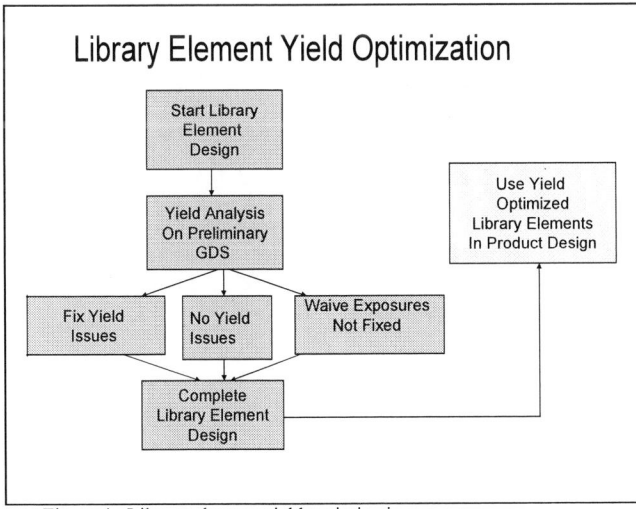

Figure 4. Library element yield optimization process.

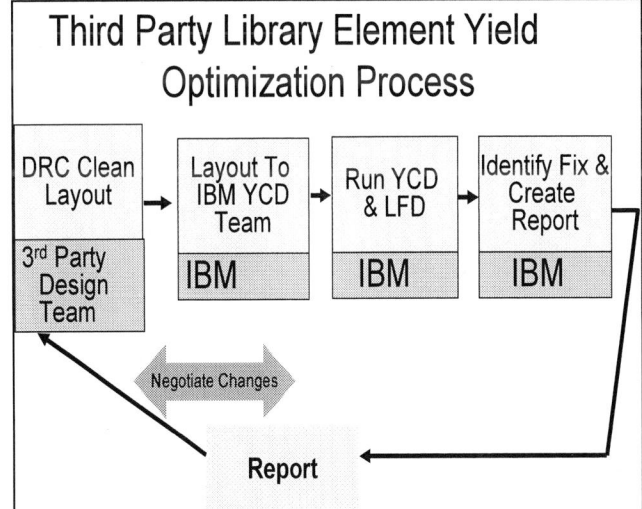

Figure 5. Third Party library element optimization process.

978-1-61284-408-4/11 $26.00 © 2011 IEEE

accept the increased risk through a waiver process. Manufacturing has a better chance to increase the robustness of the manufacturing process for these elements if warned early.

III. THIRD PARTY YIELD OPTIMIZATION CASE STUDIES

A. Overview

This process has been used with two 32nm third party core library element providers. One engagement started after much of the core had been designed. A second engagement started early in the design cycle. In both cases the process outlined in Fig. 5 was followed. This was done prior to the final core design so that changes could be incorporated in planned updates to the core. Case studies for each are presented in the next two sections.

B. Late Design YCD and LFD Engagement Case Study

YCD and LFD were run on a large third party macro in the core to evaluate compliance to the recommended rules. While the design was robust from a manufacturing standpoint, the score for a number of recommended rules was lower than the threshold established for this type of core. The severity of the fail (0 represents the highest priority for a fix and 1 represents a fix that is desirable) and number of each type of fail were identified.

A report was prepared for the third party library supplier containing a list of recommended rules that failed the threshold set to meet yield, examples of shapes that fail the recommended rules, and recommendations to improve the YCD score. Proposed design fixes for each fail type were identified. Fig. 6 – Fig. 8 show examples of areas identified as major yield exposure (redundant contacts, redundant vias, and minimum metal area). Changes were requested for each of these potential exposures.

Fig. 9 shows a source and drain area (Rx) to gate spacing concern. Figure 10 flags a concern with overlap of metal. Shapes in Fig. 9 and Fig. 10 remain unchanged because of negligible impact to core performance and/or area.

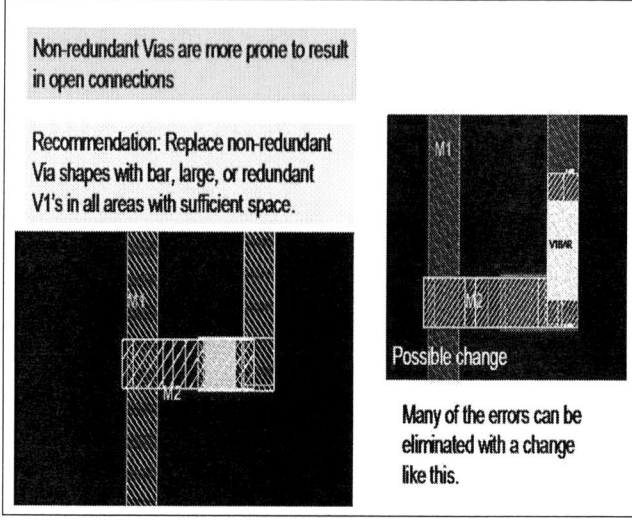

Figure 7. Redundant via improvement

Reasons for the lower than desired redundant via coverage were explored. The number and type of via choices available in the technology file supplied to the third party library supplier was identified as the source of the problem. Fixing these issues required that a new technology file with expanded via options be supplied to the third party library suppliers. Prior to this study, the contents of the technology file supplied to the third party library supplier for use in their router had not been examined with sufficient scrutiny.

C. Early Design YCD and LFD Engagement Case Study

After the experience with the engagement described above, new criteria were established for engagement with third party library element suppliers. Library element macros are evaluated as soon as available so that router options and quality of base library elements used could be identified as early as possible. Macros and the overall library element are then revaluated later in the design process.

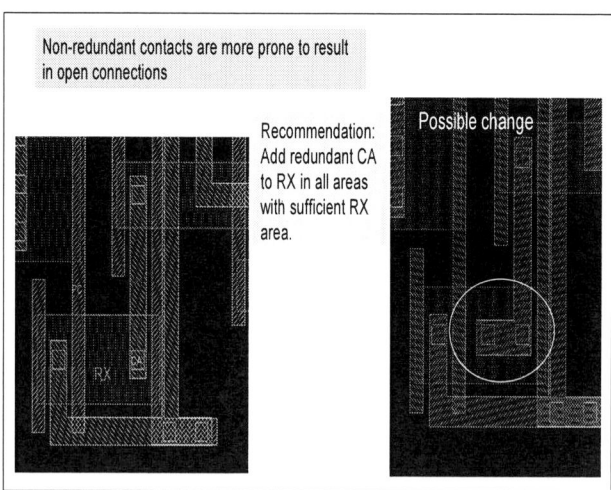

Figure 6. Redundant contact improvement.

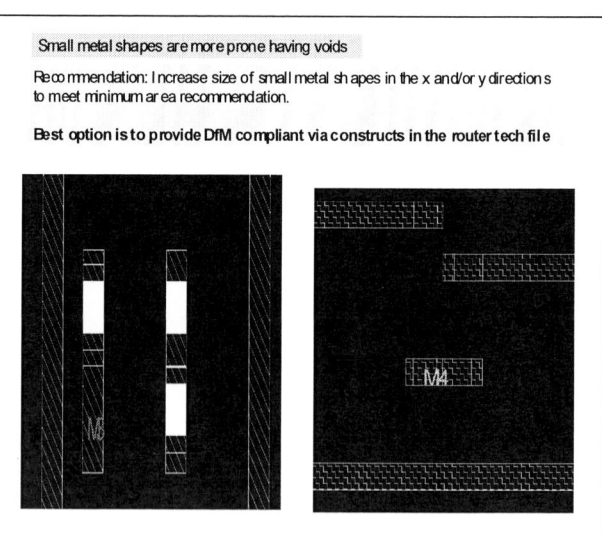

Figure 8. Metal minimum area improvement.

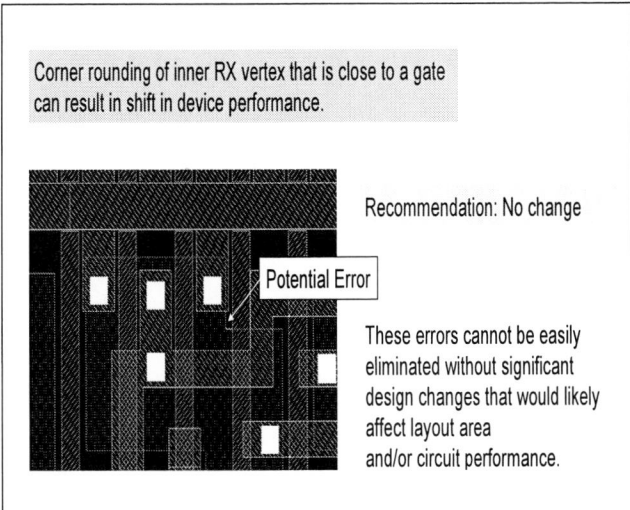

Figure 9. Spacing change was not requested (area and performance impact).

A second third party library element was evaluated with LFD and YCD. LFD analysis indicated no problems. Overall the core macro met YCD requirements. Evaluation of the macro revealed that most of the violations were present in circuits designed by IBM and supplied to the third part library suppliers. Design of library element building blocks for third party library suppliers used a separate workflow manager that had been missed when the process described in Fig. 4 was created. Because the circuit was not released directly into the library, design checks to verify that YCD and LFD had been run were not present. After this problem was identified, the workflow manager for library elements provided to third party library elements was updated with the same requirements and checks used for the internally designed and released library elements. Update of this process ensures that good design practices are used at appropriate points in the library element design cycle. This

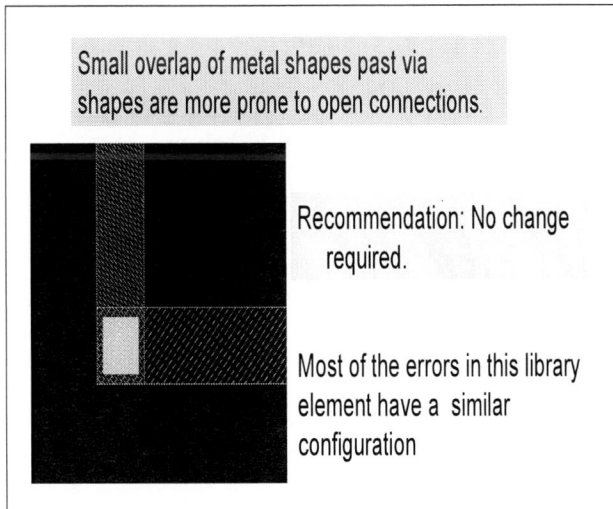

Figure 10. Metal overlap change was not requested

underscores the importance of examining the workflow process for all tools and rules used to design library elements internally as well as at third party suppliers.

D. Observations

Both case studies show the need for careful scrutiny of the tools, rules, and base library elements supplied to third party library suppliers. All library elements supplied must meet YCD and LFD requirements. It is particularly important that the technology file used by third party library suppliers provides the proper via options and trade-offs so that core routing is optimized for yield.

IV. CONCLUSIONS

Yield checking tools can be successfully used to optimize yield in third party library elements. Successful yield optimization with third party library providers requires close work between the third party core library supplier and the contracting group. In some cases the resource needed or the time required to implement the change may make updates too expensive in terms of cost or time. Contracts with third party library suppliers should include resources to perform yield optimization updates. In cases where recommended updates cannot be made, early communication to manufacturing through the use of a waiver process provides the additional time for manufacturing to prioritize process learning activities to enable support for products using the core library element. Changes identified early in the design process are more likely to be implemented because they can be included in other planned updates.

V. OUTLOOK

Currently many different yield checking tools are available. While each has the ability to address particular layout concerns, the variety of different tools and lack of standardization in mechanisms and rules between different manufacturing lines makes implementation of yield checking difficult when library elements or products are developed by a variety of companies that may not all have access to the yield checking tool selected by a manufacturing line. Development of common rule formats and standardized tools will allow third party designers to run these checks effectively on their own.

ACKNOWLEDGMENT

The authors would like to acknowledge Jennifer Lynch, Richard Hee, Garry Hughes, Todd Bailey, George Hefferon, and David Medeiros for management support of this project.

REFERENCES

[1] C.H. Stapper, F.M. Armstrong, and K. Saji, "Integrated Circuit Yield Statistics", Proceedings of IEEE, April 1983, pp. 453-470.

[2] E. Papadopoulou, D.T. Lee, "Critical Area Computation via Voronoi Diagrams", IEEE Trans. On Computer-Aided Design of Integrated Circuits and Systems, vol.18, issue 4, 1999, pp. 463-474.

978-1-61284-408-4/11 $26.00 © 2011 IEEE

[3] J. Bickford, M. Buhler, J. Hibbeler, J. Koehl, D. Muller, S. Peyer, C. Schulte, "Yield Improvement by Local Wiring Redundancy", Proceedings of the 2006 7th International Symposium on Quality Electronic Design, March 2006, pp. 6 pp. |CD-ROM.

[4] T. S. Barnett, J. Bickford, and A. J. Weger, "Product Yield Prediction System and Critical Area Database", 2007 IEEE/SEMI Advanced Semiconductor Manufacturing Conference, May 2007, pp. 351-355.

[5] R. Kanj, R.V. Joshi, and J. Sivagnaname, "Design Considerations for PD/SOI SRAM: Impact of Gate Leakage and Threshold Voltage Variation," IEEE Transactions on Semiconductor Manufacturing, vol. 21, no. 1, Feb. 2008, pp 33-40.

[6] J. Bordelon and P. Manair, "Improving Yield Through Parametric Variability Characterization and Modeling," Solid State Technology, Nov 2007, pp. 38-40.

[7] J. Bickford, J. Goss, R. McMahon, R.V. Joshi, and R. Kanj, "Use of Scalable Parametric Measurement Macro to Improve Semiconductor Technology Characterization, and Product Test," 11th International Symposium on Quality Electronic Design, March 2010, pp. 315-319.

[8] J. A. Torres, I. Graur, M.C. Simmons, and S. Kanodia, "Layout Verification in the Era of Process Uncertainty: Requirements for Speed, Accuracy, and Process Portability," BACUS, 2007, pp. 67300U-1-9.

[9] A. M. Thayer, "Geonotoxic Impurities," Chemical and Engineering News, Sept. 27, 2010, pp. 16-26.

[10] I. Graur, T. Wagner, D. Ryan, D. Chidambarrao, A. Kumaraswamy, J. Bickford, M. Styduhar, L. Wang, "Methodology for balancing design and process tradeoffs for deep-subwavelength technologies," SPIE, 2011, in press.

9781612844084